Memory in Science for Society

Memory in Science for Society

There is Nothing as Practical as a Good Theory

Edited by

Robert H. Logie, Zhisheng (Edward) Wen,
Susan E. Gathercole, Nelson Cowan,
and Randall W. Engle

OXFORD
UNIVERSITY PRESS

Great Clarendon Street, Oxford, OX2 6DP,
United Kingdom

Oxford University Press is a department of the University of Oxford.
It furthers the University's objective of excellence in research, scholarship,
and education by publishing worldwide. Oxford is a registered trade mark of
Oxford University Press in the UK and in certain other countries

Published in the United States of America by Oxford University Press
198 Madison Avenue, New York, NY 10016, United States of America

British Library Cataloguing in Publication Data

Data available

Library of Congress Control Number: 2022952036

ISBN 978–0–19–284906–9

DOI: 10.1093/oso/9780192849069.001.0001

Printed and bound by
CPI Group (UK) Ltd, Croydon, CR0 4YY

Oxford University Press makes no representation, express or implied, that the
drug dosages in this book are correct. Readers must therefore always check
the product information and clinical procedures with the most up-to-date
published product information and data sheets provided by the manufacturers
and the most recent codes of conduct and safety regulations. The authors and
the publishers do not accept responsibility or legal liability for any errors in the
text or for the misuse or misapplication of material in this work. Except where
otherwise stated, drug dosages and recommendations are for the non-pregnant
adult who is not breast-feeding

Links to third party websites are provided by Oxford in good faith and
for information only. Oxford disclaims any responsibility for the materials
contained in any third party website referenced in this work.

For Alan Baddeley with thanks for decades of inspiration.

Preface

Developing theory through applications and developing applications through theory: the influence of Alan Baddeley

Professor Alan Baddeley is a world-leading authority on human memory and cognition, best known for his multicomponent model of working memory (Baddeley & Hitch, 1974). In general psychology and cognitive sciences, there is no psychologist more central than Alan Baddeley to the progress that has been made in the last 50 years in understanding the mind, brain, and behaviour. His wide-ranging influence penetrates the diverse domains of human memory, cognition, development, and neuropsychology, and many practical applications of cognitive psychology. Besides his many distinguished honours and awards (CBE, FBA, FRS, FMedSci), and honorary degrees from the universities of Bristol, Edinburgh, Essex, Stirling, and Umea, Alan has published many influential monographs, textbooks, edited volumes, and numerous (theoretical and empirical) papers that have been cited very frequently, reaching a staggering Google Citation Score of over 205,000 (as of 11 April 2022). In recognition of his landmark contributions to psychological sciences and human cognition, this volume is presented both as a book on theory and application of memory research for a broad readership, and as a tribute for Alan to honour and celebrate his enormous contribution and influence across the topics discussed in the individual chapters.

This volume also is a personal thank you to Alan from many of the scholars who have been inspired by him and benefited from his insights and intellectual curiosity over many decades. We have invited chapters from leading international scholars, including Alan's close colleagues and associates, collaborators, and friends as well as others whose research has been influenced directly by his pioneering and groundbreaking work. More detail on the different areas of Alan's work is provided in his own chapter (Chapter 2), and within the other chapters. Here, each of the volume editors, and one of his closest colleagues (Graham Hitch), briefly share some thoughts on the considerable scholarly and personal influences gained from knowing Alan personally as well as from working with him and reading his work.

Graham Hitch

I regard myself exceptionally fortunate to have Alan as a mentor, a colleague, and a good friend for many more years than I care to remember. My first contact with him was in 1967 through a series of lectures he gave on human memory and performance. I was then a student taking a taught MSc in Experimental Psychology at the University of Sussex after completing a degree in physics at the University of Cambridge. The experimental psychology department in Sussex was an exciting place but even so, Alan's lectures stood out, inspiring me by their clarity and accounts of discoveries made through ingenious experiments, especially some of his own on short-term memory. Alan supervised my MSc project on the same topic. It didn't come to much but despite this he encouraged me to follow it up with a PhD, recommending R. Conrad as supervisor, who had been his mentor when both were at the then Medical Research Council (MRC) Applied Psychology Unit (APU) in Cambridge. By the time I started, Conrad had left the APU for Oxford but I was fortunate to be taken on by Donald Broadbent. I kept in touch with Alan throughout, feeling we had something in common, and when my 3 years were up, he invited me to return to Sussex and work with him as a postdoc. The plan was to continue investigating short-term memory. However, we soon realized it was rapidly becoming an unfashionable research topic. We wondered why and I still can recall having lunch together and asking the question 'What is short-term memory for?' We decided to set about answering by experiment. The results are described in our 1974 chapter (Baddeley & Hitch, 1974), together with our proposal of working memory as a multicomponent system. We have kept in touch ever since, wherever we have each been based, sometimes working together, sometimes not. What has kept it going from my point of view is the simple pleasure of sharing ideas with someone so good at turning them into practice, helped by regular meetings of the informal working memory discussion group to which Robert Logie and Sue Gathercole refer in their comments below.

My chapter in this book is intended to illustrate just one example of Alan's many penetrating insights over the years, and a small way of saying thank you for his encouragement, continued interest, and support throughout. The phonological loop is a great example of his trademark approach of developing straightforward, highly transmissible theoretical accounts of robust and replicable phenomena, capturing key insights and capable of ready application to practical problems. When neural network modelling came on the scene in the 1980s, Alan was quick to see its potential and strongly encouraged me and others to explore its application to working memory where the phonological loop seemed ripe for treatment. I am pleased to be able to report that this has proved highly fruitful with broad implications extending far beyond the loop itself (see Hitch, Chapter 8, and other chapters in this volume).

Alan's outstanding contribution to our science has been driven by a natural and persistent curiosity, a collaborative and collegiate approach, and an amazing ability to draw fruitfully on previous research, all combined with a genius for communication. It has been a real privilege to be one of the many who have had the opportunity to work alongside him and it gives me great pleasure to be able to say we are still collaborating as I write this in 2022, including yet further experiments on the phonological loop!

Graham J. Hitch
University of York, UK

Robert Logie

As a first year PhD student at University College London in 1976, my supervisor presented me with a preprint of *The Psychology of Memory* by Alan D. Baddeley, which I proceeded to read cover to cover (I later bought a copy of the book). Its accessible, scholarly style was a major inspiration for my interest in research on the relationship between human memory and visual mental imagery. I first met Alan in 1980 when I was thrilled to be offered a job at the MRC APU (now the Cognition and Brain Sciences Unit) in Cambridge, UK, where I spent an incredibly stimulating, happy, and productive 6 years working with him on a wide range of research projects that reflected Alan's broad interests, including the cognitive impact of simulated deep sea diving (e.g. Logie & Baddeley, 1985), childhood dyslexia, working memory impairments in Alzheimer's disease, cognitive processes in counting, and face recognition. He was instrumental in igniting my interest in working memory that led directly to my own theoretical and empirical work on visual and spatial working memory (Logie, 1986, 1995). His ideas on working memory gave me a theoretical framework and direction for that research. Pursuing the synergy between rigorous theoretical and experimental cognitive psychology, and the application of theory and findings to cognition outside the laboratory was what researchers did at the APU, and it reflected the highly successful approach and mentorship of Alan Baddeley as unit director. Alan's influence also led to my own strong motivation to conduct theoretically driven applied research on a range of topics, including cognition in burglars, the design of digital memory systems, the design of computerized patient monitoring systems in intensive care, and the assessment and understanding of cognitive impairments associated with brain damage (e.g. Logie et al., 1992, 2018; Van der Meulen et al., 2010; and see Cubelli et al., Chapter 15, this volume).

Particularly memorable were the small, annual, informal working memory meetings that started in the early 1980s, led by Alan, and organized by Graham Hitch in a self-catering cottage owned by Manchester University near Satterthwaite in the English Lake District. We slept in bunk beds, took turns lighting the wood fire and cooking breakfast, climbed the local hills, and enjoyed supper in the local pub after a wonderful day of discussing experiments, data, and theory. After I left the APU for a lectureship in Aberdeen, we tried various locations in Scotland for the meeting before Alan and Graham came across Parcevall Hall, a wonderful 16th-century manor house (with 17th- and 20th-century additions) in rural Yorkshire. This has hosted those annual meetings since the early 1990s. Each year, Alan's deeply insightful comments offer solutions to puzzling data patterns, and his own presentations of data and theory inspire new directions for the research of those who come along from all stages of career. It is my favourite meeting and, at the time of writing, I am really looking forward to returning to Parcevall Hall in June 2022 after 2 years of video conferences during a global pandemic.

Throughout the rest of my career, I have continued to work with Alan from time to time, even when our theoretical views started to diverge, and greatly enjoyed being involved in organizing the three international working memory conferences in Cambridge (1994), Kyoto (2004), and Cambridge (2014), each timed to mark an additional decade following the seminal Baddeley and Hitch (1974) chapter. Now in his late eighties, he continues to be inspirational, publishing books, chapters, and papers, and generating ideas for new experiments, while maintaining his hallmark combination of scholarship, clear and creative thinking, and humour. He currently (2022) provides external supervision for PhD students working on long-term forgetting with me and Sergio Della Sala in Edinburgh. It is an honour and an enormous pleasure to have led the preparation of this tribute to a friend and colleague who remains a major figure in our field.

Robert H. Logie
University of Edinburgh, UK

Sue Gathercole

Without Alan's influence I simply wouldn't have had the rewarding academic career I have experienced. My lecturer and his former student Neil Thomson introduced the (at that time, pretty much new) concept of working memory, which Alan developed with Graham Hitch, in my undergraduate studies at York University in 1977. It was eye-opening, providing a perfect demonstration of how cognitive theory can transform understanding and illuminate human experience with relatively simple hypothetical structures.

The most transformative point in my career was joining the APU as a postdoc, providing the opportunity to work closely with the director, Alan. Until then, my research had been restricted to standard paradigms such as serial recall and the Eriksen flanker task to address theoretical issues in short-term memory and attention. I was increasingly feeling uneasy about restricting myself only to 'how' questions about mechanisms when even bigger 'why' questions were going unanswered. The most pressing one for me was why we even have a temporary memory system when neuropsychological patients with dramatically reduced memory spans can do just about anything—produce fluent speech, understand ordinary language, read—bar recalling lengthy shopping-list-like sentences. Alan had been intrigued by findings from the developmental linguist and theorist David Crystal, and speculated that short-term memory problems were the source of the disordered language development of many children that David had described. He suggested that we might work together on this and kickstarted a research programme, shared by many internationally, revealing the key role played by working memory in supporting not those skills already acquired but in learning new ones (as demonstrated by short-term memory patients), particularly in acquiring the sound structure of language. My later work exploring the broader consequences of working memory impairments on children's classroom function was directly informed by Alan's continuing curiosity and insights into the role played by working memory in what actually happens at the coalface of learning rather than just in the psychology laboratory. Some of his greatest admirers are to be found in the international community of education practitioners. From Australia to South Africa and throughout Europe, I have met teachers and speech and language therapists whose practical work in the classroom has been transformed by Alan's thinking.

Alan's personal and intellectual generosity and his equal interest in colleagues from all stages of career development mark him out as exceptional. Some of my greatest 'work' memories are of the annual working memory meetings created by Alan and Graham Hitch, in which literally anyone interested in the broad field congregated annually. The discussions were great but it is the hiking, garden chats, and long walks in the dark to the pub in their company that are truly memorable for anyone who has ever had the good fortune to be at the meeting in Parcevall Hall in North Yorkshire. I'm already looking forward to the next one this summer. Thank you for it all, Alan.

Susan E. Gathercole
University of Cambridge, UK

Nelson Cowan

My chapter includes a first-hand account of some important ways in which Alan Baddeley influenced me throughout my career. Here, I would like to provide a broader context. One of the interesting and exciting aspects of living through a part of history is that nobody had come along yet to sort out what was the most important in that history, or what events came first, second, and third. From my graduate school training in the 1970s in an infant speech perception laboratory, what I remember best was not the well-worked-out, impressively researched system of working memory from Baddeley and Hitch (1974). (I am not sure when I first read it but my sufficiently thorough reading was later, and quite compelling when I finally undertook it.) What I remember reading in graduate school was the exciting new ideas, and clever new methods, from Baddeley et al.'s (1975) article on the word length effect. Alan's work in that era was highly relevant to my interests: partly my interest in understanding the conscious mind, and partly my related interest in language processing. In that regard, Alan followed an esteemed tradition of applying memory research in a penetrating manner in the domain of language, and with special importance for language understanding. That tradition can be seen also in Ebbinghaus (1885/1913) and his description of what he called a participant's *first fleeting grasp* of a short series of nonsense syllables; in the work by Miller and Selfridge (1950) on memory for word series of various length and approximation to English; and in the title and content of Broadbent's (1958) book, *Perception and Communication*, which also focused on temporary memory.

For someone like me with a key interest in understanding consciousness, the topic of working memory helped to define the boundaries of what can reside in the human mind at one moment. In the field, there had been an ambivalence between indications that the limit is in *capacity*, specifically the number of items or chunks (Miller, 1956), and alternatively, indications that the limit is in the rate of *decay*, the duration for which an item can be held in temporary storage (Peterson & Peterson, 1959). Baddeley et al. (1975) deftly highlighted the important role of time and decay. Baddeley and Hitch (1974) also recognized a general, attention-related component of working memory storage that might well be capacity limited. The inspiration of a good question, trying to reconcile these two views, helped lead me to a theory (Cowan, 1988) in which these limits are both present in different ways, with a decay of activated features of long-term memory (or, from a more modular perspective, the phonological and visuospatial storage buffers of Baddeley, 1986) as well as a capacity limit in the focus of attention, a key subset of activated memory. Baddeley has remained intellectually open and flexible and, when he saw a need for a capacity limit, he adopted it also in the form of a possible characteristic of an episodic buffer (Baddeley, 2000, 2001). Without Alan Baddeley's work, I am not sure I would have perceived some of the most important frontiers of the field. Throughout many years, I have also been fortunate enough to enjoy a dialogue with him to discuss some of the most important issues from his work, based on both behavioural research and the neuropsychological evidence that has been so important, for which he has played a crucial role bridging disparate fields.

Alan has a remarkable way of combining penetrating scientific analysis, persistent interest in important questions, and warmth and civility. When I met him on a visit to the APU in Cambridge, UK, in the autumn of 1990, we went on a walk and he explained the latest intrigue in the neuropsychological findings about working memory, and talked about my own work too. That was an exhilarating experience for a youngish investigator, as I was then. I learned from his 1986 book that a habit of his was walking and working (tape recorder in hand). Also on that visit, I found Alan's dissertation in the unit's library and was astounded at how he had already established, in the 1960s, concepts, themes, and methods that he continued to pursue steadily, to great effect in the many following years. Further reflecting his civility was an unusual incident at a conference in Valencia, Spain, in 2001. I was scheduled to talk in a symposium just after Alan. The room was exceedingly packed in anticipation of Alan's talk. However, Alan didn't appear. It turned out that he was attending a previous talk that ended late and he was too polite to leave before it ended, and then also too polite to push through the crowd to enter the room to give his own talk. I ended up having to switch slots with him and benefited from the captive audience. These are only a few of the direct and roundabout ways that knowing Alan has helped me throughout the years.

Nelson Cowan
University of Missouri-Columbia, USA

Zhisheng (Edward) Wen

I have always cherished the great fortune to have been able to enter academia in the most exciting time of all (at the dawn of the 21st century) when the great work of Baddeley was already well established across the board. My PhD research (2004–2008) conducted at the Chinese University of Hong Kong put Baddeley's multicomponent model into practice in second language learning and processing. Since then, I have become so obsessed with working memory in my research, talks, publications, and even my daily life, that one of my senior colleagues (Bill Littlewood) has virtually nicknamed me 'WME' (aka: 'working memory Edward!'). My very first organized international roundtable held at the Hong Kong University of Science and Technology in 2012 was on working memory and second language acquisition, which later gave rise to my first edited volume entitled *Working Memory in Second Language Acquisition and Processing* (Wen et al., 2015) to which Alan Baddeley (2015) kindly contributed the first chapter, though the original plan was only to write a short preface! My first academic research monograph titled *Working Memory in Second Language Learning: Towards an Integrated Approach* (Wen, 2016) was also conceived and intended to emulate Baddeley's seminal work with Susan Gathercole on *Working Memory and Language* (1993). Alan's back-cover endorsement to that volume meant a lot to me!

My very first and formal encounter with Alan in person only came in July 2014 during the Third International Conference of Working Memory held in Cambridge, UK (organized by Sue Gathercole and Robert Logie). That brief encounter significantly injected further fuel into my already decade-long devotion to working memory and second language research. Baddeley's opening speech presented at that conference (entitled 'Working memory at 40') predicted that the topic 'Working memory and second language learning' could be one of the most promising areas of research in the decades to come. Further encouraged and boosted by Alan's vision, I continued my efforts to explore the role of working memory in learning and speaking a second language, and later named my foreign language aptitude theory the phonological/executive (or P/E) model, in which the 'P' part is ascribed to Baddeley's phonological loop (Baddeley et al., 1998). The ensuing 8 years of exciting endeavours culminated in the recent compendium entitled *The Cambridge Handbook of Working Memory and Language* (2022) in which Alan contributed the introductory chapter. Currently, I am working with Alan and Nelson Cowan on a shorter monograph text, *Working Memory in First and Second Language*, to be published by Cambridge University Press. Overall, I feel extremely blessed and grateful to have known and conversed with Alan personally and read his pioneering work over these years. Alan's insights and vision have always been instrumental in sustaining and boosting my academic pursuit of the working memory–second language 'enterprise' (as Alan has aptly called it). That short stint at Cambridge when I spent the most relaxed afternoon and evening chatting and drinking with Alan has constituted the most memorable highlight of my entire academic life.

Zhisheng (Edward) Wen
Hong Kong Shue Yan University, Hong Kong SAR, China

Randy Engle

Alan Baddeley has been hugely influential on me and my work over my entire career. The first time I met him on a visit to the APU in Cambridge, UK, it was like meeting a rock star. I would have been less nervous meeting Freddie Mercury. However, Alan was most gracious and welcoming. He has always had that combination of being positive and affirming about my work along with a tremendous ability to see the flaws and weaknesses. Even then he has a great ability to communicate a negative review by making it about the work and not about the worker. At a dinner in his honour many years ago, Pat Rabbitt said of Alan that he 'has a very wide bottom'. He then explained to us non-Brits that this was a nautical term used to compliment a boat that was very stable and capable of covering much area. I have always thought that Rabbitt characterized Baddeley well. Alan has had, and is still having, a remarkable career.

Randall W. Engle
Georgia Institute of Technology, USA

References

Baddeley, A. D. (1986). *Working memory* (Oxford Psychology Series No. 11). Clarendon Press.

Baddeley, A. (2000). The episodic buffer: a new component of working memory? *Trends in Cognitive Sciences*, 4(11), 417–423.

Baddeley, A. (2001). The magic number and the episodic buffer. *Behavioral and Brain Sciences*, 24(1), 117–118.

Baddeley, A. D. (2015). Working memory in second language learning. In Z. Wen, M. Mota, & A. McNeill (Eds.), *Working memory in second language acquisition and processing* (pp. 17–28). Multilingual Matters.

Baddeley, A. D. (2022). Working memory and challenges of language. In J. Schwieter & Z. Wen (Eds.), *The Cambridge handbook of working memory and language* (pp. 19–30). Cambridge University Press.

Baddeley, A., Gathercole, S. E., & Papagno, C. (1998). The phonological loop as a language learning device. *Psychological Review, 105*(1), 158–173.

Baddeley, A. D., & Hitch, G. J. (1974). Working memory. In G. H. Bower (Ed.), *The psychology of learning and motivation: Advances in research and theory* (Vol. VIII, pp. 47–89). Academic Press.

Baddeley, A. D., Thomson, N., & Buchanan, M. (1975). Word length and the structure of short-term memory. *Journal of Verbal Learning and Verbal Behavior, 14*(6), 575–589.

Broadbent, D. E. (1958). *Perception and communication.* Pergamon Press.

Cowan, N. (1988). Evolving conceptions of memory storage, selective attention, and their mutual constraints within the human information-processing system. *Psychological Bulletin, 104*(2), 163–191.

Ebbinghaus, H. (1913). *Memory: A contribution to experimental psychology* (H. A. Ruger & C. E. Bussenius, Trans.). Teachers College, Columbia University. (Originally in German, *Ueber das gedächtnis: Untersuchen zur experimentellen psychologie*) (Original work published 1885)

Gathercole, S., & Baddeley, A. (1993). *Working memory and language.* Lawrence Erlbaum Associates.

Logie, R. H. (1986). Visuo-spatial processing in working memory. *Quarterly Journal of Experimental Psychology, 38*(2), 229–247.

Logie, R. H. (1995). *Visuo-spatial working memory.* Lawrence Erlbaum Associates.

Logie, R. H., & Baddeley, A. D. (1985). Cognitive performance during simulated deep-sea diving. *Ergonomics, 28*(5), 731–746.

Logie, R. H., Wolters, M., & Niven, E. H. (2018). Preserving and forgetting in the human brain. In V. Mezaris, C. Niederee, & R. H. Logie (Eds.), *Personal multimedia preservation: Remembering or forgetting images and video* (pp. 9–45). Springer.

Logie, R. H., Wright, R., & Decker, S. (1992). Recognition memory performance and residential burglary. *Applied Cognitive Psychology, 6*(2), 109–123.

Miller, G. A. (1956). The magical number seven, plus or minus two: Some limits on our capacity for processing information. *Psychological Review, 63*(2), 81–97.

Miller, G. A., & Selfridge, J. A. (1950). Verbal context and the recall of meaningful material. *American Journal of Psychology, 63*(2), 176–185.

Peterson, L. R., & Peterson, M. J. (1959). Short term retention of individual verbal items. *Journal of Experimental Psychology, 58*(3), 193–198.

Schwieter, J., & Wen, Z. (Eds.). (2022). *The Cambridge handbook of working memory and language.* Cambridge University Press.

Van der Meulen, M., Logie, R. H., Freer, Y., Sykes, C., McIntosh, N., & Hunter, J. (2010). When a graph is poorer than 100 words: A comparison of computerised natural language generation, human generated descriptions and graphical displays in neonatal intensive care. *Applied Cognitive Psychology, 24*(1), 77–89.

Wen, Z. (2016). *Working memory and second language learning: Towards an integrated approach.* Multilingual Matters.

Wen, Z., Baddeley, A., & Cowan, N. (in press). *Working memory in first and second language.* Cambridge University Press.

Wen, Z., Mota, M. G., & McNeill, M. (2015). *Working memory in second language acquisition and processing.* Multilingual Matters.

Acknowledgements

The editors would like to thank the authors of chapters in this volume for accepting the invitation to contribute, and Martin Baum at Oxford University Press for support, encouragement, and patience from inception to completion of this volume.

The editors are happy to confirm that net royalties from sales of this book will be donated to the charity Water Aid.

Contents

11. **Parent–child autobiographical reminiscing as a foundation for literacy, memory, and science education**   273
Robyn Fivush, Catherine A. Haden, and Elaine Reese

12. **Working memory in language learning and bilingual development**   295
Michael F. Bunting and Zhisheng (Edward) Wen

PART 3   IMPAIRMENTS OF MEMORY

13. **Age-related changes in everyday prospective memory**   325
Fergus I. M. Craik and Julie D. Henry

14. **Mental imagery: using working memory theory to design behaviour change interventions**   355
Jackie Andrade

Contributors

Richard J. Allen, PhD
Associate Professor
School of Psychology
University of Leeds
Leeds, UK

Jackie Andrade, PhD
Professor
School of Psychology
University of Plymouth
Plymouth, UK

Alan Baddeley, PhD, CBE, FRS, FBA,
FMedSci
Emeritus Professor
University of York
York, UK

Vicki Bruce, PhD, DBE, FBA, FRSE
Emeritus Professor
Department of Psychology
University of Newcastle
Newcastle, UK

Adam Bulley, PhD
Postdoctoral Fellow
Department of Psychology
Harvard University
Cambridge, MA, USA
School of Psychology and Brain and
Mind Centre
The University of Sydney
Sydney, New South Wales, Australia

Michael F. Bunting, PhD
Senior Scientist
Centre for the Advanced Study of Language
University of Maryland
College Park, MD, USA

A. Mike Burton, PhD, FBA, FRSE
Professor
Department of Psychology
University of York
York, UK

Alexander P. Burgoyne, PhD
Postdoctoral Researcher
School of Psychology
Georgia Institute of Technology
Atlanta, GA, USA

Nelson Cowan, PhD
Curators' Distinguished Professor
Department of Psychological Sciences
University of Missouri-Columbia
Columbia, MO, USA

Fergus I. M. Craik, PhD
University Professor Emeritus in Psychology
University of Toronto
Toronto, ON, Canada

Roberto Cubelli, PhD
Professor
Department of Psychology and Cognitive
Sciences
University of Trento
Trento, Italy

Sergio Della Sala, MD, MSc, PhD, FRSA,
FRSE, FBPsS
Professor
Department of Psychology
University of Edinburgh
Edinburgh, UK

Randall W. Engle, PhD
Professor
School of Psychology
Georgia Institute of Technology
Atlanta, GA, USA

Robyn Fivush, PhD
Samuel Candler Dobbs Professor of
Psychology
Department of Psychology
Emory University
Atlanta, GA, USA

Susan E. Gathercole, PhD, FBA, OBE
Emeritus Professor
University of Cambridge
Cambridge, UK

Agnieszka J. Graham, PhD
Lecturer
School of Psychology
Queen's University Belfast
Belfast, UK

Catherine A. Haden, PhD
Professor of Developmental Psychology
Loyola University
Chicago, IL, USA

Rebecca K. Helm, PhD
Associate Professor
University of Exeter Law School
Exeter, UK

Julie D. Henry, PhD
Professor
Department of Psychology
University of Queensland
Brisbane, Queensland, Australia

Graham J. Hitch, PhD
Emeritus Professor
Department of Psychology
University of York
York, UK

Robert H. Logie, PhD, FBPsS, FPsyS, FRSE
Professor of Human Cognitive Neuroscience
Department of Psychology
University of Edinburgh
Edinburgh, UK

Cody A. Mashburn, BS
Graduate Student
School of Psychology
Georgia Institute of Technology
Atlanta, GA, USA

Elaine Reese, PhD
Professor
Department of Psychology
University of Otago
Otago, New Zealand

Valerie F. Reyna, PhD
Lois and Melvin Tukman Professor of Human Development
Department of Psychology
Cornell University
Ithaca, NY, USA

Henry L. Roediger III, PhD
James S. McDonnell Distinguished University Professor
Washington University in St. Louis
St. Louis, MO, USA

Daniel L. Schacter, PhD
William R. Kenan, Jr. Professor of Psychology
Department of Psychology
Harvard University
Cambridge, MA, USA

Amanda H. Waterman, PhD
Professor in Cognitive Development
School of Psychology
University of Leeds
Leeds, UK

Zhisheng (Edward) Wen, PhD
Professor
Department of English Language and Literature
Hong Kong Shue Yan University
Hong Kong SAR, China

Barbara A. Wilson, PhD, OBE
Founder
Oliver Zangwill Centre for Neuropsychological Rehabilitation
Ely, UK
St George's Hospital
London, UK

John T. Wixted, PhD
Distinguished Professor
Department of Psychology
University of California, San Diego
San Diego, CA, USA

Tian-xiao Yang, PhD
Associate Professor
Institute of Psychology
Chinese Academy of Sciences
Beijing, China

1
Introduction

When applying memory theory does, and does not work

Robert H. Logie, Zhisheng (Edward) Wen, Susan E. Gathercole, Nelson Cowan, and Randall W. Engle

Memory is at the core of functioning in everyday life, and understanding the principles of how it functions has benefited from substantial theoretical developments based on well-controlled laboratory studies. However, there are major challenges when it comes to direct application of theories and principles of memory function to practical problems in society. This is in part because there are multiple ongoing debates between seemingly rival memory theories. Even more challenging is that problems outside the laboratory are complex: key factors are difficult to identify, and often are impossible to control. A theory might be robust in a laboratory setting, but fail to capture all that is important when taken out of the laboratory. Conversely, the solution to a practical problem might not offer generalized principles of memory that can be deployed to address completely different problems in a different context. The study of when and why memory succeeds and fails outside the laboratory typically is focused on specific practical problems. Moreover, laboratory-based theory development often has been carried out by different researchers from those who tackle problems in society. The good news is that times are changing. Laboratory-based memory theory increasingly is being applied to aid solutions to major challenges in society, with the same researchers developing the theory and seeking practical solutions. We have come a very long way since the first Practical Aspects of Memory Conference when Neisser (1978) observed that 'If *X* is an interesting or socially significant aspect of memory, then psychologists have hardly ever studied *X*'. Since those remarks, a wide range of socially significant aspects of memory have been studied in considerable detail, and many substantial advances in theories of the general principles of memory have benefited from investigation outside of the laboratory, in addition to those theories helping to address practical problems.

This book offers chapters by leading international researchers who have demonstrated how substantial understanding of important practical problems can be gained by drawing on memory theories derived from empirical studies in the laboratory, and in turn have shown how testing the applicability of memory theories to a range of practical problems can lead to important refinements and generalizability of the theory. Although the synergy between theory and application in

Robert H. Logie, Zhisheng (Edward) Wen, Susan E. Gathercole, Nelson Cowan, and Randall W. Engle, *Introduction* In: *Memory in Science for Society*. Edited by: Robert H. Logie, Zhisheng (Edward) Wen, Susan E. Gathercole, Nelson Cowan, and Randall W. Engle, Oxford University Press. © Oxford University Press 2023. DOI: 10.1093/oso/9780192849069.003.0001 memory research was inspired by Neisser from his comments in 1978, it has been a long tradition in the UK from the work of Bartlett (e.g. 1932), Broadbent (e.g. 1958), and notably Baddeley (e.g. 1966, 1990, 2019, Chapter 2, this volume). Alan Baddeley is the world's current leading authority on the interaction between human memory theory development and its application, with a record of such research spanning more than five decades and still ongoing. His approach to research serves as an inspiration for the title and contents of the current book, which also draws on the work of multiple researchers worldwide who adopt the approach of developing memory theory in science with a passion for applying that science to a wide range of challenges in society. The book demonstrates that scientific passion has found its way into addressing challenges in the areas of education, intelligence and life attainment, second language learning, autobiographical memory, memory for faces and eyewitness testimony, future planning and thinking, lifespan cognitive development and age-related cognitive decline, following instructions, and assessment and rehabilitation of cognitive impairment following brain damage. Although there have been many successes, individual chapters also discuss both when memory theory has and when it has not (at least as yet) been helpful in addressing practical issues beyond the memory laboratory.

Chapter 2 is by Alan Baddeley. It describes his approach of undertaking both theory-driven laboratory and applied research, influenced strongly by the work of researchers at the Medical Research Council Applied Psychology Unit in Cambridge, UK. Two of the editors (Logie and Gathercole) and several authors of chapters in this book (Andrade, Della Sala, Hitch, and Wilson) also were strongly influenced by early and mid-career experience at that Unit, while many of the other contributors to this book had extended visits. Baddeley has offered an extensive, informative, and entertaining historical discussion of his work elsewhere (Baddeley, 2019). In the current volume, he briefly describes those historical influences, then goes on to discuss three more recent projects on applying memory theory to memorizing the Qur'an, to second language learning, and to what determines rates of long-term forgetting in Alzheimer's disease and epilepsy. In so doing, he highlights some important lessons from historical work that are crucial for successful contemporary application of memory theory. One of these is the importance of working closely with practitioners and expert researchers in the area of application, both to thoroughly understand the practical problems and to facilitate the implementation of any useful insights, tests, or procedures that might arise from carrying out potentially applicable research. This is also a major theme for the chapters in this volume by Bruce and Burton (Chapter 3) on face identification, by Wixted and Roediger (Chapter 4) on eyewitness identification, by Cubelli, Logie, and Della Sala (Chapter 15) on memory impairments following brain damage, and by Wilson (Chapter 16) on memory rehabilitation, and appears consistently as a theme in many other chapters. As Wilson notes, it is crucial to know when memory theory might help, and when it might not.

A second lesson from the historical perspective is to avoid 'reinventing the wheel'. It is of course important to show that findings replicate, but Baddeley notes that contemporary research does not always recognize that it is replicating highly relevant previous research published before the age of the internet. Sometimes what is described as novel and cutting-edge in contemporary research effectively is offering a new perspective or even simply a relabelling of ideas and findings from a previous era without recognizing the original source. This theme also arises in Chapter 15 by Cubelli, Logie, and Della Sala, although all chapters discuss the historical context as well as the leading contemporary theories and findings for the research work that they describe. Therefore, the major thrust throughout the book is to build substantially on what has gone before, both to advance scientific understanding and to discover how that understanding can help the wider society.

The remaining chapters are arranged under three broad headings: 'Memory challenges in adults', 'Memory development', and 'Impairments of memory'.

To kick off the 'Memory challenges in adults' section, the errors in eyewitness identification are addressed in Chapter 3 from Vicki Bruce and Mike Burton, and in Chapter 4 from John Wixted and Henry Roediger, III. Bruce and Burton note that most people have a particular difficulty in recognizing unfamiliar faces when they are subsequently encountered, both in daily life and when an eyewitness is asked to identify whether they recognize the perpetrator of a crime. This has very wide implications for the use of photographs of faces for official identification by a human observer, such as in passports and driving licences. Bruce and Burton also note the importance of this finding for development of theories of face identification that tend to be focused on the relative ease with which the familiar faces of family, friends, and well-known public figures are identified. Crucially, such theories also should account for the inability to recognize that multiple photographs of an unfamiliar face depict the same person in different lighting conditions, poses, or at different ages.

Wixted and Roediger note that the difficulty with recognizing unfamiliar faces can be compounded if a witness to a crime is asked repeatedly to identify a potential suspect (whose face is typically unfamiliar) in live or photograph line ups, and then again in court: confidence in their judgement can be increased by the repeated experience, even if their initial identification was very uncertain. They argue that only the initial identification coupled with the stated confidence at that time should be used in criminal investigations, and low initial confidence should raise concerns about the strength of this initial identification as evidence in the subsequent court proceedings. This challenges the widely held view in the memory research literature that the confidence of a witness is poorly correlated with accuracy, but shows that confidence can be manipulated by the process of investigation, thereby reducing what may be an initial very strong link between confidence and identification accuracy. Multiple attempts at identification can be seen as analogous to contaminating other evidence from a case. Wixted and Roediger refer to some striking and deeply troubling real-life cases, where this form of 'contamination' of the witness evidence from multiple attempts at identification has led to conviction even when there is other clear evidence of the innocence of a suspect.

In Chapter 5, Rebecca Helm and Valerie Reyna discuss findings suggesting that when people make decisions in everyday life, they might rely on information that is detailed and accurate, referred to as verbatim representations, or, perhaps more frequently, rely on the general meaning or key details, referred to as gist representations. This distinction is set in the context of fuzzy trace theory, suggesting that a 'fuzzy' or gist memory trace is a core feature of both memory and decision-making. In many cases, reliance on a gist representation can be efficient and helpful when making decisions. However, in a range of scenarios, this reliance on gist can result in poor decisions about personal health, a lack of understanding of medical information, and behaviour that risks negative consequences. Helm and Reyna argue that it may even result in innocent suspects pleading guilty to a crime that they did not commit. One finding that might seem counterintuitive is that young children appear to rely less on gist memory than do adults: this may result in more accurate testimony from a child witness, and a greater chance of an inaccurate or false memory from an adult.

The impact of memory representations on decision-making also is addressed by Adam Bulley and Daniel Schacter in Chapter 6. Here, the focus is on future thinking, or to what extent possible future scenarios affect decisions in daily life. In many cases there is 'delay discounting' or a focus on immediate consequences, while largely ignoring how a decision might affect an individual in the future. Examples are drawn from financial, health, relationships, politics, and impact on the environment, along with interventions that might help people reduce delay discounting and make better decisions. However, an important caveat is the need to identify specific decisions that might benefit from delay discounting, so that applications of laboratory-based theory can be targeted effectively.

In Chapter 7, Cody Mashburn, Alexander Burgoyne, and Randall Engle continue the theme of cognition in everyday life, by exploring in depth the variation across individuals in working memory capacity, control of attention, and fluid intelligence or the ability to think flexibly and solve novel problems. The chapter first describes the theoretical and empirical basis of each of these cognitive abilities, how they are measured, and how they are defined. Particularly striking are the findings that cognitive abilities are stable over the lifespan: someone with high fluid intelligence at age 11 will also have high fluid intelligence in later life. Mashburn and colleagues discuss important and conclusive evidence for how each of these cognitive abilities is related to academic success, career development, and both mental and physical health, while recognizing that other factors also contribute to life success such as socioeconomic status.

For the final chapter in this section, Chapter 8, Graham Hitch addresses the problem of how we can retain and recall the order of items that we encounter in our daily lives, whether this is remembering the sound sequences when learning new words, sequences of numbers for a security code, a telephone contact, the address of a new friend, or sequences of pictures. Although this seems a very simple task, how we do it is very challenging to explain. Hitch demonstrates why obvious explanations don't work, and how the development of computational models of serial order memory have offered significant insights. For example, it could be that serial order is remembered as a chain of associations, so the first item acts as a cue to the second item, that acts as a cue to the third item, and so on. However, if this were to be the case, then, for a list of six items, if we forget item 4, then we should also forget items 5 and 6, because the cue for item 5 has been forgotten. However, typically, people can recall later items, even if they forget earlier ones in a list. Also, a common error is to swap the order of adjacent items during recall, for example, recalling item 4 before recalling item 3 in the list. So, this simple explanation, known as 'chaining' simply does not work. Hitch describes some alternative theories, and describes in some detail one that he has developed with colleagues in which the order is retained by linking each list item with its position in the list, providing a context for each item. As each item is recalled, the context shifts to maintain the serial order. This kind of computational model can generate errors that are similar to those obtained in recall of serial order by human volunteers. However, limitations remain and it may be that humans have more than one way in which they remember serial order, so elements of more than one theory may offer a more complete solution.

Nelson Cowan opens the next section on 'Memory development' with a discussion in Chapter 9 of how his own research on working memory theory was inspired by the work of Donald Broadbent and Alan Baddeley in the UK, and motivated by understanding consciousness. At the core of his theory is the control of limited capacity attention coupled with information that is currently activated, or readily available from stored knowledge. A further important feature is the concept of 'chunking' or combining several items into a single temporary memory representation. This can lead to substantial increases in the amount of material that can be recalled, particularly if the chunk is associated with some meaning, such as the sequence of letters U–S–A, or a sequence of words in a coherent sentence. Cowan shows how these basic theoretical assumptions have been used to understand aspects of cognitive development during childhood and adolescence, with implications for adult ageing, and for the impact of memory impairments that arise in Down and Williams syndromes, autism spectrum disorders, language disorders, and dyslexia.

Chapter 10 continues the theme of remembering sequences, in this case, the ability of children and adults to remember and carry out a set of instructions. Richard Allen, Amanda Waterman, Tian-xiao Yang, and Agnieszka Graham note how ubiquitous this requirement is in everyday life, whether it involves instructions from a kindergarten teacher, constructing flat-pack furniture, or astronauts carrying out tasks on the International Space Station. Allen and colleagues work within the theoretical working memory framework most closely associated with Baddeley, comprising a central executive, an episodic buffer, and specialist temporary stores for verbal sequences or temporary visual representations. They point to other theoretical work that suggests an additional component that acts as a temporary store for sequences of movements that is also involved in action planning. One important finding is that if presentation of the instructions is accompanied by the participant miming the actions to be taken, then recall of the action sequence is better than if only verbal instructions are given. There are also intriguing findings that children who have difficulty in following sequences of instructions in class appear also to have poor working memory capacity. It is not simply that they are failing to pay attention to the teacher. A practical intervention from these insights is to suggest that such children are given shorter sequences of instructions, or to wait for the child to perform one instructed action before giving the next instruction. Also, asking typically developing children or healthy adults mentally to imagine performing each action as each instruction is given could enhance subsequent recall.

Shared recollection of events and its role in child development is the focus for Chapter 11 by Robyn Fivush, Elaine Reese, and Catherine Haden. The emphasis here is on real-life remembering in conversations about a shared experience, with lucid examples of children's learning of literacy and scientific concepts through dialogue with adults, notably their parents, and particularly mother–child reminiscing. They note the core role of autobiographical remembering in maintaining a sense of continuity of self across the lifetime, along with the emotions felt during a past event with the strong recollection that 'the event happened to me' rather than to someone else. This also gives a sense of being influenced by and holding a set of cultural values, and to construct a personal, coherent narrative about our lives. This then provides a theoretical framework to understand how and why personally experienced events are recalled and shared with others through language.

Another important aspect of the links between memory and language development is discussed by Michael Bunting and Zhisheng Wen in Chapter 12. They show how the concept of a phonological loop within Baddeley's (see Baddeley, Chapter 2, this volume; Baddeley et al., 2021) view of working memory was shown to be important, not only for short-term verbal memory, but also for acquiring language. The ability to remember a verbal sound sequence had been shown by Gathercole (e.g. Gathercole & Baddeley, 1993) to be important for remembering the sound sequence within a new word. However, language is broader than learning vocabulary, and Bunting and Wen go on to show how other aspects of language learning and development of bilingualism can be set in the context of Cowan's embedded processes model (see Cowan, Chapter 9, this volume) and Engle's view on attentional control (see Mashburn et al., Chapter 7, this volume).

The section on 'Impairments of memory' opens with a discussion by Fergus Craik and Julie Henry in Chapter 13 on remembering to do things, known as prospective memory, and how this ability changes with adult ageing. As with many cognitive abilities, including memory, there is a notable decline as people get older, and prospective memory is no exception, particularly when remembering to carry out some intended action is self-initiated. Older people can use external cues such as notes or even specific times as very effective reminders, so can get help from their environment to offset underlying cognitive decline. This, together with high motivation for older people to do as well as they can in research studies, helps to explain why older people often perform rather better than younger people in prospective memory research studies that take place in their own familiar environment, but perform more poorly in laboratory-based studies. This theoretically driven research on what influences prospective memory leads on to interventions, such as environmental support, that could help older people avoid forgetting to deal with finances, take medication (or avoid taking it twice by mistake), keep appointments, or lock the door when going out of their home.

A very different set of challenges has been addressed in Chapter 14 by Jackie Andrade who has conducted extensive research on the use of visual working memory, and particularly visual imagery as an intervention to deal with cravings for food, alcohol, or other behaviours that can have negative consequences on physical and mental health. There is a detailed discussion of the theoretically motivated research on visual mental imagery, set within the context of the multiple component framework for working memory. This is followed by the theoretical notion of elaborated intrusion, referring to intrusive thoughts linked, for example, with food cravings. Andrade goes on to describe evidence for the successful use of functional imagery training to control the intrusive thoughts that in turn leads to a reduction in the cravings. This represents highly successful applications of research on mental imagery to the underlying causes of specific impairments of thoughts and behaviours.

In the discussion of the neuropsychology of working memory in Chapter 15, Robert Cubelli, Robert Logie, and Sergio Della Sala argue that any memory theory should be able to predict, and account for, the range of different memory impairments that are observed following brain damage, as well as provide an understanding of memory in healthy individuals. They provide highlights of over a century of research on the impact of brain injury on cognition, including the focal lesions resulting from a stroke or tumour, and the more widespread, progressive damage associated with Alzheimer disease and other forms of dementia and diseases that affect the brain. They then offer a more detailed tour of almost five decades of research showing how the multiple component theoretical framework for working memory has provided insights into the wide range of different, but very specific patterns of deficit and sparing associated with different forms of brain damage. This enhanced understanding can help target the help needed for the specific deficit identified. For example, a patient might have a specific problem in remembering sequences of numbers, words, or instructions, but have no difficulty in finding their way around or remembering what they have done during the day. Other patients might show the opposite pattern of impaired and intact abilities. Cubelli and colleagues show further how the study of brain-damaged individuals has led to developments in the theoretical framework that have also been useful for understanding healthy working memory. They express a concern that many researchers who develop theories of memory tend to focus on healthy individuals, and tend not to consider evidence from neuropsychological studies that challenges theoretical assumptions. So, applying theory to neuropsychology can benefit clinical management and support for patients, with a reciprocal benefit for theoretical development.

In the final chapter, Chapter 16, Barbara Wilson describes her extensive clinical and research experience in developing techniques to support individuals with memory impairments, and new skills that such individuals might learn to reduce the impact of brain damage on their memory abilities. She takes a very pragmatic approach, identifying when memory theory has and when it has not been helpful. One example of success is the use of errorless learning with patients who have largely intact working memory, but who suffer from long-term memory impairments, known as amnesia. Because this technique avoids mistakes in recall, this minimizes the possibility of interference from any errors with subsequent accurate recall. However, some aspects of memory support or rehabilitation have required a very practical solution that was not theoretically motivated. Wilson emphasizes the importance of treating each patient as an individual, and the need to directly address the practical problems that each encounters in their daily lives. Laboratory-based theory can help in some circumstances, but does not replace the practical knowledge and experience of the clinician or the care provider. Wilson sums up her recipe for success in neuropsychological rehabilitation with a quote from Yogi Berra: 'In theory there is no difference between theory and practice. In practice there is.'

Finally, we note with the greatest sadness a missing chapter. Martin Conway (1952–2022) was yet another researcher whose work began in the early 1980s at the Applied Psychology Unit and who was inspired and supported by Alan Baddeley. Throughout his career Martin was intrigued by autobiographical memory, the complex memory system that supports our engagement with everyday life. He showed that this is achieved through integration with cognitive systems that abstract meaningful and personally relevant material from our past and integrate it with our current and future experience, stabilizing the individual and their sense of self and supporting life goals across all time frames. At the beginning of Martin's career almost nothing was known about autobiographical memory and there were few methods to guide its research. Combining flair, steadfastness, and a natural ability for theoretical abstraction, Martin laid the foundations for current understanding of the many facets of autobiographical memory—its cognitive structure (Conway & Pleydell-Pearce, 2000), neurobiological underpinnings, psychoanalytic role and links to the self (Conway, 2005), clinical disturbances, and cultural manifestations (Conway, 2013). Through creative collaborations with poet Bill Herbert and artist Shona Illingworth, his view of autobiographical memory came alive and was made accessible to the public in a range of artistic media (https://www.city.ac.uk/news-and-events/news/2017/06/the-art-and-science-of-memory). Martin was himself a poet, a wise and witty communicator who will be greatly missed by all those who knew and have been influenced by him. His contributions to psychological science will, of course, live on.

The quote from Berra, in Chapter 16 by Wilson, emphasizes a key point made by Baddeley in Chapter 1 that any researcher carrying out what might now be referred to as 'knowledge exchange' should engage practitioners in the area of application. This is essential both for understanding the practical problems and to facilitate the implementation of any useful insights, tests, or procedures that might arise from conducting potentially applicable research. We hope that like the contributors to this book, readers will be persuaded of the reciprocal benefits, for theory and practice, of work that seeks to apply research.

References

Baddeley, A. D. (1966). Influence of depth on the manual dexterity of free divers: A comparison between open sea and pressure chamber testing. *Journal of Applied Psychology, 50*(1), 81–85.

Baddeley, A. D. (1990). *Human memory: Theory and practice*. Lawrence Erlbaum.

Baddeley, A. D. (2019). *Working memories: Postmen, divers, and the cognitive revolution*. Routledge.

Bartlett, F. C. (1932). *Remembering: A study in experimental and social psychology*. Cambridge University Press.

Broadbent, D. E. (1958). *Perception and communication*. Pergamon Press.

Conway, M. A. (2005). Memory and the self. *Journal of Memory and Language, 53*(4), 594–628.

Conway, M. (2013). *Flashbulb memories*. Psychology Press.

Conway, M. A., & Pleydell-Pearce, C. W. (2000). The construction of autobiographical memories in the self-memory system. *Psychological Review, 107*(2), 261–288.

Gathercole, S., & Baddeley, A. D. (1993). *Working memory and language*. Lawrence Erlbaum Associates, Inc.

Neisser, U. (1978). Memory: What are the important questions? In M. M. Gruneberg, P. E. Morris, & R. N. Sykes (Eds.), *Practical aspects of memory* (pp. 3–24). Academic Press.

2

On applying cognitive psychology

Alan Baddeley

I began my research career in the late 1950s, at the Medical Research Council Applied Psychology Unit (APU) in Cambridge, UK, aiming for a PhD on theoretical aspects of the design of postal codes with my boss Conrad dealing with the practical aspects. I tried to do both, combining theory with practice and produced a splendid set of codes based on information theory and the letter structure of English. Alas, these were politely declined by the Post Office who had already decided on a code structure; a useful lesson but not too discouraging as I have been attempting to combine basic and applied psychology ever since. Hence, I was pleased to be asked to write this introduction to the present volume. My delight cooled somewhat when the question arose as to just what to include. It is easy to say why combining basic and applied research is rewarding. It allows our theories and techniques to be extended beyond the constrained limits of the laboratory, throwing light on their generality and potentially raising new and interesting questions. It also offers the possibility of working with new and interesting people in previously unfamiliar areas. And finally, it might even be useful! My challenge was how to capture and illustrate all of these.

I remember well a series of talks whereby six members of the APU each talked about the question of how best to apply psychology. I was last and feared there would be nothing left to say, only to discover all the talks were informative and convincing but quite different from each other, as will, I suspect, be the chapters that follow this introduction. Furthermore, I have already attempted to cover my own approach to combining basic and applied work at book length (Baddeley, 2019) and do not feel up to the challenge of compressing the book into a brief introduction. I have decided instead to describe the research styles that have continued to influence me throughout my career, typified by the work of three people encountered during my early years. I illustrate this continued influence by briefly describing three projects that have excited me during the last 5 years, before concluding with some general thoughts on the problems and rewards of applying cognitive psychology.

I spent the first 9 years of my research career at the APU in Cambridge, which had the remit of bridging the gap between theory and application in psychology. It was founded in 1944 on the basis of a range of wartime projects, and directed initially by Kenneth Craik, the brilliant young originator of the concept of theories as models (Craik, 1943) who developed and applied a computational model to the practical issue of gun aiming. He was, however, tragically killed in

Alan Baddeley, *On applying cognitive psychology* In: *Memory in Science for Society*. Edited by: Robert H. Logie, Zhisheng (Edward) Wen, Susan E. Gathercole, Nelson Cowan, and Randall W. Engle, Oxford University Press. © Oxford University Press 2023. DOI: 10.1093/oso/9780192849069.003.0002 a cycling accident in 1944, leaving Frederick Bartlett to take over. By the time I joined the APU, Donald Broadbent had just become director and his classic book *Perception and Communication* had just been published (Broadbent, 1958). In it he attempted to apply a broad information processing theoretical framework to a wide range of experimental data, much of it collected by the APU during the preceding years.

The book's general spirit had much in common with the model-based approach to theorizing advocated by Craik and at its centre was Broadbent's model of attention, a topic that was ignored as experimentally intractable by the neo-behaviourist approaches to theory that dominated US psychology at the time. Broadbent showed that it could be a very fruitful topic by linking the experimental study of practical problems with theoretical development based on modelling within a broad information processing approach. Similar developments were evolving in North America and these were subsequently reviewed and presented to a wider readership in Ulric Niesser's (1967) influential book *Cognitive Psychology* which gave its name to this rapidly developing field. Broadbent's early model which combined research on attention with that on short-term memory (STM), subsequently influenced the 'modal model' proposed by Atkinson and Shiffrin (1968). This in turn aimed to integrate the extensive memory research that had been carried out within the information processing framework over the previous decade. In due course, this influenced the Baddeley and Hitch (1974) multicomponent model of working memory, as we described in the symposium marking the 50th anniversary of the classic Atkinson and Shiffrin paper (Baddeley, Hitch, & Allen, 2019). My own approach reflects Broadbent's influence in attempting to create and develop a model that has at its core the interaction of memory and attention and as its aim, the capacity to extend beyond the laboratory and to be applicable to a range of practical problems.

A very different influence on my approach to psychology came from one of the assistant directors, Christopher Poulton, a very active experimentalist who studiously avoided theorizing. He responded to many requests for advice of a practical nature and was happy to tackle a wide range of issues from researching and advising *The Times* newspaper on a new typeface to assessing the effect of anoxia on Everest climbers and from the role of anticipation in skilled performance to optimizing the multiscreen displays that were being introduced in order to monitor the motorways (freeways) that were just being constructed in the UK at the time. Although he was rather dismissive of theory, he was a highly innovative methodologist who appeared to relish the challenge of collecting good and reliable data under a wide range of challenging situations beyond the laboratory. He was particularly critical of the failure of many published studies to guard against 'asymmetrical transfer effects' in experiments containing two or more conditions involving within-participant designs. Counterbalancing is the usual way of dealing with the effects of one condition, A, on a subsequent condition, B. Poulton pointed out that the overall result may be quite different when the order of conditions is reversed (Poulton & Freeman, 1966). For example, an effective strategy applied in A may carry over to B, whether appropriate or not. Similarly, frustration with a difficult condition A may lead to de-motivation for easier condition B which may then lead to underperformance. A year spent with S. S. Stevens, a very influential sensory theorist at Harvard University, led him to criticize his host's conclusions on the grounds that they were based on psychophysical methods in which repeated judgements were made by the same participant (Poulton, 1979). He went on to demonstrate that quite different results were obtained when each observation was based on a different single judgement by each person, a theoretical point that is I believe now generally accepted, though the demanding single observation method is itself rarely followed (though see Baddeley & Scott, 1971, for its application to STM).

The second assistant director, Conrad (he resisted the use of his first name) was, in my own view, an outstanding applied psychologist who also made important contributions to basic theory. Having completed a PhD based on the practical task of monitoring a series of dials each operating at a different rate, he was invited to apply his multitasking findings to the manning of telephone exchanges. He succeeded in doing so in a way that impressed not only management but also telephonists who welcomed a work rate comfortably set between boredom and overload. As a result, he became an influential advisor to the General Post Office, at that time a government department responsible for both post and telephones. In return, the Post Office agreed to fund someone to work on a project of mutual interest. This resulted in my own appointment to work on the design of postal codes and to register my work for a PhD.

Much of Conrad's effectiveness was based on the extent to which he was embedded in the system, encouraging the awareness of the importance of human factors and adding advice that was recognized by both unions and management. One instance of this comes through the development of letter sorting machines where he first persuaded the Post Office to avoid a machine-paced system where one letter is delivered at a standard timed rate, as developed by the US postal service. He convinced the engineers to develop an unpaced system in which the response to one item calls up the next. This proved to be good advice since not all letters are equally easy to process; hence with a fixed pace system a slow response can lead to a cluster of errors and a breakdown of processing, as was later found with the paced US system. The resulting unpaced machine was successful in avoiding this problem, but in observing the output, Conrad noticed that the distribution of sorting times was not characteristic of that found in the basic reaction time literature. He then realized that the sorters were inserting a brief pause until they heard a particular auditory signal from the machine. He went on to discover that if the response was too rapid the operator could beat the machine, resulting in a jam. The sorters had learnt to use the auditory cue to avoid jams but at the cost of unnecessary unfilled time. A discussion with the engineers resolved the need for this, leading to a smoother and faster sorting. Conrad went on to calculate that the average time saved per letter was very little, but, as he pointed out, multiplied by the number of letters per year would result in a very substantial financial saving.

Conrad was able to operate so effectively because he was accepted and respected within the organization controlling the postal system. Embedding within an institution has a number of advantages, not least that of thoroughly understanding the background, achieving mutual trust, and, from the psychologist's viewpoint, providing a valuable way of validating existing concepts and developing new ones. However, this depends on having a suitable host together with flexible long-term support, in our case provided by the Medical Research Council, an enviable but all too rare situation.

Conrad later switched to another applied environment, that of the education of the deaf, initially noting the association between congenital deafness and problems in learning to read, linking this to the potential role of verbal STM during the early stages of reading. Of the 450 deaf children within the system, Conrad found that only four were above average on reading, and three of these used sign language rather than lipreading, the standard mode of education at the time. His work led to an influential book, *The Deaf School Child: Language and Cognitive Function* (Conrad, 1979), which played a crucial role in the demanding task of convincing the deaf education establishment that signing had considerable advantages over lip reading, a task he eventually achieved, resulting in the very changed scene that is found today.

Although it was not obvious at the time, looking back over my research career (Baddeley, 2019), I seem to have been influenced by all three styles of research. Like Broadbent, I have tried to develop a broad and applicable model though it is less ambitious than that attempted in Broadbent's 1958 book. The multicomponent working memory model does indeed derive from his original model with its attempt to combine STM and attention (Baddeley, Hitch, & Allen, 2019). Our model was also devised with a view to its application across a range of cognitive activities, hence the term working memory. It has now been applied across a wide range of areas, most successfully where the model is part of a sustained programme of research extending over several years as reflected, for example, in the chapters in this volume by Cubelli et al. (Chapter 15), and by Andrade (Chapter 12).

My own attempts at application have tended to be more sporadic and dependent on collaboration as in the case of neuropsychology, where my research has been very dependent on the opportunity to work with clinically active research-oriented colleagues. My clinically based research on working memory has typically focused on theoretical rather than applied issues (Baddeley, 2021). Work concerned with practical applications has by contrast typically been directed towards deficits in long-term memory, applying existing theory to issues such as test design and neuropsychological rehabilitation (e.g. Baddeley & Wilson, 1994; Wilson et al., 1989; see also Wilson, Chapter 16, this volume).

Lasting influences

Broadbent

My clearest link with Broadbent's work is of course the development of the multicomponent model of working memory as described earlier. A less direct involvement, however, stemmed from a shared interest in the impact of emotion on cognition, a topic that absorbed his concern over the latter part of his career. Following a substantial review of the field entitled *Decision and Stress* (Broadbent, 1971), he moved to Oxford and devoted his attention to the very practical study of stress in the context of workers in the car industry. My own interest, however, sprang initially from research on divers where I discovered that the effects of nitrogen narcosis on performance was substantially greater when they were tested under open-sea conditions than when tested in a dry pressure chamber (Baddeley, 1966; Baddeley et al., 1968). Further work demonstrated that this stemmed largely from the effects of anxiety experienced at depth in the open sea (see Baddeley, 2019, Chapter 6).

As running experiments under open-sea conditions is complex and demanding, I attempted to extend the work on fear to what initially appeared to be a more controlled though still anxiety-provoking environment by testing novice parachutists about to make their first jump or people about to give a talk at the APU (Idzikowski & Baddeley, 1983; 1987). However, although both of these generated anxiety effects, adequate control proved problematic and failed to deliver effects that were substantial and reliable enough to allow the testing of detailed hypotheses. Most people who volunteer for such experiments seemed to be very good at resisting the effects of ethically acceptable short-term stressors. This led to the idea that it might prove more fruitful to move to the more directly relevant clinical field where long-term stress clearly is an important factor.

We were fortunately able to convince the Medical Research Council to allow us to build up a small but very effective group of clinical psychologists with an interest in combining their work with the methods of cognitive psychology. Finding interested clinicians was not initially easy and when our next 5-year assessment came around, the success of the enterprise was far from clear. Fortunately, we were given the benefit of the doubt and by the next assessment we had developed a very effective team comprising Fraser Watts, Mark Williams, John Teasdale, and later Andrew Matthews. Not only were they publishing important papers but they were instrumental in setting up a new journal, *Cognition and Emotion*, publishing a very influential book (Williams et al., 1988), and later doing classic early work demonstrating the efficacy of mindfulness training in the treatment of depression (Segal et al., 2002).

Conrad

Of my three mentors, I feel I have learnt most from Conrad, although we never published together as the assumption at the time in Cambridge psychology was that the work of the PhD student should be entirely separate from that of the supervisor, a view that I now regard as misguided from the viewpoint of both the student and the supervisor. I suspect Conrad might well have agreed on this point as I believe I was his first and last PhD student. I did, however, learn a great deal from him both by observing his approach and by his comments on my own work. Initially, I would finish writing and immediately pass it on for comment, whereupon his suggestion was that I should leave it for a week, go back and read it, and only then pass it on—wise words! I learned from him the value of being embedded in the field of application, in his case initially the Post Office and subsequently the world of deaf studies, although I cannot claim the same degree of continuity in my own applied work. My most sustained area of applied research has been in neuropsychology where my involvement has been somewhat fragmented and dependent on collaboration with clinically active neuropsychologists with a joint interest in the potential value of cognitive psychology. In this, I have been particularly fortunate, first from my introduction to neuropsychology by Elizabeth Warrington, continuing with rehabilitation-related collaboration with Barbara Wilson (see Wilson, Chapter 16, this volume), and the subsequent opportunity to study Jon, a fascinating developmental amnesic patient with Faraneh Vargha-Khadem. Another very long-standing series of collaborations has been with Hans Spinnler, Sergio Della Sala, Beppe Vallar, and Costanza Papagno, originating in the link between the APU and the group developed by De Renzi in Milan (see Cubelli et al., Chapter 15, this volume). It still continues through Sergio, who is now at the University of Edinburgh.

I have always liked to have two or more research activities operating at the same time so that if one is moving rather slowly, the other is often more active. One of these is typically concerned with working memory and since my arrival at the University of York almost 20 years ago, it has involved collaboration with Graham Hitch and Richard Allen, together with a range of other colleagues, notably including visiting scientists. This is typically accompanied by a somewhat different stream of research, often involving studies of long-term memory with a more applied focus, which I suggest still show the influence of my early years at the APU. I have selected three projects carried out largely over the last 5 or so years that I think illustrate the persistence of the early influence of the APU and its concern to link theory with its application beyond the laboratory.

The last 5 years: a continuing influence

Learning the Qur'an

My first example is not strictly applied in that it does not address a specific practical problem, but rather demonstrates the applicability beyond the laboratory of the basic concepts and methods of cognitive psychology. As such, it is close in spirit to the Poulton approach to research. It stemmed from the suggestion of an old friend, the clinical psychologist Narinder Kapur, that it might be interesting to study Hafiz, the term for people who have learnt by heart the whole of the Qur'an. His interest was prompted by existing work demonstrating the extensive spatial learning required by London taxi drivers who are required to memorize and then demonstrate knowledge of the streets and roads across the whole of Greater London. This demanding task results in modification of the hippocampus in contrast to equivalent measures on bus drivers who have driven for a similar period, but along standard routes. I was sceptical of the practicality of the neuroimaging aspect of his suggestion of a verbal equivalent, but have always been intrigued by the question of why anyone would undertake such a massive verbal learning task and what its consequences might be. I agreed that, if feasible, it might be interesting to find out if this had any impact on verbal memory more generally. We were joined in our venture by Faisal Mushtaq in the School of Psychology at University of Leeds and Rashaun Black, a master's student supervised by Faisal. As a team, we should at least merit top marks for religious and cultural diversity with Narinder who, from a Hindu background, went to Roman Catholic school in Northern Ireland, Faisal from a Muslim family with a sister training to be a Hafiz who could explain what was involved, Rashaun whose parents were Christian immigrants from Jamaica, and myself a long-lapsed Church of England protestant. So, what were our questions? They were firstly, whether mastering the 77,449 words of the Qur'an would improve or perhaps even interfere with the capacity for new verbal learning. Secondly, I was surprised to discover that not all Hafiz actually understand the Arabic in which it is written. Would this influence performance?

We managed to locate ten Hafiz volunteers, all with university degrees and compared them with 12 non-Hafiz of similar religious and ethnic background and ten non-Muslim graduates. There was little existing literature on our question although there were some claims for positive effects on learning and cognition more generally that were, however, based on rather weak evidence. As expected, our results suggested that the Hafiz had retained the contents of the Qur'an to a reasonably high degree. However, this was not accompanied by better performance on verbal memory tasks in their native English language, a result broadly in line with early attempts to increase learning capacity through rote learning (Reed, 1917; Sleight, 1911). I was, however, surprised to find that half our sample did not understand Arabic and yet appeared to have mastered the Qur'an just as well as their Arabic speaking colleagues, with no clear evidence that they had taken longer (Black et al., 2020). This was very unexpected given the emphasis in most verbal learning studies from Bartlett onwards on the importance of meaning. In discussing this with a friend from a Muslim background, she suggested I listen to someone reciting the Qur'an and sent me a link. It then became clear that the recitation of the Qur'an has a very musical aspect. This reminded me of the fact that I myself need only a few notes to anticipate the continuation of a surprisingly large number of pieces, despite the fact that I do not regard myself at all musically sophisticated. Is our view of long-term memory perhaps excessively dominated by meaning, and just how different would our theories be if they had been based on music rather than language?

So what did I myself learn from this small but intriguing project? First, I learnt something about the Qur'an, that it originated at a time when literacy was rare and was a means whereby the whole community could potentially be fully aware of its contents. This contrasts with the early years of the Roman Catholic Church when the bible was translated only into Latin, a language understood by the priesthood but not by their flock, with considerable resistance to its being translated into the local language. Second, I also discovered that learning the Qur'an begins with the simpler more straightforward sections and gradually builds up to the more complex issues. This would mean that Hafiz could be expected to have a better knowledge of the Qur'an and hence a Hafiz might have more influence within the community. Becoming a Hafiz still commands respect, although the lack of familiarity with its language presumably reduces its direct practical value. While our study suggests that it does not produce wider cognitive enhancement it may well have benefits of a more social and perhaps spiritual nature. Meanwhile, the lack of difference we found between those who do and don't understand Arabic, should be treated with caution given the small numbers involved, but would seem to merit further investigation.

Chinese language learners

My next series of studies stemmed from a very practical question though it was not expected to lead to a specific practical outcome and as such should probably be categorized as applicable rather than applied. It came from an enquiry by an American in Paris, a language teacher who thought that his method of improving the French pronunciation of his learners might be of relevance to the concept of a phonological loop. It was a very simple method, involving hearing and immediately repeating back spoken French phrases. It was the time of year for initiating students' projects and together with Sven Mattys, who has considerably more linguistic knowledge then myself, I decided to investigate. Given the constraints of an undergraduate project, we thought it might be too ambitious to use a foreign language, deciding instead to teach one of two contrasting local dialects 'Scouse' from Liverpool and 'Geordie' from Newcastle. We compared three groups of participants, one requiring overt repetition which we assumed would involve both the acoustic input and articulatory output components of the loop, a second condition involved silent subvocal rehearsal, input but no articulatory output, while a third comprised a baseline involving no relevant training. In each case a relatively brief learning phase was followed by a final test in which the accuracy of imitating each phrase was evaluated by a panel of raters unaware of the condition.

The results suggested an advantage to overt repetition, with no benefit from simply hearing the phrases and attempting to maintain them acoustically. While our study had clear limitations, the results were sufficient to obtain a small grant to extend it to a study in which Chinese students in York as part of their training as second language teachers were tested. The results supported the idea that only repetition led to enhanced accent learning, although the degree of training we were able to provide was far more limited than would be expected on a language teaching course and the gains far from dramatic (Mattys & Baddeley, 2019).

We interpreted our results in terms of the previously suggested distinction between two aspects of the phonological loop, namely a phonological input store and its articulatory output equivalent, sometimes referred to as the 'inner ear' and the 'inner voice', respectively. This distinction was made a number of years ago to account for the fact that while articulatory suppression (repeating an irrelevant word) impairs memory span and eliminates the phonological similarity effect, participants are nonetheless able to make correct phonological homophony judgements (such as 'way-weigh' versus 'hay-high') (Baddeley & Lewis, 1981; Besner et al., 1981).

Given the very modest funding available, and that we were working against a lifetime's experience of pronunciation in the students' native Chinese, we were pleased to get a reasonably clear result, although lacking the power that would be expected from a full training study. However, in the apparent absence of other work in the area we thought it worth publishing as an example of using psychological theory to bridge the gap between theory and practice in second language learning. Publishing in the language learning literature did not prove easy however, not because the data were not strong enough, but because it was deemed either too theoretical by some reviewers or too applied by others. Both theorists and practitioners seemed to see our well-meaning attempt to link the two as a potentially dangerous attempt at infiltration into their specialist field. This surprised us, given clear evidence of a broad interest in the role of working memory in some approaches to second language learning (Schwieter & Wen, 2022; see Bunting & Wen, Chapter 12, this volume). We eventually gave up on our attempt to contribute directly to the world of second language learning and published in a psychology journal (Mattys & Baddeley, 2019). Our accent learning project was fun, but hardly a major applied success, so what went wrong? Basically, that I completely neglected the aspect of Conrad's approach that made him such a successful applied psychologist, namely ensuring a basic knowledge of the field that he was investigating and establishing contact with people actively practising within it.

Our project did however have one silver lining. As part of our training study, we measured the digit span of Mandarin and native English speakers. Digit span in Chinese is known to be substantially greater than that in English speakers which in turn is greater than a range of other languages in which the digits have two or even more syllables (Naveh-Benjamin & Ayres, 1986). This pattern of results fits neatly into the concept of a phonological loop as longer digits take longer to rehearse, allowing more trace decay to occur before each digit is refreshed again (Baddeley, Thomson, & Buchanan, 1975; Naveh-Benjamin & Ayres, 1986), although it is important to note that this interpretation remains controversial (Caplan et al., 1992; Cowan, 1992; Nairne, 2002). Digits in English and Chinese are both monosyllabic although the consonant vowel nature of Chinese digits potentially allows faster rehearsal than the consonant vowel consonant nature of English. This seemed to be an opportunity to explore the substantial but still unresolved questions of the Chinese advantage in more detail.

We were fortunate in being able to collaborate with a York colleague who was responsible for a course for Chinese teachers of English (Mattys et al., 2018). We compared the span for both digits and words of our Chinese students with that of native English speakers and related this to articulation rate for the relevant material. As expected, span was greater and articulation rate faster for the Chinese group, and digits were repeated more rapidly and recalled better than words for both groups. Furthermore, at an individual level, speed of articulation correlated with span, with fast talkers remembering more items. However, articulation speed was less good at predicting memory performance in Mandarin Chinese speakers for both digits and words ($r = 0.33$ and 0.034 respectively), than was the case for English speakers ($r = 0.52$ and 0.43 respectively), suggesting the possible importance of a further non-articulatory factor or factors responsible for the Chinese advantage. A similar conclusion was drawn in a classic study by Zhang and Simon (1985) who found that their model linking articulation speed to verbal STM in Chinese speakers was clearly supported but suggested the need for one or more further contributing factors. What might these be?

A quite separate project with a Chinese sabbatical visitor, John Xu, suggested a possible answer. The study aimed simply to compare the characteristics of verbal STM in English and Chinese by extending the work of Zhang and Simon (1985) on the role of phonological similarity in STM. We did, however, discover an unexpected difference between the two languages. Memory span for sequences of visually presented letters or words is reduced when the items are phonologically similar to each other, an effect that is totally eliminated when subvocal rehearsal is disrupted by articulatory suppression in English speakers. This is also true for a range of languages, notably including Japanese, which like Chinese, has a logographic script (Saito et al., 2008). We found that this was not the case, however, for speakers of either Mandarin or Cantonese Chinese where the effect of phonological similarity is reduced but clearly remains. This contrasts with the virtually complete absence of such an effect across multiple experiments in speakers of English, suggesting that Chinese speakers are able to maintain a phonological trace while simultaneously articulating a sequence of irrelevant sounds (Baddeley et al., 2023).

We suggest that Chinese speakers may be able to maintain the acoustic trace while ignoring the potentially disruptive articulatory feedback, as is the case for simultaneous interpreters, for example (Chincotta & Underwood, 1998). We suggest that this capacity may be linked to an emphasis on rote learning in Chinese which is, we propose, associated with the need to learn to associate several thousand pictograms with their pronunciation and meaning, though this is somewhat assisted by the tonality of the language which, as we found, offers a modest supplement to the articulatory features of the words. If correct, our proposal suggests a potential capacity to use the perceptually based 'inner ear' component of the phonological loop and perhaps a way of extending the verbal span of English speakers by teaching them to use this to supplement their normal reliance on articulation—perhaps rather a long shot but worth exploring.

It might at this point be useful to use the two projects just described to revisit my initial claim for the advantages of applying cognitive psychology theory beyond the laboratory. These were that it tested the generality of basic theory, it raises new questions and opens up new areas, involves interacting with new people with different ideas, and finally that it might be useful. I would claim that my work over the last 5 years has done all of these but the last. Our work on memorizing the Qur'an certainly broadened the field and although its conclusion, that memory is not like a muscle that responds to training, is hardly new, it still constitutes an all too popular belief. The success of non-Arabic speakers was, however, an interesting surprise, though given the sample size, clearly needs replication. The practical relevance of our work is rather limited as I suspect few try to learn the Qur'an simply to improve their memory. Our study of STM in Chinese language speakers began with an applied aim, to evaluate a method of training pronunciation, but was clearly some distance from direct applicability although it did produce results of potential theoretical significance.

Accelerated long-term forgetting

At this point, I aimed to introduce my final topic which, although begun more than a decade ago, had seemed to blossom in the last 5 years, resulting in a clear solution to a clinical problem offering a happy conclusion to my chapter. However, as ever, life is not so simple. Here is what happened. About a decade ago I was invited to join a group of UK neuropsychologists, neurologists, and neuropsychiatrists who were all interested in tackling a clinical problem, how to detect accelerated long-term forgetting. Perhaps surprisingly, almost all memory disorders involve problems in acquiring new episodic memories while at the same time showing normal rates of subsequent forgetting. Claims of faster forgetting have been made from time to time but from the well-known amnesic patient HM to Alzheimer's disease, evidence for faster forgetting is typically minimal at best and methodologically controversial. This is clinically convenient since it means that memory can be assessed in a single session with no need for further testing. However, over the years exceptions have emerged, notably in patients with temporal lobe epilepsy who can present with apparently normal memory when tested over a single session while subsequently showing dramatic forgetting. Clinically this is important, particularly for the patient whose problems may be discounted in the light of test results, but also theoretically because of its implications for understanding forgetting. The group I joined had come together to try to develop a means of detecting such cases.

I was happy to join as I enjoy developing tests. I assumed that the problem would be easily solved, given that psychologists have been studying forgetting since Ebbinghaus in 1885. I was wrong. Not only did tools not exist, but nor did adequate theory. The basic problem was that of testing the same person after several delays given that each successive test was liable to further modify the original memory trace. The effect of repeated testing has been well studied and can be shown to have either a positive effect, as found in the retrieval practice effect where retrieval can be more effective in enhancing later retention than a further learning trial (Roediger & Karpicke, 2006), while under other conditions an opposite effect may occur, where retrieving one item can actively inhibit the later accessibility of other non-retrieved items (Anderson, 2003; Anderson et al., 1994). A further crucial basic problem is that of how to measure amount of forgetting. Should it be in absolute number of items lost, percentage loss, or something more complex? The answer seems to depend on one's theory of forgetting, which in turn alas may depend on how forgetting is measured. These two problems led to separate but related research streams. The more basic theoretical issues were pursued in collaboration with Sergio Della Sala, Robert Logie, and their students in Edinburgh and are reported elsewhere (Rivera-Lares et al., 2022; Stamate et al., 2020). The practical problem of developing an adequate clinical test was tackled mainly in York and Leeds and will be described next.

One solution to the problems resulting from retesting the same material is to test only a subsection of encoded material with a different sample tested at each delay, as used by Cassel et al. (2016) who presented four separate stories, testing a separate story after each of four delays. This appeared to work, showing accelerated forgetting in their epilepsy patients. It did, however, have drawbacks including a heavy load on initial learning, even with relatively simple stories that allowed few questions per story, together with possible noise from differences between stories depending on their serial order of presentation and test. At this point I remembered creating the 'Wrecks Test', a test that seemed likely to avoid these problems, being easy to learn and yielding a large number of questions.

The Wrecks Test was devised in the 1960s to test the performance of trainee divers in a study on the effects of cold carried out in a refrigerated tank in a Los Angeles heat wave. It comprised brief accounts of four wrecks each specified in terms of the type of ship, its name, its depth, the nature of the seabed, and its surroundings. It proved easy to learn since each wreck was easily imageable while the matrix structure facilitated retrieval (Broadbent et al., 1978) and allowed a large number of probed recall questions (Baddeley, Cuccaro, et al., 1975). Since we could not assume a deep interest in wrecks in patients, we turned to crime which seems to be of universal interest. The test involves four crimes, each involving a range of associated features, for example: 'An elderly Russian lady had her handbag snatched outside the cathedral by a young girl who then ran off.' Questions could then ask, for example, 'What was the nationality of the victim of the crime outside the cathedral?' or 'What crime was experienced by the Russian person?' The four crimes allowed a total of 80 such probe questions. An initial pilot comparing older and young participants across delays of up to 6 weeks looked promising, as did a study carried out in Oxford (Drane, 2012) which found much faster forgetting in a sample of temporal lobe epilepsy patients, although unfortunately the control group showed a ceiling effect. Further studies carried out in Leeds and elsewhere suggested that such ceiling effects were atypical but did report very little forgetting.

This raised the first of several problems. Could it be the case that testing one component of a crime might activate the whole episode, hence serving as a reminder?

We tested this by comparing the delayed performance of people tested at each of several delays with that of people tested only after the longest delay. It was clear that the intermediate tests were indeed slowing the rate of forgetting. A similar pattern of results was found for a parallel test of visual memory based on retaining features of four door scenes (Baddeley, Atkinson, et al., 2019). We then asked the question of whether this reminder effect from intervening tests resulted from the integrated nature of the material used for both the crimes episodes and door scenes. We tested this by moving to a design involving the recognition of individual words or scenes. Again, we sampled a quarter of the items at each delay but since they were not associated, we predicted that intervening tests would not reduce forgetting. This proved to be the case. The group tested at each of four delays remembered no more than that tested only at the longest delay. We found a similar result when we used the integrated material from the Crimes and Four Doors Tests provided we tested a different episode at each delay as was employed by Cassel et al. (2016), indicating that the crucial factor is the independence of material within the successive tests (Baddeley et al., 2021). The Crimes and Four Doors Tests had sadly not escaped the retrieval facilitation effect after all.

However, all might not be lost as there are two possible mechanisms underlying the enhanced retention effect. One of these assumes that each test serves as a new learning trial. If this is the case, patients with an acquisition deficit will mistakenly be categorized as showing accelerated long-term forgetting. A second possibility, however, is that the intervening tests of a subsample of the learned material simply reactivates the existing trace, hence serving as a valuable marker of its overall decline. If this is the case then patients who are known to show clear learning impairment with little evidence of more rapid forgetting should also show a clear benefit when a different subsample of integrated material is tested at each delay. Help was at hand in addressing this issue through the Edinburgh link where Andreea Stamate was using a similar sampled testing approach to studying forgetting in patients with Alzheimer's disease (Stamate et al., 2020). Such patients are known to have a problem in acquisition, but not in forgetting. If the interpolated testing effect resulted from relearning, then the patients should show less benefit from the interpolated retrieval tests. She found the expected deficit in initially learning the material when patients were compared to age-matched controls, with both groups showing a similar rate of forgetting over an unfilled week's delay. However, both benefitted to an equivalent extent from interpolated testing, even though different features were sampled on each occasion. It appeared to be the case that the interpolated retrieval advantage stemmed from priming of the existing trace rather than from further learning.

It was now time to apply our tests to the temporal lobe epilepsy patients who were most likely to show accelerated long-term forgetting. This formed part of the doctoral thesis of Tom Laverick who compared a group of patients and matched controls using both the Crimes Test and its visual equivalent, the Four Doors Test. Retention was tested in each participant after delays of 20 minutes, 24 hours, and 1 week. He found performance to be equivalent at the 20-minute delay but with clear evidence of faster forgetting in the patients at the longer delays (Laverick et al., 2021), thus replicating the earlier Oxford result but without the problematic ceiling effects. We had finally validated our test—or had we?

The fact that a test works under carefully controlled experimental conditions does not mean that it is ready to use in the much more complex clinical environment. The fact that effects could be found in both our own and other experimental studies suggested that faster forgetting might be much more widespread than originally assumed. The next step therefore seemed to be to move to a multicentre clinical study involving collaboration with other members of the original informal clinically engaged group. Plans seemed to be coming to fruition when the COVID-19 epidemic arrived.

We decided to use the pause in patient access to develop a version of the two tests that could be administered indirectly and obtained a small grant to cover this. We were also fortunate to have the help of a patient group, Epilepsy Action, and abundant volunteers from patients and their partners. Preliminary results arrived during the writing of this chapter and they were certainly mixed! With evidence from a more conventional multi-trial verbal learning task suggesting accelerated forgetting in the patient group while data from the Crimes and Four Doors Tests were disappointing. First, performance levels on the Crimes and Four Doors Tests were low, unlike virtually all of our earlier studies. Furthermore, the patients showed little evidence of forgetting on these tests over delays previously found effective. Finally, informal questioning suggested that some patients at least were trying to rehearse during the delay intervals, something that had occasionally been reported by older groups but was not typical of the student participants used for much of the development work. We think we can tackle all of these. Increasing the initial level of acquisition may simply require more trials although tackling the problem of potential rehearsal between tests is likely to be possible, but harder to address.

On translating theory into practice

So what conclusions can be drawn from the three very diverse applied topics that have occupied me over the last few years, and how characteristic are they of the rewards and frustrations of applying cognitive psychology? It could certainly be argued that in terms of concrete practical achievements they are rather limited, producing results that are potentially applicable rather than providing an answer to a specific applied question.

The demonstration that learning the Qur'an does not lead to a general increase in memory capacity is hardly surprising given the evidence collected over a century ago, that memory is not like a muscle that is strengthened by practice, although I suspect this may still be a common belief. Our work on acquiring a foreign accent is potentially of more practical significance but would certainly need further developments with much larger samples of participants before considering its educational application. Even if successful, establishing it against a wide range of existing commercial programmes, with some indeed advocating a similar approach, would be challenging and probably best approached via an established and reputable publisher within the foreign language instruction field. Finally, as shown by our initially successful attempt to develop a measure of accelerated long-term forgetting, moving from the relatively well-controlled clinical environment with extensively studied patients to the indirect testing of a much less selective group of volunteer patients can introduce a range of further complications.

However, I would argue that applied research will typically develop at least initially through a gradual accumulation of evidence before it can be securely translated into practice. Examples are Conrad's very extensive work with the deaf, based first on his basic research on STM, then applied to the acquisition of reading skills. When combined with a knowledge of the problem of literacy in the deaf community it led to the demonstration that sign language can replace verbal STM as a means of maintaining serial order and hence facilitate reading acquisition. Having established this experimentally he then had to show that his results could be applied in the field and only then could he convince those in a position to introduce change. A similar story could be told about the general acceptance of phonics within the UK educational system following many years of theoretical and practical development from research carried out across a wide international range of investigators (Stuart et al., 2008).

The role of institutional memory

Of course, a great deal of application occurs in situations where a decision must be made relatively rapidly with no time for experimentation. Perhaps the most immediately effective piece of advice provided by the APU during my tenure as director followed a query from the then prime minister, Margaret Thatcher, as to whether the government should follow the US example and introduce lie detectors into government security services. We produced a report advising against their introduction which she duly accepted, citing our advice. We ourselves were not working on the issue but were able to rely on the extensive literature carried out by others. Similarly, Conrad's previously described detection of a glitch in the operation of the letter sorting machines allowed a rapid solution but was based on a general knowledge of the characteristics of reaction time distributions acquired from a wide range of studies. These, like much successful application, depend on the capacity of a person or indeed a field to maintain and continuously update and retrieve its knowledge either through the individual or a continuing institutional memory. In Conrad's case he was the institutional memory within the Post Offices human factors system. When he left, this was lost. Hence, an early project I myself carried out under Conrad's direction on massed and distributed practice in training postmen to type (Baddeley & Longman, 1978) was followed some years later after my return as director by an identical request for research on this issue. I re-sent the report resisting the temptation to charge them twice!

The problem of institutional forgetting can unfortunately be a much more widespread problem as in the case of the study of vigilance, the capacity to sustain attention while performing an extended monitoring task. As illustrated in Broadbent's (1958) classic, this was a major research topic during the Second World War resulting in an extensive literature with theories ranging from Skinnerian through cognitive expectancy approaches to arousal-based physiological approaches. Leading figures in those areas all attempted to apply their theoretical expertise to account for the rich array of empirical evidence focused on this important practical problem of which failure to report the multiple Japanese aircraft approaching Pearl Harbour was just one example. After the war, the theorists largely returned to their earlier concerns and the equivalent practical problem of industrial inspection became increasingly automated, and as a result the topic largely vanished from the literature. Consequently, when vigilance again became important with the rise of terrorism and the need for airport inspection, all the earlier knowledge appeared to have vanished. I became aware of this when I was contacted by Jeremy Wolfe who had unknowingly replicated an earlier study of ours (Baddeley & Colquhoun, 1969) on the effect of signal frequency on search efficiency. He had learned of our work from a much older colleague who happened to remember it. Why had it not been picked up via a search engine such as Google? I tried to do so myself, finding that the word 'vigilance' evoked large numbers of studies on the advantages of flocking in birds and herding in deer but nothing on watch keeping. It seems that Google forgets too in the sense that retrieval becomes increasingly difficult over time. By prioritizing what is currently fashionable, it tends to hide work from earlier times though using background knowledge did eventually allow me to access at least some of the old work.

Lessons from the last 5 years

So what have I learned from the last 5 years of applied or at least applicable research? One potential benefit of applied projects is that they can test the generality of existing theory by extending it beyond the carefully constrained limits of the laboratory. In doing so they are able to generate new ideas while potentially challenging the status quo. This was certainly the case with our Qur'an study which provided a reminder of the much earlier evidence against the muscle training view of memory. It also provided the striking finding that the failure to understand and speak the Arabic in which it is written does not appear to be a major problem in this massive learning task although it should be stressed that this was an unexpected observation based on a small sample and certainly requires replication. Our research on acquiring a new accent also had unexpected theoretical implications in pointing to the need to distinguish between two aspects of phonological rehearsal. One aspect is based on articulation, the 'inner voice', which appears to play an important role in accent learning as well as articulatory rehearsal. The other aspect, the perceptually based 'inner ear', by contrast may play a role in the superior digit span found in speakers of Chinese.

Our third applied study, attempting to detect accelerated long-term forgetting, has also been theoretically productive. One aspect has been the prompting of a parallel series of studies carried out with colleagues in Edinburgh on the characteristics of long-term forgetting which is already proving productive (Stamate et al., 2020; Rivera-Lares et al., 2022) and is changing our theoretical views on long-term memory. However, as this aspect of the project was explicitly theoretically motivated, it differs from the incidental theoretical benefits from a typical applied project. This does not apply to our attempt to create a useful clinical test. This raised the need to address two basic problems: first, how to design an easily learned memory task that contained multiple potential questions and second, how to minimize interference when testing the same person repeatedly. The first had been solved inadvertently many years before with the Wrecks Test which combined readily integrated imageable scenes with a matrix retrieval structure (Broadbent et al., 1978). We thought we had solved the repeated retrieval problem by probing a separate sample of each scene on each test. We were wrong in the case of the Crimes Test, but made the incidental discovery that the repeated testing advantage appeared to operate via priming of the existing trace rather than relearning, a potentially theoretically important result but one that clearly requires replication and further investigation. We also found that the intervening testing of a subsample of material does not appear to affect performance when the items learned are independent words or scenes. This has both practical clinical implications and in addition throws further light on the general theoretical understanding of when and how successive retrievals interact, an issue of considerable theoretical interest (Anderson, 2003; Anderson et al., 1994; Roediger & Karpicke, 2006).

Of the various merits of applying psychology, the one that has most obviously been achieved is that of extending my experience of wider fields and the pleasure of making new friends. This is perhaps most obvious in the Qur'anic memory study from which I learned more about other cultures as well as broadening my knowledge of long-term memory by studying it within a context that would be difficult and laborious to study within the laboratory. The attempt to test methods of acquiring a second language accent introduced me firstly to the complex and fragmented field of second language teaching and secondly to the intriguing differences between English and Chinese languages and their implications for both the practical and theoretical implications of such differences understanding verbal working memory. My attempt to apply memory theory within the clinical field of epilepsy has both extended the range of my clinical involvement while also familiarizing me with the surprisingly neglected field of long-term forgetting. Finally, in each case, I have met new and interesting colleagues with a range of different but complementary skills.

In conclusion, I have tried to give an impression of my own approach to applying cognitive psychology, conscious of the fact that I am certainly not typical, having tried to combine the theoretical with practical for over 60 years during which psychology and the world in general have changed dramatically. However, I hope the description of some of the topics that have interested me over the last 5 years may have convinced you that it is still possible to take our knowledge of cognitive psychology beyond the laboratory in ways that are theoretically fruitful, intellectually stimulating, and sometimes perhaps practically useful.

Acknowledgements

I am grateful to Graham Hitch and Richard Allen for their helpful suggestions on an earlier draft.

References

Anderson, M. C. (2003). Rethinking interference theory: Executive control and the mechanisms of forgetting. *Journal of Memory and Language*, 49(4), 415–445.

Anderson, M. C., Bjork, R. A., & Bjork, E. L. (1994). Remembering can cause forgetting: Retrieval dynamics in long-term memory. *Journal of Experimental Psychology: Learning, Memory, and Cognition*, 20(5), 1063–1087.

Atkinson, R. C., & Shiffrin, R. M. (1968). Human memory: A proposed system and its control processes. In K. W. Spence & J. T. Spence (Eds.), *The psychology of learning and motivation: Advances in research and theory* (Vol. 2, pp. 89–195). Academic Press.

Baddeley, A. D. (1966). Influence of depth on the manual dexterity of free divers: A comparison between open sea and pressure chamber testing. *Journal of Applied Psychology*, 50(1), 81–85.

Baddeley, A. D. (2019). *Working memories: Postmen, divers and the cognitive revolution*. Routledge.

Baddeley, A. D. (2021). Developing the concept of working memory: The role of neuropsychology. *Archives of Clinical Neuropsychology*, 36(6), 861–873.

Baddeley, A. D., Atkinson, A. L., Hitch, G. J., & Allen, R. J. (2021). Detecting accelerated long-term forgetting: A problem and some solutions. *Cortex*, *142*, 237–251.

Baddeley, A., Atkinson, A., Kemp, S., & Allen, R. (2019). The problem of detecting long-term forgetting: Evidence from the Crimes Test and the Four Doors Test. *Cortex*, *110*, 69–79.

Baddeley, A. D., & Colquhoun, W. P. (1969). Signal probability and vigilance: A reappraisal of the
'signal-rate' effect. *British Journal of Psychology*, *60*(2), 169–178.

Baddeley, A. D., Cuccaro, W. J., Egstrom, S. H., Weltman, G., & Willis, M. A. (1975). Cognitive efficiency of divers working in cold water. *Human Factors*, *17*(5), 446–454.

Baddeley, A. D., de Figueredo, J. W., Hawkswell Curtis, J. W., & Williams, A. N. (1968). Nitrogen narcosis and performance underwater. *Ergonomics*, *11*(2), 157–164.

Baddeley, A. D., & Hitch, G. J. (1974). Working memory. In G. A. Bower (Ed.), *Recent advances in
learning and motivation* (Vol. 8, pp. 47–89). Academic Press.

Baddeley, A. D., Hitch, G. J., & Allen, R. J. (2019). From short-term store to multicomponent working
memory: The role of the modal model. *Memory & Cognition*, *47*(4), 575–588.

Baddeley, A. D., & Lewis, V. J. (1981). Inner active processes in reading: The inner voice, the inner ear
and the inner eye. In A. M. Lesgold & C. A. Perfettie (Eds.), *Interactive processes in reading* (pp. 107–
129). Lawrence Erlbaum.

Baddeley, A. D., & Longman, D. J. A. (1978). The influence of length and frequency of training sessions
on the rate of learning to type. *Ergonomics*, *21*(8), 627–635.

Baddeley, A. D., & Scott, D. (1971). Short-term forgetting in the absence of proactive interference.
Quarterly Journal of Experimental Psychology, *23*(3), 275–283.

Baddeley, A. D., Thomson, N., & Buchanan, M. (1975). Word length and the structure of short-term
memory. *Journal of Verbal Learning and Verbal Behavior*, *14*(6), 575–589.

Baddeley, A. D., & Wilson, B. A. (1994). When implicit learning fails: Amnesia and the problem of error
elimination. *Neuropsychologia*, *32*(1), 53–68.

Baddeley, A. D., Xu, Z., Ho, S. T., & Hitch, G. J. (2023). *Journal of Memory, Language, 2023*, 129.

Besner, D., Davies, J., & Daniels, S. (1981). Phonological processes in reading: The effects of concurrent
articulation. *Quarterly Journal of Experimental Psychology*, *33*(4), 415–438.

Black, R., Mushtaq, F., Baddeley, A. D., & Kapur, N. (2020). Does learning the Qur'an improve memory
capacity? Practical and theoretical implications. *Memory*, *28*(8), 1014–1023.

Broadbent, D. E. (1958). *Perception and communication*. Pergamon Press.

Broadbent, D. E. (1971). *Decision and stress*. Academic Press.

Broadbent, D. E., Cooper, P. J., & Broadbent, M. H. (1978). A comparison of hierarchical retrieval
schemes in recall. *Journal of Experimental Psychology: Human Learning and Memory*, *4*(5), 486–497.

Caplan, D., Rochon, E., & Waters, G. S. (1992). Articulatory and phonological determinants of word-
length effects in span tasks. *Quarterly Journal of Experimental Psychology*, *45*(2), 177–192.

Cassel, A., Morris, R., Koutroumanidis, M., & Kopelman, M. (2016). Forgetting in temporal lobe epilepsy: When does it become accelerated? *Cortex*, *78*, 70–84.

Chincotta, D., & Underwood, G. (1998). Simultaneous interpreters and the effect of concurrent articulation on immediate memory. *Interpreting*, *3*(1), 1–20.

Conrad, R. (1979). *The deaf school child: Language and cognitive function*. Harper & Row.

Cowan, N. (1992). Verbal memory span and the timing of spoken recall. *Journal of Memory and
Language*, *31*(5), 668–684.

Craik, K. J. W. (1943). *The nature of explanation*. Cambridge University Press.

Drane, E. S. M. (2012). *An exploration of the experience of living with epilepsy in later life* [Unpublished
doctoral dissertation]. University of Oxford.

Idzikowski, C., & Baddeley, A. D. (1983). Fear and dangerous environments. In R. Hockey (Ed.), *Stress
and fatigue in human performance* (pp. 123–144). John Wiley & Sons Ltd.

Idzikowski, C., & Baddeley, A. D. (1987). Fear and performance in novice parachutists. *Ergonomics*,
30(10), 1463–1474.

Laverick, T., Evans, S., Freeston, M., & Baddeley, A. (2021). The use of novel measures to detect accelerated long-term forgetting in people with epilepsy: The Crimes Test and Four Doors Test. *Cortex*,
141, 144–155.

Mattys, S., & Baddeley, A. D. (2019). Working memory and second language accent acquisition. *Applied Cognitive Psychology*, *33*(6), 1113–1123.

Mattys, S., Baddeley, A. D., & Trenkic, D. (2018). Is the superior verbal memory span of Mandarin speakers due to faster rehearsal? *Memory & Cognition*, *46*(3), 361–369.

Nairne, J. S. (2002). Remembering over the short-term: The case against the standard model. *Annual Review of Psychology*, *53*, 53–81.

Naveh-Benjamin, M., & Ayres, T. J. (1986). Digit span, reading rate, and linguistic relativity. *Quarterly Journal of Experimental Psychology*, *38*(4), 739–751.

Neisser, U. (1967). *Cognitive psychology*. Appleton-Century Crofts.

Poulton, E. C. (1979). Models for biases in judging sensory magnitude. *Psychological Bulletin*, *86*(4), 777–803.

Poulton, E. C., & Freeman, P. R. (1966). Unwanted asymmetrical transfer effects with balanced experimental designs. *Psychological Bulletin*, *66*(1), 1–8.

Reed, H. B. (1917). A repetition of Ebert and Meumann's practice experiment on memory. *Journal of Experimental Psychology*, *2*(5), 215–346.

Rivera-Lares, K., Logie, R., Baddeley, A., & Della Sala, S. (2022). Rate of forgetting is independent of initial degree of learning. *Memory & Cognition*, 1–13. Advance online publication.

Roediger, H. L., III, & Karpicke, J. D. (2006). Test-enhanced learning: Taking memory tests improves long-term retention. *Psychological Science*, *17*(3), 249–255.

Saito, S., Logie, R. H., Morita, A., & Law, A. (2008). Visual and phonological similarity effects in verbal immediate serial recall: A test with Kanji materials. *Journal of Memory and Language*, *59*(1), 1–17.

Schwieter, J. W., & Wen, Z. (2022). *The Cambridge handbook of working memory and language*. Cambridge University Press. ,

Segal, Z. V., Williams, J. M. G., & Teasdale, J. D. (2002). *Mindfulness-based cognitive therapy for depression: A new approach to preventing relapse*. Guilford Publications.

Sleight, W. G. (1911). Memory and formal training. *British Journal of Psychology*, *4*(3–4), 386–457.

Stamate, A., Logie, R., Baddeley, A. D., & Della Sala, S. (2020). Forgetting in Alzheimer's disease: Is it fast? Is it affected by repeated retrieval? *Neuropsychologia*, *138*, 1073511.

Stuart, M., Stainthorp, R., & Snowling, M. (2008). Literacy as a complex activity: Deconstructing the simple view of reading. *Literacy*, *42*(2), 59–66.

Williams, J. M. G., Watts, F. N., MacLeod, C., & Mathews, A. (1988). *Cognitive psychology and emotional disorders*. John Wiley & Sons.

Wilson, B. A., Cockburn, J., Baddeley, A. D., & Hiorns, R. (1989). The development and validation of a test battery for detecting and monitoring everyday memory problems. *Journal of Clinical and Experimental Neuropsychology*, *11*(6), 855–870.

Zhang, G., & Simon, H. A. (1985). STM capacity for Chinese words and idioms: Chunking and acoustical loop hypotheses. *Memory & Cognition*, *13*(3), 193–201.

PART 1
MEMORY CHALLENGES IN ADULTS

3

The problem of face identification

Vicki Bruce and A. Mike Burton

Research into human face memory really got going in the 1970s, stimulated by the real-world problem of eyewitness misidentifications coupled with laboratory demonstrations of phenomenal memory for once-viewed unfamiliar faces. Why was face recognition so good in the laboratory, but apparently so error-prone in real-world eyewitnessing? This paradox drew a number of psychologists to the problem of face recognition (including the first author) and has continued to drive aspects of research to this day.

Alan Baddeley and his colleagues at the Medical Research Council Applied Psychology Unit in Cambridge, UK, investigated aspects of face memory in the late 1970s, and they were remarkably prescient in identifying important and practically relevant problems that the field has returned to, productively, more recently. In this chapter we will describe how concern with errors in eyewitness memory for faces both stimulated applied research and generated theoretical understanding in the 1980s and 1990s—though not, we will argue, understanding that helped much with the practical problems we started with. We will argue that well-informed and ecologically valid applied research does not just aid practice, but directly helps us to understand the problem space within which to develop better theories.

The problem of eyewitness misidentification

It is sad to note that many hundreds of miscarriages of justice have been found to rest wholly or mainly on the mistaken testimony of eyewitnesses (see some of the cases in the Innocence Project (https://innocenceproject.org/) for some disturbing recent ones in the US, and Davies & Griffiths, 2008, for a review of cases in English courts). Although many problems associated with eyewitness testimony continue to this day, they have been documented for at least 50 years. In the UK in the 1970s, Lord Devlin (Devlin, 1976) led a committee of enquiry that described in detail two such cases—those of Mr Dougherty and Mr Virag.

Mr Dougherty had been identified from police photographs by two of the employees of a store from which a shoplifter had attempted to steal three pairs of curtains in August 1972. For various reasons there was no line-up held, but the prosecution witnesses both identified Dougherty when he was placed among the jurors

Vicki Bruce and A. Mike Burton, *The problem of face identification* In: *Memory in Science for Society*. Edited by: Robert H. Logie, Zhisheng (Edward) Wen, Susan E. Gathercole, Nelson Cowan, and Randall W. Engle, Oxford University Press.
© Oxford University Press 2023. DOI: 10.1093/oso/9780192849069.003.0003

on the day of the trial. Dougherty had many previous convictions for petty crimes and this led the police to be doubtful about the alibi he offered for the day of the crime. But he did have a very strong alibi for that day. He and his girlfriend had taken a coach trip from Sunderland to Whitley Bay accompanied by six children (his four plus her two). Up to 40 people might have been called as alibi witnesses but only a couple were. The jury chose to believe the identification evidence over the alibi—as has happened in many cases since.

In the second case considered by Devlin's committee, the identification evidence appeared overwhelming. A total of eight different witnesses identified Laszlo Virag from line-ups as being the man they had seen committing one or other of the armed offences in Liverpool and Bristol committed in January and February 1969. There were fingerprints in a vehicle connected with one of the robberies that did not match Virag's, but the conclusion drawn was that he must have been acting with an associate. Virag had an alibi for the night of the Bristol offences, because he was playing cards, with his wife, at a gambling club. There were others at the club who could have strengthened his alibi, but as Devlin's report makes clear, there were a number of problems in getting written statements from them.

One of the witnesses who identified Virag from a line-up was PC Smith, who was wounded in the attempt to apprehend the driver of a car which was pursued following reports of suspicious activity around parking meters. Smith got a very good view of his assailant. (He also heard him speak, reporting his noticeable foreign accent, which was one of the factors that led to the Hungarian Virag, who had previous convictions for theft, coming under suspicion.) In court, Smith announced, when affirming that the person in the dock—Virag—was the man who had injured him 'his face is imprinted in my brain'. After Virag was convicted and imprisoned for the offences, a Ukrainian, Georges Payen, was found to be responsible and Virag was eventually exonerated.

Baddeley (2019) recalls that, as an expert in memory, he was contacted in 1975 by the solicitor acting for George Davis, another man protesting that he was being wrongly accused of robbery with violence in the East End of London. The strongest evidence against Davis was that a police officer recognized him, almost a year after the robbery, from a line-up. Baddeley was unable to provide a statement that recognition was impossible at such a delay, however he did point out something he found more problematic: that the witness had already recognized Davis from photographs, prior to the line-up. Davis's conviction led to public protests about his innocence, including the vandalization of a test cricket pitch, and eventually the conviction was overturned as being 'unsafe' (i.e. based on insufficient evidence). Davis was a career criminal, and soon after his release was arrested for his part in other armed offences.

What these three cases have in common is that the people who were wrongly convicted all had some previous history of criminal conviction, leading police and prosecutors to be disinclined to believe them when they protested their innocence. Their alibis were not believed either by the prosecutors or by the juries. In Dougherty's case, for example, one of the witnesses to his bus excursion had herself served time for previous convictions. While it is not permissible in a trial to raise the accused's prior record, the integrity of witnesses can be questioned and challenged. As happened with Virag, alibi witnesses in such cases were suspicious of the law, and reluctant to testify.

The identification procedures in almost all cases involved witnesses who had been shown photographs before also attending a line-up—a practice that even in the late 1960s was prohibited. Devlin (1976, paragraph 25) cites the following two paragraphs from Home Office Circular no. 9 (1969):

> Photographs of suspects should never be shown to witnesses for the purpose of identification if circumstances allow of a personal identification. Even where a mistaken identification does not result, the fact that the witness has been shown a photograph of the suspect before his ability to identify him has been properly tested at an identification parade will considerably detract from the value of his evidence.

And:

> If a witness makes a positive identification from photographs, other witnesses should not be shown photographs but should be asked to attend an identification parade.

Implicit in these paragraphs of guidance is that viewing photographs might in some way influence the witness facing an identification parade (as Baddeley pointed out to Davis's solicitor), though this isn't spelled out as clearly as it could be. The important point is that the person chosen in the line-up could have seemed familiar because of being remembered from the photographs. Knowing that a face is familiar is easier than knowing why it is.

But perhaps most troubling of all, is that what these three cases also have in common is that police, prosecutors, witnesses, and jurors took the identification evidence as over-ruling all other. Inexplicably (given the science we know now), someone who says that the person is the criminal is assumed to be correct rather than fallible.

Changes in appearance

At the time when the Devlin report was published, psychologists were marvelling at the recently discovered powers of visual memory. Studies by Shepard (1967) and Standing et al. (1970) had both shown that recognition memory for once-viewed pictures was phenomenally good when pairs of old and new pictures were shown for forced choice recognition memory following exposure to hundreds of items in the study phase. For example, Standing et al. (1970) showed participants 2500 pictures for 10 seconds each, and found recognition accuracy of a subset of these was 90% correct after a delay averaging 1.5 days. Studies with more homogeneous materials were also conducted at about the same time, and those using faces typically found recognition memory for once-viewed unfamiliar faces was good too (Hochberg & Galper, 1967) and considerably higher than for other materials such as snowflakes and inkblots (Galper & Hochberg, 1971) or houses and airplanes (Yin, 1969). And herein lies the paradox of face memory. If we are so good at remembering faces, why do eyewitnesses so often make such costly errors?

One of the many problems that a witness to a crime must deal with is that the face of a suspect shown in photographs or in a line-up is not identical to the face as witnessed at the crime itself. The perpetrator might have been wearing a disguise, or other paraphernalia that might conceal some part of the face. Patterson and Baddeley (1977) published the first study that looked systematically at the effects of changing appearance on recognition memory for faces. In their first experiment, some of the faces shown to participants were those of actors whose faces in the test phase were shown playing a different role, often involving quite major changes in hairstyle, facial hair, and so forth. The experiment revealed dramatic effects of these changes in appearance—performance dropping from a d' of 3 (98% correct hits and 29% false positives) to a d' of just 0.58 (45% hits and 29% false positives)—close to chance.

In their second experiment, the effect of disguise was investigated more systematically, but now the task was changed. Participants were asked to learn the names of a relatively small set of faces shown in full-face views and, once learned, tried to identify these when shown in three-quarter view or profile with different elements of disguise. From performance of 89% correct when the targets were shown without disguise or with a change in glasses only, performance dropped to below 60% when there was a change in the beard or beard plus other features in addition. With beard, wig, and glasses changed, identification was just 39% correct.

This early study clearly demonstrates that changing the appearance of a face impairs performance, and the large effects of disguise have been replicated and extended since (e.g. see Noyes & Jenkins, 2019). However, there was another aspect of Patterson and Baddeley's study which was often cited too. In their first experiment, another set of studied faces was tested either using the same picture as previously used in the study, or with a change in viewpoint and expression. They found no effect of this manipulation, and they were not alone. Davies et al. (1978) also showed little effect of pose change in a study principally investigating the effects of changing mode of presentation from photograph to line drawing and vice versa. The absence of any overall effect on d' of the pose manipulations in these early studies may be some consequence of the experimental designs within which face pose changes were incorporated alongside more dramatic variations in disguise or mode. Certainly these null effects of pose and expression variation have not survived replication.

We now know that changes in expression or viewpoint can dramatically affect recognition memory for once-viewed unfamiliar faces. For example, Bruce (1982) investigated the effect of a change in viewpoint (from full-face to three-quarter view or vice versa), expression (neutral to smiling), or both, on recognition memory for a set of male faces who sported a variety of facial hair and hairstyles (this was in the late

1970s, long before buzz cuts). They were shown in head and shoulders shots with no attempt to conceal their clothing. Despite this, a change in viewpoint or expression reduced the hit rate in recognition memory from 90% to 76%, and a change in both brought it down to only 60% correct. These are large effects given that the distinctive hairstyles and necklines were visible and remained constant across these pose and expression transformations. Moreover, and more surprising still, we know now that superficial changes in image quality, as may happen when pictures are taken from different cameras, also reduce performance considerably—*even when memory isn't involved at all*—but we will come back to that.

In real investigations, the witness is asked to indicate whether the person in front of them now matches their memory for a person they saw in visually different circumstances. Add to that the myriad of further 'estimator' variables (Wells, 1978) that might affect their ability to remember the face (lighting, exposure duration, difference in ethnicity, stress, age, etc.; see Ryder et al., 2015), it is more surprising that witnesses are ever correct than that they frequently make mistakes.

However, even though participants on average may find it difficult to correctly recognize a face that has been seen briefly and whose appearance might have changed, we can never know for certain that *this* witness has got it wrong. And in the case of Laszlo Virag, there were eight witnesses who identified him. Two of these were police officers, whose testimony might have seemed particularly credible. After all, police undergo training that should enhance their observational prowess. Perhaps professional training is enough to compensate for any intrinsic fallibility in face recognition? And perhaps some witnesses are just better than others?

Individual differences in face recognition and training

The research that Patterson and Baddeley (1977) conducted on the effects of disguise and pose change was one part of a larger programme of work on face recognition that was conducted at the Applied Psychology Unit quite soon after Baddeley became director there in 1974 (see Baddeley, 2019, for a full and entertaining account of this and other research conducted during this period). During the course of the research, some 400 volunteers were tested in various experiments and the researchers noted that there appeared to be large individual differences in their abilities. Woodhead and Baddeley (1981) retested a group of 19 of the best performers and 19 of the worst performers and gave them a further recognition memory test of faces, plus new tests of recognition memory for paintings and for words. The difference in face recognition performance between these groups was confirmed, with the 'good' group recognizing 96% and the 'bad' group 85% of previously viewed faces correctly from a set of 50 pairs of old with new items. There was no difference between the two groups in recognition memory for words, for which both scored 96% correct. Interestingly, there was some differentiation between the groups in the task using paintings, with the 'good' face recognizers 99% correct, significantly better than the bad ones, who scored 95% correct. This was consistent with the idea that 'visual' memory involved rather different skills from 'verbal' memory (cf. Paivio, 1971). As can been seen from the scores, however, performance in all conditions was near ceiling, and participant numbers were relatively small, making it difficult to explore the overall results further with correlations. Recent studies of individual differences in face recognition ability have confirmed and considerably extended our understanding (see Lander et al., 2018, for a review) and we return to this later in the chapter. Nonetheless, it seems clear from this early investigation that our eyewitnesses could have different underlying abilities and that this could be a further factor affecting their performance.

Perhaps, though, it is possible to train people to recognize faces more successfully? Surely the police should have special skills as a result of such training? In the 1970s, Baddeley and his team set out to evaluate a face recognition training system that was being used in the British army, where participants were taught to analyse the constituent features of faces, noting particular distinctive characteristics. Woodhead et al. (1979) reported three studies in which they compared performance of trained participants after the 3-day course with that of a group who were untrained. In one of the studies, the task was to remember target faces and in the other studies, they were to match targets against arrays. They found no benefit of the training course in any of the studies, and in one of them it actually made people significantly worse. This suggests that training based on analysis of individual features of faces is not useful—at least for a task which involves matching whole faces. Indeed, more recent work has confirmed how challenging it is to find any system of training that significantly enhances matching performance in naïve participants or even security professionals. Towler et al. (2019) studied 11 training courses currently offered to police, security, and border agencies internationally, including in the UK, the US, and Australia. Despite almost all participants in these courses believing that they helped, the training actually produced almost no benefit on formal tests of identification and matching.

There have been many studies which underline the fact that people whose jobs involve face recognition and who may have been trained, or at least encouraged to examine faces carefully, may not be any better than the average observer. For example, Burton, Wilson, et al. (1999) compared three groups of participants' abilities to recognize from photographs the faces of people they had just viewed on video footage from a security camera. The first group of participants were psychology students who should have been familiar with the people who had been captured on video, the second group comprised students from other courses who would not have been familiar with these targets, and the final group were serving police officers with an average of 13.5 years of experience. The group familiar with the targets were almost perfect at the task, but there was no difference at all in the accuracy of the two other groups who performed only slightly above chance.

More recently, White et al. (2014) examined Australian passport officers, whose jobs include checking that faces match passport photos. When asked to carry out a set of photo-to-photo face matches, the passport officers performed no better than a group of untrained students—about 80% accurate in both groups. The passport officers were also tested with 'live' volunteers, who approached a desk for a check against a passport-style photo, taken just a few weeks earlier. Even in this situation, performance was surprisingly inaccurate, at about 90% accuracy. Given the thousands of people undergoing passport checks every day, a 10% error rate is very high. However, it is important to point out that these average error rates conceal very large individual differences. White et al. (2014) report some of the passport officers performing at very high levels, while some performed surprisingly poorly. They argue that this highlights the importance of selection for jobs involving face identification. Given that training seems to have little benefit, natural variation in ability should probably be emphasized when selecting people for these roles.

On average, it seems that professionals are no better than the rest of us at tasks of face recognition or matching. White et al. (2021) reviewed 29 published studies in which various professionals (border officers, passport issuance officers, police officers, etc.) were compared to the general population. While most professionals were indistinguishable from any other viewer, there were exceptions. A specialist group called 'forensic examiners', who are used in the USA to scrutinize facial evidence in criminal investigations, do perform better than the general public on most tasks, having been trained in a detailed feature-comparison strategy (Phillips et al., 2018). However, they still do not perform at perfect levels, even when comparing photos taken in good light and in the same pose. Interestingly, Phillips et al. also show that untrained people with naturally very high face recognition abilities (so-called super-recognizers; Russell et al., 2009) perform almost as well as professional 'forensic examiners' on many tasks. This natural variability in performance has been exploited in some organizations; for example, the Metropolitan Police (London, UK) created a 'super-recognizer unit' to work on difficult cases of identification. Robertson et al. (2016) showed that these officers clearly outperform most viewers from the general public. So, while there are a few exceptions, in general it seems that most groups of professionals who have been trained to examine faces carefully, and/or whose jobs involve doing this, are no better, on average, than untrained observers.

Searching for target faces in the real world

As has already been mentioned, in some of Alan Baddeley's early research he and his team alternated between using recognition memory tasks and using tasks where participants had to match test faces against a set of targets that were in front of them—removing the burden of memory. This method was taken to the extreme when—in an attempt to see whether three-quarter view exemplars were superior to other poses for recognition—Logie et al. (1987) asked, via newspaper adverts, for Cambridge citizens to look out for a particular individual whose picture was publicized (in different poses in different suburbs) and to phone in for a prize. But only one person spotted the target in the first run of that particular experiment. They then paid members of their panel of volunteers to look out for targets who they were told would be in Cambridge city centre on a particular day and time. Almost 80% of their 43 volunteers spotted no targets at all, and over 90% of the sightings that were reported were false alarms. Finally, Logie et al. recruited a further 145 volunteers who were each given either a full-face view, a three-quarter view, or a profile photo of a single target whom they were told they would meet coming towards them at some point on a proscribed circular route. In these very constrained conditions, 70% of the volunteers spotted a target, but there was no difference in success depending on the pose of the target that they scrutinized.

However, the most significant results of the project were probably those that led to such disappointment by the research team. Looking out for specified unfamiliar faces—without any burden of memory—is a difficult task, and even under very constrained conditions, it is far from perfect. More recently, researchers have begun to examine the problems of searching CCTV footage for target individuals, for example, missing persons or known criminals. Mileva and Burton (2019) showed people real footage from a London transport hub and asked them to look for target individuals for whom photos were available in a number of forms (e.g. passports, social media, and arrest-style images). Even in crowded scenes, the viewers were able to find target individuals at around 50–60% accuracy, though there were many false identifications. Interestingly, viewers were better at finding people in these real crowded scenes if they had three quite different images of the target person, rather than a single image, such as used in the Logie et al. (1987) experiments. However, providing even more images (up to 16) or moving videos of the search targets did not help.

Where the early research led

So, back in the 1970s, we knew a number of things about everyday face recognition. We knew that witnesses to crimes were fallible. We knew that there were individual differences in face recognition ability. We knew that training didn't seem to help. And we knew that even a relatively simple task of looking for a specific target person in the world was surprisingly difficult too.

This was some of the background around the time that both applied and theoretical research into face recognition really took off. Some of the applied research was aimed at improving the tools used by the police to help witnesses recall faces using 'composite' systems such as Photofit. This research was led in the UK by a group in Aberdeen, funded by the UK Home Office (the government department in charge of policing). At the time, many police forces in the UK and further afield had been persuaded that a feature-based 'Photofit' kit would be a useful way to build a likeness of a wanted criminal that could help in investigations. Photofit was an ingenious invention of one 'Jacques Penry'—not his real name—a photographer and entrepreneur. (For a fascinating historical account of how this system came to be adopted by the police, see Lawrence, 2020.)

The system required witnesses to select, from a large number of photographed face features, the individual features of the face they were trying to remember. The composite was then built from these selected features—hair, eyes, nose, and so on. The research team in Aberdeen showed that likenesses built using Photofit in fact bore very little resemblance to the target faces (H. D. Ellis et al., 1978) and this led to productive research that helped improve composite systems. The difficulty experienced by witnesses using older and some of the newer composite systems demonstrates a theoretically important point too. It seems that faces are not represented in a way that makes it natural to describe or recognize their parts. Some of the more successful recent systems have moved away from a system requiring the witness to recall individual parts of faces and now use holistic recognition-based systems instead, such as EFIT-V and EvoFIT (see Frowd, 2015, for a detailed review of the evaluation and development of composite systems).

The applied problems identified early on also motivated wider attempts to build and test a theoretical model of how faces were recognized. One of us (Bruce) argued strongly at the time that we would not be able to understand face recognition errors and difficulties unless we had an understanding of how people generally succeeded in the more usual everyday task of recognizing familiar faces. And a number of us in the UK and beyond soon developed functional models of face recognition which placed the identification of familiar faces at their core. The model which became most well known in the field at that time was that of Bruce and Young (1986) (Figure 3.1), a synthesis and extension of other models published at about the same time (e.g. H. D. Ellis, 1986; Hay & Young, 1982; Rhodes, 1985; see Young & Bruce, 2011, for a reflection on the strengths and weaknesses of the model 25 years on).

Everyday errors and difficulties directly informed the development of this functional model. For example, Young et al. (1985) asked volunteers to record, over an 8-week period, all the difficulties and errors that they noticed in their daily lives—such as forgetting someone's name, or mistaking a person at the bus stop for a neighbour. The pattern of errors was consistent with the three-stage sequence at the heart of the Bruce and Young model of identification. The first stage was the visual analysis of the face that allowed it to be recognized as familiar. This was then followed by the access of knowledge of why the face was familiar. The final stage was retrieval of the name. The sequence was a rigid one. People may think a known person is unfamiliar (failure of the first stage), or know why s/he is familiar (a politician, perhaps), but not be able to recall their name. But people never behave in ways that violates that sequence. We do not look at a face and say 'That's Donald Trump—why is he familiar?', except in jest.

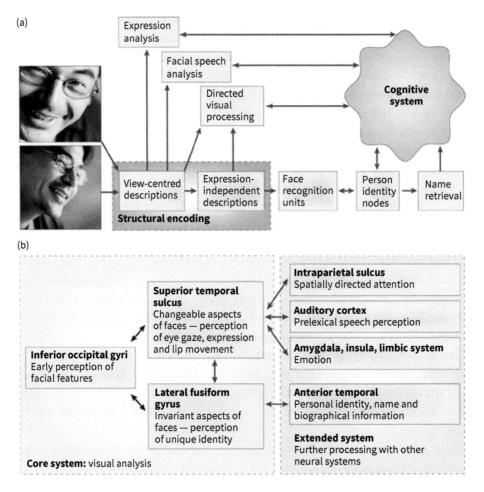

Figure 3.1 Comparison of the Bruce and Young (1986) (a) and Haxby et al. (2000) (b) models. Reproduced from Box 1 of Calder, A. J., & Young, A. W. (2005). Understanding the recognition of facial identity and facial expression. *Nature Reviews Neuroscience, 6*(August), 641–651. © Nature Publishing Group. Used with permission from Andy Young.

Young et al. (1985) also noted a number of incidents—not themselves errors—where their diarists were struck by a *resemblance* between a person and someone else. They noted that these experiences were ones that might have led to an error, if the context had not ruled this out. For example, one noted that they saw someone who looked very like a famous comedian, but as this happened in their local shop, they ruled that it must be resemblance, rather than identity. It's clear that there must be more than just a sequence of stages that succeed or fail—resemblance information must be evaluated within a wider context.

This issue of facial resemblance, and the role of context in our facial recognition decision-making, is also of critical importance in the practical problems of face recognition and face matching we started with. We will come back to discuss this at length later in this chapter.

Neuropsychological evidence from case studies of people with brain injuries affecting face perception (Damasio et al., 1982; de Haan et al., 1991; Tranel & Damasio, 1985) informed the development of our essentially functional models at this time. But the rapid progress from neural imaging studies of face processing in intact human brains allowed the development of models in which neural rather than functional pathways were specified (e.g. Calder & Young, 2005; Gobbini & Haxby, 2007; Haxby et al., 2000; and beyond) (Figure 3.1).

Figure 3.1 shows the explicit comparison drawn by Calder and Young (2005) between the original Bruce and Young (1986) functional model and Haxby et al.'s (2000) neurological one. To make the comparison easier, the Bruce and Young model was reoriented horizontally in this figure.

In both the early functional model, and the later neurological one, there is an explicit separation between processes involved in deriving the identity of a familiar person from the invariant aspects of their face (in the lower of the two main pathways shown), and those involved in deriving other kinds of meaning from changeable aspects such as facial expressions in the upper of these two main pathways. Haxby et al. (2000) identify the lateral fusiform gyrus as the site for analysis of stable ('invariant') aspects of a face, including its identity. Other researchers refer to this as the fusiform face area (Kanwisher et al., 1997; Tong et al., 2000). In contrast, it is the superior temporal sulcus that mediates the visual analysis of facial changes. Bruce and Young (1986) had failed to specify how the analysis of eye gaze fitted within their framework, a major omission rectified by Haxby et al. (2000) and other more recent neurally informed models (Carlin & Calder, 2013; Hooker et al., 2003).

While the neuroscientific progress was significant, it did not itself address other questions about detailed mechanisms and representations that were posed but not answered by the functional models. To do this, in our own group, we developed the early 'stage' model of face recognition as a computational model (Burton et al., 1990; Burton, Bruce, & Hancock, 1999) which did remarkably well at accounting for a number of aspects of human face recognition performance. Figure 3.2 shows a depiction of the IAC model of face recognition, based on a simple connectionist architecture, Interactive Activation and Competition (McClelland, 1981).

While models such as the one shown in Figure 3.2 are neutral with respect to neural coding, they are constrained by the requirement that they must 'work' computationally. In the IAC model shown here, we explored the interface between an image-analysis at the front end (implemented as principal component analysis (PCA) on pictures of faces) and a more conceptual recognition system which can also be accessed by other routes—for example, we can recognize people by their names as well as their faces. In this way, we made predictions about the sequence in which certain processes are carried out, and the relations between the representational building blocks that come together to support recognition.

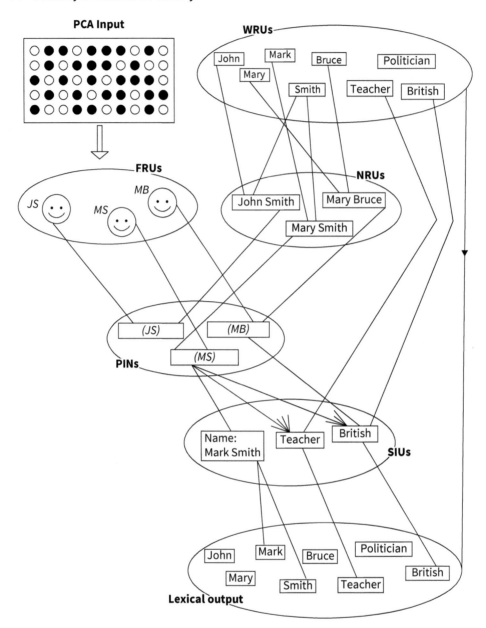

Figure 3.2 An outline of the IAC model of face recognition. Reproduced from Burton, A. M., Bruce, V., & Hancock, P. J. B. (1999). From pixels to people: a model of familiar face recognition. *Cognitive Science*, *23*(1), 1–31.

As we developed the computational model, we learned a lot more about the minutiae of some of the tasks such as repetition priming that we were using to test the model (Bruce & Valentine, 1985; A. W. Ellis et al., 1996, 1997) than we did about understanding why eyewitnesses make so many mistakes. Bruce and Young (1986) wrote at some length about contextual effects, and the fundamental role likely to be played by context in memory for relatively unfamiliar faces, but in our modelling efforts, 'context' was limited to the facilitation of familiarity decisions by associatively related 'prime' faces, such as the faster recognition of Stan Laurel's face after priming with Oliver Hardy's (Bruce & Valentine, 1986). We knew that familiar face recognition (usually successful) was different from unfamiliar face recognition (often incorrect) but we did not realize how important this observation was to getting to grips with the practical problems faced by the witness, the security services, and the passport officer.

The problem of face matching

It was a serendipitous observation that reset our ambitions. We were further developing our computational model of face recognition and trying to build a 'front-end' visual analysis system that behaved the same way as human recognition (Burton, Bruce, & Hancock, 1999). So we set out to compare a number of computer vision recognition algorithms with human performance, particularly to see whether these produced the same drop-off in performance we expected humans to show when viewpoint and/or expression changed between to-be-matched items. We created a 'gold standard' task of human face matching to compare with computer systems to see if performance decrements were similar as poses varied, and if similar confusions between faces were made (e.g. see Burton et al., 2001).

The task we asked humans to perform was to examine a face that had been captured as a still image from a high-quality video and decide whether the person was present in an array of ten photographs shown in Figure 3.3 (Bruce et al., 1999). The test array pictures were taken on a different, static camera. In one condition of the experiment the faces were shown in the same, full-face, pose and neutral

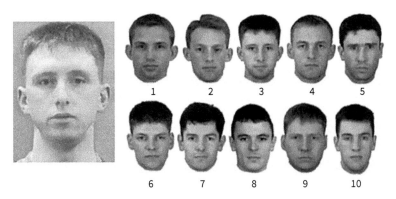

Figure 3.3 Example of the matching task used by Bruce et al. (1999). Is the target (left) present in the array and if so, which one is he? Reproduced with permission from Bruce, V., Henderson, Z., Greenwood, K., Hancock, P. J. B., Burton, A. M., & Miller, P. (1999). Verification of face identities from images captured on video. *Journal of Experimental Psychology: Applied, 5*(4), 339–360. Copyright © 1999, American Psychological Association.

expression. In another condition, the target face was smiling and in a third, he was shown in three-quarter pose. There was no memory load in this experiment and we expected it to be very easy when viewpoint and expressions were the same—it was really the drop-off when there were pose changes that interested us. We constrained the arrays to contain faces of superficially similar appearance to try to create some possible errors in human vision. We need not have worried. We discovered very early on that even in the pose and expression matching condition, errors were frequent rather than rare, with performance (averaged over target present and absent trials) just 70% correct.

As expected, performance was poorer still when there was a change in pose to contend with as well, dropping to 64% when the target face was shown smiling and 61% when the head angle was changed to 30 degrees. In a second experiment, we constrained the task. Instead of asking whether the person was present or not in the array, we used only target present arrays and told our participants he would always be there. All they needed to do was to choose the person who most closely resembled the target. Again, the error rate surprised us, since errors were made on 20% of the arrays on average. (In Figure 3.3, the correct response is number 3.)

We should not really have been surprised. Richard Kemp and colleagues (1997) had already published a remarkable, ecologically valid study of face matching using supermarket cashiers in a simulated training session involving photo identification (ID) credit cards which needed to be checked. Performance was poor. When someone was carrying a 'fake' photograph containing a foil which was similar in appearance, 64% of the cards were accepted as valid. Indeed, about a third of cards went unchallenged even when the photographs were of people quite dissimilar in appearance. Performance on the valid identity cards was higher—suggesting a strong bias to accept rather than reject an identity—but even here 14% of the matches were challenged when the photo showed a person with different paraphernalia to the target's current appearance.

In retrospect, Kemp et al.'s (1997) study was telling us something very important—matching unfamiliar faces is difficult. In fact, it took some time for this lesson to become accepted, and there was some initial reluctance to agree that face matching is hard. Perhaps the demand characteristics of the supermarket were somehow odd, perhaps the photos were too small or the time for matching too short. It appeared relatively easy to explain away a single experiment in a setting outside the laboratory. Recall that it was very well understood in the 1990s that unfamiliar face *memory* was fallible—years of eyewitness research had shown that (see Wixted & Roediger, Chapter 4, this volume, for important recent evidence on this). However, the assumption was that this fallibility, in large part, reflected our imperfect *memories*. Furthermore, psychological attempts to overcome eyewitness errors focused almost entirely on memory, either through reinstatement of encoding context or the development of memory-focused interview techniques (Memon & Bull, 1991; Memon et al., 2010). However, if Kemp et al.'s findings were to be taken seriously (and eventually they were), then it would require a rethink about where the problem lay in failures of unfamiliar face recognition—certainly memory could not be wholly to blame when very high errors were observed in the absence of any memory load at all.

Since these early reports, there have been numerous studies showing the difficulty of unfamiliar face matching. This is not tied to specific viewing conditions or experimental protocols; unfamiliar faces are hard to match whether the viewer is asked to pick a target from an array of ten candidates (Bruce et al., 1999) or simply to indicate whether two photos are the same or different identities (Burton et al., 2010). Unfamiliar faces are hard to match in poor image quality, but also in very high image quality (Henderson et al., 2001). They are hard to match when photos were taken months apart or on the same day (Bindemann & Sandford, 2011; Megreya et al., 2013). Furthermore, it is just as hard matching live individuals to their photos as it is matching pairs of photos (Megreya & Burton, 2008; Ritchie et al., 2020). The message is now clear: different images of the same person can look dissimilar enough to be classified differently. And two different faces can look similar enough to be wrongly classified as the same person.

It is worth briefly reflecting on why this was such a difficult lesson to learn. One possibility is that our everyday experience of recognizing the people we know, usually with no difficulty, leads us to think that face recognition is *generally* accurate—for most people looking at most faces. If you can match your brother to his passport photo easily then it is hard to understand why someone else cannot. Over 30 years ago, Alan Baddeley and colleagues observed that there is, in fact, a general propensity to overgeneralize one's own knowledge (e.g. Nickerson et al., 1987). For example, if someone happens to know 'the island on which Napoleon was born', then they tend to assume that other people know that too. This general 'egocentric bias' has been observed in social psychological research for many years (e.g. Holmes, 1968; Krueger & Clement, 1994; Ross et al., 1977) and applies not just to knowledge, but also to behaviour. When given a choice of two possible actions, we tend to predict that others will choose the same action as us.

The egocentric bias has been observed in face perception too. Ritchie et al. (2015) showed viewers pairs of faces, asking them to make 'same person' or 'different person' decisions. They recruited their participants from three different universities, and used photos of staff from these universities as the materials. As expected, participants were much more accurate when matching familiar faces (i.e. those from their own institution) than unknown faces. However, what was most interesting was their response when asked to guess other people's performance on the same matching test. These participants believed that the faces they themselves knew, would also be easier for other viewers—regardless of familiarity. In short, they overgeneralized their own familiarity, falsely believing that the faces they found easy to match were somehow just easier for everyone. This idea is further supported by recent work showing that people overgeneralize their own abilities across a range of face perception tasks (Zhou & Jenkins, 2020).

Why use photo identification?

A further reason for our reluctance to embrace the difficulty of unfamiliar face matching is that it seems to imply a need for societal change. The use of photo ID is very common in the modern world, and questioning its efficacy is somewhat challenging to established security procedures. Of course, the most high-profile use of photo ID comes in passports or other official identity documents, and we have already described the fallibility of some highly trained staff in these security-critical settings. Partly in acknowledgement of this vulnerability, government agencies are currently investing considerable effort in computer-supported face recognition (Phillips et al., 2018) and other biometric sources such as fingerprints and iris scans (Labati et al., 2016). However, the use of photo ID is now highly prevalent across many day-to-day settings. For example, people are often asked to prove their identity for the purchase of age-restricted goods, accessing workplaces, or entry to leisure venues such as sports facilities and nightclubs. Photo ID cards for these purposes do not carry the same national security risks as passports, but they are routinely used to prove identity by almost all citizens of Western countries.

Modern research on the use of photo ID continues to confirm the work of Kemp et al. (1997), described above. Viewers are very poor at matching unfamiliar individuals to their photo ID, and this problem persists even for those who carry out these checks routinely, such as cashiers, bar staff, or bouncers (Robertson & Burton, 2021; Weatherford et al., 2021). These results seem to echo those found in the laboratory using standard face matching tests such as the Glasgow Face Matching Test (Burton et al., 2010) or the Kent Face Matching Test (Fysh & Bindemann, 2018). However, in some ways, these laboratory tests are unlike matching in the real world. For example, they typically contain a 50/50 mix of matching and mismatching trials, whereas the proportion of people trying to use fraudulent ID (i.e. photos that do not match the carrier) is much lower than this in virtually all real-world settings. Furthermore, face matching tests typically show pairs of isolated faces, with no surrounding context, whereas real ID checks take place in the context of an identity *document* that contains other information, such as a passport or driving licence.

In fact, recent research has shown that many factors can influence viewers' accuracy in a matching task. The simple frequency of mismatch items in a sequence of trials does not seem to affect overall accuracy unless item-by-item feedback is given (Bindemann et al., 2010; Papesh et al., 2018). However, increased time pressure does have an effect: restricting participants' viewing time to a few seconds per face pair can considerably reduce their accuracy (Bindemann et al., 2016; Fysh & Bindemann, 2017; O'Toole et al., 2007). Interestingly, this time pressure manipulation seems to have a specific effect on detecting mismatches.

Some recent work, inspired by real-world ID checking, suggests that the context in which a face is embedded, that is, a card or document, can affect viewers' matching decisions. Robertson and Burton (2021) asked participants to decide whether someone should be served alcohol on the basis of a proffered ID—a judgement that requires both an age check (from the ID) and a face matching check (to establish that the ID photo matches the carrier). They found that the age check dominated this decision. If the card showed someone to be too young to purchase alcohol, then they were almost always rejected. However, if the ID showed a legitimate age, then viewers were highly likely to accept the ID and serve the 'customer', even when the faces mismatched. The pattern was consistent across different forms of ID commonly used in the UK (driving licences and UK PASS+ cards) and experienced cashiers showed the same tendency, albeit to a lesser extent than those with no relevant experience.

In fact, simply embedding a photo in an ID document turns out to affect identity decisions, even in cases where the information on the ID is task irrelevant. Figure 3.4 shows two conditions in a series of experiments reported by McCaffery and Burton (2016). Viewers were simply asked to make match/mismatch decisions for pairs of faces, though in some conditions one of the faces was presented within a passport frame, which viewers were told to ignore. The results consistently showed that the addition of a passport frame did not alter the overall accuracy of participants' decisions, but it did alter their bias. Viewers were more likely to accept a pair of faces as being a match in the presence of a passport. Why might such a bias exist? Perhaps the authority of the passport frame inclines people to believe that it is being used

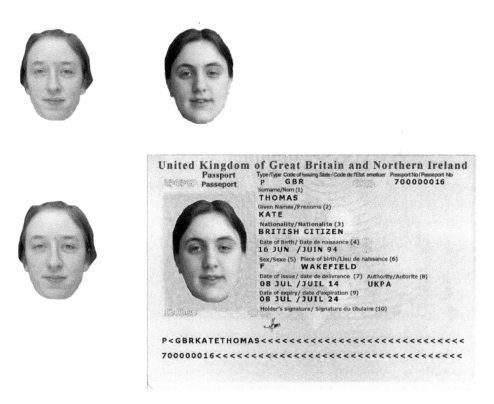

Figure 3.4 Matching faces presented in isolation, top, or embedded in a passport frame, bottom. Reproduced from McCaffery, J. M., & Burton, A. M. (2016). Passport checks: Interactions between matching faces and biographical details. *Applied Cognitive Psychology*, 30(6), 925–933.

legitimately. In fact, this turns out not to account for the effect—exactly the same pattern is observed using a range of other ID cards, some with official authority, such as UK driving licences, and others bearing less authority, such as student ID or social club cards (Feng & Burton, 2019). However, removing or blurring the text from these cards, or rendering them in a script unfamiliar to the viewers (Bulgarian) eliminates the bias (Feng & Burton, 2021). It seems that this effect relies on some kind of interference between face processing and readable text on the ID—a picture/text interference which is very commonly observed in cognitive psychology, but was unexpected in this simple face matching case. At a more general level, these studies point to the role of context (in this case, meaningful document text) influencing the task of face matching, just as it affects performance in tasks involving face memory.

Note that the consistent 'same person' response bias introduced to face matching tests when presented in ID is exactly the bias one would seek to avoid in security-critical situations. By lowering the checker's criterion for accepting a match, the potential for fraudulent use of photo ID is increased. However, there are also some important points to learn here about the relationship between laboratory-based and applied research. First, the very large number of unfamiliar face matching studies that now exist in the literature could give the impression that we now have a good understanding of the task. Many of these studies (including our own) make reference to practical issues as motivation for the experiments they contain. However, the studies discussed previously serve as a reminder that laboratory tasks always represent simplifications. It is fair to say that the results of McCaffery and Burton (2016) were initially surprising, but they highlight the fact that even simple tasks like face matching can behave very differently in artificial and real settings. It may also be interesting theoretically to note the details of the effect. Laboratory tests for face matching tend to be designed to avoid bias, allowing for simple accuracy to be measured. However, here we have a situation in which overall accuracy is not affected by the manipulation intended to capture real-world aspects of the task—instead, the effect shows up only as a response bias. This points to the possibility that our laboratory designs can sometimes obscure critical aspects of the phenomena under study—a possibility that will be important for the next section.

What should we study when we study faces?

As face perception has become a more popular topic, technical advances have given us more control over the materials we use. Early papers can seem odd to the modern reader, with their example faces cut out from magazines or photographed and printed by a university technician. Alan Baddeley's work on spotting real people as they walk through towns (Logie et al., 1987) seems similarly anachronistic. Surely there are too many sources of variance in these experiments—too many differences in lighting, view, or transient expressions—to allow proper scientific experimentation. Later, it became possible to control for many of these variables, and laboratory-based experiments began to present images with 'nuisance' variation eliminated. So, for example, many experiments used faces that were cropped to eliminate people's hair, make-up, and facial adornments and attempts were made to standardize low-level visual characteristics such as luminance and contrast.

The ability to control facial images using computer graphics manipulations in fact carries two associated problems. First, the aim to eliminate apparently spurious variation has no end point: just when you think you have controlled all the important factors, a new aspect of image description is pointed out that did not form part of the stimulus preparation. Second, the more standardized the experiments, the further they get from the faces we recognize every day—which do have hair, facial adornments, and complex reflective properties that change moment to moment. In fact, a desire to control low-level properties of faces has sometimes led to the development of databases comprising face-like materials derived purely from graphical manipulations and preserving only the visual characteristics relevant to the questions posed in particular experiments (e.g. see databases reported in Leopold et al., 2001, and Loffler et al., 2005). There is no doubt that highly controlled, often artificial, images allow one to address very specific scientific questions. However, their applicability to real face recognition can become moot under these circumstances.

In response to this tendency for tight control, an opposing position has developed based on the idea that studies of face recognition should, where possible, use *ambient images* (Burton, 2013; Jenkins & Burton, 2011; Jenkins et al., 2011). We recognize faces every day from a huge variety of sources—some in person, some on TV or social media, in many different poses, lighting conditions, and expressive states. Eliminating this variation is not only difficult (perhaps impossible) but, we have argued, may specifically obscure the very phenomena we are aiming to study. To understand why, we need to consider the issue of within-person variation in face images.

Figure 3.5 shows different images of different people, but how many people? Your response is likely to depend on whether you are familiar with anyone here: in fact, the answer is two. Jenkins et al. (2011) showed sets of images like this, comprising different photos of just two different people—in one case, two Dutch TV presenters. Asking participants to sort the photos into piles, one for each person, they made an important observation. Dutch students, who recognized the people shown, were able to sort the photos accurately into two piles. But viewers unfamiliar with the faces (UK students) were much worse at the task, sorting the photos into 7.5 piles on average (modal response = 9; range = 3–16). So, a task that was very easy for viewers familiar with the faces, was very hard and performed very inaccurately by unfamiliar viewers. What is interesting here is that this particular task does not rely on telling faces apart—even UK viewers rarely mixed up the two presenters. Instead, the problem was *telling faces together*. The natural variation between ambient images, uncontrolled but easily recognized by familiar viewers, means that unfamiliar viewers cannot easily see which face images belong together. Could this mean that a key part of understanding familiar face recognition is understanding the variability

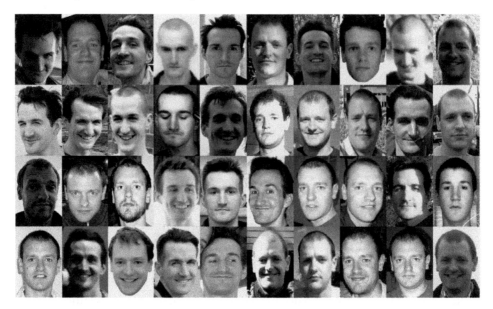

Figure 3.5 Multiple images of multiple people, but how many different people are shown here? Reproduced from Jenkins, R., White, D., van Montfort, X., & Burton, A. M. (2011). Variability in photos of the same face. *Cognition, 121*(3), 316. Copyright 2011 by Elsevier. Reprinted with permission.

itself? And if so, does it imply that eliminating the variability from experiments might obscure the process?

The idea that variability might be informative for face recognition, and not just a nuisance, was inspired by the work of the British psychologist, Maggie Vernon (1901–1991). Vernon had written extensively on a key problem of visual perception, the fact that our visual world appears stable, despite variations from moment to moment in viewpoint, lighting, and other changes in the projections from our visual field. Indeed, Vernon had suggested that the 'possible and permissible variations of appearance' might actually be utilized by the visual system in defining boundaries between representations (Vernon, 1952, cited in Bruce, 1994). In a lecture honouring Vernon, Bruce (1994) noted that faces represent a class of stimuli that vary in particular ways—particularly in the non-rigid deformations produced by speech and expression that change the stimulus pattern quite markedly *within-identity*. Computing these changes would allow the viewer to discard them for the purposes of computing somebody's identity, while at the same time retaining them for other crucial face perception tasks, such as decoding facial speech or judging emotional states.

These ideas were subsequently extended into more general theories of within- and between-person variability, and their relative contribution to face identification (Burton, 2013; Jenkins & Burton, 2011; Jenkins et al., 2011). It has now become quite well established that in tasks such as these, unfamiliar viewers have particular difficulty in 'telling faces together', and a core component of this appears to be the level of variability over which one has experienced a face. So, when asked whether two photos show the same person, a familiar viewer has a broader experience on which to draw, effectively being able to sample from the range of experience with this person. In contrast, an unfamiliar viewer has no evidence about the ways that a particular individual's face can vary, and must resort to simple image comparison, a process which might or might not deliver the correct answer (Hancock et al., 2000).

Of course, the serious problem of eyewitness misidentification arises when a witness erroneously 'tells together' his or her memory of a perpetrator's face with that of an innocent suspect. We note that in many (though not all) cases of eyewitness misidentification, such as the Virag example we described earlier, there was a clear resemblance between the innocent suspect and the person eventually convicted of the crime. What the witness or witnesses must do is to interpret such 'resemblance' in a legal and psychological context that may strongly bias them to identify someone rather than no one.

The importance of within-face variability suggests a particular stance on face learning—how faces come to be familiar in the first place. We learn new faces throughout life, and it has been a major challenge in this field to understand how this can happen, particularly in the context of the large differences between familiar and unfamiliar face processing. An emphasis on variability leads one to ask whether face learning might actually reflect learning the variability underlying a particular face. Though seemingly innocuous, this is quite a radical suggestion—previous approaches to face learning have suggested that we need to encode a key set of distances and measurements uniquely defining a person's face—a 'configural processing' account of familiarity (Maurer et al., 2002; Tanaka & Gordon, 2011). Instead, an approach based on variability suggests that one should abandon the search for some key, Platonic set of descriptors of an individual's face, available in all recognizable instances, and instead simply observe the range of instances encoding a person-specific distribution for each known face.

Support for this idea comes from an analysis of face images themselves. Burton et al. (2015) sampled multiple images of Hollywood actors. They then performed a PCA separately on each actor's photos. This allowed them to characterize the ways in which each person varied. The results were compelling. While the photos of all actors shared some early image variation in common (representing viewpoint, brightness, and so on) they also displayed a great deal of *idiosyncratic* variation. So, the difference between Harrison Ford and Brad Pitt was not only in their location in 'face space', but also in the shape of their distributions—photos of Brad Pitt vary along different dimensions to photos of Harrison Ford. This seems to be strong evidence for the notion of idiosyncratic variation—and it implies that to learn a new face we need to sample over the distribution of that specific face. General principles will not get us very far, and knowing how one face varies is of limited use for understanding how another face varies.

This emphasis on person-specific learning is borne out in experimental tests. So, we have known for a long time that face learning can be linked to number of encounters and duration of exposure, just like everything else. However, it is now clear that face learning is improved by exposure to a greater range of photos of a person: independently of exposure duration, learners benefit from seeing more variability (Murphy et al., 2015; Ritchie & Burton, 2017). This also begins to explain why training in *general* face recognition tends not to work (Towler et al., 2019; Woodhead et al., 1979)—if familiarity depends on learning a broad distribution of examples for an individual, and if this range is idiosyncratic to that individual, then it is difficult to see how learning to recognize one face will help very much when trying to recognize or match another.

One extreme form of within-person variation can be seen as people age. While we all see friends and family members across a long duration, there are also a few public figures who are recognizable across their entire adult lifespan. Figure 3.6 shows an example, the Beatle Paul McCartney, who has been famous across his entire adult life, and is easily recognizable by many people. Coping with this large variation has puzzled psychologists for some time—for example, do we need to maintain separate representations for young, middle-aged, and elderly Pauls? Extending the idea of within-person idiosyncratic variation described above, Mileva et al. (2020) performed PCA on images of people for whom multiple photos were available across 60 years (mostly 1960s pop stars). They were able to demonstrate the remarkably generalizable character of within-person variability. So, while individual photos of a person aged 20 and 60 were almost impossible to match on the basis of image comparison, this was relatively easy for a model trained on within-person variability *for a single age*. So, somebody unfamiliar with a pop star would find it very difficult to match photos taken 40 years apart. But familiarity with the star at 20, along with seeing multiple varying images at that age, makes them relatively easy to recognize at 60. This seems to be an example in which variability itself is a very powerful tool to support the recognition system.

Our discussion of familiar face recognition over these extreme levels of variability—for example, over a whole adult lifespan—highlights the difficulty faced by any perceptual system. Research on the 'front-end' visual image processing component of face recognition has much in common with the practical problem of automatic face recognition for security and surveillance. The relatively simple statistical image analysis provided by techniques such as PCA, preserves a transparency that is useful for building cognitive models (Kramer et al., 2018; Mileva et al., 2020). However, more powerful techniques, many based on deep convolutional neural networks, now perform at much higher accuracy than models built on PCA, and even exceed human unfamiliar face recognition (O'Toole et al., 2018; Phillips et al., 2018). There is debate about the utility of these very powerful computational systems for building perceptual theory (Blauch et al., 2021; Young & Burton, 2021) and the recent resurgence of interest in connectionist models has raised challenges across a

Figure 3.6 Paul McCartney from his 20s to his 60s. Readers of a certain age will find it easy to recognize all of these photos (Mileva et al., 2020). Image attributions are given in the 'Acknowledgements' section.

range of topics in human cognition. More practically, it seems that automatic face recognition systems, often shown to be inadequate in real-world settings, may soon become available in highly accurate form.

The psychological research described in this section relies on the use of ambient images—uncontrolled, naturally occurring photos of the type that people recognize every day. The use of such images remains somewhat controversial in laboratory-based studies, but we argue that it is necessary to understand within-person variability (how photos of the same person can differ) as well as between-person differences. In fact, this is a lesson learned from applied research in face recognition. If our theories are to be of any utility to those using face recognition for practical purposes, then we are obliged to address the issue of generalizability—to what extent are our experiments eliminating potentially important variables? However, the relationship between research and practice is two-way. Without the constraints imposed by thinking about practical problems (eyewitness testimony, photo ID, etc.) it would be relatively easy to withdraw into a world of artificial, well-controlled stimuli, which are satisfying to use experimentally, but bear little relation to human face perception as it happens. Of course, the use of naturally occurring, ambient stimuli make experimental design even more challenging than normal. While we have argued that large natural variance is necessary for understanding the problem, we remain obliged, for example, to ensure that factors we regard as 'noise' do not vary systematically with the independent variables we manipulate. This can be challenging, but the evidence from the last decade or two seems to be that the greatest theoretical gain can be made when we are working as closely as possible to the real-world phenomena we are seeking to understand.

Conclusion

We began with the problem of eyewitness memory, describing some historical cases in which witnesses, in good faith, made misidentifications that led to miscarriages of justice. Today, the forensic landscape is somewhat different. While witness memory remains a key part of many legal cases, the role of security surveillance has increased dramatically—eliminating fallible memory but bringing problems of its own. Furthermore, photo ID, once confined almost exclusively to passports, is now ubiquitous across many societies. In the context of these changes, what has been the benefit of the almost 50 years of research we have described?

One pessimistic view is that psychological research has produced rather little practical benefit. We have documented extensively the fallibility of eyewitnesses, face matching, and image comparison, even by those deemed expert. While there has been some acknowledgement of this in various jurisdictions around the world, there continue to be many miscarriages of justice based on compelling but inaccurate evidence presented by witnesses and experts (Edmond et al., 2017; Garvie, 2019; see Wixted & Roediger, Chapter 4, this volume).

On the more positive side, it is clear that our understanding of the fundamental processes of face recognition is now considerably more advanced than when the applied problem of witness memory drew psychologists to the issue. We understand much more about the fallibility of face recognition, not only due to the imperfections of human memory, but also through the inherent differences between familiar and unfamiliar faces. We also understand, prompted by the practical issues surrounding photo ID, that there is a lot more to face recognition than image matching, and that familiarity is much more than simple exposure. In these regards, it seems fair to argue that theoretical work in face processing has gained more from the study of practical issues than vice versa.

One increasingly promising route for understanding the relationship between theoretical and applied research in face recognition lies in the study of metacognition (Zhou & Jenkins, 2020). Why is it that we can know that eyewitness testimony is fallible and yet find a specific testimony so compelling? Why is it that we can understand that unfamiliar face matching is hard, and yet consistently choose a photo for our ID that is poorly recognized by other people (White et al., 2017)? These are questions that reflect at the personal level, issues that arise also at a societal level. The answers may lie in our own inherent biases in face recognition. Our everyday experience supports the idea that face recognition is very good—it is easily possible to pass a day, for example in the workplace, recognizing dozens of people with no apparent errors at all. And it could be this very success that causes the misperception that face recognition is a generally reliable system—simply because we fail to take into account the powerful benefits of familiarity. If this is true, then the transfer of knowledge from the laboratory to the real world is itself a psychological problem—because doing so requires addressing our own egocentric biases. While this suggests one way forward, it may be an even harder problem to crack than face recognition.

Acknowledgements

Image attributions for Figure 3.6 (Paul McCartney): top row from left to right: Mikael J. Nordstrom (Public Domain), Pixabay (Public Domain—CC0), Michael Cooper (Public Domain); middle row from left to right: Terry George (CC BY-NC-SA 2.0), Nationaal Archief (CC BY-SA 3.0 NL), Fresh on the net (CC BY 2.0); bottom row from left to right: Oli Gill (CC BY-SA 2.0), Comediadeflowers (CC BY-SA 4.0), Angie Mills (CC BY-SA 3.0).

References

Baddeley, A. (2019). *Working memories: Postmen, divers and the cognitive revolution*. Routledge.

Bindemann, M., Avetisyan, M., & Blackwell, K.-A. (2010). Finding needles in haystacks: Identity mismatch frequency and facial identity verification. *Journal of Experimental Psychology: Applied, 16*(4), 378–386.

Bindemann, M., Fysh, M. C., Cross, K., & Watts, R. (2016). Matching faces against the clock. *i-Perception, 7*(5), 1–18.

Bindemann, M., & Sandford, A. (2011). Me, myself, and I: Different recognition rates for three photo-IDs of the same person. *Perception, 40*(5), 625–627.

Blauch, N. M., Behrmann, M., & Plaut, D. C. (2021). Computational insights into human perceptual expertise for familiar and unfamiliar face recognition. *Cognition, 208*, 104341.

Bruce, V. (1982). Changing faces: Visual and non-visual coding processes in face recognition. *British Journal of Psychology, 73*(1), 105–116.

Bruce, V. (1994). Stability from variation: The case of face recognition. *Quarterly Journal of Experimental Psychology, 47*(1), 5–28.

Bruce, V., Henderson, Z., Greenwood, K., Hancock, P. J. B., Burton, A. M., & Miller, P. (1999). Verification of face identities from images captured on video. *Journal of Experimental Psychology: Applied, 5*(4), 339–360.

Bruce, V., & Valentine, T. (1985). Identity priming in the recognition of familiar faces. *British Journal of Psychology, 76*(3), 373–383.

Bruce, V., & Valentine, T. (1986). Semantic priming of familiar faces. *Quarterly Journal of Experimental Psychology Section A, 38*(1), 125–150.

Bruce, V., & Young, A. (1986). Understanding face recognition. *British Journal of Psychology, 77*(3), 305–327.

Burton, A. M. (2013). Why has research in face recognition progressed so slowly? The importance of variability. *Quarterly Journal of Experimental Psychology, 66*(8), 1467–1485.

Burton, A. M., Bruce, V., & Hancock, P. J. B. (1999). From pixels to people: A model of familiar face recognition. *Cognitive Science, 23*(1), 1–31.

Burton, A. M., Bruce, V., & Johnston, R. A. (1990). Understanding face recognition with an interactive activation model. *British Journal of Psychology, 81*(3), 361–380.

Burton, A. M., Kramer, R. S. S., Ritchie, K. L., & Jenkins, R. (2015). Identity from variation: Representations of faces derived from multiple instances. *Cognitive Science, 40*(1), 202–223.

Burton, A. M., Miller, P., Bruce, V., Hancock, P. J. B., & Henderson, Z. (2001). Human and automatic face recognition: A comparison across image formats. *Vision Research, 41*(24), 3185–3195.

Burton, A. M., White, D., & McNeill, A. (2010). The Glasgow face matching test. *Behavior Research Methods, 42*(1), 286–291.

Burton, A. M., Wilson, S., Cowan, M., & Bruce, V. (1999). Face recognition in poor-quality video: Evidence from security surveillance. *Psychological Science, 10*(3), 243–248.

Calder, A. J., & Young, A. W. (2005). Understanding the recognition of facial identity and facial expression. *Nature Reviews Neuroscience, 6*(8), 645–651.

Carlin, J. D., & Calder, A. J. (2013). The neural basis of eye gaze processing. *Current Opinion in Neurobiology, 23*(3), 450–455.

Damasio, A. R., Damasio, H., & Van Hoesen, G. W. (1982). Prosopagnosia: Anatomic basis and behavioral mechanisms. *Neurology, 32*(4), 331–341.

Davies, G. M., Ellis, H. D., & Shepherd, J. W. (1978). Face recognition accuracy as a function of mode of representation. *Journal of Applied Psychology, 63*(2), 180–187.

De Haan, E. H., Young, A. W., & Newcombe, F. (1991). Covert and overt recognition in prosopagnosia. *Brain, 114*(6), 2575–2591.

Devlin, P. (1976). *Report to the Secretary of State for the Home Department of the Departmental Committee on Evidence of Identification in Criminal Cases.* Her Majesty's Stationery Office.

Edmond, G., Towler, A., Growns, B., Ribeiro, G., Found, B., White, D., Ballantyne, K., Searston, R. A., Thompson, M. B., Tangen, J. M., Kemp, R. I., & Martire, K. (2017). Thinking forensics: Cognitive science for forensic practitioners. *Science & Justice, 57*(2), 144–154.

Ellis, A. W., Burton, A. M., Young, A. W., & Flude, B. M. (1997). Repetition priming between parts and wholes: Tests of a computational model of familiar faced recognition. *British Journal of Psychology, 88*(4), 579–608.

Ellis, A. W., Flude, B. M., Young, A. W., & Burton, A. M. (1996). Two loci of repetition priming in the recognition of familiar faces. *Journal of Experimental Psychology: Learning, Memory and Cognition, 22*(2), 295–308.

Ellis, H. D. (1986). Processes underlying face recognition. In R. Bruyer (Ed.), *The neuropsychology of face perception and facial expression* (pp. 1–27). Lawrence Erlbaum Associates.

Ellis, H. D., Davies, G. M., & Shepherd, J. W. (1978). A critical examination of the Photofit system for recalling faces. *Ergonomics, 21*(4), 297–307.

Feng, X., & Burton, A. M. (2019). Identity documents bias face matching. *Perception, 48*(12), 1163–1174.

Feng, X., & Burton, A. M. (2021). Understanding the document bias in face matching. *Quarterly Journal of Experimental Psychology*, *74*(11), 2019–2029.

Frowd, C. (2015). Facial composites and techniques to improve image recognizability. In T. Valentine & J. P. Davis (Eds.), *Forensic facial identification: Theory and practice of identification from eyewitnesses, composites and CCTV* (pp. 43–70). Wiley Blackwell.

Fysh, M. C., & Bindemann, M. (2017). Effects of time pressure and time passage on face-matching accuracy. *Royal Society Open Science*, *4*(6), 170249.

Fysh, M. C., & Bindemann, M. (2018). The Kent face matching test. *British Journal of Psychology*, *109*(2), 219–231.

Galper, R. E., & Hochberg, J. (1971). Recognition memory for photographs of faces. *American Journal of Psychology*, *84*(3), 351–354.

Garvie, C. (2019). *Garbage in, garbage out: Face recognition on flawed data*. Georgetown Law Center on Privacy and Technology. https://www.flawedfacedata.com/

Gobbini, M. I., & Haxby, J. V. (2007). Neural systems for recognition of familiar faces. *Neuropsychologia*, *45*(1), 32–41.

Hancock, P. J. B., Bruce, V., & Burton, A. M. (2000). Recognition of unfamiliar faces. *Trends in Cognitive Sciences*, *4*(9), 330–337.

Hay, D. C., & Young, A. W. (1982). The human face. In A. W. Ellis (Ed.), *Normality and pathology in cognitive functions* (pp. 173–202). Academic Press.

Haxby, J. V., Hoffman, E. A., & Gobbini, M. I. (2000). The distributed human neural system for face perception. *Trends in Cognitive Sciences*, *4*(6), 223–233.

Henderson, Z., Bruce, V., & Burton, A. M. (2001). Matching the faces of robbers captured on video. *Applied Cognitive Psychology*, *15*(4), 445–464.

Hochberg, J., & Galper, R. E. (1967). Recognition of faces: I. An exploratory study *Psychonomic Science*, *9*(12), 619–620.

Holmes, D. S. (1968). Dimensions of projection. *Psychological Bulletin*, *69*(4), 248–268.

Hooker, C. I., Paller, K. A., Gitelman, D. R., Parrish, T. B., Mesulam, M. M., & Reber, P. J. (2003). Brain networks for analyzing eye gaze. *Cognitive Brain Research*, *17*(2), 406–418.

Jenkins, R., & Burton, A. M. (2011). Stable face representations. *Philosophical Transactions of the Royal Society B: Biological Sciences*, *366*(1571), 1671–1683.

Jenkins, R., White, D., Van Montfort, X., & Burton, A. M. (2011). Variability in photos of the same face. *Cognition*, *121*(3), 313–323.

Kanwisher, N., McDermott, J., & Chun, M. M. (1997). The fusiform face area: A module in human extrastriate cortex specialized for face perception. *Journal of Neuroscience*, *17*(11), 4302–4311.

Kemp, R., Towell, N., & Pike, G. (1997). When seeing should not be believing: Photographs, credit cards and fraud. *Applied Cognitive Psychology*, *11*(3), 211–222.

Kramer, R. S. S., Young, A. W., & Burton, A. M. (2018). Understanding face familiarity. *Cognition*, *172*, 46–58.

Krueger, J., & Clement, R. W. (1994). The truly false consensus effect: An ineradicable and egocentric bias in social perception. *Journal of Personality and Social Psychology*, *67*(4), 596–610.

Lawrence, P. (2020). Policing, 'science', and the curious case of Photo-FIT. *Historical Journal*, *63*(4), 1007–1031.

Labati, R. D., Genovese, A., Muñoz, E., Piuri, V., Scotti, F., & Sforza, G. (2016). Biometric recognition in automated border control: A survey. *ACM Computing Surveys*, *49*(2), 1–39.

Lander, K., Bruce, V., & Bindemann, M. (2018). Use-inspired basic research on individual differences in face identification: Implications for criminal investigation and security *Cognitive Research: Principles and Implications*, *3*, 26.

Leopold, D., O'Toole, A. J., Vetter, T., & Blanz, V. (2001). Prototype-referenced shape encoding revealed by high-level aftereffects. *Nature Neuroscience*, *4*(1), 89–94.

Loffler, G., Yourganov, G., Wilkinson, F., & Wilson, H. R. (2005). fMRI evidence for the neural representation of faces. *Nature Neuroscience*, *8*(10), 1386–1391.

Logie, R. H., Baddeley, A. D., & Woodhead, M. M. (1987). Face recognition, pose and ecological validity. *Applied Cognitive Psychology*, *1*(1), 53–69.

Maurer, D., Le Grand, R., & Mondloch, C. J. (2002). The many faces of configural processing. *Trends in Cognitive Sciences, 6*(6), 255–260.

McCaffery, J. M., & Burton, A. M. (2016). Passport checks: Interactions between matching faces and biographical details. *Applied Cognitive Psychology, 30*(6), 925–933.

McClelland, J. L. (1981). Retrieving general and specific information from stored knowledge of specifics. In *Proceedings of the third annual meeting of the Cognitive Science Society* (pp. 170–172). Cognitive Science Program.

Megreya, A. M., & Burton, A. M. (2008). Matching faces to photographs: Poor performance in eyewitness memory (without the memory). *Journal of Experimental Psychology: Applied, 14*(4), 364–372.

Megreya, A. M., Sandford, A., & Burton, A. M. (2013). Matching face images taken on the same day or months apart: The limitations of photo ID. *Applied Cognitive Psychology, 27*(6), 700–706.

Memon, A., & Bull, R. (1991). The cognitive interview: Its origins, empirical support, evaluation and practical implications. *Journal of Community & Applied Social Psychology, 1*(4), 291–307.

Memon, A., Meissner, C. A., & Fraser, J. (2010). The cognitive interview: A meta-analytic review and study space analysis of the past 25 years. *Psychology, Public Policy, and Law, 16*(4), 340–372.

Mileva, M., & Burton, A. M. (2019). Face search in CCTV surveillance. *Cognitive Research: Principles and Implications, 4*(37), 1–21.

Mileva, M., Young, A. W., Jenkins, R., & Burton, A. M. (2020). Facial identity across the lifespan. *Cognitive Psychology, 116*, 101260.

Murphy, J., Ipser, A., Gaigg, S. B., & Cook, R. (2015). Exemplar variance supports robust learning of facial identity. *Journal of Experimental Psychology: Human Perception and Performance, 41*(3), 577–581.

Nickerson, R. S., Baddeley, A., & Freeman, B. (1987). Are people's estimates of what other people know influenced by what they themselves know? *Acta Psychologica, 64*(3), 245–259.

Noyes, E., & Jenkins, R. (2019). Deliberate disguise in face identification. *Journal of Experimental Psychology: Applied, 25*(2), 280–290.

O'Toole, A. J., Castillo, C. D., Parde, C. J., Hill, M. Q., & Chellappa, R. (2018). Face space representations in deep convolutional neural networks. *Trends in Cognitive Sciences, 22*(9), 794–809.

O'Toole, A. J., Phillips, P. J., Jiang, F., Ayyad, J., Penard, N., & Abdi, H. (2007). Face recognition algorithms surpass humans matching faces over changes in illumination. *IEEE Transactions on Pattern Analysis and Machine Intelligence, 29*(9), 1642–1646.

Paivio, A. (1971). *Imagery and verbal processes.* Holt, Rinehart & Winston.

Papesh, M. H., Heisick, L. L., & Warner, K. A. (2018). The persistent low-prevalence effect in unfamiliar face-matching: The roles of feedback and criterion shifting. *Journal of Experimental Psychology: Applied, 24*(3), 416–430.

Patterson, K. E., & Baddeley, A. D. (1977). When face recognition fails. *Journal of Experimental Psychology: Human Learning and Memory 1977, 3*(4), 406–417.

Phillips, P. J., Yates, A. N., Hu, Y., Hahn, C. A., Noyes, E., Jackson, K., Cavazos, J. G., Jeckeln, G., Ranjan, R., Sankaranarayanan, S., Chen, J. C., Castillo, C. D., Chellappa, R., White, D., & O'Toole, A. J. (2018). Face recognition accuracy of forensic examiners, superrecognizers, and face recognition algorithms. *Proceedings of the National Academy of Sciences of the United States of America, 115*(24), 6171–6176.

Ritchie, K. L., & Burton, A. M. (2017). Learning faces from variability. *Quarterly Journal of Experimental Psychology, 70*(5), 897–905.

Ritchie, K. L., Mireku, M. O., & Kramer, R. S. (2020). Face averages and multiple images in a live matching task. *British Journal of Psychology, 111*(1), 92–102.

Ritchie, K. L., Smith, F. G., Jenkins, R., Bindemann, M., White, D., & Burton, A. M. (2015). Viewers base estimates of face matching accuracy on their own familiarity: Explaining the photo-ID paradox. *Cognition, 141*, 161–169.

Rhodes, G. (1985). Lateralized processes in face recognition. *British Journal of Psychology, 76*(2), 249–271.

Robertson, D. J., & Burton, A. M. (2021). Checking ID-cards for the sale of restricted goods: Age decisions bias face decisions. *Applied Cognitive Psychology, 35*(1), 71–81.

Robertson, D. J., Noyes, E., Dowsett, A. J., Jenkins, R., & Burton, A. M. (2016). Face recognition by Metropolitan Police super-recognisers. *PLoS One, 11*(2), e0150036.

Ross, L., Greene, D., & House, P. (1977). The 'false consensus effect': An egocentric bias in social perception and attribution processes. *Journal of Experimental Social Psychology*, *13*(3), 279–301.

Russell, R., Duchaine, B., & Nakayama, K. (2009). Super-recognisers. People with extraordinary face recognition ability. *Psychonomic Bulletin & Review*, *16*(2), 252–257.

Ryder, H., Smith, H. M. J., & Flowe, H. D. (2015). Estimator variables and memory for faces. In T. Valentine & J. P. Davis (Eds.), *Forensic facial identification: Theory and practice of identification from eyewitnesses, composites and CCTV* (pp. 159–183). Wiley Blackwell.

Shepard, R. N. (1967). Recognition memory for words, sentences and pictures. *Journal of Verbal Learning and Verbal Behavior*, *6*(1), 156–163.

Standing, L., Conezio, J., & Haber, R. N. (1970). Perception and memory for pictures: Single-trial learning of 2500 visual stimuli. *Psychonomic Science*, *19*(2), 73–74.

Tanaka, J. W., & Gordon, I. (2011). Features, configuration and holistic face processing. In A. J. Calder, G. Rhodes, M. H. Johnson, & J. V. Haxby (Eds.), *The Oxford handbook of face perception* (pp. 15–30). Oxford University Press.

Tong, F., Nakayama, K., Moscovitch, M., Weinrib, O., & Kanwisher, N. (2000). Response properties of the human fusiform face area. *Cognitive Neuropsychology*, *17*(1–3), 257–280.

Towler A., Kemp, R. I., Burton, A. M., Dunn, J. D., Wayne, T., Moreton, R., & White, D. (2019). Do professional facial image comparison training courses work? *PLoS One*, *14*(2), e0211037.

Tranel, D., & Damasio, A. R. (1985). Knowledge without awareness: An autonomic index of facial recognition by prosopagnosics. *Science*, *228*(4706), 1453–1454.

Vernon, M. D. (1952). *A further study of visual perception*. Cambridge University Press.

Weatherford, D. R., Roberson, D., & Erickson, W. B. (2021). When experience does not promote expertise: Security professionals fail to detect low prevalence fake IDs. *Cognitive Research: Principles and Implications*, *6*(1), 1–27.

Wells, G. L. (1978). Applied eyewitness-testimony research: System variables and estimator variables. *Journal of Personality and Social Psychology*, *36*(12), 1546–1557.

White, D., Kemp, R. I., Jenkins, R., Matheson, M., & Burton, A. M. (2014). Passport officers' errors in face matching. *PLoS One*, *9*(8), e103510.

White, D., Sutherland, C. A. M., & Burton, A. L. (2017). Choosing face: The curse of self in profile image selection. *Cognitive Research: Principles and Implications*, *2*(1), 23.

White, D., Towler, A., & Kemp, R. I. (2021). Understanding professional expertise in unfamiliar face matching. In M. Bindemann (Ed.), *Forensic face matching: Research and practice* (pp. 62–88). Oxford University Press.

Woodhead, M. M., & Baddeley, A. D. (1981). Individual differences and memory for faces, pictures, and words. *Memory & Cognition*, *9*(4), 368–370.

Woodhead, M. M., Baddeley, A. D., & Simmonds, D. C. V. (1979). On training people to recognize faces. *Ergonomics*, *22*(3), 333–343.

Yin, R. K. (1969). Looking at upside-down faces. *Journal of Experimental Psychology: General*, *81*(1), 141–145.

Young, A. W., & Bruce, V. (2011). Understanding person perception. *British Journal of Psychology*, *102*(4), 959–974.

Young, A. W., & Burton, A. M. (2021). Insights from computational models of face recognition: A reply to Blauch, Behrmann and Plaut. *Cognition*, *208*, 104422.

Young, A. W., Hay, D. C., & Ellis, A. W. (1985). The faces that launched a thousand slips: Everyday difficulties and errors in recognizing people. *British Journal of Psychology*, *76*(4), 495–523.

Zhou, X., & Jenkins, R. (2020). Dunning–Kruger effects in face perception. *Cognition*, *203*, 104345.

4

Signal detection theory and eyewitness identification

John T. Wixted and Henry L. Roediger III

Something is surprisingly wrong with an idea long endorsed by eyewitness identification experts and long accepted by most educated people. The wrongheaded idea is that when eyewitnesses claim to recognize the person who committed a crime, their decisions are not only often wrong but their confidence when making those decisions is not particularly diagnostic of accuracy. According to this way of thinking, it is a mistake to believe that high confidence in recognizing the perpetrator implies high accuracy and low confidence implies low accuracy. Yet signal detection theory, which has effectively guided thinking about recognition memory in the basic science literature for decades, inherently predicts a strong confidence–accuracy relationship (Egan, 1958; Green & Swets, 1966; Roediger et al., 2012; Wixted, 2020). Should we therefore assume that the theory applies to the countless recognition memory tasks that are used in the basic science laboratory—including face recognition memory—but becomes completely irrelevant the moment we consider recognition memory in the forensic/legal context? That proposition seems unlikely.

What, then, explains the apparent discrepancy? In a typical recognition memory experiment conducted in the basic science laboratory, participants are first presented with a list of items (e.g. words, faces, or objects) to memorize. On the subsequent memory test, the *targets* from the list are randomly intermixed with similar *foils* or fillers that are new, and each test item is presented (one at a time) for an 'old/new' recognition decision. Typically, the experimenter's goal is to test the participant's ability to discriminate old items from new. Therefore, the test items differ only in their status as having appeared on the list or not. On all other dimensions, they are equated (on average).

To ensure they are equated, the experimenter does not prompt the participant to declare certain test items to be 'old' ('Are you sure this one wasn't on the list?') and does not provide feedback following some 'old' decisions ('You got that one right!') before assessing confidence. In addition, the experimenter does not manipulate or otherwise tamper with the strength of the memory signals that will be generated by selected foils. Such manipulations are performed to enhance our theoretical understanding of memory, such as in the Deese–Roediger–McDermott

John T. Wixted and Henry L. Roediger III, *Signal detection theory and eyewitness identification* In: *Memory in Science for Society*.
Edited by: Robert H. Logie, Zhisheng (Edward) Wen, Susan E. Gathercole, Nelson Cowan, and Randall W. Engle, Oxford University Press.
© Oxford University Press 2023. DOI: 10.1093/oso/9780192849069.003.0004

false memory procedure (Deese, 1959; Roediger & McDermott, 1995), but not to assess memory for items that are equated except for their prior occurrence on a list.

Under objective testing conditions like these, typical of the basic science laboratory, participants must rely solely on the unbiased memory signal generated by a test item to determine how confident they should be that it is old or new. These are the conditions under which the a priori predictions of signal detection theory about the strong confidence–accuracy relationship apply, and those predictions are almost invariably confirmed (e.g. see Delay & Wixted, 2021; Mickes et al., 2007, 2011; Tekin et al., 2018, 2021; Tekin & Roediger, 2017).

In years gone by, police line-ups often provided anything but an objective test of recognition memory (where a line-up consists of one suspect and several 'fillers'). For example, line-ups were constructed in such a way that the suspect stood out by using fillers who looked nothing like the suspect (providing a clue as to who should be identified), and/or the line-up administrator would praise a witness who happened to tentatively identify the suspect ('Good job, you picked the right guy'; Wells & Bradfield, 1998). Such confirming feedback would inflate confidence above what it would have been if the rating were based solely on the strength of the memory signal associated with the identified face. Worse, the very act of testing memory for a suspect strengthens the witness's memory of that suspect; the power of testing or retrieval practice on memory has been shown in numerous experiments (see McDermott, 2021). Thus, if another test of memory is performed following the first, the suspect's face will generate a stronger memory-match signal than it did before, and the witness may misattribute the source of that strong signal to the initial crime scene rather than to the previous test, thereby inflating confidence that the suspect is the perpetrator (Wixted et al., 2021). Signal detection theory does not predict a strong confidence–accuracy relationship under compromised circumstances like these because, again, it assumes that confidence is determined by the strength of an unbiased memory signals generated by an item presented on an objective test of recognition memory.

With that background in mind, it is worth taking a moment to reconsider the core problem in the real-world debacle that has, perhaps more than anything else, cemented the impression that eyewitness memory is unreliable. Since the early 1990s, the Innocence Project has used DNA evidence to reverse more than 375 exonerations of wrongful convictions of the innocent (Innocence Project, 2021). How did these innocent people end up being convicted by a jury? The answer is now well known: eyewitness misidentification was a contributing factor in approximately 70% of those convictions, and in every case, the witness misidentified the innocent defendant at trial with high confidence (Garrett, 2011, p. 49). In the eyes of many, the interpretation of this finding is obvious: eyewitness memory is unreliable. Moreover, the solution seems equally obvious: courts of law should deemphasize eyewitness identification evidence no matter how confident the witness might be. In theory, deemphasizing eyewitness confidence will protect innocent suspects from being wrongfully convicted.

On closer inspection, this widely accepted interpretation may not be the best diagnosis of the problem, and the proposed solution may not serve the cause of justice. For example, if on an initial objective test of uncontaminated memory using a line-up, eyewitness identification evidence is reliable, but then the criminal justice system itself renders it unreliable by using improper tests (e.g. involving unfair line-ups or confirming feedback) or by testing a witness's memory more than once, then discounting all eyewitness evidence would not be justified. Only the contaminated evidence should be considered with suspicion, which is something that is true of all types of forensic evidence.

Consider an analogy. If police investigators (1) did not like the results of an initial forensic DNA test conducted on a murder weapon (e.g. the results came back inconclusive), then (2) contaminated the evidence with the DNA of the suspect, then (3) conducted another test on the contaminated evidence (obtaining a conclusive match this time), and finally (4) used the results of that conclusive test to win a conviction, would the problem have anything to do with the reliability of forensic DNA evidence? No. The proper diagnosis of the problem in this hypothetical scenario would focus on the conduct of police investigators and prosecutors, not on the inherent reliability of forensic DNA evidence. The same diagnosis would apply if, on the initial DNA test, improper handling of the evidence or laboratory procedures resulted in a conclusive (but false) match.

Fortunately, the DNA scenario outlined above is purely hypothetical. Unfortunately, in the case of eyewitness identification, it is quite literally the norm (even when a proper line-up is properly administered). If the first test using a line-up yields something other than an immediate, high-confidence identification of the suspect, the police often test memory again using another line-up involving the same suspect (e.g. see eyewitness misidentification cases summarized in The National Registry of Exonerations (https://www.law.umich.edu/special/exoneration)). They might do so after the witness reports having calmed down or after having had a 'revelation' or because the police obtain a better photo of the suspect than the one used in the first test or because they decide that a live line-up would be better than the photo line-up they used on the first test (e.g. Jones et al., 2008). It might not matter very much if, on a second line-up test, all of the same faces were used again. In that case, all of the faces would have been familiarized by the first test, so the suspect would not stand out, and the confidence–accuracy relationship would remain largely intact (Lin et al., 2019). However, in practice, when a second line-up is administered, the same suspect typically appears along with a new set of fillers. Therefore, the suspect's face (innocent or guilty) will have been selectively familiarized by the first test, potentially wreaking havoc on the information value of confidence.

Unfortunately, even if none of that happens, memory is eventually tested again, in court and in front of the jury. Judges often believe that the courtroom test can be conducted independent of prior tests (Garrett, 2012), as if testing memory using a line-up leaves the witness's memory unchanged. This idea, although catastrophically wrong, justifies the courtroom identification. In any case, whatever the rationale, a witness's memory for a suspect is almost always tested again, thereby ensuring that the confidence expressed by the eyewitness will be based on factors other than the uncontaminated memory signal generated by the suspect's face on the first test.

Returning to the key question: is the core problem here the fact that eyewitness memory is unreliable or is it instead the fact that other actors in the criminal justice system engage in behaviour that is (unintentionally) designed to elicit a high-confidence misidentification of an innocent suspect? As it turns out, it is the latter, not the former. After all, eyewitnesses are unaware of the fact that police investigators irretrievably contaminated their memory when they administered the first test. A more accurate diagnosis of the problem leads to a more effective solution: simply stop testing a witness's memory for a suspect more than once, and use an objective test of memory when a line-up is administered to the witness the first time, early in the police investigation (Wixted & Mickes, 2022; Wixted et al., 2021).

The procedures for conducting an objective test have been painstakingly worked out over the years and are now well known (Wells et al., 2020). In terms of construction, a line-up should consist of one suspect and at least five fillers who physically resemble the suspect (Figure 4.1). Ideally, the fillers would all match the description of the suspect provided by the eyewitness to the police. On a test like this, the faces are generally equated except (possibly) for their old/new status. That is, one of the faces (the face of the suspect) may have been encountered once before at the crime scene.

A properly constructed line-up also needs to be properly administered. In particular, the line-up should be administered by an officer who is blind to the suspect's identity, the instructions to the witness should stipulate that the perpetrator may or may not be in the line-up (to induce a neutral response bias), and an immediate statement of confidence should be obtained (Wells et al., 2020). It is reasonable to

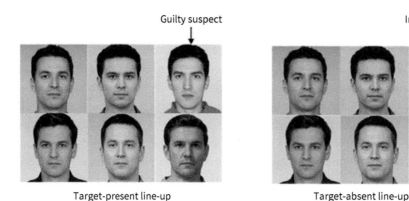

Target-present line-up   Target-absent line-up

Figure 4.1 The left panel illustrates a target-present line-up containing a photo of a guilty suspect and five similar fillers. The right panel illustrates a target-absent line-up containing a photo of an innocent suspect and five similar fillers.

wonder how confidence should be assessed, exactly, but it turns out that almost any rating scale works just fine (Tekin et al., 2018; Tekin & Roediger, 2017), and even verbal confidence ratings provide essentially the same information (Smalarz et al., 2021; Tekin et al., 2018). The important point is that confidence should be immediately assessed after an identification is made; the question of exactly how to do that is much less important.

Line-ups constructed and administered in this manner are often referred to as 'pristine' (Wixted & Wells, 2017), and researchers generally use pristine procedures in their investigations of eyewitness identification performance. How reliable is eyewitness identification performance when a witness's uncontaminated memory is initially tested using a pristine line-up procedure? In assessing the reliability of forensic memory evidence, *that* is the question. It is analogous to the question we ask when judging the reliability of any kind of forensic evidence, such as forensic DNA: how reliable is an uncontaminated forensic DNA sample when it is tested using proper procedures?

We turn now to the next chapter of this story, which is that, for decades, researchers concluded that even under pristine initial testing conditions involving uncontaminated memory evidence, eyewitness identification is still a highly error-prone process, with confidence, at best, only weakly related to accuracy. If that were true, then (1) signal detection theory would not apply, (2) eyewitness memory would be inherently unreliable, and (3) forensic memory evidence would be properly discounted by the legal system. However, as we shall see, it is not true (Roediger et al., 2012; Wixted et al., 2015, 2018; Wixted & Wells, 2017).

The confidence–accuracy relationship on an initial, pristine test

To meaningfully assess the relationship between eyewitness confidence and accuracy, it is important to understand that the details matter. Intuition suggests that measuring this relationship should be simple and straightforward, but the opposite is true. Because the issue is vitally important (lying, as it does, at the heart of the question of whether eyewitness memory is reliable or unreliable), in this section, we consider the intricate details that have resulted in decades of confusion. To begin, consider the following summary of the state of scientific knowledge about this issue in 1995:

> A major source of juror unreliability is their reliance on witness confidence, which (a) is a weak indicator of eyewitness accuracy even when measured at the time an identification is made and under relatively 'pristine' laboratory conditions and (b) appears to be highly malleable and influenced by post-identification factors such as repeated questioning, briefings in anticipation of cross-examination, and feedback about the behavior of other witnesses. (Penrod & Cutler, 1995, p. 830)

The importance of the distinction in the above quotation between confidence expressed at the time of an initial, pristine test of memory and confidence expressed in a later post-identification test of memory (e.g. at trial) cannot be overemphasized. The long-standing scientific consensus about the unreliability of high-confidence identifications made *subsequent to* the initial identification remains in place, as in part (b) of the above quote. No subsequent test provides the sought-after information because testing recognition memory of the suspect's face contaminates memory of that suspect (Wixted et al., 2021). As the above quotation illustrates in part (a), the field long believed that confidence is largely non-diagnostic of accuracy even on an *initial* uncontaminated and properly administered test of memory. However, that conclusion has been completely reversed in recent years. Contrary to what was long understood to be true, confidence on an initial test of memory is highly diagnostic of accuracy in most situations (Roediger et al., 2012; Wixted et al., 2015; Wixted & Wells, 2017).

What changed? The methods used to experimentally investigate the issue have not changed much at all, so that is not what accounts for the change in thinking. In a typical line-up study dating back to the 1970s, a participant-witness observes a mock crime (e.g. a perpetrator stealing a purse) and then makes a decision about whether a suspect is present from either a target-present line-up or a target-absent line-up (Figure 4.1). On such a test, the participant can either identify someone (the suspect or one of the fillers) or reject the line-up. After deciding, the participant is asked to rate their confidence, often using a 0–100 scale.

Usually, a participant completes only one study-test trial in mock crime studies because, in the real world, eyewitnesses usually observe only one crime and are tested with only one line-up. Thus, for each participant, only two pieces of information are involved in the analysis of the relationship between confidence and accuracy: (1) the accuracy of their singular decision (correct or incorrect) and (2) their confidence in that decision (e.g. measured using a 0–100 scale). Both an identification of the suspect from a target-present line-up and a rejection of a target-absent line-up are correct responses and might therefore be coded as '1'. All other decisions (identifying a suspect from a target-absent line-up, identifying a filler from either a target-present or a target-absent line-up, and rejecting a target-present line-up) are incorrect and might therefore be recorded as '0'.

What should we do with numbers like these to measure the confidence–accuracy relationship? What would *you* do if asked to measure that relationship? It is an important question to consider because if the data are analysed incorrectly, the resulting conclusion about the confidence–accuracy relationship will be correspondingly off the mark.

An intuitively appealing approach is simply to compute the correlation between the confidence and accuracy measures across participants. This is the approach that was almost invariably used from the 1970s through the late 1990s and is still used sometimes today. To perform this analysis, one simply computes the correlation between a column of accuracy scores (e.g. 0, 0, 1, 0, 1, 1, 1, . . .) and their corresponding confidence ratings (e.g. 80, 45, 99, 25, 50, 85, 65, . . .). This measure is technically known as the point-biserial correlation coefficient because one measure is dichotomous (i.e. accuracy) and the other is more-or-less continuous (i.e. confidence), but it is computationally identical to the familiar formula for computing the Pearson product-moment correlation (r).

The point-biserial correlation (r_{pb}) is computed once across all levels of confidence using the following formula:

$$ r_{pb} = \frac{\left(\bar{Y}_1 - \bar{Y}_0\right)}{s_n} \sqrt{\frac{n_1 n_0}{n}} \tag{4.1} $$

where $\bar{Y}_1$ is the average confidence rating for the n_1 witnesses who were correct, $\bar{Y}_0$ is the average confidence rating for the n_0 witnesses who were incorrect, s_n is the standard deviation of all confidence ratings (correct and incorrect combined), and $n = n_1 + n_0$. This approach attempts to boil down the confidence–accuracy relationship to a single number.

In a typical study, r_{pb} was found to be approximately 0.20 and was often close to 0 (e.g. Deffenbacher, 1980). Therefore, the conclusion was that the information provided by eyewitness confidence about accuracy is negligible. Later studies computed the correlation separately for 'choosers' (i.e. excluding participants who rejected the line-up) and non-choosers (i.e. those who reject the line-up). Choosers yielded a higher correlation of 0.37 (Sporer et al., 1995). However, this is still rather unimpressive because it means that confidence explains only about $0.37^2 \times 100\%$ = 14% of the variance in the accuracy of choosers. For non-choosers, the correlation was noticeably lower (0.12), accounting for $0.12^2 \times 100\%$ = 1% of the variance in accuracy.

Based on findings like these, courts across the US have come to distrust eyewitness confidence, often specifically pointing out that the *correlation* between confidence and accuracy is low. Although factually true, unbeknown to the judges raising this concern, it has little to do with the question of interest. Despite its intuitive appeal, the virtual irrelevance of the information provided by the correlation coefficient will become clear as we proceed with our analysis. Although it turns out to be the wrong measure, on the positive side, the correlation was computed using data from an initial test of memory using pristine testing procedures. Thus, at least the focus is where it should be (i.e. on the initial, properly conducted line-up).

What is wrong with using the correlation coefficient to measure the relationship between confidence and accuracy when eyewitness memory is tested using a line-up? As Juslin et al. (1996) first pointed out, it simply does not answer the question of interest. As they put it: 'The more precise argument we make is (a) that the correlational measure can be severely misleading and may often fail to disclose a forensically useful confidence–accuracy relation, and (b) the correlation coefficient does not provide the court with the kind of information that is needed' (p. 1305). They further argued that the information of interest is provided by a *calibration* analysis. In a calibration analysis, the subjective confidence expressed by a witness corresponds to the estimated chances of being correct. Thus, a witness who identifies a suspect with 75% confidence is estimating that their chances of being correct are 75%. This estimate is compared to the actual percentage correct over all participants who expressed 75% confidence. If witnesses who express 50% confidence are 50% correct, witnesses who express 75% confidence are 75% correct, and witnesses who express 100% confidence are 100% correct, then confidence would be perfectly calibrated with accuracy.

Unlike the point-biserial correlation coefficient, calibration accuracy scores are computed separately for each confidence rating. For a given rating (R), percentage correct (PC_R) is given by:

$$PC_R = 100\% \times \frac{C_R}{C_R + E_R} \tag{4.2}$$

where C_R represents the number of witnesses who made a correct line-up decision with confidence rating R, and E_R represents the number of witnesses who made an erroneous line-up decision with confidence rating R. As noted earlier, a correct decision consists of a suspect identification from a target-present line-up and the rejection of a target-absent line-up, with all other possible decisions being incorrect. If there are ten levels of confidence ratings (10, 20, 30, and so on through to 100), there would be ten corresponding accuracy scores. Instead of trying to boil the confidence–accuracy relationship down to a single number, one would instead simply plot the ten accuracy scores on the y-axis against the ten levels of confidence on the x-axis, yielding a calibration plot.

Juslin et al. (1996) asked a pertinent question that we paraphrase here: if a witness expressed 75% certainty in their identification of a suspect, which would better inform a judge and a jury, the fact that $r_{pb}^2 = 0.14$ or the fact that $PC_{75\%} = 70\%$? It is hard to imagine what jurors would do with the knowledge that confidence accounts for 14% of the variance in accuracy, but it is easy to imagine that they could make effective use of the knowledge that witnesses who express 75% confidence are, on average, correct 70% of the time. Moreover, and counterintuitively, the point-biserial correlation coefficient can be close to 0 even when confidence is *perfectly* calibrated to accuracy (Juslin et al., 1996). For these reasons, Juslin et al. argued that calibration provides much more useful information than the correlation coefficient does.

In the years following the introduction of this insight, some researchers switched to using the calibration approach, and they confirmed what Juslin et al. (1996) found: confidence is a lot more predictive of accuracy than had been previously appreciated (e.g. Brewer et al., 2002; Brewer & Wells, 2006; Weber & Brewer, 2004). As was eventually done in studies using the correlation approach, calibration is usually computed separately for choosers (those who identify either a suspect or a filler) versus non-choosers (those who reject the line-up). The results were similar to what had been found using the correlation coefficient in that the confidence–accuracy relationship was noticeably stronger for choosers (e.g. Brewer & Wells, 2006). However, for choosers, at least, the confidence–accuracy relationship could no longer be reasonably characterized as weak. We illustrate the calibration approach using actual data shortly, but first we address a technical detail associated with calibration analyses that has not often been considered.

When Equation 4.2 is applied to choosers, the value in the numerator consists only of correct identifications from target-present line-ups for a given level of confidence because that is the only correct response a chooser can make. The denominator includes those correct identifcations plus incorrect chooser identifications for the same level of confidence, of which there are three kinds: (1) incorrect suspect identifications from target-absent line-ups, (2) incorrect filler identifications from target-absent line-ups, and (3) incorrect filler identifications from target-present line-ups.

In a common experimental design, no one in the target-absent line-up is designated as an innocent suspect. In that case, one can estimate innocent suspect identifications by dividing all target-absent filler identifications by line-up-size (k) because, from the witness's point of view, an innocent suspect is effectively just another filler. Thus, for example, if $k = 6$ and there were 60 filler identifications from a target-absent line-up made with 75% confidence across all participants, then the estimated number of innocent suspect identifications made with that level of confidence would be 60/6 = 10, and the estimated number of target-absent filler identifications made with that level of confidence would be 60 − 10 = 50. For this type of design, the estimated number of innocent suspect identifications plus the estimated number of filler identifications from target-absent line-ups (10 + 50 = 60) is simply equal to the observed number of filler identifications from target-absent line-ups (60 in this hypothetical example). Thus, one need not actually go to the trouble of estimating the number of innocent suspect identifications to compute a calibration accuracy score, but we mention it here because the issue will come up again before we finish this story.

It is important to note that, in practice, calibration has typically been computed in a different way. The computational formula that seems truest to the spirit of the calibration approach is what we have described thus far (Equation 4.2). Equation 4.3 spells out Equation 4.2 in more detail for the case in which target-absent line-ups contain a designated innocent suspect:

$$PC_R = 100\% \times \frac{G_R}{G_R + \left(I_R + F_{TP_R} + F_{TA_R} \right)} \tag{4.3}$$

where G_R represents the number of guilty suspect identifications from target-present line-ups (which are correct decisions) made with confidence rating R, I_R represents the number of (incorrect) innocent suspect identifications from target-absent line-ups made with confidence rating R, and F_{TP_R} and F_{TA_R} represent the number of (incorrect) filler identifications from target-present and target-absent line-ups made with confidence rating R, respectively. For the case in which target-absent line-ups do not contain a designated innocent suspect, the equation is the same except that I_R is not included.

Equation 4.3 unpacks three kinds of incorrect decisions in the denominator, which are grouped by parentheses, to make the point that most calibration studies have not actually included all of these errors when accuracy is computed for each level of confidence. Instead, most have simply excluded filler identifications from target-present line-ups (i.e. they have excluded F_{TP_R}) on the grounds that, in the real world, those are known errors (Brewer & Wells, 2006). However, in the real world, filler identifications from target-absent line-ups are also known errors and should therefore *also be excluded* (i.e. F_{TA_R} should also be excluded). Yet they are invariably included in calibration studies.

By including filler identifications from target-absent line-ups while excluding filler identifications from target-present line-ups, the calibration approach results in overall higher accuracy scores than would otherwise be the case. Importantly, that fact has not been lost on those who translated calibration findings for the legal system. For example, Figure 4.2a shows calibration data reported by Sauer et al. (2008) computed using the usual method of including filler identifications from target-absent line-ups while excluding target-present filler identifications, whereas Figure 4.2b shows what happens when all errors made by choosers are included in the analysis (as it seems they should be if one takes the calibration approach). Notice in particular the large effect on the accuracy of identifications made with high confidence.

In an amicus brief filed in support of a defendant's request to instruct the jury that 'witnesses who are highly confident of their identifications are not therefore necessarily reliable', the American Psychological Association (APA, 2014, p. 3) considered these very data. In that brief, the APA pointed out that 'there is overwhelming consensus as to the core findings of that research' (p. 10). With regard to the reliability of eyewitness identification assessed under pristine testing conditions, the amicus brief had this to say:

> Importantly, error rates can be high even among the most confident witnesses. Researchers have performed studies that track, in addition to identification accuracy, the subjects' estimates of their confidence in their identifications. In one article reporting results from an empirical study, researchers found that among witnesses who made positive identifications, as many as 40 percent were mistaken, yet they declared themselves to be 90 percent to 100 percent confident in the accuracy of their identifications ... This confirms that many witnesses are over-confident in their identification decisions. (pp. 17–18)

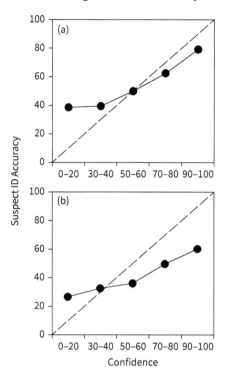

Figure 4.2 Confidence–accuracy calibration plot for choosers based on data reported by Sauer et al. (2008). The dashed diagonal line represents perfect calibration. (a) For this plot, the common method of excluding target-present filler identifications was used. Some degree of overconfidence is evident at the high end of the scale because the accuracy of identifications made with a 90–100 confidence score was only about 80% correct. According to this approach, high-confidence identifications are accurate but accuracy falls well below perfect calibration of 95% correct. (b) The same data using responses made by all choosers, including target-present filler identifications. Now, an extreme degree of overconfidence is evident at the high end of the scale because the accuracy of identifications made with a 90–100 confidence score was only about 60% correct (far below perfect calibration of 95% correct).

Thus, as recently as 2014, the APA informed the legal system that the error-prone nature of high-confidence identifications was empirically confirmed by the data shown in Figure 4.2b.

Although calibration analyses represent an improvement over the point-biserial correlation coefficient for assessing the confidence–accuracy relationship, as just illustrated, they have nevertheless continued to badly mislead the legal system, just as the correlation coefficient did in prior years. As noted above, the problem with calibration analyses is that the standard rationale for excluding target-present filler identifications (namely, such as identifications are known errors) applies equally to target-present *and* target-absent line-ups. Thus, filler identifications should be excluded from the calculations altogether, and doing so finally makes it clear that confidence is highly predictive of accuracy, with high-confidence identifications being highly reliable.

Starting over

Next, we reconsider how to measure the confidence–accuracy relationship, starting from the beginning. The first step is to create fair target-present and target-absent line-ups. After these line-ups are constructed but before a witness makes a decision, there is some baseline (prior) probability that it is a target-present line-up (in which case the suspect in the line-up is guilty). Most laboratory studies use an equal number of target-present and target-absent line-ups (equal base rates), so the *prior* probability that the suspect in the line-up is guilty (i.e. one's best estimate of guilt prior to the witness making a decision) is 0.50.[1] After the witness decides, this prior probability is updated, yielding the *posterior* probability of guilt (i.e. the best estimate that the suspect is guilty in light of the witness's decision).

As noted earlier, when a witness is tested using either a target-present or a target-absent line-up, the decision outcome falls into one of three categories: (1) a suspect identification, (2) a filler identification, or (3) a line-up rejection (no identification). Critically, each of these decision outcomes provides unique information about the innocence or guilt of the suspect in the line-up. In other words, the *posterior* probability of the suspect's guilt differs depending on the decision outcome. Therefore, to compute the posterior probability that it is a target-present line-up, an analysis must be performed on three separate groups of eyewitnesses: (1) those who identified a suspect, (2) those who identified a filler, and (3) those who rejected the line-up.

To appreciate why the analysis should not be performed on all eyewitnesses combined or on choosers versus non-choosers, as has been done for decades, consider the example presented in Figure 4.3. In this example, there are 50 target-present plus 50 target-absent line-ups (100 line-ups in all), each of which is associated with one of three possible decision outcomes. From those 100 line-ups, imagine that witnesses identified the suspect in 35 of them, 25 from a target-present line-up (the witness identified the guilty suspect) and 10 from a target-absent line-up (the witness identified the innocent suspect). Thus, the posterior *odds* that it is a target-present line-up (in which case the suspect is guilty) given that the witness landed on the suspect in the line-up would be 25/10 = 2.5 and the posterior *probability* would be 25/(25 + 10) = 0.71. Thus, given a suspect identification, the estimated probability that it is a target-present line-up (so the suspect is guilty) increased from a prior probability of 0.50 before the test to a posterior probability of 0.71 after the test.

The same analysis can be performed on the remaining two outcomes. Imagine that 23 of the 100 witnesses identified a filler, nine from a target-present line-up and

---

[1] In the real world, the prior probability of guilt likely varies by jurisdiction, depending on how much evidence of guilt the police require to place a suspect's photo in a line-up. In Houston, Texas, which requires no evidence at all, Wixted, Mickes, et al. (2016) used a signal detection model to estimate that 35% of suspects were guilty and 65% innocent. However, other jurisdictions require that the investigating officer be able to articulate at least some evidence of guilt before placing a suspect in the line-up. Those jurisdictions will presumably have higher base rates of guilt in the line-ups they administer to eyewitnesses.

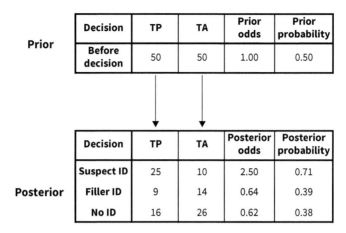

Decision	TP	TA	Prior odds	Prior probability
Before decision	50	50	1.00	0.50

Prior

Decision	TP	TA	Posterior odds	Posterior probability
Suspect ID	25	10	2.50	0.71
Filler ID	9	14	0.64	0.39
No ID	16	26	0.62	0.38

Posterior

Figure 4.3 A concrete example of 100 line-ups (50 target-present (TP) and 50 target-absent (TA)) illustrating the prior odds and probability that it is a TP line-up (in which case the suspect is guilty) as well as the posterior odds and probability that it is a TP line-up depending on the decision outcome.

14 from a target-absent line-up. Thus, the posterior odds that it is a target-present line-up (i.e. the posterior odds that the suspect in the line-up is guilty) given that the witness landed on a filler would be 9/14 = 0.64 and the posterior probability would be 9/(9 + 14) = 0.39. In other words, because our estimated probability that it is a target-present line-up decreased from 0.50 before the test to 0.39 after the test, a filler identification would be *probative of innocence* (not guilt).

Finally, imagine that the remaining 42 witnesses rejected the line-up, 16 from a target-present line-up and 26 from a target-absent line-up. Thus, the posterior odds that it is a target-present line-up (i.e. the posterior odds that the suspect in the line-up is guilty) given that the witness rejected the line-up would be 16/26 = 0.62 and the posterior probability that the suspect is guilty would be 16/(16 + 26) = 0.38. Again, as when a filler identification occurs, a line-up rejection is also probative of innocence. The numerical values used in these examples are representative of empirical estimates obtained from laboratory research (Wells et al., 2015).

Usually, when addressing the question of eyewitness reliability in a court of law, the question of interest is how accurate the identification was *given that a suspect identification was made* during the initial line-up test (Mickes, 2015). This is the question that judges and jurors should want to know the answer to because (1) a filler who was mistakenly identified does not end up being put on trial (i.e. fillers are known to be innocent), and (2) a witness who initially rejected the line-up containing the suspect should not testify against that suspect at trial (because a line-up rejection is probative of innocence).

Assuming a fair line-up, the fact that the witness landed on the suspect moves the needle in the direction of guilt (from 0.50 to 0.71 in the example presented above). Does the degree to which the needle moves vary as a function of confidence expressed on the initial (uncontaminated) test of memory? That is our main question, and as we have mentioned from the beginning, signal detection theory predicts that the answer is 'Yes' (Roediger et al., 2012; Wixted et al., 2015). We now consider why that is.

In the context of eyewitness identification, the latent variable of interest is the strength of the memory-match signal between a face in the line-up and the witness's memory of the perpetrator. The closer the match, the stronger the memory-match signal will be. Theoretically, a guilty suspect will generate a stronger memory-match signal than an innocent suspect. As illustrated in Figure 4.4, a core assumption of signal detection theory is that the strength of the memory-match signals generated by, for example, many innocent suspects are not identical to each other but are instead variable. In other words, they generate a distribution of memory signals across witnesses and the corresponding suspects. The simplest version of signal detection theory assumes a Gaussian distribution, but that is not an essential aspect of the theory. Virtually any continuous unimodal distribution would suffice. The key assumption is that because the innocent suspects do not correspond to the perpetrator (and therefore do not usually provide a strong match to the witness's memory of the perpetrator), the mean of the innocent suspect distribution is relatively low. For guilty suspects, everything is the same except that the mean of the distribution is higher because the guilty suspect provides a better match to the witness's memory of the perpetrator. But even in the case of a guilty suspect, the match may not be perfect because, for example, the perpetrator's face may have been encoded incorrectly, or the correctly encoded face may have degraded over time due to forgetting, or the face may have changed since the crime was committed (e.g. a new beard), and so on.

As illustrated in Figure 4.5, decisions are made in relation to various criteria arrayed along the memory-match axis. A six-point confidence scale ranging from

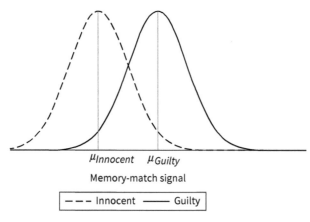

$\mu_{Innocent}$ μ_{Guilty}

Memory-match signal

- - - Innocent ——— Guilty

Figure 4.4 Distribution of underlying (latent) memory signals generated by innocent and guilty suspects according to the basic signal detection model. The model assumes that, across line-ups, the average strength of the memory-match signal generated by guilty suspects (μ_{Guilty}) is higher than the average strength of the memory-match signal generated by innocent suspects ($\mu_{Innocent}$).

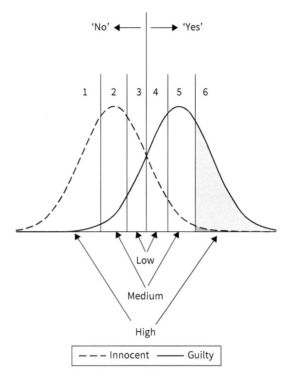

Figure 4.5 An illustration of the signal detection interpretation of confidence ratings made using a six-point confidence scale. The shaded regions to the right illustrate the proportion of innocent suspects (dark shading) and guilty suspects (light shading) receiving a high-confidence (6) 'Yes' decision.

1 = 'Sure No' to 6 = 'Sure Yes' is illustrated here, but we could just as easily imagine the scale ranging from −100 ('Sure No') to + 100 ('Sure Yes'). Figure 4.5 illustrates why high-confidence 'Yes' decisions should be reliable. The reason is that innocent suspects rarely generate memory signals strong enough to exceed the criterion for deciding 'Yes' with high confidence (6), but guilty suspects frequently generate memory signals that strong.

Figure 4.6 illustrates this issue in more detail. The proportion of guilty suspects generating memory signals yielding confidence ratings of 6, 5, and 4 are 0.31, 0.33, and 0.20, respectively (left column). The proportion of innocent suspects generating memory signals yielding confidence ratings of 6, 5, and 4 are 0.01, 0.04, and 0.11, respectively (right column). Thus, for high-confidence suspect identifications (confidence = 6) in this illustrative example, the posterior probability of guilt is predicted to be 0.31/(0.31 + 0.01) = 0.98. In other words, high confidence implies high accuracy. For medium-confidence suspect identifications (confidence = 5), the posterior probability of guilt is predicted to be 0.33/(0.33 + 0.04) = 0.88, and for low-confidence suspect identifications (confidence = 4), the posterior probability of guilt is predicted to be 0.20/(0.20 + 0.11) = 0.65. In other words, low confidence implies lower accuracy; high confidence indicates higher accuracy.

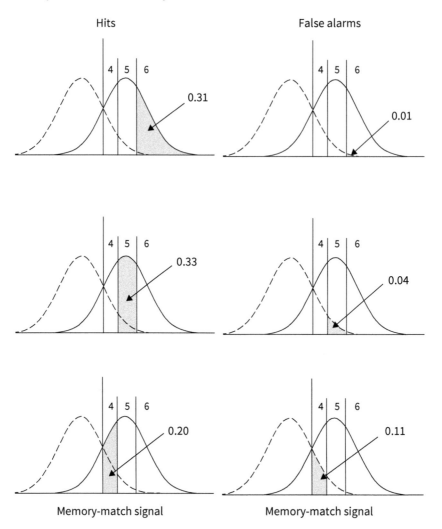

Figure 4.6 An illustration of the predicted relationship between confidence and accuracy according to signal detection theory for 'Yes' decisions. The shading in the left column shows the proportion of guilty suspects receiving a correct 'Yes' decision with high (6), medium (5), or low (4) confidence, top to bottom. The shading in the right column shows the proportion of innocent suspects receiving an incorrect 'Yes' decision with high (6), medium (5), or low (4) confidence, top to bottom.

Using this approach, the confidence–accuracy relationship for suspect identifications made on an initial test using a line-up is revealed by plotting the posterior probability of guilt as a function of confidence. This is much like a calibration plot, but because it differs from what the field has conceptualized as a calibration plot since Juslin et al. (1996), Mickes (2015) referred to it as a *confidence–accuracy characteristic* (CAC) plot. When the base rates of target-present and target-absent line-ups are equal (as they usually are in the laboratory), a CAC plot depicts the posterior probability of guilt given that a suspect identification was made. If the base rates happen not to be equal, it is best to adjust the data such that the CAC plot reflects what the results would be in the equal base-rate scenario (Wixted, 2020). That is, it reflects eyewitness performance in the 'maximum entropy' situation (i.e. making the fewest assumptions about what is unknown). If the defence and prosecution believe they have information about the base rate of guilt in police line-ups (unlikely, to be sure), they can adjust the CAC data to fit the relevant jurisdiction accordingly. Assumptions about base rates are separate and distinct from the question of the reliability of eyewitnesses per se.

Do empirical CAC data show the strong confidence–accuracy relationship predicted by signal detection theory? To find out, Wixted and Wells (2017) reanalysed data from many studies that had previously been analysed using the point-biserial correlation coefficient or by plotting calibration curves. The CAC results are shown in Figure 4.7a. Obviously, the confidence–accuracy relationship is almost as strong as it could be. Keep in mind that the data that were analysed were the same kind of data that the APA used in its amicus briefs to declare that high-confidence identifications from an initial, pristine line-up are highly error prone.

As shown in Figure 4.7b, similar results were obtained in a police department field study in which actual eyewitnesses to crimes reported their confidence (Wixted, Mickes, et al., 2016). When real police line-ups are used, it is not known if a given suspect identification is correct or not, but, with certain assumptions in place (an equal-variance signal detection model), it is possible to estimate aggregate accuracy (see Cohen et al., 2020).

The data from actual eyewitnesses came from $N = 348$ photo line-ups administered by the Houston Police Department (Wixted, Mickes, et al., 2016). In analysing the data, we assumed that memory was uncontaminated. This is a reasonable assumption because (1) the line-ups were conducted early in the police investigation (before much

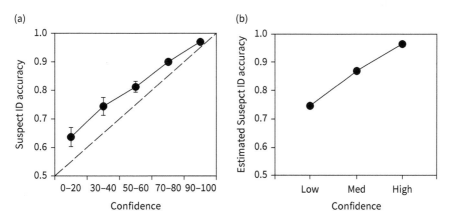

Figure 4.7 (a) CAC plot showing suspect identification accuracy (proportion correct) averaged across 15 studies with comparable scaling on the confidence (*x*-) axis (Wixted & Wells, 2017). (b) Estimated suspect identification accuracy (proportion correct) as a function of confidence for the data from the Houston Police Department field study assuming equal base rates (Wixted, Mickes, et al., 2016).

time for contamination to occur); (2) the perpetrators were strangers to the witnesses (i.e. they had never been encountered before the crime, so the odds of another encounter before the line-up were presumably low); (3) the odds of a chance encounter with the same innocent person the police would happen to place in the line-up were also presumably low; and (4) when asked, the witnesses rarely indicated having seen the suspect they identified other than at the crime scene (only 8% of the time).

We also assume that the witnesses based their decisions on the strength of the memory signals generated by the faces in the line-up because the line-up procedure was properly administered. That is, the line-ups were blindly administered (so the police officer could not selectively steer the witness to the suspect in the line-up), and a random sampled subset of 50 line-ups was assessed to be fair (i.e. mock witnesses were unable to identify the suspect at greater than chance accuracy based on the description of the perpetrator provided by the eyewitness). Witnesses were also informed that the perpetrator may or may not be in the line-up. Thus, we assume that, given an identification, the only path to landing on the suspect at greater than chance probability (1/6 for these six-person line-ups) is via a match between the suspect's face and the face of the perpetrator held in memory.

With those assumptions in place, the goal is to estimate the number of suspect identifications that were made to guilty suspects (G_R) and the number that were made to innocent suspects (I_R), separately for each level of confidence (R). Yet, for suspect identifications, all we have is S_R (the number of suspect identifications made with confidence rating R), such that $S_R = G_R + I_R$. In other words, G_R and I_R are unknown. Critically, however, we also have the number of filler identifications made with each level of confidence (F_R). When a signal detection model is fit to the observed data (i.e. to the values of S_R, F_R, plus the overall number of line-up rejections), the estimated parameters are μ_{Target} (mean of the guilty suspect distribution) and c_1, c_2, and c_3 (the locations of the low-, medium-, and high-confidence criteria). The values of μ_{Lure} (the mean of the innocent suspect/filler distribution) and $\sigma_{Target} = \sigma_{lure} = \sigma$ (the standard deviations of the target and lure distributions) were always fixed at 0 and 1, respectively. Once those parameters are estimated, they can be used to directly estimate G_R and I_R, from which suspect identification accuracy can be computed for each level of confidence using this equation: $G_R / (G_R + I_R)$.

It seems like magic. How can the model unpack guilty suspect identifications and innocent suspect identifications when, in the raw data to which the model was fit, those values were collapsed? Moreover, we do not even know what proportion of line-ups contain an innocent suspect. To ensure that it is nonetheless possible to estimate G_R and I_R, Wixted, Mickes, et al. (2016) fitted the model to laboratory data collapsed over target-present and target-absent line-ups (to mimic the fit to real data). We then compared the model-based estimates of G_R and I_R to their known values, and the estimates were very close.

Recently, Cohen et al. (2020) investigated how the model could achieve seemingly impossible feats like this and found that the key is the use of the same filler distribution for target-present and target-absent line-ups. When responses are broken down by confidence under those conditions, the model is constrained to estimate a single set of optimal parameter values that can then be used to recover G_R, I_R, and even the base rate of target-present line-ups from the collapsed data (the prior probability of a TP line-up for the Houston line-ups was estimated to be approximately 0.35). They validated its ability to do so across multiple laboratory studies, each time fitting the model to collapsed data and then verifying the suspect identification and base rate estimates. However, it only works if the line-ups are fair such that it makes sense to assume that the memory signals generated by innocent suspects and fillers for both target-present and target-absent line-ups are drawn from the same distribution.

In Figure 4.7b, which presents the model-based confidence–accuracy relationship from the Houston Police Department field study, note the remarkably high estimated accuracy of suspect identifications made with high confidence (just as in the laboratory data in Figure 4.7a). This result is in stark contrast to the idea that eyewitness identifications are inherently unreliable. That almost universal impression fails to distinguish eyewitness identifications in general (on the first test, second test, third test, in court) from eyewitness identifications made on an initial uncontaminated test of memory from a properly constructed and administered line-up. On a test like that, the confidence–accuracy relationship is often very strong, as predicted by signal detection theory. The same strong relationship is typically observed on list-memory tasks conducted in the basic science laboratory (e.g. Mickes et al., 2007, 2011). Thus, a long-standing contradiction between basic applied research on recognition memory is not a contradiction after all.

Now consider again the data reported by Sauer et al. (2008). These are the data that were characterized in an APA amicus brief as indicating that witnesses who expressed a confidence score of 90–100 were only 60% correct. Figure 4.8 shows what those same data actually indicate when properly analysed (i.e. Figure 4.8 is a CAC plot). Clearly, accuracy is high across the board, and is especially high for

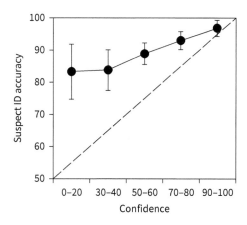

Figure 4.8 CAC plot showing suspect identification accuracy (percentage correct) for the data reported by Sauer et al. (2008). The data were collapsed across their Thief and Waiter conditions.

identifications made with a 90–100 confidence score. For those identifications, accuracy in this study was 96.8% correct.

Perceptions of eyewitness reliability

Unfortunately, the idea that eyewitness memory is unreliable is now almost universally accepted. However, it was not always that way. In a survey conducted some time ago, 76% of American university undergraduates (Deffenbacher & Loftus, 1982) and 73% of American police officers (Brigham & Wolfskiel, 1983) believed that there is a positive confidence–accuracy relation. Were they wrong? They were if they were thinking of the confidence–accuracy relationship when memory is tested in the courtroom (by which time memory has been contaminated, at a minimum by earlier tests but also perhaps by news stories, police feedback, etc.), but they were right if they were thinking about uncontaminated memory initially tested using a proper line-up procedure. Presumably, they were not drawing this distinction at all, so they were half right and half wrong.

Since that time, opinions have changed. For example, Brewin et al. (2019) recently surveyed psychology students and others about whether they agreed with phrases like 'An eyewitness's confidence is never a good predictor of his or her identification accuracy'. Remarkably, among psychology students, 82% agreed. We assume these students would be surprised by the data summarized in Figure 4.7, as many people are.

A more nuanced perspective is needed. When considering the relationship between confidence and accuracy (which is a question about the reliability of eyewitness memory), one needs to be clear about the conditions of the test. Does the question pertain to the conditions that are used to assess the reliability of other kinds of forensic evidence (i.e. when the evidence is not contaminated and is properly tested)? Under those conditions, eyewitness identification evidence is clearly highly reliable. Or does the question instead pertain to conditions that would render any kind of forensic evidence unreliable (i.e. when the evidence is contaminated and/or improperly tested)? Under those conditions, eyewitness identification evidence is clearly unreliable for reasons having nothing to do with the reliability of eyewitnesses (and everything to do with the behaviour of other actors in the legal system).

An intriguing exception to the story outlined above applies to line-up rejections. As noted earlier, line-up rejections are probative of innocence (illustrated earlier in Figure 4.3). However, curiously, confidence in a line-up rejection often provides little additional information beyond that (e.g. Brewer & Wells, 2006). This is in contrast to list-memory laboratory data, where confidence is indicative of accuracy for both positive ('old') decisions and negative ('new') decisions (e.g. Tekin et al., 2021). Indeed, Tekin et al. (2021) found that in virtually every variation on the CAC theme, confidence informs accuracy.

Why the difference for line-up rejections? The answer is not clear, but it might have to do with the fact that the decision to reject a line-up is made to a group of faces, not to a single face, as when a positive identification is made (e.g. Lindsay et al., 2013). More research is needed to explain this intriguing difference between basic laboratory studies and line-up studies, but for positive identifications, it is clear that a strong confidence–accuracy relationship is observed either way. In addition, contrary to the long-standing idea that eyewitness memory is inherently unreliable, high confidence often implies high accuracy.

Reactions to the claim that eyewitness memory is reliable

High-confidence accuracy in the real world

For a field that has long cautioned the legal system about the dangers of relying on witnesses who express high confidence, the recent developments summarized here have generated an understandably mixed reaction. Some continue to argue that high-confidence identifications are likely to be error prone in the real world (Berkowitz et al., 2022). Relatedly, others have argued that the impressive high-confidence accuracy evident in CAC curves may not generalize to the real world because biased line-ups might be created despite the best of intentions (Sauer et al., 2019). Still others have searched for conditions where, even according to CAC analysis, high-confidence identifications are unreliable (Lockamyeir et al., 2020).

All of these studies emphasize the potential for high-confidence identifications being unreliable in the real world (the same concern the field has focused on for decades) despite the recent revelations about laboratory data that were once thought to support such concerns but are now understood to support the exact opposite perspective. However, unaccountably, none of them mentioned (or even cited) the only study involving actual eyewitnesses to a crime that specifically investigated the confidence–accuracy relationship in the real world (Wixted, Mickes, et al., 2016). That study confirmed results from the laboratory studies (Figures 4.7a,b). Thus far, there is scant evidence from actual eyewitnesses supporting the speculative possibility that the same message the field has sent for decades (distrust initial high-confidence identifications from a pristine line-up) still applies.

Non-pristine procedures

Another common reaction, enshrined in APA amicus briefs, is that although confidence might be predictive of accuracy under perfectly pristine conditions, any deviation from a perfect line-up test automatically renders confidence unreliable. However, a more accurate summary of the current state of scientific knowledge is as follows: (1) confidence is strongly predictive of accuracy when a pristine procedure is used (and, in our view, such procedures should always be used), and (2) *certain* non-pristine procedures have been empirically shown to compromise the confidence–accuracy relationship. In particular, an unfair line-up has that effect (Mickes, 2015). However, the effect of other non-pristine procedures on the CAC plot (e.g. failing to warn the witness that the perpetrator may or may not be in the line-up, using a non-blind administrator) remain unknown. We hasten to add that this is not an argument for line-up administrators being non-blind because the potential for un-conscious influence clearly exists under those conditions. However, that compelling argument is distinct from the (not yet empirically established) claim that research has demonstrated that, in general, non-blind administrators result in a comprom-ised confidence–accuracy relationship.

Estimator variables

Yet another claim (also recently enshrined in APA amicus briefs; see APA, 2019) is that various unfavourable 'estimator variables' (e.g. short exposure duration, long distance between the witness and the perpetrator, high stress, cross race, long reten-tion interval, etc.) compromise the reliability of eyewitness identification. However, when the data are plotted as a CAC curve, as they should be, most of the evidence weighs against this claim (Semmler et al., 2018). Poor estimator variables make it less likely that a high-confidence identification will be made (because they impair memory), but they do not appreciably affect the relationship between confidence and accuracy (which is an entirely separate issue). Apparently, witnesses realize their memories are poor, and they adjust their criterion for deciding 'high confidence' ac-cordingly. Indeed, for the police department field study mentioned earlier (yielding the results depicted in Figure 4.7b), more than half the cases were cross-race, and for most of them, a weapon was present (presumably eliciting high stress and weapon focus).

The fact that a strong confidence–accuracy relationship was still observed in real eyewitnesses despite unfavourable estimator variables is consistent with a substan-tial and growing body of laboratory-based research on the subject (Carlson et al., 2017; Nguyen et al., 2017; Palmer et al., 2013; Pezdek et al., 2021; Smalarz et al., 2021; Wixted, Read, & Lindsay, 2016). Indeed, even when memory is tested by recall using an interview, recent evidence suggests that the relationship between confidence and accuracy is strong. In one study (Spearing & Wade, 2022), participants reported confidence either immediately after each response or at the end of the memory test. As the authors put it: 'The timing of the confidence judgement did not affect the confidence–accuracy relationship, and the confidence–accuracy relationship re-mained strong even when participants encoded the event under poor visibility con-ditions' (Spearing & Wade, 2022, p. 1).

To be clear, our claim is not that it is impossible to find a condition where a pristine test of uncontaminated memory yields unimpressive high-confidence accuracy. Our claim, instead, is that the evidence from laboratory studies that were thought to be representative of the real world (and which were cited in APA amicus briefs to establish the unreliability of high-confidence identifications) actually show that the accuracy of high-confidence identifications often approaches 95% correct or higher. As reported by Semmler et al. (2018), this is true even in many studies that test the effect of unfavourable estimator variables. A few studies conducted since then tell a similar but more nuanced story. However, they continue to show how impressively resistant high-confidence identifications are to impairment.

In a large-N study involving 1588 participants each tested four times on different eyewitness crime scenarios (6233 responses), Nyman et al. (2019) manipulated viewing distance between the witness and the target from 5 m to 110 m. They found that, as performance decreased to chance levels (approximately 60 m or more), the number of high-confidence identifications decreased dramatically. Thus, people appropriately adjust confidence as memory conditions worsened (a key point). When tested in the 5–40 m range using simultaneous line-ups, adults were approximately 95% correct when expressing 81–100% confidence. The authors summarized their results as follows: 'A CAC analysis (Mickes, 2015) confirmed that high confidence is associated with high accuracy at distances up to 40 m. This was true for all age groups. After 40 m there were too few high-confidence observations to reliably analyze the results' (p. 538).

Virtually identical results had been reported in an earlier study of distance using once-tested participants. More specifically, based on a reanalysis of results reported by Lindsay et al. (2008), in terms of CAC, high-confidence accuracy for adults was 95% correct for the short-distance (5 m) condition and 94% correct long-distance (37 m) condition (Semmler et al., 2018). This was true even though overall memory performance was much better in the 5 m condition.

In another study of the effect of distance between the witness and perpetrator, Lockamyeir et al. (2020) simulated different viewing distances (from 3 m to 20 m) in a mock crime video (i.e. they did not manipulate the actual distance between a live witness and target, as in the studies discussed above). Unlike the studies discussed above, the long-distance condition resulted in near-chance performance: whereas the correct and false identification rates in the 3 m condition were 0.54 and 0.07, respectively, the corresponding values in the 20 m condition were 0.19 and 0.12 (i.e. the hit rate barely exceeded the false alarm rate). Although accuracy for identifications made with 90–100% confidence exceeded 95% in the 3 m condition (a typical result), it was only 63% correct in the 20 m condition (albeit with wide error bars because high-confidence identifications were rare). Thus, there may be limits to how resistant high-confidence identification accuracy is to poor encoding conditions. That is, as performance approaches random chance levels, high-confidence identifications may no longer be highly accurate.

Giacona et al. (2021) also pushed performance down to low levels, though not quite as low as Lockamyeir et al. (2020) did. The poor estimator variable condition was created by combining multiple suboptimal estimator variables (distance, weapon focus, retention interval). The correct and false identification rates under good viewing conditions were 0.65 and 0.06, respectively, whereas the corresponding values in the poor viewing condition were 0.21 and 0.09 (somewhat better than chance). High-confidence accuracy exceeded 95% correct under good viewing conditions (again, a typical result). It was lower under very poor viewing conditions but was still surprisingly accurate (90% correct).

In summary, extremely poor estimator variable conditions appropriately lead to many fewer high-confidence suspect identifications (Nyman et al., 2019) and sometimes, but not always, reduce the accuracy of high-confidence identifications as well. In the real world, where estimator variable conditions vary widely, high-confidence identifications associated with extremely poor estimator variable conditions will likely rarely be observed (precisely because witnesses rarely make high-confidence identifications under those conditions). This would explain why, for the real-world data summarized in Figure 4.7b, estimated high-confidence accuracy was very high despite what would ordinarily be considered poor estimator variable conditions, such as cross-race, high stress, weapon focus, and so on. These variables significantly impair performance, but the effect sizes are usually modest (e.g. Carlson & Carlson, 2014; Dodson & Dobolyi, 2016; Pezdek et al., 2021).

Individual differences

Just as differences in estimator variables can affect accuracy across conditions (e.g. short vs long exposure duration), individual differences in face recognition accuracy can affect accuracy across eyewitnesses (see Bruce & Burton, Chapter 3, this volume). Do those individual differences disappear when identifications are made with high confidence, as they often do for estimator variables? A recent study investigated that very question and found that the answer is 'No' (Grabman et al., 2019). That is, high-confidence identification errors were more likely for individuals who are worse face recognizers.

Despite being more error prone, were the poor face recognizers nevertheless still reasonably accurate when they made a high-confidence identification? The answer to that question remains unknown because the results were reported in terms of calibration (Equation 4.3), not in terms of the more forensically relevant measure, suspect identification accuracy. As noted earlier, calibration scores as low as 60% can translate to suspect identification accuracy scores of 97% correct. Thus, going forward, it seems important to assess suspect identification accuracy for high-confidence indentifications made by poor face recognizers.

Test a witness's memory for a suspect only once

We conclude by reiterating what might be the most important message of all. For decades, and continuing to this day, the concern has been that eyewitnesses who express high confidence in an identification are often wrong. However, it is important to understand that this concern applies to expressions of high confidence in a court of law, not to expressions of high confidence on a proper, initial, test of uncontaminated memory conducted in a police investigation. Both in the laboratory and in the real world, evidence of high-confidence misidentifications on a first test is hard to come by. Indeed, the available evidence suggests that the main problem may be that the police often ignore inconclusive identifications on the initial test—the only test of uncontaminated memory (but see Helm & Reyna, Chapter 5, this volume). An initial test in which a witness rejects a target-present line-up or picks a filler is evidence of the innocence of the suspect (Wells et al., 2015).

In his book *Convicting the Innocent: Where Criminal Prosecutions Go Wrong*, Garrett (2011) analysed trial materials for 161 DNA exonerees who had been misidentified by one or more eyewitnesses in a court of law. In every case, the eyewitness testified with high confidence at trial that the defendant was the perpetrator. However, identifications that occur at trial are not initial identifications, far from it. Memory for the suspect was at least tested at the beginning of the police investigation and possibly several times since. The defendant's face may also have been seen in news coverage, and police and prosecutors may have told the witness about other evidence pointing to the defendant's guilt. The eyewitness may have mentally replayed the crime many times with the suspect's face after having picked him in the line-up. Therefore, the fact that eyewitnesses make confident misidentifications in the courtroom shows only that contaminated memory evidence (like any kind of contaminated forensic evidence) is unreliable. The question of interest is what happened on the *first* test (conducted early in the police investigation), not the *last* test (conducted in front of the judge and jury, at trial). The first test changes memory, and often makes it worse.

Critically, in 57% of those cases, information was available about the level of confidence expressed on the initial and presumably uncontaminated memory test. As Garrett (2011) put it:

> I expected to read that these eyewitnesses were certain at trial that they had identified the right person. They were. I did not expect, however, to read testimony by witnesses at trial indicating that they earlier had trouble identifying the defendants. Even if there had been problems, eyewitnesses might not recall their hesitance at the time they first identified the defendant. Yet in 57% of these trial transcripts (92 of 161 cases), the witnesses reported that they had *not* been certain at the time of their earlier identifications. (p. 49, emphasis in original)

Indeed, some testimony indicated that the witnesses were not even certain enough to pick anyone from the line-up, or they picked a filler.

Recently, Garrett disavowed his original conclusions, arguing that 'we do not have any idea how confident these eyewitnesses were in their initial identifications because there is no record' (Berkowitz et al., 2020, p. 12). But there is a record, and it consists of the trial testimony he analysed. Obviously, recollections at trial can be wrong, but there is every reason to believe that they were likely accurate. After all, witnesses are biased to recall their initial low-confidence identifications as having been made with high confidence (e. g. Wells & Bradfield, 1998), not the other way around.

In the debate over the reliability of eyewitness memory, which has generally conflated initial and later tests, one surprising (and overlooked) fact is that documented cases of initial misidentifications having been made with high confidence are hard to find. We have been searching for an exception since our 2015 paper (Wixted et al., 2015), and have found only a few (see The National Registry of Exonerations (https://www.law.umich.edu/special/exoneration)). Therefore, it seems that most of the misidentifications in the DNA exoneration cases (perhaps the large majority) began with something other than a high-confidence misidentification of the suspect. Examples abound.

In one of the most famous cases of eyewitness misidentification—which has long been used to illustrate the unreliability of eyewitness memory—Jennifer Thompson (a rape victim) misidentified Ronald Cotton from a photo line-up early in the police investigation. By the time of trial, she was absolutely certain that he was the rapist, but how certain was she on the initial test? The answer to that question bears on the reliability of eyewitness memory. On that test, she hesitated for nearly 5 minutes, narrowing the possibilities to two people, before finally tentatively identifying Cotton. In other words, on the only test of uncontaminated memory, the result was inconclusive. The fact that her (now contaminated) memory was tested again, ultimately leading to a wrongful conviction, reflects a mistake made by police and prosecutors, not the eyewitness. The eyewitness made it clear that her identification of Cotton was inconclusive; it is the police and prosecutors who ignored that fact and relied on her contaminated memory (which they themselves contaminated) to convict Cotton.

Are there more stories like this? There are many more, and we provide a few more examples to drive home the point. In another rape case, the witness identified DNA exoneree Habib Abdal at trial, claiming that she was absolutely certain he was the rapist. However, when she first saw Abdal in a photo at the police station (the initial, uncontaminated memory test), she could not identify her assailant (Garrett, 2011, p. 45). The same sequence of events played out in the case of DNA exoneree Alan Crotzger. That is, the witness failed to identify him as the perpetrator on the only test that matters (the initial line-up) and ended up identifying him with absolute certainty at trial (Garrett, 2011, p. 56). In the case of DNA exoneree Neil Miller, the witness narrowed the choices down to two photos but was not sure either was the perpetrator, yet her uncertainty vanished once inside the courtroom, at trial (Garrett, 2011, p. 57).

DNA exoneration cases like these are particularly informative, but there are also other troubling cases that are similar except that innocence cannot be established so conclusively. Consider the case of *Watson v. Commonwealth of Virginia* (APA, 2019):

> The case stemmed from the robbery of Joseph Jackson and Paul Abbey and the murder of Abbey by three men in a parking lot. Jackson's identification of Watson as one of the assailants was a key issue at trial. Jackson initially identified Watson in a photo lineup, but indicated that he was 'not sure.' At the preliminary hearing, however, Jackson expressed certainty that Watson was one of the assailants (p. 3).

It is the same story yet again. Or consider the even more troubling case of Charles Don Flores. In a murder investigation, the witness described one of the perpetrators as a white male in his 30s with shoulder length hair. Even though he is a heavyset Hispanic man with a crew cut, the police suspected Flores. Therefore, they placed his photo in a line-up with other Hispanic men and presented it to the witness. Quite understandably, the witness did not identify anyone (i.e. she rejected the line-up). This makes sense because it is hard to see why photos of large Hispanic males with short hair would generate a strong memory-match signal when compared against the memory of a white male with long hair stored in the witness's brain. Thus, on the initial test, her failure to identify Flores provides evidence of innocence, not guilt. Yet a year later, the witness identified Flores at his trial with high confidence. Although some circumstantial evidence connected Flores to the crime, the eyewitness testimony was the only evidence tying him to the crime scene. Flores was convicted of the crime and sentenced to death. He has been on death row for 23 years and has exhausted all of his appeals. Although it has not happened yet, there are no remaining legal barriers to the setting of his execution date.

Conclusion

The upshot of our message is this: instead of trying to convince the legal system to ignore expressions of high confidence in the courtroom, we should try to convince the legal system *not* to ignore expressions of confidence (high or low) on the first test. The first test contaminates memory, so there are no do overs. Had this simple rule been followed from the beginning, many of the wrongful convictions in the DNA exoneration cases may not have happened in the first place. The witnesses who returned an inconclusive forensic memory test result on the initial test did their jobs, and it is time to exonerate them, too. The criminal justice system made the mistake of testing a witness's memory for a suspect more than once. Perhaps the time has finally come for that to stop.

References

American Psychological Association (2014). *Commonwealth v. Walker*. http://www.apa.org/about/offi ces/ogc/amicus/walker.aspx

American Psychological Association (2019). *Watson v. Commonwealth of Virginia*. https://www.apa. org/about/offices/ogc/amicus/watson-virginia

Berkowitz, S. R., Garrett, B. L., Fenn, K. M., & Loftus, E. F. (2022). Convicting with confidence? Why we should not over-rely on eyewitness confidence. *Memory, 30*(1), 10–15.

Brewin, C. R., Andrews, B., & Mickes, L. (2020). Regaining consensus on the reliability of memory. *Current Directions in Psychological Science, 29*(2), 121–125.

Brewer, N., Keast, A., & Rishworth, A. (2002). The confidence-accuracy relationship in eyewitness identification: The effects of reflection and disconfirmation on correlation and calibration. *Journal of Experimental Psychology: Applied, 8*(1), 44–56.

Brewer, N., & Wells, G. L. (2006). The confidence-accuracy relation in eyewitness identification: Effects of lineup instructions, foil similarity, and target-absent base rates. *Journal of Experimental Psychology: Applied, 12*, 11–30.

Brigham, J. C, & Wolfskiel, M. P. (1983). Opinions of attorneys and law enforcement personnel on the accuracy of eyewitness identification. *Law and Human Behavior, 7*(4), 337–349.

Carlson, C. A., & Carlson, M. A. (2014). An evaluation of lineup presentation, weapon presence, and a distinctive feature using ROC analysis. *Journal of Applied Research in Memory and Cognition, 3*(2), 45–53.

Carlson, C. A., Dias, J. L., Weatherford, D. R., & Carlson, M. A. (2017). An investigation of the weapon focus effect and the confidence-accuracy relationship for eyewitness identification. *Journal of Applied Research in Memory and Cognition, 6*(1), 82–92.

Cohen, A. L., Starns, J. J., Rotello, C. M., & Cataldo, A. M. (2020). Estimating the proportion of guilty suspects and posterior probability of guilt in lineups using signal-detection models. *Cognitive Research: Principles and Implications, 5*(1), 21.

Deffenbacher, K. A. (1980). Eyewitness accuracy and confidence: Can we infer anything about their relationship? *Law and Human Behavior, 4*(4), 243–260.

Deese, J. (1959). On the prediction of occurrence of particular verbal intrusions in immediate recall. *Journal of Experimental Psychology, 58*(1), 17–22.

Deffenbacher, K. A., & Loftus, E. F. (1982). Do jurors share a common understanding concerning eyewitness behavior? *Law and Human Behavior, 6*(1), 15–30.

Delay, C. G., & Wixted, J. T. (2021). Discrete-state vs. continuous models of the confidence-accuracy relationship in recognition memory. *Psychonomic Bulletin & Review, 28*(2), 556–564.

Dodson, C. S., & Dobolyi, D. G. (2016). Confidence and eyewitness identifications: The cross-race effect, decision time and accuracy. *Applied Cognitive Psychology, 30*(1), 113–125.

Egan, J. P. (1958). *Recognition memory and the operating characteristic* (Tech Note AFCRC-TN-58-51). Indiana University, Hearing and Communication Laboratory.

Garrett, B. L. (2011). *Convicting the innocent: Where criminal prosecutions go wrong*. Harvard University Press.

Garrett, B. L. (2012). Eyewitnesses and exclusion. *Vanderbilt Law Review, 65*, 451–506.

Giacona, A. M., Lampinen, J. M., & Anastasi, J. S. (2021). Estimator variables can matter even for high-confidence lineup identifications made under pristine conditions. *Law and Human Behavior, 45*(3), 256–270.

Grabman, J. H., Dobolyi, D. G., Berelovich, N. L., & Dodson, C. S. (2019). Predicting high confidence errors in eyewitness memory: The role of face recognition ability, decision-time, and justifications. *Journal of Applied Research in Memory and Cognition, 8*(2), 233–243.

Green, D. M., & Swets, J. A. (1966). *Signal detection theory and psychophysics* [Reprinted, with corrections to the original 1966 ed.]. Robert E. Krieger Publishing Co.

Innocence Project (2021). *Understand the causes: The causes of wrongful conviction*. Innocence Project. https://www.innocenceproject.org/eyewitness-identification-reform/

Jones, E. E., Williams, K. D., & Brewer, N. (2008). 'I had a confidence epiphany!': Obstacles to combating post-identification confidence inflation. *Law and Human Behavior, 32*(2), 164–176.

Juslin, P., Olsson, N., & Winman, A. (1996). Calibration and diagnosticity of confidence in eyewitness identification: Comments on what can be inferred from the low confidence-accuracy correlation. *Journal of Experimental Psychology: Learning, Memory, and Cognition, 22*(5), 1304–1316.

Lin, W., Strube, M. J., & Roediger, H. L. (2019). The effects of repeated lineups and delay on eyewitness identification. *Cognitive Research: Principles and Implications, 4*(1), 1–19.

Lindsay, R. C. L., Kalmet, N., Leung, J., Bertrand, M. I., Sauer, J. D., & Sauerland, M. (2013). Confidence and accuracy of lineup selections and rejections: Postdicting rejection accuracy with confidence. *Journal of Applied Research in Memory and Cognition, 2*(3), 179–184.

Lindsay, R. C. L., Semmler, C., Weber, N., Brewer, N., & Lindsay, M. R. (2008). How variations in distance affect eyewitness reports and identification accuracy. *Law and Human Behavior, 32*(6), 526–535.

Lockamyeir, R. F., Carlson, C. A., Jones, A. R., Carlson, M. A., & Weatherford, D. R. (2020). The effect of viewing distance on empirical discriminability and the confidence-accuracy relationship for eyewitness identification. *Applied Cognitive Psychology, 34*(5), 1047–1060.

McDermott, K. B. (2021). Practicing retrieval facilitates learning. *Annual Review of Psychology, 72,* 609–633.

Mickes, L. (2015). Receiver operating characteristic analysis and confidence-accuracy characteristic analysis in investigations of system variables and estimator variables that affect eyewitness memory. *Journal of Applied Research in Memory & Cognition, 4*(2), 93–102.

Mickes, L., Hwe, V., Wais, P. E., & Wixted, J. T. (2011). Strong memories are hard to scale. *Journal of Experimental Psychology: General, 140*(2), 239–257.

Mickes, L., Wixted, J. T., & Wais, P. E. (2007). A direct test of the unequal-variance signal-detection model of recognition memory. *Psychonomic Bulletin & Review, 14*(5), 858–865.

Nguyen, T. B., Pezdek, K., & Wixted, J. T. (2017). Evidence for a confidence-accuracy relationship in memory for same- and cross-race faces. *Quarterly Journal of Experimental Psychology, 70*(12), 2518–2534.

Nyman, T. J., Lampinen, J. M., Antfolk, J., Korkman, J., & Santtila, P. (2019). The distance threshold of reliable eyewitness identification. *Law and Human Behavior, 43*(6), 527–541.

Palmer, M., Brewer, N., Weber, N., & Nagesh, A. (2013). The confidence–accuracy relationship for eyewitness identification decisions: Effects of exposure duration, retention interval, and divided attention. *Journal of Experimental Psychology: Applied, 19*(1), 55–71.

Penrod, S., & Cutler, B. (1995). Witness confidence and witness accuracy: Assessing their forensic relation. *Psychology, Public Policy, and Law, 1*(4), 817–845.

Pezdek, K., Abed, E., & Cormia, A. (2021). Elevated stress impairs the accuracy of eyewitness memory but not the confidence–accuracy relationship. *Journal of Experimental Psychology: Applied, 27*(1), 158–169.

Roediger, H. L., & McDermott, K. B. (1995). Creating false memories: Remembering words not presented in lists. *Journal of Experimental Psychology: Learning, Memory, and Cognition, 21*(4), 803–814.

Roediger, H. L., Wixted, J. H., & DeSoto, K. A. (2012). The curious complexity between confidence and accuracy in reports from memory. In L. Nadel & W. Sinnott-Armstrong (Eds.), *Memory and law* (pp. 84–118). Oxford University Press.

Sauer, J., Brewer, N., & Wells, G. L. (2008). Is there a magical time boundary for diagnosing eyewitness identification accuracy in sequential line-ups? *Legal and Criminological Psychology, 13*(1), 123–135.

Sauer, J. D., Palmer, M. A., & Brewer, N. (2019). Pitfalls in using eyewitness confidence to diagnose the accuracy of an individual identification decision. *Psychology, Public Policy, and Law, 25*(3), 147–165.

Semmler, C., Dunn, J., Mickes, L., & Wixted, J. T. (2018). The role of estimator variables in eyewitness identification. *Journal of Experimental Psychology: Applied, 24*(3), 400–415.

Smalarz, L., Yang, Y., & Wells, G. L. (2021). Eyewitnesses' free-report verbal confidence statements are diagnostic of accuracy. *Law and Human Behavior*, *45*(2), 138–151.

Spearing, E. R., & Wade, K. A. (2022). Providing eyewitness confidence judgements during versus after eyewitness interviews does not affect the confidence-accuracy relationship. *Journal of Applied Research in Memory and Cognition*, *11*(1), 54–65.

Sporer S. L., Penrod S., Read D., & Cutler B. (1995). Choosing, confidence, and accuracy: A meta-analysis of the confidence–accuracy relation in eyewitness identification studies. *Psychological Bulletin*, *118*, 315–327.

Tekin, E., DeSoto, K. A., Wixted, J. H., & Roediger III, H. L. (2021). Applying confidence accuracy characteristic plots to old/new recognition memory experiments. *Memory*, *29*(4), 227–243.

Tekin, E., Lin, W., & Roediger, H. L. (2018). The relationship between confidence and accuracy with verbal and verbal+ numeric confidence scales. *Cognitive Research: Principles and Implications*, *3*(1), 41.

Tekin, E., & Roediger, H. L. (2017). The range of confidence scales does not affect the relationship between confidence and accuracy in recognition memory. *Cognitive Research: Principles and Implications*, *2*(1), 49.

Weber, N., & Brewer, N. (2004). Confidence-accuracy calibration in absolute and relative face recognition judgments. *Journal of Experimental Psychology: Applied*, *10*(3), 156–172.

Wells, G. L., & Bradfield, A. L. (1998). 'Good, you identified the suspect': Feedback to eyewitnesses distorts their reports of the witnessing experience. *Journal of Applied Psychology*, *83*(3), 360–376.

Wells, G. L., Kovera, M. B., Douglass, A. B., Brewer, N., Meissner, C. A., & Wixted, J. T. (2020). Policy and procedure recommendations for the collection and preservation of eyewitness identification evidence. *Law and Human Behavior*, *44*(1), 3–36.

Wells, G. L., Yang, Y., & Smalarz, L. (2015). Eyewitness identification: Bayesian information gain, base-rate effect equivalency curves, and reasonable suspicion. *Law and Human Behavior*, *39*(2), 99–122.

Wixted, J. T. (2020). The forgotten history of signal detection theory. *Journal of Experimental Psychology: Learning, Memory, and Cognition*, *46*(2), 201–233.

Wixted, J. T., & Mickes, L. (2022). Eyewitness memory is reliable, but the criminal justice system is not. *Memory*, *30*(1), 67–72.

Wixted, J. T., Mickes, L., Clark, S. E., Gronlund, S. D., & Roediger, H. L. (2015). Initial eyewitness confidence reliably predicts eyewitness identification accuracy. *American Psychologist*, *70*(6), 515–526.

Wixted, J. T., Mickes, L., Dunn, J., Clark, S. E., & Wells, W. (2016). Relationship between confidence and accuracy for eyewitness identifications made from simultaneous and sequential police lineups. *Proceedings of the National Academy of Sciences of the United States of America*, *113*(2), 304–309.

Wixted, J. T., Read, J. D., & Lindsay, D. S. (2016). The effect of retention interval on the eyewitness identification confidence-accuracy relationship. *Journal of Applied Research in Memory and Cognition*, *5*(2), 192–203.

Wixted, J. T., Mickes, L., & Fisher, R. P. (2018). Rethinking the reliability of eyewitness memory. *Perspectives on Psychological Science*, *13*(3), 324–335.

Wixted, J. T., & Wells, G. L. (2017). The relationship between eyewitness confidence and identification accuracy: A new synthesis. *Psychological Science in the Public Interest*, *18*(1), 10–65.

Wixted, J. T., Wells, G. L., Loftus, E. F., & Garrett, B. L. (2021). Test a witness's memory of a suspect only once. *Psychological Science in the Public Interest*, *22*(Suppl 1), 1S–18S.

5

Fuzzy trace theory

Memory and decision-making in law, medicine, and public health

Rebecca K. Helm and Valerie F. Reyna

The way that information is encoded and processed in working memory, and then stored and retrieved from long-term memory, has implications not only for memories themselves but also for decision-making. Fuzzy trace theory (FTT), a dual-process theory of memory and decision-making, makes predictions about the ways information is mentally represented in working and long-term memory and how the type of representation relied on when making decisions influences decision-making and the outcomes that flow from those decisions.

Specifically, FTT posits that when most adults hear information, they encode both the literal information (referred to as the verbatim information) (e.g. the precise statistical risks and benefits associated with having a particular surgery or the height of a person observed committing a crime) and the 'gist' of the information. Gist is the bottom-line meaning that people extract from information, and can be encoded at varying levels of abstraction, from more precise ordinal distinctions (e.g. I am more likely to avoid a negative outcome if I have surgery than if I do not; the person I saw was taller than me) to less precise categorical distinctions (e.g. there is a non-trivial risk of serious long-term harm if I do not have the surgery that can be avoided by having the surgery; the person I saw was huge).

The form of mental representation used when recounting experiences and when drawing on information to make decisions has implications for both what is remembered and how decisions are made. For example, the representations that a decision-maker relies on regarding outcomes and their likelihoods (e.g. 'it only takes once' to contract the human immunodeficiency virus (HIV) from unprotected sex), combined with the values they apply to those representations (e.g. contracting HIV is bad), have been shown to influence their decisions regarding risk (Reyna, 2021). Note that, in the spirit of Baddeley (2000), who has influenced our work, the concept of 'memory' representations applies to situations in which people fully remember presented information or even when the information is displayed in front of them.

In this chapter, we begin by introducing the basic tenets of FTT, describing how these tenets have been supported in rigorous experimental and mathematical tests, highlighting how these tenets distinguish FTT from other related theories, and

Rebecca K. Helm and Valerie F. Reyna, *Fuzzy trace theory* In: *Memory in Science for Society*. Edited by: Robert H. Logie, Zhisheng (Edward) Wen, Susan E. Gathercole, Nelson Cowan, and Randall W. Engle, Oxford University Press. © Oxford University Press 2023.
DOI: 10.1093/oso/9780192849069.003.0005

outlining some general implications of these tenets for policy in applied contexts. We then discuss how insight provided by FTT can improve the way that we address a range of specific practical problems in society. First, we examine the FTT-informed conjoint recognition and other 'phantom' memory paradigms and their importance in assessing memory accuracy and in designing procedures to enhance memory accuracy in criminal investigations. Next, we review FTT's predictions relating to how people evaluate evidence (including numerical evidence) and use this evidence to form conclusions. We review the ways in which this work can help us to understand and improve evaluations of evidence and related judgements in a range of contexts including enhancing understanding of medical test results, helping civil juries appropriately translate judgements of harm severity into damage awards, and helping people more effectively identify and discount misinformation and disinformation. Finally, we examine FTT's predictions relating to decisions under risk. We describe how FTT predicts and explains rational and technically irrational (but typically adaptive) decisions, specifically the risky-choice framing effect, and show how understanding the influence of mental representations on decisions under risk can help us to understand and improve a variety of decisions including decisions as to whether to plead guilty in criminal trials, decisions as to whether to engage in crime, and decisions as to whether to have unprotected sex. We conclude by discussing why evidence-based theory is more informative regarding practical solutions to applied problems than are empirical results that are not motivated by questions of mechanism.

Introducing fuzzy trace theory

Origins and the independence of gist and verbatim memory

FTT is a dual-process theory of memory and decision-making. Its core concepts are informed by work in psycholinguistics, where the distinction between verbatim and gist, discussed above, was first established (e.g. Kintsch, 1974). However, that literature considered gist memory as being derived from verbatim memory, that people extracted gist from verbatim memory and then discarded verbatim information (e.g. Clark & Clark, 1977). FTT adopts the distinction between gist and verbatim, but rather than assuming that gist is extracted from verbatim, FTT predicts—and evidence supports—that gist and verbatim are encoded, stored, and retrieved separately (Reyna, 2012).

The claim that gist and verbatim memories are encoded, stored, and retrieved separately has been tested and supported across multiple experiments seeking to isolate verbatim and gist memory (e.g. Reyna & Brainerd, 1995). For example, research examining memory for narrative sentences has shown that memory for presented sentences (verbatim memory) and for inferences that can be drawn from those sentences (gist) are stochastically independent from each other. In other words, memory for gist does not depend on memory for verbatim (see Reyna & Kiernan, 1994; 1995; Singer & Remillard, 2008). For example, in experiments with children and with adults, participants were asked to remember specific sentences (e.g. the bird is in the cage, the cage is under the table, the bird has yellow feathers) and then later were asked to indicate which exact sentence's they remembered. Recognition of the sentences (e.g. the bird is in the cage) was found to be independent of systematic misrecognition of true inferences (e.g. the bird is under the table) (e.g. Reyna & Kiernan, 1994).

Research has also shown that gist can be encoded in the absence of full verbatim encoding (e.g. participants can remember themes from a word list even when it is presented too fast for individual words to be encoded; see Brainerd & Reyna, 2005) and verbatim can be encoded in the absence of gist (e.g. when meaningless syllables are presented; Brainerd, et al., 1995). The independence of gist and verbatim memories has also been tested and supported through the use of mathematical models (Reyna & Brainerd, 2011; Stahl & Klauer, 2008, 2009).

This independence result shows that verbatim and gist representations differ and both are required to explain behaviour. However, the effect can be manipulated per FTT by varying the cues in questions (e.g. whether presented information is provided on the test) and the context (e.g. delay between study and test): When participants perform a short 'buffer' task between sentence presentation and test, *independence* is observed. When the test occurs immediately and verbatim stimuli have been repeatedly studied, *negative dependence* is observed; gist-consistent inferences are rejected based on recollecting verbatim stimuli. When a longer delay occurs between study and test (e.g. a week), *positive dependence* is observed; both gist-consistent inferences and verbatim stimuli are accepted based on consistency with gist (semantically *in*consistent sentences are still mainly rejected). All three relationships—independent, negative, and positive—have been observed for the identical stimuli and even for the same participants under theoretically specified conditions.

In fact, FTT predicts that gist-consistent false memories (never-presented stimuli) will be better 'remembered' than true memories (presented stimuli) under specific conditions, a counterintuitive result that challenges widespread assumptions about memory (Brainerd & Reyna, 1998). In this connection, Wixted and Roediger (Chapter 4, this volume) assert that 'the key assumption of signal detection theory' is that 'because the innocent suspects do not correspond to the perpetrator (and therefore do not usually provide a strong match to the witness's memory of the perpetrator), the mean of the innocent suspect distribution is relatively low. For guilty suspects, everything is the same except that the mean of the distribution is higher because the guilty suspect provides a better match to the witness's memory of the perpetrator.' In other words, false memories—what was not witnessed—cannot be stronger than true memories of what was witnessed. However, per FTT, the mean of the false memory distribution significantly exceeded the mean of the true memory distribution for 106 of 617 data sets on recognition memory (most reported in Brainerd & Reyna, 2018). The effect varied predictably as a function of the accessibility of gist memories. This theoretically predicted violation of a key assumption indicates that the scope of memory theories must be expanded to include gist as well as verbatim memories.

Other core concepts

Other tenets of FTT concern the characteristics of gist and verbatim memory, and associated decision-making. Understanding these tenets can help us understand both what people are likely to remember and how they are likely to draw on memories when making decisions. Four core tenets with particular relevance to policy are discussed below: intuitive decision-making is distinct from 'hot' or 'fast' decision-making, there is a developmental trajectory from reliance on more verbatim memory to more gist memory, task characteristics can determine reliance on gist or verbatim memory, and individual differences can determine reliance on gist or verbatim memory.

Intuitive decision-making is distinct from 'hot' or 'fast' decision-making

Although FTT is characterized as a dual-process theory, it differs from traditional dual-process theories. Traditional dual-process theories rely on a distinction between 'type 1' and 'type 2' processes to explain decision-making. Type 1 processes are fast and intuitive whereas type 2 processes are slow and deliberative (e.g. Evans & Stanovich, 2013; Kahneman, 2011). These theories are termed 'default interventionist' since type 1 processing is often seen as the default method of processing, that can be overridden by higher-order type 2 processing (Kahneman, 2011; but see also Barbey & Sloman, 2007). FTT distinguishes intuitive thinking from 'fast' thinking, separating the role of impulsivity (or lack of inhibition) from the role of intuitive (as opposed to detailed) cognition, which is determined, according to FTT, by the type of memory representation relied on. Thus, according to FTT, inhibition per se is not a reasoning mode, but instead acts to withhold thoughts or actions (Hare et al., 2008; Reyna & Brainerd, 2011).

Work drawing on FTT has therefore recognized three distinct components important in the decision-making process—rewards and other motivational benefits (which might be termed a 'hot' influence), inhibition (which might be termed a 'cold' influence), and memory representation relied on (gist or verbatim) (Reyna, Wilhelms, et al., 2015).

There is a developmental trajectory from reliance on more verbatim memory to reliance on more gist memory

The recognition that reliance on gist is distinct from 'hot' or 'fast' cognitive processes relates to another central tenet of FTT. While traditional dual-process theories typically expect reasoning to become more analytical and less intuitive with age (e.g. Stanovich et al., 2008), FTT recognizes reliance on gist memory as being developmentally advanced (Reyna & Brainerd, 2011). Both verbatim- and gist-based abilities have been found to develop during childhood, alongside an increasing preference for reliance on gist (Brainerd et al., 2011; Reyna & Farley, 2006).

According to FTT, reliance on gist memory is advanced and increases with age (i.e. experience) and with expertise, meaning that decision-makers increasingly rely on simpler but more meaningful distinctions (see, e.g. Reyna & Lloyd, 2006). This recognition allows FTT to predict systematic developmental reversals in both memory (Brainerd, et al., 2011) and decision-making (e.g. Klaczynski & Felmban, 2014; Morsanyi et al., 2017; Reyna et al., 2014; Reyna & Ellis, 1994). Put simply, where the development of false memories or systematic biases are the result of reliance on gist, FTT predicts developmental reversals.

Thus, for example, FTT predicts developmental reversals in risky-choice biases, namely, that children are less biased than adults: that is, children do not reverse their preferences for risk when the same net outcomes are described as gains versus losses (Reyna & Farley, 2006). However, adults do show such a bias that has been linked to gist thinking, preferring to avoid risk for gains (e.g. winning prizes) but to seek risk for losses (e.g. losing prizes from an initial endowment of 'house money' such that outcomes feel like losses but they are actually equivalent net gains). Moreover, showing these gist-based 'framing' biases is associated with healthier real-world risk-taking and lower levels of criminal behaviour (Reyna, Estrada et al., 2011; Reyna, Helm et al., 2018). Although traditional approaches to rationality emphasize trading off the magnitudes of reward (number of prizes) against magnitudes of risk—precise and objective verbatim analysis—research using eye-tracking data suggests that *younger* people (adolescents) are more likely to process decision options in this precise and balanced way than adults; they acquired more information than adults in a more thorough manner than adults, engaging in trade-offs prior to making a decision (Kwak et al., 2015). As predicted by FTT, adults are less likely to trade-off magnitudes of risk and reward, instead relying on categorical gist, such as 'it only takes once' or 'winning something is better than maybe winning nothing'. In an independent but dramatic demonstration of this theoretical principle that more advanced thinkers are less likely to rely on risk–reward trade-offs, Decker et al. (1993) showed that criminals' willingness to offend varied 'rationally' as a function of level of risk and reward (though not penalty), whereas matched controls were unwilling to offend regardless of the magnitudes of risk and reward.

As a corollary of the developmental trajectory from reliance on verbatim to reliance on gist, adults are thought to have what is known as a fuzzy-processing preference in decision-making, meaning that they will rely on the simplest gist possible to make a decision (e.g. if two options are distinguished from one another on a categorical level, they will rely on this distinction; if not, they will move to consider any ordinal distinction, and so on; see figures in Reyna, 2012). This preference is the opposite of that which would be predicted by information processing theories, which assume that elaborate reasoning proceeds until excessive cognitive load forces simpler processing (Reyna & Brainerd, 2011).

Task characteristics can determine reliance on gist or verbatim memory

Although FTT predicts a developmental trajectory from reliance on more verbatim memory to reliance on more gist memory, the memory relied on can also be influenced by specifics of a task that a decision-maker is presented with, at least when gist and verbatim representations of information are both accessible. This prediction is important from a policy perspective, since policymakers may have the ability to control the specifics of the task that a person is facing, and in this way shift the type of memory that this person is relying on (Thaler & Sunstein, 2008). First, a task can require reliance on either gist or verbatim mental representations. For example, if a person is asked to remember exact words or numbers that do not have particular meaning (e.g. remember the number 137), this will push them to rely more on verbatim representations.

A person will also be pushed more towards verbatim processing when asked to choose between two options that have the equivalent gist. For example, a person faced with a decision of whether to plead guilty or go to trial might know that both plea and trial involve a short custodial sentence. Therefore, to decide whether to plead guilty or go to trial, a person would have to consider more fine-grained information (e.g. 2 months vs 3 months). Decision-makers considering narrative information (e.g. jurors considering a legal case) can be encouraged to rely on gist by presenting information in a 'story' format, giving clearer meaning to information and facilitating the extraction of gist, as opposed to a scrambled or arbitrary order (see Dewhurst et al., 2007; Pennington & Hastie, 1992; Reyna & Brainerd, 1995). When plea options differ qualitatively (e.g. non-felony vs felony conviction), as with other decisions, this, too, elicits gist-based processing (e.g. Helm & Reyna, 2017). Although we speak of verbatim-based and gist-based processing, FTT assumes all processes occur roughly in parallel and that different processes predominate based on the task (e.g. whether options can be discriminated based on gist) and on the people performing the task (Reyna & Brainerd, 1995).

Finally, reliance on gist or verbatim may be dependent on delay. Gist is more stable over time, and a person will be forced to rely on gist when verbatim memory is no longer accessible (Kintsch, 1974; Reyna & Kiernan, 1994; 1995). However, depending on cues in questions (verbatim or gist content) and the granularity required in responses (ranging from simple dichotomous choice to exact numerical judgements), reliance on gist predominates even when exact information remains visible or when delays are short.

Individual differences can determine reliance on gist or verbatim memory

In addition to age and task characteristics, reliance on gist or verbatim can be determined by individual differences in decision-makers in several ways. First, individual differences may determine the extent to which an individual is able to extract gist from information. (As noted above, where information has no meaning to people, they are pushed to rely on verbatim information.) So, for example, more skilled readers with more background knowledge are likely to be better able to extract gist from a narrative even when information lacks a clear structure (Van den Broek, 2010).

Second, individual differences in metacognitive monitoring and need for cognition help a person to recognize that two options are equivalent from a verbatim perspective and to override the natural tendency to rely on gist (e.g. Stanovich & West, 2008). (Need for cognition is the desire to engage in effortful cognition, as distinguished from the ability to do so.) For example, a person with a high level of need for cognition is more likely to recognize that gain and loss versions of framing problems are equivalent from a verbatim perspective (because they engage in spontaneous computations and comparisons) and to inhibit their tendency to rely on gist as a result (Broniatowski & Reyna, 2018).

Third, certain individual differences influence the tendency to rely on gist or verbatim representations despite the general fuzzy-processing preference. For example, research suggests that some individuals with autism may be more likely to rely on verbatim memory and less likely to rely on gist memory than their peers (see Reyna & Brainerd, 2011). This prediction is supported by findings showing that autistic individuals are less prone to the risky-choice framing effect (De Martino et al., 2008) and conjunction fallacies (Morsanyi et al., 2010), effects associated with gist-based processing (e.g. Kühberger & Tanner, 2010), and are less likely to draw gist-based inferences (Jolliffe & Baron-Cohen, 2000) and to exhibit gist-based false memories (Griego et al., 2019).

Importance for policy

As described above, FTT makes predictions involving how people encode information in memory, and how people then retrieve this information in order to make decisions. This understanding is important in informing policy in applied contexts (Reyna, 2021). Among other things, it allows policymakers to consider what information people are likely to be relying on and what biases people might be susceptible to when making decisions and to ensure that the way decisions are being made in practice adheres with normative goals in society. It can also help to highlight individuals who may make decisions in a way that is different from that envisioned by policymakers, and to provide necessary interventions to ensure appropriate outcomes for those people and others affected by decisions. In the remainder of this chapter, we discuss specific implications of FTT for identifying and addressing problems in society across a range of applied contexts.

Meaning-consistency, suggestion, and susceptibility to false memory

Being able to assess where a memory is particularly likely to be false (as opposed to real) is important, particularly in the legal context where research suggests that

(1) witness testimony is often important evidence in legal cases (Brainerd, 2013), (2) witness testimony is one of the most convincing types of evidence for legal decision-makers (Semmler et al., 2011), and (3) witness testimony is a key contributor to wrongful convictions (Evidence-Based Justice Lab, n. d.; Helm, 2021c; National Registry of Exonerations, n. d.; see also Toglia & Berman, 2021; Wixted & Roediger, Chapter 4, this volume). Effectively assessing the likelihood that a witness's memory is false (or contains false elements) has the potential to increase the accuracy of convictions and acquittals, and the allocation of responsibility in civil lawsuits. Through introducing the distinction between gist and verbatim memory, FTT provides key insight into the cognitive processes underlying false memory, and, relatedly, cues that are probative in assessing whether a particular memory is likely to be false.

Introducing conjoint recognition

FTT's predictions relating to memory have been tested using what is known as the 'conjoint recognition' paradigm (Reyna et al., 2016). (Recall models have also been developed; see Reyna, 2012.) This paradigm allows the separation and identification of distinct memory processes through a multinomial measurement model.

The model separates three distinct memory processes: identity, similarity, and recollection rejection. First, an identity judgement is essentially recognition of the exact thing seen or heard through retrieving the original verbatim trace and matching it to the test item. Second, a similarity judgement is of meaning consistency with the thing seen or heard, based on the gist trace; although not identical in surface form, the test item agrees with the substance of what was seen or heard. Finally, recollection rejection involves recognizing that an item that is familiar or meaning-consistent with a viewed item is not the viewed item itself; retrieving the original verbatim trace reveals a mismatch between the test item and what was seen or heard (e.g. Lampinen et al., 2006). Instructional conditions allow the isolation of these processes: asking respondents to say yes only to exact items they have seen, asking respondents to say yes to items that are true regardless of whether seen, and asking respondents to say yes only to items that are true that are not the exact item they have seen (Reyna & Kiernan, 1994; Stahl & Klauer, 2008). Examining responses to varying cues across these conditions allows researchers to estimate how memory operates under different conditions, for example, given different cues (recognition probes) and given the accessibility of different kinds of representations. Examining memory in this way is important, since simply analysing whether a memory is true or false does not unambiguously prove its psychological origin (see Reyna et al., 2016) and therefore cannot provide reliable and generalizable insight. This insight is key in understanding false memory; the ability to influence policy is enhanced by rigorous experimental work testing the predictions and conclusions outlined below in a variety of settings.

Mechanisms underlying false memory

According to FTT, false memory can arise in two primary ways: from meaning-consistency in the absence of verbatim retrieval and as a result of external suggestion.

False memory resulting from meaning-consistency

According to FTT, retrieval of both gist and verbatim traces, provided that what is encoded is accurate, support true memory. Both traces help individuals remember events, and both will lead to the correct identification of targets. Although gist and verbatim reinforce each other for true items, they work in opposition to each other for meaning-consistent but unpresented items on verbatim tests (Brainerd & Reyna, 2005; Reyna & Kiernan, 1994). For example, in witness testimony, it is essential to identify an exact individual as opposed to someone who resembles the perpetrator (see Bruce & Burton, Chapter 3, this volume, although the dimensions of meaningful similarity are not fully understood for faces, but see Bartlett et al., 2009). While verbatim memory can generally suppress false memory on such tests, gist memory can promote it. Specifically, retrieval of gist traces supports false memories for similar or meaning-consistent events (although note that in some cases gist memory may be necessary to provide probative information, for example, if a witness is asked whether a person was behaving strangely; see Reyna et al., 2016).

A person remembering the gist of an event may accept a meaning-consistent event (called a 'related distractor' in the false recognition paradigm) as what they have seen as a result of retrieving encoded gist (or reconstructively processing gist in recall), especially when the person does not access the verbatim trace that can suppress this type of false memory through recollection rejection. So, for example, a witness might remember a person they saw as a young blonde woman (gist). If they are then asked whether a different young blonde woman is the person that they saw, they might be susceptible to making a misidentification (i.e. saying yes) as a result of gist-based similarity. However, if they remember the verbatim face of the person that they saw they can use this to recognize that the new woman is not the person they saw, despite gist consistency (Wixted & Wells, 2017). When verbatim memory is not retrieved, gist-based similarity can be strengthened, for example, through repeatedly cueing gist (Reyna, 2000; Reyna et al., 2016). When this happens, false memories that are clear and vivid can arise as a result of strengthened gist memory. This phenomenon is known as *phantom recollection* and it can occur either simply through repeated retrieval of gist (as in repeated discussion or interrogation about a crime) or it can occur because when people process gist, they can recover realistic contextual details that make the memories appear real (Arndt, 2012; Brainerd & Reyna, 2019; Reyna, 2000). For example, the image of a gist-consistent person as an offender might be accompanied by memory for real or imagined details of the crime such as the location, the weather, and the behaviour of a victim or bystanders. The redintegration of veridical or plausible details with vivid gist memories has been well documented (Reyna & Brainerd, 1995). This process can also be exacerbated by asking what seem to be neutral recognition or recall questions; analogous to the Heisenberg uncertainty principle, merely measuring memory alters it (Reyna, 2000; Reyna et al., 2007). Thus, there is no such thing as testing memory without manipulating it (cf. Wixted & Roediger, Chapter 4, this volume), but there are measurement approaches, such as conjoint recognition, that disentangle these processes.

False memory resulting from suggestion

Verbatim memory therefore generally has a protective effect against false memory for meaning-consistent information, through the process of recollection rejection. One exception to this rule occurs when verbatim memory has been corrupted (Reyna et al., 2016). When external suggestion occurs, people may retrieve verbatim traces of suggestions rather than true events (Brainerd & Reyna, 2019). In this case, verbatim memory would support rather than suppress false memory. Therefore, FTT predicts two distinct types of false memory. First, 'spontaneous' false memory, caused by meaning connections and reliance on gist and, second, false memory arising from suggestion. By recognizing that false memory depends on verbatim and gist retrieval as well as surrounding circumstances, FTT makes a number of predictions that are important for practice and policy.

Eyewitness identifications are predictably unreliable

As noted above, FTT predicts that false memory is likely to occur either as a result of meaning consistency, where gist memory is relied on in the absence of verbatim retrieval, or where verbatim memory is corrupted as the result of suggestion. Memory is therefore compromised in situations where a person other than the offender is meaningfully similar to the offender and verbatim memory is not retrieved, or where verbatim memory has been corrupted.

Whether verbatim memory is retrieved is likely to depend on both the individual decision-maker and the circumstances in which they are being asked to remember witnessed events. Certain characteristics of investigations can make it less likely that witnesses will retrieve verbatim memory, and therefore make it more likely that they will develop false memory for meaning-consistent others. For example, as noted above, verbatim memory is less stable than gist memory (Kintsch, 1974; Reyna & Kiernan, 1994, 1995). As a result, increasing time periods between an event and the retrieval of memory increase the chance that gist will be retrieved in the absence of verbatim, and relatedly the risk of false memory for meaning-consistent others. In terms of the corruption of verbatim memory, corruption may occur where a witness is presented with suggestive information about a case. One common type of suggestion used is suggestive questioning in legal interviews. FTT shows that such questioning increases the risk of false memory and decreases the integrity and reliability of legal investigations.

However, it is also important to note that FTT predicts that spontaneous false memories occur routinely even in the absence of suggestion or time delay. This inaccuracy can arise purely as the result of a line-up procedure, in which a witness is asked to pick out the person they saw from a set of similar looking people. When one person in the line-up is a better match for the gist of the target than the others, the person making the identification will be susceptible to falsely and confidently recognizing that person as the offender. This susceptibility provides an explanation for the low levels of accuracy seen in some experimental research examining memory for faces (e.g. Haber & Haber, 2001) and also for the role of faulty identifications in wrongful convictions in practice (see National Registry of Exonerations, n. d. For an example of a wrongful conviction resulting from inaccurate witness identification, see the case of Ronald Thompson; O'Neill, 2001).

Although fair line-ups and using 'pristine' procedures for memory interrogation can reduce gist-based misidentifications (Wixted & Wells, 2017), it is important to recognize that meaning-based processing occurs from the onset of police investigations (e.g. witnesses use knowledge and prejudices to point a finger at plausible suspects) to later testimony under oath, often after substantial delays. Testimony encompasses many memory reports beyond face recognition judged with line-ups, such as what, when, and where a crime happened and what victims, bystanders, and suspects did or said. To exhort the legal system to ignore any evidence gathered after an initial memory test and then include evidence only from fair line-ups composed of unfamiliar participants (see Wixted & Roediger, Chapter 4, this volume), is not only unrealistic but it can lead to miscarriages of justice. To take just a handful of examples, an 'initial' memory test can occur days, weeks, months, or years after a crime, all of which are not immediate and thus likely to draw on memory for gist. Witnesses can recant initial statements under cross-examination at trial because it is revealed that their confidence was never high or they had an axe to grind with respect to the defendant or they misinterpreted the gist of events, as examples, that a victim was 'attacked' or that a defendant had a fearful expression; such judgements are rife with gist-based processing that cannot be dismissed simply because they contradict initial statements.

Witnesses with false memory can be confident

The phenomenon of phantom recollection explains findings in the literature whereby people report memories confidently and in detail that are known to be false (e.g. Ceci et al., 1994; Loftus, 2003). The fact that people can have confident and vivid false memories is important in the legal system because some jurors still believe that eyewitness identifications are especially likely to be accurate when accompanied by statements of strong confidence (Brainerd & Reyna, 2019). Research has shown that laypeople are significantly less likely than experts to endorse the idea that confidence can be influenced by factors other than memory accuracy (Benton et al., 2006; Helm, 2021a; Kassin et al., 2001). One study found that almost 40% of laypeople interviewed (but no memory experts) believed that the testimony of one confident eyewitness should be enough to convict a defendant of a crime (Simons & Chabris, 2011). There is therefore a risk that those making important judgements about how to weight memory evidence in the legal system are equating confidence with accuracy and convicting defendants on memory evidence that is weak from a scientific perspective. Although legal procedure can require a warning to be given to jurors that confident witnesses are not always accurate (e.g. a Turnbull direction in England and Wales), further work needs to be done to ensure such instructions are effective in influencing juror knowledge (Dillon et al., 2017; Helm, 2021a).

Recent research has challenged the assumption that high confidence witness iden-tifications are inaccurate, drawing on impressive evidence from laboratory and field studies (see Wixted & Roediger, Chapter 4, this volume). However, as suggested above, the conditions under which confidence is a highly reliable cue are limited. Although those conditions can be enforced to some degree for face recognition, memory in everyday life is frequently based on true and false memories of the gist of events or information. Therefore, understanding how gist memories influence judgements and decisions, and what factors contribute to confidence and accuracy, are all essential for practical applications, such as those in the legal system.

Children are not always less reliable witnesses than adults

Understanding developmental trends in false memory has particular relevance for deciding how to treat reports made by children both generally and when children's accounts conflict with accounts given by adults. FTT suggests that current legal ap-proaches that focus on the comparative reliability of adult memory and unreliability of child memory (e.g. McAuliff et al., 2007) are oversimplified as a result of general-izations across different types of false memory (see also Otgaar et al., 2017).

Although increased responsiveness to suggestion in children is likely to make children more susceptible to false memory arising as a result of suggestion (see, e.g. Bruck & Ceci 1997; Ceci & Friedman 2000), FTT suggests a different pattern in false memories arising spontaneously as a result of meaning-consistency in the absence of verbatim retrieval. Specifically, FTT predicts that false memories of this type (more specifically, false memories resulting from relying on meaningful gist; see Reyna et al., 2016) will generally increase with age (Brainerd et al., 2011). Less reliance on gist as opposed to verbatim representations with age (e.g. Reyna & Ellis, 1994) means that children are less likely to rely on meaning connections among objects that would lead them to allocate them with a common gist (Ceci et al., 2010; Hritz et al., 2015). For example, given a list of common fruits using age-normed vocabulary, children do not spontaneously rely on the gist of the list as being about fruit; their false memories for unpresented fruit words are lower than those of older participants. A simple instruction noting that the words are examples of fruit is enough to increase false memories for unpresented fruits. Age increases in false memory have been demonstrated in a wide range of experimental work across various contexts (e.g. Brackman et al., 2019; Dewhurst et al., 2007; Fisher & Sloutsky, 2005; Ross et al., 2006).

Drawing on FTT to recognize the more nuanced nature of developmental trends in false memory is important in the forensic context. Children's memory can be the central evidence in investigations and court cases, and is often especially important in cases involving domestic abuse, in which children are frequently complainants or key witnesses. Ensuring that children's memory is not peremptorily dismissed and is properly considered even in the face of competing adult memory is important in maximizing accuracy and fairness in these cases. Adopting a nuanced approach based on theory supported by experimental work is key in avoiding both believing children who are likely to be unreliable and dismissing children making genuine reports. Research has also begun to draw out the policy implications of FTT for ageing witnesses, specifically highlighting the risk that older people remember gist in the absence of verbatim and will therefore be susceptible to false memory based on meaning-consistency (Reyna & Brainerd, 2011). Gist-based processing in ageing may also contribute to sound judgements and decisions in the legal system in the many instances in which literal verbatim thinking is inappropriate (e.g. juries judging whether a 'reasonable' person would have used deadly force to stop a fleeing unarmed shoplifter).

False memory research: benefits to society

Through providing insight into the cognitive mechanisms underlying false memory, FTT can help legal systems more accurately assign probative value to witness accounts and relatedly to achieve greater accuracy in legal convictions. To take one example of many legal cases in which FTT has been applied, Reyna et al. (2002) illustrate how repeated questioning that induced gist processing produced self-incriminating testimony from a defendant who was subsequently convicted of manslaughter. The defendant began with high confidence that the allegations were untrue but ended up believing with high confidence that they were true, apparently contrary to fact. FTT shows that false memory can be complex and can result from distinct processes involving distinct cognitive mechanisms. This complexity highlights the need to move away from procedures allowing reliance on 'common-sense' principles in assessing memory, and towards procedures where decision-makers assessing witness testimony are properly informed about false memory (Brainerd & Reyna, 2019). FTT can help to inform these procedures, including improved methods for interrogating memory (Reyna et al., 2007).

Evaluation of evidence

Through describing how information is encoded in memory and retrieved from memory, FTT also makes predictions about how people encode evidence and draw on it to make decisions, for example, in medical and legal contexts. Understanding how evidence is utilized in decisions increases understanding of these decisions, facilitates evaluation of whether those decisions are consistent with policy goals, and informs interventions where necessary. Below we explore areas in which FTT has informed research that has examined the influence of memory representations (gist or verbatim) on the evaluation of evidence and related decision-making.

Interpretation of medical information

First, through recognizing the distinction between gist and verbatim and the role of gist in developmentally advanced decision-making, FTT has informed interventions to improve comprehension of information relating to health, and associated health judgements (e.g. Reyna, Broniatowski, & Edelson, 2021; for a systematic review identifying 94 studies testing FTT's predictions; see Blalock & Reyna, 2016).

As noted above, FTT predicts that people encode both gist and verbatim representations of information, and that adults tend to rely on gist. However, reliance on accurate gist will only be possible where people are able to understand and extract meaning from information. Where information is meaningless to them, people may fall back on fragile verbatim information. For example, when patients are deciding whether to undergo an unfamiliar procedure, informed choice requires that they accurately understand meaningful differences between their options rather than purely being able to recall precise verbatim statistics (Reyna, 2008). It is therefore important that interventions to help patients are aimed at enhancing their ability to attach meaning to information, as opposed to just providing information. Where patients are not able to extract accurate gist, they will not be able to rely on it, despite the natural tendency to do so. Facilitating reliance on gist through helping patients attach meaning to information is also predicted to be important in avoiding errors in the interpretation of information relating to medical risks, such as probabilities of adverse outcomes.

Note that getting the gist of information in FTT goes far beyond notions of plain language, basic literacy, and numeracy (Reyna et al., 2009). Literate and numerate patients can encode the risk of a disease or treatment when presented with plain language that includes numbers, but that is not the same thing as getting the gist—for example, whether a 20% lifetime risk of invasive breast cancer is high or low (Reyna, 2008). Errors in interpreting numbers can be caused by a lack of knowledge or a failure to encode information accurately. Some of these errors might be reduced through greater numeracy but not errors that involve failing to get the gist (e.g. Peters, 2020).

In addition, there is another class of errors that is more straightforward to address—errors arising from what is known as processing interference, confusion when information about overlapping classes of events is presented, such as a genetic risk of developing breast cancer and the base rate, or unconditional, risk of developing breast cancer. Base rate neglect is an example of the tendency to give too little weight to the denominator in a probability (e.g. judging 40/100 as being more likely than 4/10; Bar-Hillel, 1980). This processing error is known as a class-inclusion error in FTT and it can be reduced significantly by using two-by-two tables or Venn diagrams to make classes and their probabilities distinct (non-overlapping; Wolfe et al., 2015).

Base rate neglect can have important implications in applied decision-making. Consider the following example:

> The pre-test probability of a disease is 10%. Eighty per cent of people with the disease will test positive and 80% of people without the disease will test negative (i.e. the test has 80% sensitivity and 80% specificity).

Given a choice between 30% and 70%, decision-makers think that the probability that a person who has a positive test result has the disease is relatively high; they overwhelmingly choose 70%. However, the correct answer is closer to 30% because among those who test positive, very few will actually have the disease due to the pretest probability of 10%. These errors are not necessarily resolved by increasing numeracy (Portnoy et al., 2010; Reyna et al., 2009). As a result, even highly numerate people can have difficulty with understanding numerical information such as conditional probabilities. FTT provides a route through which this difficulty can be ameliorated by recognizing the distinction between verbatim and gist, as well as disentangling numerator and denominator information. According to FTT, these errors can be corrected by encouraging decision-makers to rely on gist as opposed to verbatim representations and by presenting diagrams or labels that separate classes of events. Reliance on gist has a protective effect against such mistakes since gist involves understanding the meaning of numbers rather than analytical quantitative calculations that are sensitive to interference (Reyna et al., 2009). Therefore, according to FTT, interventions in this area (e.g. base-rate neglect and conjunction or disjunction fallacies), should not necessarily target precise verbatim details and mathematical ability to reduce class-inclusion biases because people who commit this error are often high in those abilities but should enhance decision-makers' understanding of and reliance on meaningful gist (see Wolfe & Reyna, 2010). This conclusion is consistent with work showing that advanced practitioners in the medical field tend to rely on gist rather than verbatim information in their area of expertise (Lloyd & Reyna, 2009).

Interventions based on FTT have been developed in order to enhance decision-makers' reliance on gist, and as a result promote informed choice and reduce errors in comprehension.

For example, the BRCA Gist (BReast CAncer and Genetics Intelligent Semantic Tutoring) system has been introduced in order to communicate genetic risk of breast cancer to those receiving test results. BRCA Gist is an intelligent web-based tutoring system that uses artificial intelligence to encourage people to form flexible gist representations of numerical information relating to breast cancer risk (Wolfe et al., 2015). BRCA Gist has been shown to be more effective than existing interventions in increasing comprehension relating to breast cancer risk, and thus can play a role in ensuring that patients make informed decisions. Another web-based decision support tool that was developed to promote reliance on gist focused on effectively informing patients with rheumatoid arthritis about complex information about the disease and the need for escalating care after failing traditional disease-modifying antirheumatic drugs. Research showed that this tool improved knowledge, willingness to escalate care appropriately, and the likelihood of making an informed and value-concordant choice relating to care (Fraenkel et al., 2012). The success of these interventions in the medical context has led to suggestions that gist-based interventions might be used in other areas where decision-makers are required to utilize complex information, such as information relating to forensic testing in criminal adjudication (Helm et al., 2017).

Juror damage awards

One area in which people are asked to evaluate evidence to reach decisions is in the justice system. In the US, jurors evaluate case evidence in both civil and criminal cases and reach determinations of responsibility. In the civil context, they are also often required to award damages to the plaintiff in the event that a defendant is found responsible for causing harm to them. Damages awarded by jurors in civil cases are designed to put a plaintiff back in the position they would be in had the harm not been done to them, and cover both pecuniary loss (e.g. loss of income or medical expenses) and non-pecuniary loss (e.g. pain and suffering). FTT has provided important insight into how jurors allocate damages for non-pecuniary loss (Hans & Reyna, 2011), a process that commentators have described as 'rudimentary and elusive' (Greene & Bornstein, 2003). Understanding this process is important in ensuring that the civil jury process is fair to both plaintiffs and defendants, particularly given widespread criticism over the unpredictability of the civil jury (e.g. Hans & Eisenberg, 2010).

The 'Hans–Reyna' model of damage award decision-making outlines FTT-informed predictions relating to how civil jurors draw on evidence, and how that evidence is used to make damage award decisions. As described above, FTT posits that people encode both gist and verbatim representations of information, and the representation relied upon in decision-making will be determined by the decision-maker (with adults having a preference for relying on the simplest gist they can) and the requirements of the task (with tasks that cannot be resolved through reliance on gist pushing even adult decision-makers towards finer grained distinctions). When considering evidence, jurors are predicted to represent and encode each piece of evidence as gist and as verbatim, and also to represent the body of evidence in a gist-based way (similar to the 'story' predicted by Pennington & Hastie, 1986, 1992) and a verbatim way (a list of specific pieces of evidence; see Kintsch, 1974). Jurors will then generally rely on the simplest gist that they can to complete a given task.

In allocating damage awards, the Hans–Reyna model predicts that jurors will rely largely on gist, making gist-based judgements relating to whether damages are warranted or not (a categorical judgement), and to categorize the level of deserved damages, for example, as low, medium, or high (Reyna, Hans, et al., 2015). To reach a precise damage award, jurors will allocate a precise number to the gist of the deserved damages (e.g. allocating a 'high' number when it is determined that an injury warrants a 'high' level of damages; Hans & Reyna, 2011). This model has implications for jury decision-making. For example, the model provides insight into the mechanisms through which anchors are predicted to influence juries. Anchors are numerical values that can bias decision-makers' judgements in the direction of the anchor value (Bystranowski et al., 2021). According to FTT, the stage of the decision-making process at which precise numbers are relevant is the stage at which a number is allocated to a gist (i.e. when numbers are mapped to low, medium, or high gists). Therefore, one influence of numerical anchors on damage awards will be in influencing which specific numbers are considered low, medium, or high. This influence should be greatest where a number is meaningful, since that meaning will allow jurors to put the size of awards in perspective so they can be understood as low, medium, or high (Reyna, Hans, et al., 2015). This greater influence of meaningful as opposed to meaningless anchors has been shown in a line of experimental work (Hans et al., 2018; Helm et al., 2020; Reyna, Hans, et al., 2015). This work can help inform interventions such as judicial instructions or attorney guidance that utilize anchors to appropriately guide jurors by helping them to contextualize numbers rather than biasing them (Helm et al., 2020).

Information consumption

Recent research has also drawn on FTT in order to further our understanding of how online media platform users decide to act on and share received information (Broniatowski & Reyna, 2020; Reyna, 2021). FTT's approach to the consumption of information online builds on the existence of gist and verbatim memory representations and the fact that, as noted above, adults are generally driven by reliance on gist representations where possible. Therefore, presented information is most likely to influence opinions when people can extract a clear gist from it (e.g. 'vaccines are safe and effective' or 'COVID-19 vaccine side effects are nil').

One way to facilitate the extraction of gist (and therefore the tendency of individuals to subsequently rely on that gist) is to ensure that information is presented in a way that is conducive to the extraction of gist. Stories that make sense to a reader allow them to extract a coherent gist and are more likely to be accepted and acted upon (Broniatowski & Reyna, 2020). Previous research has shown that creating a coherent order for pieces of information, as in a 'story' format, helps participants to extract and rely on gist by increasing comprehensibility and allowing the extraction of meaning (e.g. Dewhurst et al., 2007; Pennington & Hastie, 1992; Reyna, 2012). Extraction of gist from narratives with poorly defined causal structures may only be possible for skilled readers or those with sufficient background knowledge (Van den Broek, 2010). As stated in one recent paper, 'more difficult texts are likely to appeal only to those subjects possessing the willingness and ability to expend the effort to comprehend them' (Broniatowski & Reyna, 2020, p. 435). This role of meaning extraction is predicted to combine with motivational factors and social values in influencing decisions to share information online (Reyna, 2021). These predictions have been supported by recent experimental work (Broniatowski et al., 2016) and have been developed into a model of online media platform users' decisions to act on and share information (see Broniatowski & Reyna, 2020).

Importantly, from a policy perspective, official communications are typically more likely to focus on literal facts (e.g. information on how vaccines work) rather than emphasizing causal relations among facts in a way that is conducive to the extraction of gist. By contrast, fake news often provides a narrative focusing specifically on causal explanations. This contrast creates a risk that fake news can be more comprehensible and memorable than official communication (Reyna, Broniatowski, & Edelson, 2021). Official communications should seek to more clearly emphasize causal relations and meaning in order to avoid being less comprehensible and less memorable and therefore less likely to be acted on than fake news.

Evaluating evidence: benefits to society

Through making the distinction between gist and verbatim representations, and showing the impact that each can have on the evaluation of evidence, FTT provides insight into the factors that influence evidence evaluation and interventions to ensure that evidence evaluation takes place in a way that is conducive to healthy outcomes. The areas considered here highlight some key areas where the distinction between gist and verbatim is likely to be important and work in these areas highlights key principles with the potential to be important to a wide range of policy: (1) generally, decision-making based on gist promotes informed consent, and reduces errors associated with processing of information; (2) meaningful cues are likely to be most helpful in assisting decision-makers allocating specific numbers to a gist; and (3) information is most likely to influence behaviour when an individual can extract gist from it.

Decisions under risk

Another area in which FTT can provide insight with the potential to benefit society is when considering decision-making under risk, for example, decisions where a person is choosing between a sure option with a certain outcome and risky option with the potential for a better outcome or a worse outcome when compared with the sure option. Through the distinction between gist and verbatim representations, FTT explains and predicts observed effects in the literature on decision-making under risk, most importantly the risky-choice framing effect, and provides insight into the way decisions are made that has the potential to be important in informing policy.

The risky-choice framing effect

The risky-choice framing effect is important to understand, since the inconsistencies in risk preference involved in the effect provide insight into the mechanisms underlying risky choice. The risky-choice framing effect is the tendency of decision-makers to pick the sure outcome when a decision is framed as a gain but the risky option when the same decision is framed as a loss (Tversky & Kahneman, 1981, 1986). In gain-framed problems, decision-makers choose between a sure option and a gamble typically of equal expected value (e.g. gaining $1000 for sure vs a 50% chance of gaining $2000 and a 50% chance of nothing). In loss-framed problems, decision-makers might be given an endowment (e.g. of $2000) and must choose between losing some money for sure and a gamble, again typically of equal expected value (e.g. losing $1000 for sure vs a 50% chance of losing $2000 and a 50% chance of losing nothing). In both frames, decision-makers are essentially choosing between keeping $1000 and a 50% chance of keeping $2000. However, adults tend to be more likely to choose the sure option in the gain frame and more likely to choose the risky option in the loss frame.

FTT explains the framing effect as the result of reliance by most adults on gist (e.g. Kühberger & Tanner, 2010). As noted above, adults have a preference for relying on the simplest level of gist possible. Thus, where there are meaningful categorical distinctions between options, adults will rely on these distinctions to differentiate options and dictate their decisions. In the gain frame, decisions boiled down to their simplest gist become gaining something for sure versus maybe gaining something and maybe gaining nothing. This gist promotes selection of the sure option. In the loss frame, decisions boiled down to their simplest gist become losing something for sure versus maybe gaining something and maybe gaining nothing. This gist promotes selection of the risky option.

Thus, FTT predicts the framing effect, which was also predicted in early theories of decision-making under risk such as prospect theory. The latter has been compared to FTT with results supporting FTT and with new effects introduced by FTT (for a review, see Broniatowski & Reyna, 2018). Experimental research has tested FTT's explanation for framing effects and has provided support for this explanation when compared to competing explanations (most notably the explanations of prospect theory; Kühberger & Tanner, 2010; Reyna, Brainerd, et al., 2021). Importantly, the fact that reliance on gist is predicted to increase with age and experience also predicts and explains counterintuitive developmental reversals that have been found in the literature, where adults and experts are more likely than children and novices to show framing effects, as discussed (Reyna et al., 2014; Reyna & Ellis, 1994). Work utilizing framing effects to test FTT predictions therefore provides support for the predicted importance of gist in risky decisions in adults, but also the greater importance of verbatim in risky decisions in children. This importance of gist has implications in applied contexts that are important in society. Two such contexts are considered here.

Deciding whether to plead guilty

When defendants are accused of crimes in criminal justice systems (including the systems in the US, England and Wales, Scotland, and Northern Ireland), they face a choice of whether to plead guilty. Research shows that some innocent defendants plead guilty and that plea decisions are determined by more than just factual guilt and innocence (Blume & Helm, 2014; Dervan & Edkins, 2013; Zimmerman & Hunter, 2018). Examining plea decisions as risky choices can be helpful in understanding those decisions. If defendants do plead guilty, they will typically receive a sentence that is less severe than the one that they would receive if convicted at trial. Therefore, defendants making guilty plea decisions are choosing between accepting guilt and a certain punishment versus pleading not guilty and facing a potentially worse punishment (if convicted at trial) or no punishment at all (if acquitted at trial). In this way, guilty plea decisions are similar to the decisions under risk described above, where decision-makers must pick between a sure and a risky option, but with the added complication of factual guilt (a gist-based consideration; see Helm, 2018). FTT can therefore provide important insight into these decisions, which can help to ensure that decisions are being made in a way that is consistent with normative legal goals.

First, FTT predicts that plea decisions in typical adults will be driven by reliance on gist. Therefore, decisions will be determined by the simplest meaningful differences between options that will resolve the decision. On the one hand, where outcomes when pleading guilty and convicted at trial are meaningfully similar (e.g. both would involve a short period in custody), guilt or innocence may be the main gist-based distinction relevant to the problem and therefore should be determinative of decisions (provided that it is important to the decision-maker). On the other hand, where outcomes when pleading guilty and convicted at trial are not meaningfully similar (e.g. a defendant would face a period in custody if convicted at trial but would not if they pled guilty), the gist-based difference between the options has the potential to compete with factual guilt or innocence and to lead factually innocent defendants to plead guilty. This prediction has been supported in experimental work (see Helm, 2022). Real cases also illustrate this phenomenon, for example, several of the now-acquitted defendants in the UK Post Office scandal have described how they pleaded guilty despite believing they were innocent since pleading guilty would mean they would (or so they thought) avoid jail (see Helm, 2021b). To protect innocent defendants from pleading guilty, FTT therefore suggests that, as far as possible, plea and trial outcomes should not incentivize pleading guilty to escape confinement. Categorical gist differences between outcomes, such as the ability to still 'have a life' despite incarceration, should be explained to defendants.

Second, FTT predicts that there are certain groups that are likely to be driven more by verbatim representations. Important groups thought to be more reliant on verbatim processing are youthful defendants and developmentally less advanced adults. As a result, children and adolescents are more likely to rely on verbatim information when deciding whether to plead guilty. Importantly, since factual guilt or innocence is a meaningful rather than quantitative dimension in the decision, FTT predicts that youth are particularly susceptible to neglecting factual guilt or innocence (and other meaningful distinctions in plea decisions) despite understanding them at a literal level. Thus, youth may plead guilty even when innocent on the basis of relatively small sentence discounts (Helm et al., 2018). This prediction has been supported by experimental work suggesting that factual guilt and innocence is less important in plea decisions in youth and that this does not reflect a difference in relevant underlying values, such as the value of not pleading guilty to a crime one has not committed (Helm et al., 2018). Adults with a tendency to rely more on verbatim processing have similarly shown a relative lack of responsiveness to factual guilt or innocence (Helm & Reyna, 2017). This FTT-informed finding has important implications for the criminal justice system, which must protect these defendants, particularly from pleading guilty when innocent (Helm & Reyna, 2017).

Risky decisions, unprotected sex, and crime

FTT can also help inform understanding of decisions to engage in risky activity, such as decisions to engage in unprotected sex, and decisions to engage in crime. Put simply,

FTT predicts that reliance on gist representations will generally have a protective effect against unhealthy risk-taking while reliance on verbatim representations can promote unhealthy risk-taking (e.g. Reyna & Farley, 2006). This protective effect of gist and negative effect of verbatim occurs specifically in situations in which risks of adverse outcomes are low and benefits are high. In such situations, precise trade-offs of risk and rewards (resulting from reliance on precise and superficial representations) can promote risky behaviour. In contrast, reliance on meaningful contextual processing (resulting from reliance on meaningful representations) results in more fuzzy processing based on categorical distinctions (e.g. based on the distinction between no chance of a serious disease and some chance of a serious disease, or no chance of criminal conviction and some chance of criminal conviction) which is driven by values and recognizes that certain outcomes are not worth risking even if rewards are high (Helm & Reyna, 2018). In this way, FTT recognizes a route to unhealthy risk-taking based on reliance on verbatim processing, in addition to existing recognized routes based on impulsivity and lack of control (e.g. Casey et al., 2011; Steinberg, 2007).

The fact that reliance on gist increases with age and thus is lower in adolescents can also explain or partly explain the increased tendency towards risk-taking observed in adolescence compared to later adulthood. In fact, a recent meta-analysis of experiments on risky decision-making showed that risk preference actually declines from childhood to adolescence to adulthood, contrary to a peak in adolescence in unhealthy risk-taking based on impulsivity and lack of control as is commonly assumed (Defoe et al., 2015).

FTT's predictions in this area have been examined and supported in the context of many health decisions, for example, whether to risk HIV or other sexually transmitted infections by engaging in unprotected sex. The risk of HIV infection from a single act of unprotected sex is relatively low (roughly 0.08% from one incident of unprotected sex; see Boily et al., 2009). As a result, reliance on verbatim processing, which promotes precise and superficial weighing up of risks and rewards, could lead a person to engage in unprotected sex. Thus, on the one hand, the low probability of infection with HIV along with high perceived benefits of sexual activity rationalize unprotected sex. On the other hand, reliance on gist representations results in more fuzzy processing based on categorical distinctions which recognizes that certain outcomes (such as HIV) are not worth risking even if rewards are high (Helm & Reyna, 2018). Experimental work provides support for this explanation for unhealthy risk-taking (Mills et al., 2008; Reyna et al., 2011), which also predicts and explains the paradoxical finding that adolescents simultaneously rate unprotected sex as being riskier than adults do and yet are more willing to engage in unprotected sex (Mills et al., 2008). The demonstrated relationship between mental representations relied on and risky behaviours in this area has informed the development of a curriculum for adolescents (Reducing the Risk Plus (RTR+)) aimed at promoting risk reduction and the avoidance of premature pregnancy and sexually transmitted infections by emphasizing gist representations (Reyna & Mills, 2014). RTR+ has been shown to be more effective than existing curricula in achieving the majority of desired outcomes

(Reyna & Mills, 2014), and its success shows the importance of cueing gist represen-tations in preventing unhealthy risk-taking.

The route from reliance on verbatim to unhealthy risk-taking also has the poten-tial to be important in understanding some types of decision to engage in crime. Importantly, many decisions to engage in criminal behaviour involve perceived high rewards and low risks. Thus, in this common situation where rewards are high and risks are low (i.e. the chances of getting caught are objectively small for each instance of criminal risk-taking), FTT predicts that reliance on verbatim representations would produce an increased risk of criminal offending (Helm & Reyna, 2018; Reyna et al., 2018). Reliance on gist representations is likely to have a protective effect against risk-taking by promoting reliance on categorical avoidance of catastrophic outcomes, such as conviction of a crime, rather than rationally trading-off risks for rewards. Note that both offenders and non-offenders typically have a strong desire to avoid conviction but gist thinkers are more likely to choose in accordance with their values. As discussed earlier, this hypothesis is supported by research indicating that residential burglars consider the risk of being caught and the potential reward (though these considerations apparently outweigh the consideration of penalties) whereas matched controls consider none of these factors and categorically avoid risk-taking (Decker et al., 1993). The hypothesis has also been supported by research in neuroscience and law, which has found an increase in neural activation in areas of the brain associated with increased cognitive effort as levels of criminal behav-iour increase, when making framing consistent choices (thought to show reluctant reliance on gist; Reyna et al., 2018). Real-world self-reported risk-taking, criminal and non-criminal, was correlated with lower levels of framing bias, that is, more ob-jective processing of risks and rewards (Reyna et al., 2018). FTT therefore adds an additional component to existing accounts of cognitive factors involved in criminal behaviour, which include differential processing of rewards/benefits (Buckholtz et al., 2010), reduced attention and inhibition (Freeman et al., 2015; Larson et al., 2013), and abnormal processing of emotional stimuli (Marsh & Cardinale, 2012).

Recognition of this additional component has the potential to inform new policy interventions aimed at the reduction of criminal behaviour. Previous interventions to reduce crime have been based primarily on encouraging high-risk individuals to think 'slowly' (i.e. to prioritize deliberation and inhibition over motivational factors; see Heller et al., 2015). However, FTT's predictions and supportive findings (e.g. Reyna et al., 2018; Reyna & Mills, 2014) suggest that while it is important to en-courage inhibition, it may also be important to encourage reliance on qualitative gist representations in order to reduce crime.

Decisions under risk: benefits to society

By recognizing that unhealthy risk-taking can be promoted by reliance on verbatim representations of risk as opposed to gist, FTT recognizes a novel path to unhealthy risk-taking. Understanding this path has the potential to inform interventions that can more effectively address unhealthy risk-taking and promote healthy behaviour, both in the areas discussed above, and more broadly.

Conclusion

Through recognizing the distinction between gist and verbatim representations, FTT provides important insight into both memory and associated decision-making. The basic tenets of FTT, which have been supported in experimental work, have informed applied work examining memory and decision-making in practice and informed specific interventions to ensure that the realities of memory and decision-making are recognized in practice and policy across a wide variety of contexts. Basing applied work on FTT in this way has had three key advantages.

First, it has generated counterintuitive predictions that have been tested and supported in empirical work, such as higher rates of 'recognition' for events that are gist-consistent but were never experienced, compared to events that were truly experienced, and developmental reversals in false memory and decision biases, that is, children are less biased than adults. These counterintuitive predictions have the potential to be important in practice in protecting certain groups, and some related policies have been implemented in practice. For example, this research has influenced guidelines for children's testimony around the world, protecting the welfare of children making legal and medical reports that conflict with those made by adults.

Second, FTT has informed work separating relevant underlying constructs and has thus facilitated nuanced predictions that effectively differentiate memories or decisions that are meaningfully different from one another. For example, by separating gist and verbatim and recognizing reliance on gist or verbatim as separable from suggestibility, FTT has allowed different types of false memory to be differentiated from one another, and different trends in false memory to be identified. In doing so, the theory can inform policy that treats memories that ought to be treated differently, differently from one another.

Third, the theory provides insight not only into decisions or memories but also into how they are generated. This insight is important where the way a decision or memory is formed is important to protect human welfare. For example, in the guilty plea context, a preference for pleading guilty in itself is not problematic (and in fact, it has been argued that it can be good that innocent people have the chance to plead guilty; see Garrett, 2015). However, it may be problematic when the decision resulted from pressures undermining the influence of guilt or innocence, rather than from the true preference of a defendant (see Helm et al., 2022). Knowing how decisions and memories are generated is also important in designing policy interventions that are likely to not only be effective, but to be effective *for the right reasons*. For example, interventions can reduce unprotected sex in young people not due to fear, but due to mature decision-making and recognition of the meaning of risk-communication messages. The three advantages outlined above make FTT well placed to consider in designing experimental work and policy interventions in order to ensure the benefits of science are felt in society in practice.

Acknowledgements

Preparation of this manuscript was supported in part by a grant from UK Research and Innovation (MR/T02027X/1) to the first author and by grants from the National Science Foundation (SES-2029420 and SES-1536238) and National Institutes of Health (RO1NR014368 and R21NR016905) to the second author.

References

Arndt, J. (2012). False recollection: Empirical findings and their theoretical implications. *Psychology of Learning and Motivation*, 56, 81–124.

Baddeley, A. (2000). The episodic buffer: A new component of working memory? *Trends in Cognitive Sciences*, 4(11), 417–423.

Barbey, A. K., & Sloman, S. A. (2007). Base-rate respect: From ecological rationality to dual processes. *Behavioral and Brain Sciences*, 30(3), 241–254.

Bar-Hillel, M. (1980). The base-rate fallacy in probability judgments. *Acta Psychologica*, 44(3), 211–233.

Bartlett, J. C., Shastri, K. K., Abdi, H., & Neville-Smith, M. (2009). Component structure of individual differences in true and false recognition of faces. *Journal of Experimental Psychology: Learning, Memory, and Cognition*, 35(5), 1207–1230.

Benton, T. R., Ross, D. F., Bradshaw, E., Thomas, W. N., & Bradshaw, G. S. (2006). Eyewitness memory is still not common sense: Comparing jurors, judges and law enforcement to eyewitness experts. *Applied Cognitive Psychology*, 20(1), 115–129.

Blalock, S. J., & Reyna, V. F. (2016). Using fuzzy-trace theory to understand and improve health judgments, decisions, and behaviors: A literature review. *Health Psychology*, 35(8), 781–792.

Blume, J. H., & Helm, R. K. (2014). The unexonerated: Factually innocent defendants who plead guilty. *Cornell Law Review*, 100(1), 157–191.

Boily, M. C., Baggaley, R. F., Wang, L., Masse, B., White, R. G., Hayes, R. J., & Alary, M. (2009). Heterosexual risk of HIV-1 infection per sexual act: Systematic review and meta-analysis of observational studies. *Lancet Infectious Diseases*, 9(2), 118–129.

Brackmann, N., Sauerland, M., & Otgaar, H. (2019). Developmental trends in lineup performance: Adolescents are more prone to innocent bystander misidentifications than children and adults. *Memory & Cognition*, 47(3), 428–440.

Brainerd, C. J. (2013). Developmental reversals in false memory: A new look at the reliability of children's evidence. *Current Directions in Psychological Science*, 22(5), 335–341.

Brainerd, C. J., & Reyna, V. F. (1998). When things that were never experienced are easier to 'remember' than things that were. *Psychological Science*, 9(6), 484–489.

Brainerd, C. J., & Reyna, V. F. (2005). *The science of false memory*. Oxford University Press.

Brainerd, C. J., & Reyna, V. F. (2018). Complementarity in false memory illusions. *Journal of Experimental Psychology: General*, 147(3), 305–327.

Brainerd, C. J., & Reyna, V. F. (2019). Fuzzy-trace theory, false memory, and the law. Policy Insights from the *Behavioral and Brain Sciences*, 6(1), 79–86.

Brainerd, C. J., Reyna, V. F., & Brandse, E. (1995). Are children's false memories more persistent than their true memories? *Psychological Science*, 6(6), 359–364.

Brainerd, C. J., Reyna, V. F., & Zember, E. (2011). Theoretical and forensic implications of developmental studies of the DRM illusion. *Memory & Cognition*, 39(3), 365–380.

Broniatowski, D. A., Hilyard, K. M., & Dredze, M. (2016). Effective vaccine communication during the Disneyland measles outbreak. *Vaccine*, 34(28), 3225–3228.

Broniatowski, D. A., & Reyna, V. F. (2018). A formal model of fuzzy-trace theory: Variations on framing effects and the Allais paradox. *Decision*, 5(4), 205–252.

Broniatowski, D. A., & Reyna, V. F. (2020). To illuminate and motivate: A fuzzy-trace model of the spread of information online. *Computational and Mathematical Organization Theory*, 26(4), 431–464.

Bruck, M., & Ceci, S. J. (1997). The suggestibility of young children. *Current Directions in Psychological Science*, 6(3), 75–79.

Buckholtz, J. W., Treadway, M. T., Cowan, R. L., Woodward, N. D., Benning, S. D., Li, R., Ansari, M. S., Baldwin, R. M., Schwartzman, A. N., Shelby, E. S., Smith, C. E., Cole, D., Kessler, R. M., & Zald, D. H. (2010). Mesolimbic dopamine reward system hypersensitivity in individuals with psychopathic traits. *Nature Neuroscience*, 13(4), 419–421.

Bystranowski, P., Janik, B., Próchnicki, M., & Skórska, P. (2021). Anchoring effect in legal decision-making: A meta-analysis. *Law and Human Behavior*, 45(1), 1–23.

Casey, B. J., Jones, R. M., & Somerville, L. H. (2011). Braking and accelerating of the adolescent brain. *Journal of Research on Adolescence*, 21(1), 21–33.

Ceci, S. J., Fitneva, S. A., & Williams, W. M. (2010). Representational constraints on the development of memory and metamemory: A developmental–representational theory. *Psychological Review*, 117(2), 464–495.

Ceci, S. J., & Friedman, R. D. (2000). The suggestibility of children: Scientific research and legal implications. *Cornell Law Review*, 86(1), 33–108.

Ceci, S. J., Loftus, E. F., Leichtman, M. D., & Bruck, M. (1994). The possible role of source misattributions in the creation of false beliefs among preschoolers. *International Journal of Clinical and Experimental Hypnosis*, 42(4), 304–320.

Clark, H. H., & Clark, E. V. (1977). *Psychology and language: An introduction to psycholinguistics.* Harcourt College Publishing.

Decker, S., Wright, R., & Logie, R. (1993). Perceptual deterrence among active residential burglars: A research note. *Criminology*, 31(1), 135–147.

Defoe, I. N., Dubas, J. S., Figner, B., & Van Aken, M. A. (2015). A meta-analysis on age differences in risky decision making: Adolescents versus children and adults. *Psychological Bulletin*, 141(1), 48–84.

De Martino, B., Harrison, N., Knafo, S., Bird, G., & Dolan, R. J. (2008). Explaining enhanced logical consistency during decision-making in autism. *Journal of Neuroscience*, 28(42), 10746–10750.

Dervan, L. E., & Edkins, V. A. (2013). The innocent defendant's dilemma: An innovative empirical study of plea bargaining's innocence problem. *Journal of Criminal Law & Criminology*, 103(1), 1–48.

Dewhurst, S. A., Pursglove, R. C., & Lewis, C. (2007). Story contexts increase susceptibility to the DRM illusion in 5-year-olds. *Developmental Science*, 10(3), 374–378.

Dillon, M. K., Jones, A. M., Bergold, A. N., Hui, C. Y., & Penrod, S. D. (2017). Henderson instructions: Do they enhance evidence evaluation? *Journal of Forensic Psychology Research and Practice*, 17(1), 1–24.

Evans, J. S. B., & Stanovich, K. E. (2013). Dual-process theories of higher cognition: Advancing the debate. *Perspectives on Psychological Science*, 8(3), 223–241.

Evidence-Based Justice Lab. (n. d.). *UK miscarriages of justice registry.* Evidencebasedjustice.exeter. ac.uk. https://evidencebasedjustice.exeter.ac.uk/miscarriages-of-justice-registry/the-cases/overv iew-graph/

Fisher, A. V., & Sloutsky, V. M. (2005). When induction meets memory: Evidence for gradual transition from similarity-based to category-based induction. *Child Development*, 76(3), 583–597.

Fraenkel, L., Peters, E., Charpentier, P., Olsen, B., Errante, L., Schoen, R. T., & Reyna, V. (2012). Decision tool to improve the quality of care in rheumatoid arthritis. *Arthritis Care & Research*, 64(7), 977–985.

Freeman, S. M., Clewett, D. V., Bennett, C. M., Kiehl, K. A., Gazzaniga, M. S., & Miller, M. B. (2015). The posteromedial region of the default mode network shows attenuated task-induced deactivation in psychopathic prisoners. *Neuropsychology*, 29(3), 493–500.

Garrett, B. L. (2015). Why plea bargains are not confessions. *William and Mary Law Review, 57*, 1415–1444.

Greene, E., & Bornstein, B. H. (2003). *Determining damages: The psychology of jury awards.* American Psychological Association.

Griego, A. W., Datzman, J. N., Estrada, S. M., & Middlebrook, S. S. (2019). Suggestibility and false memories in relation to intellectual disability and autism spectrum disorder: A meta-analytic review. *Journal of Intellectual Disability Research, 63*(12), 1464–1474.

Haber, R. N., & Haber, L. (2001, November). *A meta-analysis of research on eyewitness line-up accuracy.* Paper presented at Psychonomic Society, Orlando, FL.

Hans, V. P., & Eisenberg, T. (2010). The predictability of juries. *DePaul Law Review, 60*(2), 375–396.

Hans, V. P., Helm, R. K., & Reyna, V. F. (2018). From meaning to money: Translating injury into dollars. *Law and Human Behavior, 42*(2), 95–109.

Hans, V. P., & Reyna, V. F. (2011). To dollars from sense: Qualitative to quantitative translation in jury damage awards. *Journal of Empirical Legal Studies, 8*(Suppl 1), 120–147.

Hare, T. A., Tottenham, N., Galvan, A., Voss, H. U., Glover, G. H., & Casey, B. J. (2008). Biological substrates of emotional reactivity and regulation in adolescence during an emotional go-nogo task. *Biological Psychiatry, 63*(10), 927–934.

Heller, S. B., Shah, A. K., Guryan, J., Ludwig, J., Mullainathan, S., & Pollack, H. A. (2015). *Thinking fast and slow? Some field experiments to reduce crime and dropout in Chicago* (NBER Working Paper No. 21178). https://www.nber.org/papers/w21178

Helm, R. K. (2018). Cognitive theory and plea-bargaining. *Policy Insights from the Behavioral and Brain Sciences, 5*(2), 195–201.

Helm, R. K. (2021a). Evaluating witness testimony: Juror knowledge, false memory, and the utility of evidence-based directions. *International Journal of Evidence & Proof, 25*(4), 264–285.

Helm, R. K. (2021b, 28 April). *False guilty pleas and the post-office scandal.* Evidence-Based Justice Blog. https://evidencebasedjustice.exeter.ac.uk/false-guilty-pleas-and-the-post-office-scandal/

Helm, R. K. (2021c). The anatomy of 'factual error' miscarriages of justice in England and Wales: A fifty-year review. *Criminal Law Review, 5*, 351–373.

Helm, R. K. (2022). Cognition and incentives in plea decisions: Categorical differences in outcomes as the tipping point for innocent defendants. *Psychology, Public Policy, and Law, 28*(3), 344–355.

Helm, R. K., Dehaghani, R., & Newman, D. (2022). Guilty plea decisions: Moving beyond the autonomy myth. *Modern Law Review, 85*(1), 133–163.

Helm, R. K., Hans, V. P., & Reyna, V. F. (2017). Trial by numbers. *Cornell Journal of Law & Public Policy, 27*(1), 107–143.

Helm, R. K., Hans, V. P., Reyna, V. R., & Reed, K. (2020). Numeracy in the jury box: Numerical ability, meaning, and damage award decision making. *Applied Cognitive Psychology, 34*(2), 434–448.

Helm, R. K., & Reyna, V. F. (2017). Logical but incompetent plea decisions: A new approach to plea bargaining grounded in cognitive theory. *Psychology, Public Policy, and Law, 23*(3), 367–380.

Helm, R. K., & Reyna, V. F. (2018). Cognitive, developmental, and neurobiological aspects of risk judgments. In M. Raue, E. Lermer, & B. Streicher (Eds.), *Psychological perspectives on risk and risk analysis: Theory, models, and applications* (pp. 83–108). Springer.

Helm, R. K., Reyna, V. F., Franz, A. A., & Novick, R. Z. (2018). Too young to plead? Risk, rationality, and plea bargaining's innocence problem in adolescents. *Psychology, Public Policy, and Law, 24*(2), 180–191.

Hritz, A. C., Royer, C. E., Helm, R. K., Burd, K. A., Ojeda, K., & Ceci, S. J. (2015). Children's suggestibility research: Things to know before interviewing a child. *Anuario de Psicología Jurídica, 25*(1), 3–12.

Jolliffe, T., & Baron-Cohen, S. (2000). Linguistic processing in high-functioning adults with autism or Asperger's syndrome. Is global coherence impaired? *Psychological Medicine, 30*(5), 1169–1187.

Kahneman, D. (2011). *Thinking fast and slow.* Penguin.

Kassin, S. M., Tubb, V. A., Hosch, H. M., & Memon, A. (2001). On the 'general acceptance' of eyewitness testimony research: A new survey of the experts. *American Psychologist, 56*(5), 405–416.

Kintsch, W. (1974). *The representation of meaning in memory.* Halstead Press.

Klaczynski, P. A., & Felmban, W. S. (2014). Heuristics and biases during adolescence: Developmental reversals and individual differences. In H. Markovits (Ed.), *The developmental psychology of reasoning and decision-making* (pp. 84–111). Psychology Press.

Kühberger, A., & Tanner, C. (2010). Risky choice framing: Task versions and a comparison of prospect theory and fuzzy-trace theory. *Journal of Behavioral Decision Making, 23*(3), 314–329.

Kwak, Y., Payne, J. W., Cohen, A. L., & Huettel, S. A. (2015). The rational adolescent: Strategic information processing during decision making revealed by eye tracking. *Cognitive Development, 36*, 20–30.

Lampinen, J. M., Watkins, K. N., & Odegard, T. N. (2006). Phantom ROC: Recollection rejection in a hybrid conjoint recognition signal detection model. *Memory, 14*(6), 655–671.

Larson, C. L., Baskin-Sommers, A. R., Stout, D. M., Balderston, N. L., Curtin, J. J., Schultz, D. H., & Newman, J. P. (2013). The interplay of attention and emotion: Top-down attention modulates amygdala activation in psychopathy. *Cognitive, Affective & Behavioral Neuroscience, 13*(4), 757–770.

Lloyd, F. J., & Reyna, V. F. (2009). Clinical gist and medical education: Connecting the dots. *JAMA, 302*(12), 1332–1333.

Loftus, E. F. (2003). Make-believe memories. *American Psychologist, 58*(11), 867–873.

Marsh, A. A., & Cardinale, E. M. (2012). When psychopathy impairs moral judgments: Neural responses during judgments about causing fear. *Social Cognitive and Affective Neuroscience, 9*(1), 3–11.

McAuliff, B. D., Nicholson, E., Ravenshanes, D. (2007, March). *Hypothetically speaking: Can expert testimony improve jurors' understanding of developmental differences in suggestibility?* Paper presented at the Biennial Meeting of the Society for Research in Child Development, Boston, MA.

Mills, B., Reyna, V. F., & Estrada, S. (2008). Explaining contradictory relations between risk perception and risk-taking. *Psychological Science, 19*(5), 429–433.

Morsanyi, K., Chiesi, F., Primi, C., & Szűcs, D. (2017). The illusion of replacement in research into the development of thinking biases: The case of the conjunction fallacy. *Journal of Cognitive Psychology, 29*(2), 240–257.

Morsanyi, K., Handley, S. J., & Evans, J. S. (2010). Decontextualised minds: Adolescents with autism are less susceptible to the conjunction fallacy than typically developing adolescents. *Journal of Autism and Developmental Disorders, 40*(11), 1378–1388.

National Registry of Exonerations. (n. d.). *Browse the National Registry of Exonerations.* Law.umich. edu. https://www.law.umich.edu/special/exoneration/Pages/browse.aspx

O'Neill, H. (2001). *The perfect witness.* Death Penalty Information Centre. https://deathpenaltyinfo. org/stories/the-perfect-witness

Otgaar, H., Howe, M. L., Brackmann, N., & van Helvoort, D. H. (2017). Eliminating age differences in children's and adults' suggestibility and memory conformity effects. *Developmental Psychology, 53*(5), 962–970.

Pennington, N., & Hastie, R. (1986). Evidence evaluation in complex decision-making. *Journal of Personality and Social Psychology, 51*(2), 242–258.

Pennington, N., & Hastie, R. (1992). Explaining the evidence: Tests of the story model for juror decision-making. *Journal of Personality and Social Psychology, 62*(2), 189–206.

Peters, E. (2020). *Innumeracy in the wild: Misunderstanding and misusing numbers.* Oxford University Press.

Portnoy, D. B., Roter, D., & Erby, L. H. (2010). The role of numeracy on client knowledge in BRCA genetic counseling. *Patient Education and Counseling, 81*(1), 131–136.

Reyna, V. F. (2000). Fuzzy-trace theory and source monitoring: An evaluation of theory and false-memory data. *Learning and Individual Differences, 12*(2), 163–175.

Reyna, V. F. (2008). A theory of medical decision making and health: Fuzzy trace theory. *Medical Decision Making, 28*(6), 850–865.

Reyna, V. F. (2012). A new intuitionism: Meaning, memory, and development in fuzzy-trace theory. *Judgment and Decision Making, 7*(3), 332–359.

Reyna, V. F. (2021). A scientific theory of gist communication and misinformation resistance, with implications for health, education, and policy. *Proceedings of the National Academy of Sciences of the United States of America, 118*(15), e1912441117.

Reyna, V. F., & Brainerd, C. J. (1995). Fuzzy-trace theory: An interim synthesis. *Learning and individual Differences, 7*(1), 1–75.

Reyna, V. F., & Brainerd, C. J. (2011). Dual processes in decision making and developmental neuroscience: A fuzzy-trace model. *Developmental Review, 31*(2), 180–206.

Reyna, V. F., Brainerd, C. J., Chen, Z., & Bookbinder, S. H. (2021). Explaining risky choices with judgments: Framing, the zero effect, and the contextual relativity of gist. *Journal of Experimental Psychology: Learning, Memory, and Cognition, 47*(7), 1037–1053.

Reyna, V. F., Broniatowski, D. A., & Edelson, S. (2021). Viruses, vaccines, and COVID-19: Explaining and improving risky decision making. *Journal of Applied Research in Memory and Cognition, 10*(4), 491–509.

Reyna, V. F., Chick, C. F., Corbin, J. C., & Hsia, A. N. (2014). Developmental reversals in risky decision making, intelligence agents show larger decision biases than college students. *Psychological Science, 25*(1), 76–84.

Reyna, V. F., Corbin, J. C., Weldon, R. B., & Brainerd, C. J. (2016). How fuzzy-trace theory predicts true and false memories for words, sentences, and narratives. *Journal of Applied Research in Memory and Cognition, 5*(1), 1–9.

Reyna, V. F., & Ellis, S. C. (1994). Fuzzy-trace theory and framing effects in children's risky decision making. *Psychological Science, 5*(5), 275–279.

Reyna, V. F., Estrada, S. M., DeMarinis, J. A., Myers, R. M., Stanisz, J. M., & Mills, B. A. (2011). Neurobiological and memory models of risky decision making in adolescents versus young adults. *Journal of Experimental Psychology: Learning, Memory, and Cognition, 37*(5), 1125–1142.

Reyna, V. F., & Farley, F. (2006). Risk and rationality in adolescent decision making: Implications for theory, practice, and public policy. *Psychological Science in the Public Interest, 7*(1), 1–44.

Reyna, V. F., Hans, V. P., Corbin, J. C., Yeh, R., Lin, K., & Royer, C. E. (2015). The gist of juries: Testing a model of damage award decision-making. *Psychology, Public Policy and Law, 21*(3), 280–294.

Reyna, V. F., Helm, R. K., Weldon, R. B., Shah, P. D., Turpin, A. G., & Govindgari, S. (2018). Brain activation covaries with reported criminal behaviors when making risky choices: A fuzzy-trace theory approach. *Journal of Experimental Psychology: General, 147*(7), 1094–1109.

Reyna, V. F., Holliday, R., & Marche, T. (2002). Explaining the development of false memories. *Developmental Review, 22*(3), 436–489.

Reyna, V. F., & Kiernan, B. (1994). Development of gist versus verbatim memory in sentence recognition: Effects of lexical familiarity, semantic content, encoding instructions, and retention interval. *Developmental Psychology, 30*(2), 178–191.

Reyna, V. F., & Kiernan, B. (1995). Children's memory and metaphorical interpretation. *Metaphor and Symbol, 10*(4), 309–331.

Reyna, V. F., & Lloyd, F. J. (2006). Physician decision making and cardiac risk: Effects of knowledge, risk perception, risk tolerance, and fuzzy processing. *Journal of Experimental Psychology: Applied, 12*(3), 179–195.

Reyna, V. F., & Mills, B. A. (2014). Theoretically motivated interventions for reducing sexual risk-taking in adolescence: A randomized controlled experiment applying fuzzy-trace theory. *Journal of Experimental Psychology: General, 143*(4), 1627–1648.

Reyna, V. F., Mills, B., Estrada, S., & Brainerd, C. J. (2007). False memory in children: Data, theory, and legal implications. In M. P. Toglia, J. D. Read, D. F. Ross, & R. C. L. Lindsay (Eds.), *The handbook of eyewitness psychology. Vol. 1. Memory for events* (pp. 479–507). Lawrence Erlbaum Associates Publishers.

Reyna, V. F., Nelson, W. L., Han, P. K., & Dieckmann, N. F. (2009). How numeracy influences risk comprehension and medical decision making. *Psychological Bulletin, 135*(6), 943–973.

Reyna, V. F., Wilhelms, E. A., McCormick, M. J., & Weldon, R. B. (2015). Development of risky decision making: Fuzzy-trace theory and neurobiological perspectives. *Child Development Perspectives, 9*(2), 122–127.

Ross, D. F., Marsil, D. F., Benton, T. R., Hoffman, R., Warren, A. R., Lindsay, R. C. L., & Metzger, R. (2006). Children's susceptibility to misidentifying a familiar bystander from a lineup: When younger is better. *Law and Human Behavior, 30*(3), 249–257.

Simons, D. J., & Chabris, C. F. (2011). What people believe about how memory works: A representative survey of the US population. *PloS One, 6*(8), e22757.

Singer, M., & Remillard, G. (2008). Veridical and false memory for text: A multiprocess analysis. *Journal of Memory and Language, 59*(1), 18–35.

Semmler, C., Brewer, N., & Bradfield-Douglass, A. (2011). Jurors believe eyewitnesses. In *Conviction of the Innocent: Lessons from Psychological Research* (pp. 185–211). American Psychological Association.

Stahl, C., & Klauer, K. C. (2008). A simplified conjoint recognition paradigm for the measurement of gist and verbatim memory. *Journal of Experimental Psychology: Learning, Memory, and Cognition, 34*(3), 570–586.

Stahl, C., & Klauer, K. C. (2009). Measuring phantom recollection in the simplified conjoint recognition paradigm. *Journal of Memory and Language, 60*(1), 180–193.

Stanovich, K. E., Toplak, M. E., & West, R. F. (2008). The development of rational thought: A taxonomy of heuristics and biases. In *Advances in child development and behavior* (Vol. 36, pp. 251–285).

Stanovich, K. E., & West, R. F. (2008). On the relative independence of thinking biases and cognitive ability. *Journal of Personality and Social Psychology, 94*(4), 672–695.

Steinberg, L. (2007). Risk taking in adolescence: New perspectives from brain and behavioral science. *Current Directions in Psychological Science, 16*(2), 55–59.

Thaler, R. H., & Sunstein, C. R. (2008). *Nudge: Improving decisions about health, wealth, and happiness.* Yale University Press.

Toglia, M. P., & Berman, G. L. (2021, 30 August). *Convicted by memory, exonerated by science.* APS Observer. https://www.psychologicalscience.org/observer/convicted-memory

Tversky, A., & Kahneman, D. (1981). The framing of decisions and the psychology of choice. *Science, 211*(4481), 453–458.

Tversky, A., & Kahneman, D. (1986). Rational choice and the framing of decisions. *Journal of Business, 59*(4), S251–S278.

Van den Broek, P. (2010). Using texts in science education: Cognitive processes and knowledge representation. *Science, 328*(5977), 453–456.

Wixted, J. T., & Wells, G. L. (2017). The relationship between eyewitness confidence and identification accuracy: A new synthesis. *Psychological Science in the Public Interest, 18*(1), 10–65.

Wolfe, C. R., & Reyna, V. F. (2010). Semantic coherence and fallacies in estimating joint probabilities. *Journal of Behavioral Decision Making, 23*(2), 203–223.

Wolfe, C. R., Reyna, V. F., Widmer, C. L., Cedillos, E. M., Fisher, C. R., Brust-Renck, P. G., & Weil, A. M. (2015). Efficacy of a web-based intelligent tutoring system for communicating genetic risk of breast cancer. A fuzzy-trace theory approach. *Medical Decision Making, 35*(1), 46–59.

Zimmerman, D. M., & Hunter, S. (2018). Factors affecting false guilty pleas in a mock plea bargaining scenario. *Legal and Criminological Psychology, 23*(1), 53–67.

6

Episodic future thinking, memory, and decision-making

From theory to application

Adam Bulley and Daniel L. Schacter

Introduction: the mechanisms and functions of future thinking

One of the fastest growing areas in the cognitive sciences over the past two decades has been the study of *prospection*—the capacity to think about and imagine
the future. This topic has sparked interest in comparative, developmental, social, and clinical psychology, cognitive neuroscience, philosophy, and in diverse
branches of the social sciences. Prospection is an umbrella term that captures
a range of future-oriented cognitive processes, from making predictions about
what might happen tomorrow, through forming intentions to learn a new skill, to
planning for retirement (Gilbert & Wilson, 2007; Suddendorf & Corballis, 2007;
Szpunar et al., 2014). It also captures the ability to specifically imagine or simulate events that might occur in one's own personal future—a prospective capacity for episodic future thinking (Atance & O'Neill, 2001; Schacter et al., 2017;
Szpunar, 2010). In this chapter, we concentrate on the role of episodic future
thinking in decision-making. Our aim is to shed light on how a deeper theoretical understanding of its underlying cognitive mechanisms can inform efforts at
application—particularly when it comes to making choices with outcomes that
play out only over time.

Various lines of research on the underlying mechanisms of prospection have
demonstrated that future-oriented cognitive processes rely on memory for past
experiences. These lines of evidence include those from developmental psychology, where common developmental trajectories have been observed for memory
and prospection (Atance & O'Neill, 2001; Busby & Suddendorf, 2005; Suddendorf
& Redshaw, 2013); neuropsychology, where common impairment has substantiated the claim for a shared neurocognitive substrate of mental time travel into past
and future (Hassabis et al., 2007; Klein et al., 2002; Tulving, 1985); and cognitive

Adam Bulley and Daniel L. Schacter, *Episodic future thinking, memory, and decision-making* In: *Memory in Science for Society*.
Edited by: Robert H. Logie, Zhisheng (Edward) Wen, Susan E. Gathercole, Nelson Cowan, and Randall W. Engle, Oxford University Press.
© Oxford University Press 2023. DOI: 10.1093/oso/9780192849069.003.0006

psychology, where various studies have revealed similarities in the episodic specificity and phenomenal characteristics of remembered past and imagined future events (Addis et al., 2008; D'Argembeau & Van Der Linden, 2004; Madore et al., 2014). In cognitive neuroscience, neuroimaging research has revealed a striking correspondence in neural activity when people are asked to remember specific past events and when they are asked to imagine possible future events (Addis et al., 2007; Benoit & Schacter, 2015; Okuda et al., 2003; Schacter et al., 2007; Szpunar et al., 2007; Thakral et al., 2020). Further research has begun to chart out the mechanisms of episodic future thinking in terms of its interface with emotion (Barsics et al., 2016; Jing et al., 2019), the nature of its reliance on the semantic memory system (Irish, 2016; Irish & Vatansever, 2020), and its foundations in representations of possibility (De Brigard & Parikh, 2019; Leahy & Carey, 2020; Mahr, 2020; Redshaw & Suddendorf, 2020) or metarepresentation (Ernst et al., 2019; Redshaw, 2014; Suddendorf, 1999). Theoretical accounts of the mechanisms involved in episodic future thinking have, in turn, informed research into what episodic future thinking does and what it is for: its numerous functions (D'Argembeau et al., 2011; Schacter, 2012, 2021).

The putative functions of episodic future thinking has become a particularly intensive area of research, both from an evolutionary perspective, and in terms of everyday life (for recent reviews, see Bulley & Irish, 2018; Bulley et al., 2020; Schacter et al., 2017; Suddendorf et al., 2018). Among others, these identified functions of episodic future thinking include emotion regulation (Jing et al., 2016; MacLeod, 2016), facilitating prospective memory (Neroni et al., 2014; Spreng et al., 2018), problem-solving and goal pursuit (Ernst & D'Argembeau, 2017; Gerlach et al., 2014; Madore & Schacter, 2014), deliberately acquiring skills and knowledge (Davis et al., 2015; Suddendorf et al., 2016, 2018), social coordination and prosociality (Boyer, 2008; Gaesser, 2020; Gaesser & Schacter, 2014), anticipating emotions (Barsics et al., 2016; Gilbert & Wilson, 2007), preparing for threats (Bulley, Henry, & Suddendorf, 2017; Miloyan et al., 2018), contingency planning for mutually exclusive possibilities (Bulley et al., 2020; Redshaw & Suddendorf, 2020), compensating for anticipated performance limits (Bulley et al., 2020), and generating a sense of temporal self-continuity or personal identity (D'Argembeau, 2016, 2020; Irish et al., 2019; Strikwerda-Brown et al., 2019). This burgeoning research has underscored that episodic future thinking is fundamental to adaptive behaviour.

While each of the aforementioned functions has attracted intensive research effort in its own right, the study of one function has emerged as a target of particularly focused attention due to its clear applied promise: the role of episodic future thinking in making flexible decisions that take delayed consequences into account. We first introduce the theoretical background relevant to decision-making involving future consequences, before turning to new research that uses this theoretical foundation to apply the science of prospection.

Making trade-offs between sooner and later: introducing intertemporal choice

Intertemporal choices are those with consequences that play out only over time, involving trade-offs between sooner and later outcomes (Kable, 2014; Loewenstein & Elster, 1992). Typically, the value of a delayed outcome is discounted relative to the value of an outcome in the present (Berns et al., 2007; Ericson & Laibson, 2019; Kable, 2014)—a phenomenon also ubiquitous in non-human animals (for reviews, see Redshaw & Bulley, 2018; Stevens & Stephens, 2008). This *delay discounting* occurs for various outcomes but is most commonly studied in the context of reward. The relationship between the time to the receipt of a delayed reward and the subjective value of that reward can be described by a *delay discounting function*, which, for a given individual, captures how rapidly rewards lose their subjective value with increasing delays. A range of research demonstrates that delay discounting differs substantially between people (Anokhin et al., 2011; Cavagnaro et al., 2016; Koffarnus et al., 2013; Peters & Büchel, 2011), even if there are general principles by which intertemporal decision-making operates. The most common tools in charting both these individual differences and general principles have been *intertemporal choice tasks*, wherein participants make a series of decisions between options (usually, but not always, monetary) that differ both in magnitude and delay (Ericson & Laibson, 2019; Lempert et al., 2019).

In classical economic theory, human decision-makers faced with intertemporal trade-offs choose in a manner that maximizes some consistent utility function (Samuelson, 1937). Following this principle, the value of a reward would be discounted to the same extent for each additional unit increase in the delay to its receipt. An *exponential discounting function* satisfies this criterion. However, economists, psychologists, and other scholars attempting to make sense of intertemporal decision-making have long contended with its inconsistencies (Strotz, 1955; Thaler & Shefrin, 1981) and anomalies (Loewenstein & Prelec, 1992), noting how frequently decision-makers deviate from principles of normative economic rationality (Ainslie, 2005). For instance, the nature of the discounting function appears to depend on how far away in time a decision-maker is from a reward. This stands in contrast to the classical utility idealization of an individual, depicted as far back as Samuelson (1937), 'whose tastes maintain a certain invariance throughout the time under consideration'.[1]

A famous example of a utility function changing based on the passage of time is of the dinner guest who initially intends to have no dessert in order to stick to a diet.

---

[1] Samuelson also warned of the psychological implausibility of assumptions underlying the exponential discounted utility conceptualization. Later thinkers also cast doubt on this framework, and perspectives on intertemporal decision-making shifted decidedly away from that view. As George Loewenstein and Jon Elster put it in the preface to the 1992 book *Choice Over Time* (Loewenstein & Elster, 1992): 'The elegance of the economic approach [to intertemporal choice] inheres in its simplicity; its main drawback is its low congruence with reality' (p. x).

The delayed value of adhering to the diet is much more substantial than the prospect of the cheesecake. After dinner, now that the cheesecake has become a more imminent possibility, its value topples that of the prior intention. In intertemporal choice tasks, an analogous pattern is observed when participants choose for example *$50 in 12 months* over *$40 in 11 months*—they are willing to wait for the extra $10. Then, in a separate decision, they choose *$40 today* over *$50 in 1 month*; they would prefer the immediate reward. In both cases, the delay between the options is identical (1 month), as is the difference in magnitude ($10). The key difference is that one question concerns distant prospects, but the other refers to a more immediate opportunity. Thus the discounting function itself appears to change depending on how far away the rewards are in time—a form of what economists call *dynamic inconsistency* (Thaler, 1981).

Dynamic inconsistency and other empirical violations of the predictions from the classical economic view eventually led to the now far more popular conception of hyperbolic discounting (Ainslie, 1975, 2005; Frederick et al., 2002; Mazur, 1987; Thaler, 1981). In *hyperbolic discounting*, the rate of discounting itself drops off for rewards further away in time. At short delays, rewards very rapidly lose their subjective value. At longer delays, the decline in value is less pronounced. In the standard hyperbolic discounting framework, a single free parameter in a discounting equation (often labelled 'k') represents the discounting rate (Odum, 2011). A larger value of this k parameter indicates a 'steeper' discounting function, with rewards more quickly losing their value with delay to their receipt. A smaller k parameter reflects 'shallower' discounting, with the subjective value of rewards being less affected by delay (Figure 6.1).

Aside from standard exponential and hyperbolic discounting functions, a wide range of other models of human intertemporal choice have proliferated in the literature (T. Ballard et al., 2020; Cavagnaro et al., 2016; van den Bos & McClure, 2013). Many of these models have arisen in an attempt to capture various empirical deviations from the classical exponential discounting view, while still others have been motivated by further theoretical or empirical considerations, such as the *quasi-hyperbolic discounting* ('*beta-delta*') model popular in economics. The quasi-hyperbolic modelling approach uses a standard exponential function (*delta*), but adds a term to capture a purported bias towards rewards available in the present (*beta*) (Laibson, 1997). In this framework, decision-makers may therefore be seen as having two competing valuation regimes, one that weighs immediate outcomes heavily and one that evaluates delayed outcomes in a time-consistent manner (McClure et al., 2004), paralleling ideas of 'hot' and 'cold' motivational processes from psychology (Metcalfe & Mischel, 1999). Economists have taken to using the 'present bias' *beta* parameter as an index of self-control because it reflects the extent to which a decision-maker is swayed towards immediately tempting rewards (for a recent meta-analysis of present-bias in decision-making, see Cheung et al., 2021).

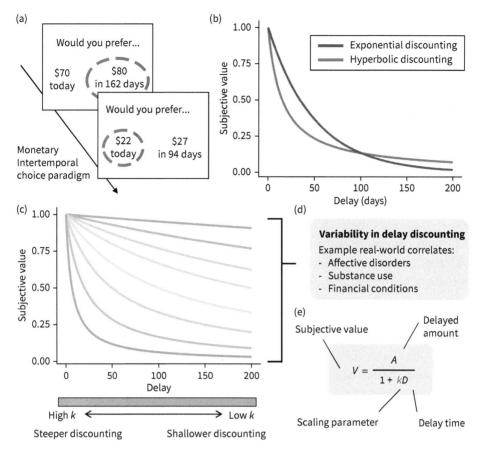

Figure 6.1 Delay discounting: general principles and individual variability. (a) In an intertemporal choice task, participants make multiple decisions about outcomes (usually monetary rewards) that vary both in magnitude and delay. (b) Exponential discounting means delayed rewards lose their value in a time-consistent manner, with each unit delay reducing subjective value to the same extent. In hyperbolic discounting, rewards lose value rapidly with initial increases in delay but then more slowly as the time horizon increases. (c) The steepness of hyperbolic discounting varies widely both between and within individuals, with a larger '*k*' parameter representing steeper discounting. (d) Variability in discounting correlates with real-world outcomes from psychopathology to retirement savings. (e) The hyperbolic discounting equation showing how the subjective value of a reward can be modelled as a function of the objective reward amount divided by the time to its receipt multiplied by the *k* scaling parameter.

The idea that two separate systems are involved in making intertemporal choices has a long history (Thaler & Shefrin, 1981). So-called *dual-process theories* have tended to contrast a fast, reactive, impulsive, automatic, or emotional system that generally prioritizes immediate gratification, with a slow, reflective, deliberative, controlled, or rational one that generally prioritizes long-term outcomes or goals

(Heatherton & Wagner, 2011; Hofmann et al., 2009; Loewenstein et al., 2015; McClure & Bickel, 2014; McClure et al., 2004). In dual-process accounts, variants of these two systems compete to control behaviour. Applied to intertemporal decision-making, this perspective has been used to cast a tendency to choose larger, later rewards (shallower discounting) as indicative of good *self-control*, while a tendency to choose smaller, sooner rewards (steeper discounting) has been seen as a failure to resist temptation.

The notion that a delay discounting parameter in intertemporal choice therefore measures self-control has become a mainstay in the psychology literature.[2] Adopting this theoretical stance leads to interpretations of an individual's discounting factor, such as their 'k' value from the hyperbolic model, as measuring something like their *capacity to control* their impulsive urges in favour of long-term outcomes. Later in this chapter, however, we will highlight a number of empirical findings that highlight why the discounting factor from an intertemporal choice task is an inadequate measure of self-control. To foreshadow, consider an individual who has no superordinate goal to delay gratification, say because they feel they have earned some fun after working hard recently towards a savings goal (Vosgerau et al., 2020). This individual is not *trying* to delay gratification, and will readily choose immediate rewards in line with their particular preference function (Keinan & Kivetz, 2008; Loewenstein, 2018; Vosgerau et al., 2020). In what sense should it be called a failure of self-control when this person chooses immediate rewards and hence has a steep discounting function? As Lempert et al. (2019) point out, 'Just as a person who is not addicted to cigarettes does not need to exert control to avoid smoking, a person with a low discounting rate may not need to exert self-control to choose a delayed reward.'

An alternative to the self-control framing of discounting in intertemporal choice can be found in value-based decision-making accounts (Berkman et al., 2017; Buckholtz, 2015; Cosme et al., 2019; Kable, 2014; Levy & Glimcher, 2012; Rangel et al., 2008). From this perspective, outcomes are evaluated in terms of their subjective value according to their various features or attributes, and a time-discounted value is thereby assigned to the options. A decision process translates the time-discounted subjective value of options into a choice, which thereby lines up, albeit imperfectly, with an underlying preference function. A value-based decision-making view of intertemporal choice does not commit to the discounting parameter from an intertemporal choice task as measuring a capacity for self-control, but rather as a measure of an individual's preference function as expressed in their decisions.

---

[2] Consider some excerpts: 'self-control [refers] only to the choice of a larger, more delayed reinforcer over a smaller, less delayed reinforcer' (Logue, 1988); 'Individuals produce small k values by forgoing small rewards now for more lucrative LLRs [larger, later rewards], and thus are considered to be self-controlled' (Critchfield & Kollins, 2001); 'Lower delay discounting (better self-control) is linked to higher intelligence' (Shamosh et al., 2008); 'some people value future consequences very highly, which, according to the discounting model of impulsive choice . . . tends to make them more self-controlled' (Kirby, 2009); 'Discounting paradigms quantify the ability to refrain from preference of immediate rewards, in favour of delayed, larger rewards' (Christakou et al., 2011); '[the] ability to delay gratification is captured by individual differences in so-called intertemporal choices' (Pornpattananangkul et al., 2017); 'self-control may be conceptualized as the set of processes that support choosing delayed rewards, particularly in instances when immediate rewards are subjectively highly valued' (Turner et al., 2019).

Adopting these different theoretical accounts of intertemporal decision-making has substantial implications for research applications into real-world decision-making domains (for further discussion, see Shenhav, 2017).[3]

Before we address those applications further, however, it is important to note that even if the delay discounting parameter does not measure self-control per se, it is still a meaningful and consequential variable. Indeed, steeper delay discounting has been associated with a wide range of critical outcomes that involve trade-offs over time (Shenhav et al., 2017). These outcomes include psychopathologies, including, perhaps most prominently, various substance use disorders (for a recent high-powered meta-analysis, see Amlung et al., 2019). A common reading of such associations is that steeper delay discounting, even as revealed in a monetary intertemporal choice task, reflects a general cross-domain propensity to devalue later consequences relative to sooner ones. In the context of a substance use disorder, for instance, the immediate benefits of consumption (e.g. hedonic pleasure, alleviating negative affect) could come to subjectively outweigh the delayed benefits of sobriety (e.g. familial repercussions, health effects). Given that steeper delay discounting has been observed not only in substance use disorders but across a wide range of mental health and psychiatric conditions, there have been calls for it to be considered a critical 'trans-disease process' (Bickel et al., 2012, 2019).

Naturally, the trans-disease and welfare implications of elevated delay discounting have in turn led intertemporal decision-making to become a key target for intervention in many fields, including in clinical psychology.[4] The success of such targeted approaches to reducing delay discounting rests on the discovery that intertemporal decision-making is highly malleable within a given individual, being sensitive to choice framing, environmental circumstances, mindset, and so forth (Lempert & Phelps, 2016). While many applied intervention approaches have taken aim at modifying monetary intertemporal decision-making, others have directly targeted behaviours supposed to involve a strong intertemporal trade-off, such as cigarette smoking. Among these various intervention approaches, episodic future thinking has emerged as a strong candidate for modifying delay discounting, and we review the promise of such interventions in the subsequent section.

---

[3] More nuanced views on the role of self-control in intertemporal choice emphasize that self-control-related processes may be selectively involved in choosing larger, later rewards specifically when smaller, sooner rewards have a higher time-discounted subjective value, as argued by Figner et al. (2010) or Turner et al. (2019). Alternatively, self-control may be involved in attempts to pre-empt or avoid dynamically inconsistent preference reversals, where an initial preference for a delayed reward flips to a preference for a smaller, sooner reward simply because the two options both become closer in time (Peters & D'Esposito, 2016; Soutschek et al., 2017). By contrast, the self-control views of the delay discounting parameter to which we primarily refer to in the text are those that broadly interpret choices of larger, later reward as indicative of a capacity for self-control, whether that be operationalized in terms of the proportion of choices of larger, later reward (Logue, 1988), a delay discounting factor such as 'k' (Critchfield & Kollins, 2001), or some other metric such as the calculated area under a discounting curve (Jimura et al., 2018).

[4] Outside of the clinical domain, delay discounting is also central to a wide range of applied societal issues including retirement saving efforts (Ericson & Laibson, 2019), benefits distribution (Shapiro, 2005), education policy (Oreopoulos, 2007), credit card borrowing (Angeletos et al., 2001), sustainability initiatives (Hardisty & Weber, 2009), and voter turnout and political participation (Schafer, 2021), among many others.

Applying episodic future thinking to modify intertemporal decision-making

Despite the fact that intertemporal decisions are about outcomes delayed in time, the involvement of episodic future thinking in intertemporal choice has remained contentious (Bulley et al., 2016). On the one hand, various non-human animal species thought to possess far less advanced prospective cognition than humans (Suddendorf, 2013) are nonetheless capable of making intertemporal decisions, albeit in tasks with often dissimilar structures to human monetary choice tasks (Redshaw & Bulley, 2018). Non-human animals can be readily trained to make intertemporal choices through associative learning, which implies a lack of model-based mental simulation in these decision processes (Stevens, 2011; Stevens & Stephens, 2008). For example, a rat might learn to associate various options with the unpleasant feelings evoked by waiting for each of them when chosen. In turn, an intertemporal decision could be executed by comparing the conditioned value associated with each of the options. This would not require any mental simulation of the delay to receipt or projection forward to a forthcoming reward context. Another independent line of evidence casting doubt on the necessity of episodic future thinking in intertemporal choice comes from studies on amnesia due to hippocampal damage. A number of individuals with hippocampal damage exhibit parallel deficits to both remembering the past and imagining the future, but appear nonetheless capable of making intertemporal decisions, and even exhibit discounting functions within the range of healthy controls (Kwan et al., 2012, 2013).

On the other hand, systems-level accounts of intertemporal decision-making now routinely include the core network involved in remembering the past and imagining the future, alongside systems responsible for valuation and executive control (Lempert et al., 2019; Peters & Büchel, 2011; Sellitto et al., 2011). These theoretical accounts are drawn in part from extensive neuroimaging investigations of intertemporal choice which implicate each of these networks. Recent models in economics (Gabaix & Laibson, 2017) and computational neuroscience (Gershman & Bhui, 2020) similarly implicate episodic future thinking as fundamental to making even simple intertemporal decisions. In each of these models, noisy prospective value estimation from simulating future outcomes produces canonical delay discounting, even for an agent who has no underlying preference for immediate rewards. Assuming a prior estimate of the value of delayed rewards near zero, the estimate of a delayed reward value can decline for rewards as those rewards move further away in time and are thus harder to anticipate. In this picture, delayed rewards become less valuable because they are harder to foresee clearly. Such models assume delay discounting relies on the capacity to estimate the value of delayed rewards via some kind of error-prone prospective simulation. Earlier models make similar assumptions about the centrality of episodic future thinking, for instance, that delay discounting occurs in part because rewards are progressively harder to find and evaluate with mental simulation when located further away in time (Kurth-Nelson et al., 2012).

In an influential review, Boyer (2008) argued that one core evolutionary function of mental time travel is that it provides a motivational counterweight to immediate opportunities, and thereby encourages more 'restrained' decisions with long-run benefits. From this view, episodic future thinking would facilitate shallower discounting by nudging decision-makers away from immediately gratifying but ultimately detrimental choices (Boyer, 2008). This view does not necessarily commit to episodic future thinking being crucial for making intertemporal decisions in general. The view does predict, however, that episodic future thinking could bias choices towards larger, later options when evoked (for related accounts, see Hoerl & McCormack, 2019; Noël et al., 2017).

A wide body of research has since tested and found support for this prediction. Starting with significant initial contributions by Peters and Büchel (2010) and Benoit et al. (2011), various studies have directly cued participants to engage in episodic future thinking while making intertemporal choices—both monetary, and those pertaining to health trade-offs (Rung & Madden, 2018b) (Figure 6.2). In the monetary choice domain, this cueing procedure leads to marked reductions in delay discounting. In the health domain, the procedure leads to analogous shifts away from immediate consumption of calorie-rich foods, cigarettes, and alcohol in favour of longer-term outcomes such as adherence to dietary goals. Illustrating the wide excitement around this procedure, in a recent meta-analysis, Rösch et al. (2022) derive 174 effect sizes from 48 independent articles, all published in the 10 years after the initial investigations mentioned above. This wide-ranging meta-analysis supports the hypothesis that the cueing effects occur on the basis of episodic future thinking, and not via some alternative mechanism such as semantic future thinking, and also finds strong evidence against a demand-characteristics explanation for the findings (Rung & Madden, 2018a).

The Rösch et al. (2022) meta-analysis reveals a number of other important results, including a significant influence of key moderating variables. For instance, the cueing effect appears to have a more pronounced effect on reducing delay discounting for emotionally positive rather than neutral or negative events (see discussion in Ballance et al., 2022; Bulley et al., 2019). This meta-analytic conclusion is derived from a range of estimates, some of which include a reversed pattern (steeper delay discounting) when people are cued to imagine negative future events (Figure 6.2c). We will return in the next section to a discussion of situations in which episodic future thinking might encourage a greater preference for immediate rewards (see also Bulley et al., 2016; Bulley & Schacter, 2020). The Rösch et al. (2022) meta-analysis further shows that the episodic cueing effect is stronger when cued future events are imagined more vividly, and when they are directly related to the delayed reward option (e.g. imagining an upcoming meal with friends while deciding whether to eat currently available chocolate chip cookies) (Dassen et al.,

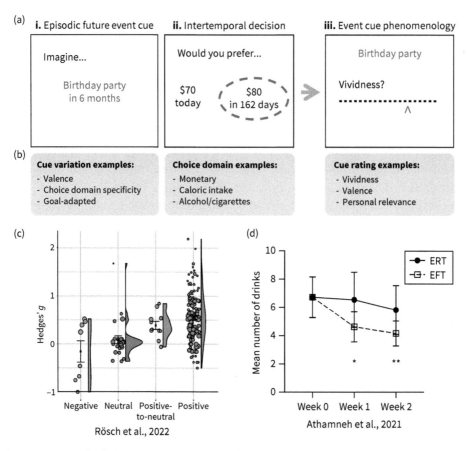

Figure 6.2 Episodic future event cueing reduces delay discounting. (a) In a standard episodic future event cueing study, participants receive cues pertaining to specific future events before (or while) making each decision in an intertemporal choice task. (b) Dozens of studies have adapted this basic paradigm with variations on the event cues, choice domain (monetary versus actual commodities), and subsequent event cue ratings. (c) A recent meta-analysis by Rösch et al. (2022) clearly reveals a graded role for the valence of cued events, with positive cues reducing delay discounting significantly more (greater Hedges' *g* effect size) than neutral or negative ones: each dot shows the observed effect size for a single study for the respective moderator level. Studies with larger samples (i.e. greater precision) are displayed with larger dots. Black squares indicate the mean, whiskers the standard error. Copyright © 2022, American Psychological Association. Reproduced with permission. (d) Athamneh et al. (2021) report that a remotely delivered episodic future cueing procedure leads to reductions in mean self-reported alcohol consumption in individuals with alcohol use disorder. Error bars represent 95% confidence intervals. * *p* <0.05 compared to week 0, ** *p* <0.01 compared to week 0. Copyright © 2021, American Psychological Association. Reproduced with permission.

2016). These features strongly support the argument that simulating future events shifts delay discounting by enabling higher-fidelity or more highly weighted representations of delayed reward values, rather than via some generic change in future orientation (Schacter et al., 2017).

Neuroimaging research has indicated a role for the hippocampus in the episodic cueing effect, and, in particular, coupling between the medial temporal lobe system and medial regions of prefrontal cortex involved in valuation (Benoit et al., 2011; Peters & Büchel, 2010; Sasse et al., 2015). The potential dependence of the episodic cueing effect on the medial temporal lobe system is supported by evidence that individuals with hippocampal damage may not be as susceptible to the effect as healthy controls (Palombo et al., 2015). Nonetheless, the cueing procedure may be effective in individuals with medial temporal lobe damage when using highly personalized cues that enable patients to draw on existing (perhaps semantic) representations of familiar contexts (Kwan et al., 2015; Palombo et al., 2016). This interpretation accords with an functional magnetic resonance imaging study that found the episodic cueing effect was less dependent on the hippocampus when participants imagined highly familiar events that could plausibly rely more on existing knowledge (Sasse et al., 2015).

The central role of the medial temporal lobes in the episodic cueing effect also accords with value-based accounts of decision-making that suggest participants sample from memory in order to make predictions about the relative value of options at a decision point (Shadlen & Shohamy, 2016). This process of evidence accumulation by internal sampling may explain the dependence of the hippocampus on decision-making tasks ostensibly unrelated to memory (Biderman et al., 2020). In the case of intertemporal decision-making, one possibility is that the cueing effect calls upon the hippocampus and other regions of the core network to bias evidence accumulation towards 'samples' (memories and future simulations) that favour the larger, later reward, increasing its relative value as represented in valuation hubs of the prefrontal cortex.

While the specific mechanisms of the episodic cueing effect have yet to be fully determined, this basic science research on the procedure has meanwhile contributed to wide-ranging efforts to apply it outside the laboratory. Indeed, perhaps the most promising aspect of the Rösch et al. (2022) meta-analysis is the finding that the episodic cueing effect is more pronounced in populations characterized by behavioural health patterns that imply steep discounting of the future, such as obesity and substance use disorders. This finding suggests the procedure can be successfully applied in populations where shifting decisions away from immediate reinforcement and towards larger, later outcomes is a high behavioural health priority (Bickel et al., 2019; Mellis et al., 2019). These applications include reducing delay discounting in individuals with health conditions such as pre-diabetes (Bickel et al., 2020), where steep delay discounting is associated with behaviours that pose a risk of disease progression to type 2 diabetes (e.g. poorer medication adherence, eating a lower-quality diet, less physical exercise) (Epstein et al., 2021). Intervention efforts have also extended beyond indices of delay discounting to directly target health behaviours in relevant populations, where it has been found that the episodic cueing effect can reduce caloric intake in overweight or obese individuals (Daniel et al., 2013; O'Neill et al., 2016), reduce alcohol demand in people with alcohol dependence (Snider et al., 2016), and reduce cigarette consumption in smokers (Stein et al., 2016).

This applied research has also generated new insights into increasing the efficacy of the episodic cueing protocol. In one recent online study, the episodic cueing procedure was adapted so that cued events were linked explicitly to participants' personal health goals (Athamneh et al., 2022), drawing on research demonstrating the centrality of episodic future thinking in supporting goal-related processes (D'Argembeau et al., 2010; O'Donnell et al., 2017). In this online study, the goal-adapted cueing procedure was more effective than the standard future thinking manipulation (which pertains to general positive future events) at reducing both cigarette demand in individuals who smoke, as well as fast-food demand in individuals with obesity (Athamneh et al., 2022).

Directly applied clinical research studies have recently begun piloting the feasibility and scalability of the episodic cueing procedure for real-world delivery, with enthusiasm growing for possible translation into the front-line treatment of delay discounting-related psychopathologies. In one recent randomized field trial, Athamneh et al. (2021) sought to deliver an episodic future thinking-based intervention for people who met the *Diagnostic and Statistical Manual of Mental Disorders* (Fifth Edition) criteria for moderate or severe alcohol use disorder, and who aimed to drink less or abstain from drinking altogether. Participants generated positive, specific, and vivid events they were looking forward to in 2 weeks', 1 month's, 6 months', and 1 year's time. During the real-world intervention phase of the experiment, remotely delivered episodic future thinking prompts (or control 'episodic recent thinking' prompts) were delivered twice daily to participants via text. The cueing procedure led to a marked reduction in both delay discounting and self-reported real-world alcohol consumption (Figure 6.2d). Interestingly, in support of the idea that episodic future thinking targets delay discounting processes common to different modalities (i.e. money vs alcohol), changes in discounting rate after the intervention predicted changes in alcohol consumption. This 'proof-of-concept' real-world alcohol use disorder trial is among a number of recent efforts to take developments in the science of prospection into clinical and other applied domains and represents an encouraging step towards the development of a trans-diagnostic episodic future thinking-based clinical tool.[5]

---

[5] There are a number of other recent lines of evidence that suggest a key role for future thinking in shifting intertemporal decisions towards larger, later rewards without directly manipulating or cueing episodic future thinking. These include economic field research showing how imposing waiting periods on decisions reduces delay discounting, perhaps via providing scope for deliberation about the delayed benefits of the larger, later choice (Brownback et al., 2019; Imas et al., 2018). Other work targets the imaginability of one's future self, for instance, by presenting participants with age-progressed images or videos of their own face, in an effort to encourage shallower discounting of future outcomes and bolster retirement saving (Hershfield et al., 2011, 2018).

Two cautions: reducing delay discounting is not always desirable and increasing farsightedness does not equate to reducing delay discounting

Despite much progress, various important open questions and caveats remain regarding the episodic cueing effect. For instance, episodic future thinking cues do not appear to affect decision-making in domains including risk-taking (Bulley et al., 2019), or probability discounting (Yin Mok et al., 2020), which suggests the effect may have a fairly narrow influence over only those trade-offs with an explicit temporal structure. Another limit pertains to the particular groups who stand to benefit from the protocol. For instance, older adults do not appear to derive as much of a shift in their delay discounting as a result of the cueing procedure (Sasse et al., 2017), possibly due to a disruption of control processes or difficulty generating episodic future thoughts when cued (for a review of age-related changes to episodic future thinking, see Schacter et al., 2018). Aside from these issues of transferability and generalizability, there are broader theoretical concerns that should be noted as the cueing procedure is developed as a clinical tool. We note two such concerns, echoing points made by Loewenstein (2018).

First, reducing delay discounting cannot be considered a desirable outcome by default. On the contrary, delaying gratification can be both maladaptive and detrimental to well-being in certain circumstances. To take some now well-worn (but often overlooked) examples, consider first an individual living in a highly volatile environment being offered some delayed return by a party thought to be untrustworthy (Jachimowicz et al., 2017; Kidd et al., 2013; Ma et al., 2018; Michaelson et al., 2013; Mischel, 1974; Pepper & Nettle, 2017). In this case, encouraging a longer-term view may indeed still be fruitful because it would enable a fuller picture of whether it is in fact worth waiting for that payoff (Bulley et al., 2016), but it can undoubtedly be the wrong decision to delay gratification for a reward that has a sufficiently low chance of materializing. For another example, take the case of an individual who is so singularly focused on delaying gratification that they continually forego pleasures in life or miss fleeting opportunities. For such an individual, shallow discounting would result in less general utility from everyday pleasures, perhaps to their overarching detriment. These two examples (an uncertain future and missing out on enjoyments) show that shallow discounting can be both maladaptive and inimical to well-being. These two examples are not unconnected, of course—the situation is all the worse for an individual who patiently waits for later outcomes, all the while missing out on small joys, and then passes away before being able to capitalize on their delayed gratification.[6]

---

[6] This is illustrated in the Academy Award-winning film *Nomadland*, where the protagonist is told a story about an office worker who has saved for decades to enjoy a retirement on the ocean, only to die having never taken his sailboat out of his driveway.

To illustrate the point in a clinical context, consider the restrictive eating subtype of anorexia nervosa (AN). While some studies find little to no differences in delay discounting between healthy controls and individuals with AN, a series of other studies have identified shallower than average discounting in AN (Decker et al., 2015; King & Ehrlich, 2020; King et al., 2016). In the majority of other psychopathologies studied, clinical populations show steeper-than-normal delay discounting (Amlung et al., 2019). One reading of these data is that AN, at least in the acutely ill under-weight stage, can develop into a kind of pathological patience: an unhealthy willing-ness to forego the immediate hedonic pleasures of eating in favour of higher-order goals like a body image ideal (Steinglass et al., 2012, 2017). While the evidence is still equivocal on the specific cognitive and neural instantiation of delay discounting in AN (King & Ehrlich, 2020), attempting to reduce delay discounting would almost certainly be the wrong approach. This reveals a blind spot: essentially every clin-ical intervention tool in the intertemporal choice literature attempts to *reduce* delay discounting, including all work we know of thus far on the episodic future thinking manipulation (Lempert et al., 2019).

The second caution about implementing the episodic future cueing procedure in applied settings is that greater farsightedness might sometimes paradoxically en-courage immediate gratification. There are numerous, well-documented reasons why increased prospection could lead to the prioritization of immediate rewards (Bulley, Pepper, & Suddendorf, 2017; McGuire & Kable, 2013; Shaddy et al., 2021), though this point is often overlooked in views of intertemporal decision-making which tend to equate greater farsightedness with a preference for delayed gratifica-tion. Take the phenomenon of anticipated regret. Consumers sometimes foresee the regret they will feel in missing out on life's pleasures if they forego immediate grati-fication, and will seek to rectify this through intentional indulgence in the present (Keinan & Kivetz, 2008). In some cases, consumers will even pre-commit to indul-gence, such as by choosing to take raffle winnings in the form of a massage vou-cher rather than as cash, thereby locking in the hedonic option and short-circuiting their own tendency to buy boring but essential items (Kivetz & Simonson, 2002). Or, having appreciated that they live in uncertain times, a decision-maker might run through their options in great, farsighted detail and thereby realize that their optimal decision is to take what they can get now (Augenblick et al., 2016; Lee & Carlson, 2015). For instance, given the substantial uncertainty around the length of delays to future payoffs common in everyday life, the farsighted strategy can be to cease waiting altogether when it becomes apparent that a delay is likely to be far longer than one originally expected (Lempert et al., 2018; McGuire & Kable, 2015, 2016).

These cases add to the reasons why steep delay discounting cannot be automat-ically treated as indicative of poor self-control. They also bring into sharp relief the reason it is misleading to treat farsightedness as synonymous with a prefer-ence for delayed rewards: there are plenty of farsighted reasons to prefer imme-diate gratification—including, but by no means limited to, anticipating later regret at missing out on small pleasures, and anticipating the risks intrinsic to deferring rewards. Nonetheless, interpretations of steeper delay discounting as a failure of future thinking remain dominant in theoretical accounts of delay discounting. Consider, for instance, the definition from the International Society for Research on Impulsivity, who characterize steep delay discounting as a form of 'choice impulsivity' reflecting a 'lack of planning and lack of regard for future consequences' (Hamilton et al., 2015). This definition falls short given the many circumstances in which an individual might have a steep discounting function precisely *because* of their long-term planning and high regard for future consequences. So, we cannot assume steep delay discounting necessarily emerges from a lack of self-control, nor can we assume it emerges from a lack of planning ahead.

One recent line of research that casts further doubt on standard framings of discounting in intertemporal choice tasks comes from work on confidence judgements (Folke et al., 2016; Soutschek & Tobler, 2020). In a recent series of studies (Bulley et al., 2021), we had participants make intertemporal choices and then report their confidence that they had made the *right decision* after each choice (see also Soutschek et al., 2021; Soutschek & Tobler, 2020). As described above, self-control theories assume that, in an intertemporal choice task, participants are attempting to maximize reward by delaying gratification but sometimes fail to choose delayed rewards because they lack the capacity to control their impulses. Considering that confidence judgements reflect the subjective belief that one's decision is correct (Desender et al., 2021; Fleming & Daw, 2017; Pouget et al., 2016), if this self-control assumption is valid, participants should be reliably more confident that they have made the right choice when they 'successfully' delay gratification for larger, later rewards. But this is not what we found. Participants were no more or less confident that they had made the right decision on average when they opted for the smaller, sooner, or the larger, later reward. Instead, participants were more confident they had made the right choice when their decisions more closely lined up with their independent valuation of the available options (Figure 6.3a, b). In other words, participants reported that the 'right' choice for them was whichever option best accorded with the subjective values that characterized their preferences. While inconsistent with the self-control failure view of steep discounting, this finding does support the aforementioned value-based decision-making account, in which decision confidence would track uncertainty in estimating and comparing the value of alternative options (Berkman et al., 2017; Kable, 2014). Further support for a value-based account of intertemporal choice comes from our finding that participants were less confident in a decision (and took longer to make it) when the two available rewards had more similar time-discounted subjective values (Figure 6.3c) (see also Konovalov & Krajbich, 2019). Plausibly, participants must retrieve memories and generate predictions to build up evidence about which of two options is better, and these processes are more difficult and take longer when the options are more similarly valuable (Bakkour et al., 2019; Shadlen & Shohamy, 2016).

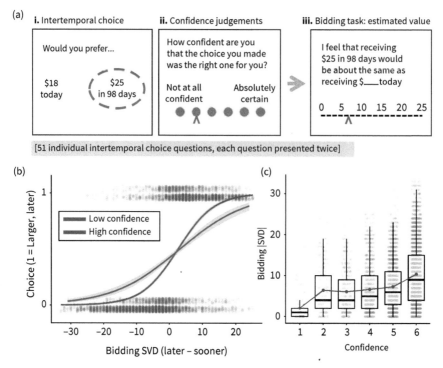

Figure 6.3 Confidence in intertemporal choice reflects value estimation and comparison.
(a) Participants made intertemporal choices (i), and then rated their confidence that the
decision they made was the right one for them (ii). In a separate 'bidding task', participants
estimated how much the delayed rewards were worth, considering the delay to their receipt
(iii). (b) Participants were reliably more confident in their decisions when their actual choices
better reflected which option had greater estimated subjective value (SVD = subjective value
difference). (c) Participants were also more confident when the absolute subjective value
difference (|SVD|) between the options increased, meaning one of the options had a more
obviously dominant time-discounted subjective value (Bulley et al., 2021).

Participants were also less confident in decisions when the larger, later reward
on offer was more delayed into the future, supporting the idea described earlier
that increasing delays to a reward can introduce uncertainty about the value of that
reward—an uncertainty that participants themselves can report explicitly (Gabaix
& Laibson, 2017; Trope & Liberman, 2010). Furthermore, supporting the claim
that this uncertainty can be modulated, we found that participants reached greater
confidence when there was more reward on offer. One plausible explanation for
this finding is that people allocate more mental effort towards making the *right*
decision when that decision is deemed more important (Hill, 2012)—possibly by
investing more mental resources towards estimating the value of future outcomes
(I. C. Ballard et al., 2017; Gershman & Bhui, 2020). Lastly, we found that decisions
made with less confidence were both more prone to changes of mind and were

more susceptible to a patience-enhancing framing manipulation, which is what one would expect if confidence were monitoring uncertainty around which option is best (Alós-Ferrer & Garagnani, 2021; Folke et al., 2016; Frömer & Shenhav, 2022). All of the above results are readily predicted by applying value-based decision-making theory to intertemporal choice. Confidence judgements appear to track uncertainty in value estimation and comparison during intertemporal choice just as they do in other kinds of value-based decisions that are traditionally viewed as unrelated to 'self-control' (De Martino et al., 2013; Folke et al., 2016). These findings reinforce the idea that conceptualizing steep discounting simply as a failure of attempted self-control reflects, at best, an incomplete theory of intertemporal choice.

The foregoing discussion implies that we should reconsider the theoretical basis for framing intertemporal decisions as 'farsighted vs shortsighted' (Rösch et al., 2022), 'optimal vs suboptimal' (Sellitto et al., 2010), 'successes vs failures' (Duckworth et al., 2018), 'better vs worse' (Shamosh et al., 2008), 'self-controlled vs impulsive' (Turner et al., 2019) or 'restrained vs opportunistic' (Boyer, 2008). The terms used to describe the discounting function reflect our theoretical assumptions, and these assumptions might be flawed—leading to poor applications in real-world contexts (Loewenstein, 2018). For example, a policy that attempts to encourage 'farsighted' choices of delayed rewards might falter because participants themselves are farsighted enough to realize the immediate options are better for them—even if simply because they represent a safer bet. Or an intervention attempting to facilitate 'self-control' might backfire when participants use that very self-control to override a habitual proclivity to save money so they can better enjoy present pleasures.

Conclusion

Multiple adaptive functions of episodic future thinking have been identified, including making flexible intertemporal decisions that take delayed consequences into account. In this chapter, we have attempted to show how theoretical advances in the study of memory and prospection have collectively borne productive applications in real-world decision-making about the future. For instance, we have reviewed the promise of attempts to harness episodic future thinking to reduce delay discounting and encourage the prioritization of future rewards. This research, while still in its early stages, offers a highly scalable, effective, and straightforward approach for modifying delay discounting in both monetary and behavioural health-related contexts. Nonetheless, important caveats remain. In order to make further progress, the field should move beyond preconceptions about impulsivity and failures of self-control, and instead focus on the diverse determinants of intertemporal decision-making—including by grappling with the complexities in how people deliberate through their trade-offs with the future.

Acknowledgements

A.B. is supported by an Australian National Health and Medical Research Council CJ Martin Biomedical Fellowship APP1162811 (GNT1162811), and an Australian Research Council Discovery Project Grant DP210101572. D.L.S. is supported by National Institute on Aging grant R01 AG008441.

References

Addis, D. R., Wong, A. T., & Schacter, D. L. (2007). Remembering the past and imagining the future: Common and distinct neural substrates during event construction and elaboration. *Neuropsychologia, 45*(7), 1363–1377.

Addis, D. R., Wong, A. T., & Schacter, D. L. (2008). Age-related changes in the episodic simulation of future events. *Psychological Science, 19*(1), 33–41.

Ainslie, G. (1975). Specious reward: A behavioral theory of impulsiveness and impulse control. *Psychological Bulletin, 82*(4), 463–496.

Ainslie, G. (2005). Précis of breakdown of will. *Behavioral and Brain Sciences, 28*(5), 635–673.

Alós-Ferrer, C., & Garagnani, M. (2021). Choice consistency and strength of preference. *Economics Letters, 198*, 109672.

Amlung, M., Marsden, E., Holshausen, K., Morris, V., Patel, H., Vedelago, L., Naish, K. R., Reed, D. D., & McCabe, R. E. (2019). Delay discounting as a transdiagnostic process in psychiatric disorders. *JAMA Psychiatry, 76*(11), 1176–1186.

Angeletos, G., Laibson, D., Repetto, A., Tobacman, J., & Weinberg, S. (2001). The hyperbolic consumption model: Calibration, simulation, and empirical evaluation. *Journal of Economic Perspectives, 15*(3), 47–68.

Anokhin, A. P., Golosheykin, S., Grant, J. D., & Heath, A. C. (2011). Heritability of delay discounting in adolescence: A longitudinal twin study. *Behavior Genetics, 41*(2), 175–183.

Atance, C. M., & O'Neill, D. K. (2001). Episodic future thinking. *Trends in Cognitive Sciences, 5*(12), 533–539.

Athamneh, L. N., Brown, J., Stein, J. S., Gatchalian, K. M., Laconte, S. M., & Bickel, W. K. (2022). Future thinking to decrease real-world drinking in alcohol use disorder: Repairing reinforcer pathology in a randomized proof-of-concept trial. *Experimental and Clinical Psychopharmacology, 30*(3), 326–337.

Athamneh, L. N., Stein, M. D., Lin, E. H., Stein, J. S., Mellis, A. M., Gatchalian, K. M., Epstein, L. H., & Bickel, W. K. (2021). Setting a goal could help you control: Comparing the effect of health goal versus general episodic future thinking on health behaviors among cigarette smokers and obese individuals. *Experimental and Clinical Psychopharmacology, 29*(1), 59–72.

Augenblick, N., Cunha, J. M., Dal Bó, E., & Rao, J. M. (2016). The economics of faith: Using an apocalyptic prophecy to elicit religious beliefs in the field. *Journal of Public Economics, 141*, 38–49.

Bakkour, A., Palombo, D. J., Zylberberg, A., Kang, Y. H., Reid, A., Verfaellie, M., Shadlen, M. N., & Shohamy, D. (2019). The hippocampus supports deliberation during value-based decisions. *eLife, 8*, 1–28.

Ballance, B. C., Tuen, Y. J., Petrucci, A. S., Orwig, W., Safi, O. K., Madan, C. R., & Palombo, D. J. (2022). Imagining emotional events benefits future-oriented decisions. *Quarterly Journal of Experimental Psychology, 75*(12), 2332–2348.

Ballard, I. C., Kim, B., Liatsis, A., Aydogan, G., Cohen, J. D., & McClure, S. M. (2017). More is meaningful: The magnitude effect in intertemporal choice depends on self-control. *Psychological Science, 28*(10), 1443–1454.

Ballard, T., Luckman, A., & Konstantinidis, E. (2020). *How meaningful are parameter estimates from models of inter-temporal choice?* PsyArXiv.

Barsics, C., Van der Linden, M., & D'Argembeau, A. (2016). Frequency, characteristics, and perceived functions of emotional future thinking in daily life. *Quarterly Journal of Experimental Psychology, 69*(2), 217–233.

Benoit, R. G., Gilbert, S. J., & Burgess, P. W. (2011). A neural mechanism mediating the impact of epi-sodic prospection on farsighted decisions. *Journal of Neuroscience*, *31*(18), 6771–6779.

Benoit, R. G., & Schacter, D. L. (2015). Specifying the core network supporting episodic simulation and episodic memory by activation likelihood estimation. *Neuropsychologia*, *75*, 450–457.

Berkman, E. T., Hutcherson, C. A., Livingston, J. L., Kahn, L. E., & Inzlicht, M. (2017). Self-control as value-based choice. *Current Directions in Psychological Science*, *26*(5), 422–428.

Berns, G. S., Laibson, D., & Loewenstein, G. (2007). Intertemporal choice—Toward an integrative framework. *Trends in Cognitive Sciences*, *11*(11), 482–488.

Bickel, W. K., Athamneh, L. N., Basso, J. C., Mellis, A. M., DeHart, W. B., Craft, W. H., & Pope, D. (2019). Excessive discounting of delayed reinforcers as a trans-disease process: Update on the state of the sci-ence. *Current Opinion in Psychology*, *30*, 59–64.

Bickel, W. K., Jarmolowicz, D. P., Mueller, E. T., Koffarnus, M. N., & Gatchalian, K. M. (2012). Excessive discounting of delayed reinforcers as a trans-disease process contributing to addiction and other disease-related vulnerabilities: Emerging evidence. *Pharmacology and Therapeutics*, *134*(3), 287–297.

Bickel, W. K., Stein, J. S., Paluch, R. A., Mellis, A. M., Athamneh, L. N., Quattrin, T., Greenawald, M. H., Bree, K. A., Gatchalian, K. M., Mastrandrea, L. D., & Epstein, L. H. (2020). Does episodic fu-ture thinking repair immediacy bias at home and in the laboratory in patients with prediabetes? *Psychosomatic Medicine*, *82*(7), 699–707.

Biderman, N., Bakkour, A., & Shohamy, D. (2020). What are memories for? The hippocampus bridges past experience with future decisions. *Trends in Cognitive Sciences*, *24*(7), 542–556.

Boyer, P. (2008). Evolutionary economics of mental time travel? *Trends in Cognitive Sciences*, *12*(6), 219–224.

Brownback, A., Imas, A., & Kuhn, M. (2019). Behavioral food subsidies. *SSRN Electronic Journal*, 1 January.

Buckholtz, J. W. (2015). Social norms, self-control, and the value of antisocial behavior. *Current Opinion in Behavioral Sciences*, *3*, 122–129.

Bulley, A., Henry, J., & Suddendorf, T. (2016). Prospection and the present moment: The role of epi-sodic foresight in intertemporal choices between immediate and delayed rewards. *Review of General Psychology*, *20*(1), 29–47.

Bulley, A., Henry, J. D., & Suddendorf, T. (2017). Thinking about threats: Memory and prospection in human threat management. *Consciousness and Cognition*, *49*, 53–69.

Bulley, A., & Irish, M. (2018). The functions of prospection: Variations in health and disease. *Frontiers in Psychology*, *9*, 2328.

Bulley, A., Lempert, K. M., Conwell, C., Irish, M., & Schacter, D. L. (2021). Intertemporal choice reflects value comparison rather than self-control: Insights from confidence judgements. *Phil. Trans. R. Soc. B.*, *377*, 20210338.

Bulley, A., Miloyan, B., Pepper, G. V., Gullo, M. J., Henry, J. D., & Suddendorf, T. (2019). Cuing both positive and negative episodic foresight reduces delay discounting but does not affect risk-taking. *Quarterly Journal of Experimental Psychology*, *72*(8), 1998–2017.

Bulley, A., Pepper, G., & Suddendorf, T. (2017). Using foresight to prioritise the present. *Behavioral and Brain Sciences*, *40*, E79.

Bulley, A., Redshaw, J., & Suddendorf, T. (2020). The future-directed functions of the imagination: From prediction to metaforesight. In A. Abraham (Ed.), *The Cambridge handbook of the imagination* (pp. 425–444). Cambridge University Press.

Bulley, A., & Schacter, D. L. (2020). Deliberating trade-offs with the future. *Nature Human Behaviour*, *4*, 238–247.

Busby, J., & Suddendorf, T. (2005). Recalling yesterday and predicting tomorrow. *Cognitive Development*, *20*(3), 362–372.

Cavagnaro, D. R., Aranovich, G. J., McClure, S. M., Pitt, M. A., & Myung, J. I. (2016). On the functional form of temporal discounting: An optimized adaptive test. *Journal of Risk and Uncertainty*, *52*(3), 233–254.

Cheung, S. L., Tymula, A., & Wang, X. (2021). *Quasi-hyperbolic present bias: A meta-analysis* (Life Course Centre Working Paper Series, 2021-15). Institute for Social Science Research, The University of Queensland.

Christakou, A., Brammer, M., & Rubia, K. (2011). Maturation of limbic corticostriatal activation and connectivity associated with developmental changes in temporal discounting. *NeuroImage, 54*(2), 1344–1354.

Cosme, D., Ludwig, R. M., & Berkman, E. T. (2019). Comparing two neurocognitive models of self-control during dietary decisions. *Social Cognitive and Affective Neuroscience, 14*(9), 957–966.

Critchfield, T. S., & Kollins, S. H. (2001). Temporal discounting: Basic research and the analysis of socially important behavior. *Journal of Applied Behavior Analysis, 34*(1), 101–122.

Daniel, T. O., Stanton, C. M., & Epstein, L. H. (2013). The future is now: Reducing impulsivity and energy intake using episodic future thinking. *Psychological Science, 24*(11), 2339–2342.

D'Argembeau, A. (2016). The role of personal goals in future-oriented mental time travel. In K. Michaelian, S. B. Klein, & K. K. Szpunar (Eds.), *Seeing the future: Theoretical perspectives on future-oriented mental time travel* (pp. 199–214). Oxford University Press.

D'Argembeau, A. (2020). Zooming in and out on one's life: Autobiographical representations at multiple time scales. *Journal of Cognitive Neuroscience, 32*(11), 2037–2055.

D'Argembeau, A., Renaud, O., & Van Der Linden, M. (2011). Frequency, characteristics and functions of future-oriented thoughts in daily life. *Applied Cognitive Psychology, 25*(1), 96–103.

D'Argembeau, A., Stawarczyk, D., Majerus, S., Collette, F., Van der Linden, M., Feyers, D., Maquet, P., & Salmon, E. (2010). The neural basis of personal goal processing when envisioning future events. *Journal of Cognitive Neuroscience, 22*(8), 1701–1713.

D'Argembeau, A., & Van der Linden, M. (2004). Phenomenal characteristics associated with projecting oneself back into the past and forward into the future: Influence of valence and temporal distance. *Consciousness and Cognition, 13*(4), 844–858.

Dassen, F. C. M., Jansen, A., Nederkoorn, C., & Houben, K. (2016). Focus on the future: Episodic future thinking reduces discount rate and snacking. *Appetite, 96*, 327–332.

Davis, J. T. M., Cullen, E., & Suddendorf, T. (2015). Understanding deliberate practice in preschool aged children. *Quarterly Journal of Experimental Psychology, 69*(2), 361–380.

De Brigard, F., & Parikh, N. (2019). Episodic counterfactual thinking. *Current Directions in Psychological Science, 28*(1), 59–66.

Decker, J. H., Figner, B., & Steinglass, J. E. (2015). On weight and waiting: Delay discounting in anorexia nervosa pretreatment and posttreatment. *Biological Psychiatry, 78*(9), 606–614.

De Martino, B., Fleming, S. M., Garrett, N., & Dolan, R. J. (2013). Confidence in value-based choice. *Nature Neuroscience, 16*(1), 105–110.

Desender, K., Donner, T., & Verguts, T. (2021). Dynamic expressions of confidence within an evidence accumulation framework. *Cognition, 207*, 104522.

Duckworth, A. L., Milkman, K. L., & Laibson, D. (2018). Beyond willpower: Strategies for reducing failures of self-control. *Psychological Science in the Public Interest, 19*(3), 102–129.

Epstein, L. H., Paluch, R. A., Stein, J. S., Quattrin, T., Mastrandrea, L. D., Bree, K. A., Sze, Y. Y., Greenawald, M. H., Biondolillo, M. J., & Bickel, W. K. (2021). Delay discounting, glycemic regulation and health behaviors in adults with prediabetes. *Behavioral Medicine, 47*(3), 194–204.

Ericson, K. M., & Laibson, D. (2019). Intertemporal choice. In B. D. Bernheim, S. DellaVigna, & D. Laibson (Eds.), *Handbook of behavioral economics: Foundations and applications 2* (pp. 1–67). North-Holland.

Ernst, A., & D'Argembeau, A. (2017). Make it real: Belief in occurrence within episodic future thought. *Memory & Cognition, 45*(6), 1045–1061.

Ernst, A., Scoboria, A., & D'Argembeau, A. (2019). On the role of autobiographical knowledge in shaping belief in the future occurrence of imagined events. *Quarterly Journal of Experimental Psychology, 72*(11), 2658–2671.

Figner, B., Knoch, D., Johnson, E. J., Krosch, A. R., Lisanby, S. H., Fehr, E., & Weber, E. U. (2010). Lateral prefrontal cortex and self-control in intertemporal choice. *Nature Neuroscience, 13*(5), 538–539.

Fleming, S. M., & Daw, N. D. (2017). Self-evaluation of decision-making: A general Bayesian framework for metacognitive computation. *Psychological Review, 124*(1), 91–114.

Folke, T., Jacobsen, C., Fleming, S. M., & De Martino, B. (2016). Explicit representation of confidence informs future value-based decisions. *Nature Human Behaviour, 1*, 1–8.

Frederick, S., Loewenstein, G., & Donoghue, T. O. (2002). Time discounting and time preference: A critical review. *Journal of Economic Literature, 40*(2), 351–401.

Frömer, R., & Shenhav, A. (2022). Filling the gaps: Cognitive control as a critical lens for understanding mechanisms of value-based decision-making. *Neuroscience and Biobehavioral Reviews, 134*, 104483.

Gabaix, X., & Laibson, D. (2017). *Myopia and discounting* (NBER Working Paper Series No. w23254). National Bureau of Economic Research.

Gaesser, B. (2020). Episodic mindreading: Mentalizing guided by scene construction of imagined and remembered events. *Cognition, 203*, 104325.

Gaesser, B., & Schacter, D. L. (2014). Episodic simulation and episodic memory can increase intentions to help others. *Proceedings of the National Academy of Sciences of the United States of America, 111*(12), 4415–4420.

Gerlach, K. D., Spreng, R. N., Madore, K. P., & Schacter, D. L. (2014). Future planning: Default network activity couples with frontoparietal control network and reward-processing regions during process and outcome simulations. *Social Cognitive and Affective Neuroscience, 9*(12), 1942–1951.

Gershman, S., & Bhui, R. (2020). Rationally inattentive intertemporal choice. *Nature Communications, 11*(3365), 1–8.

Gilbert, D. T., & Wilson, T. D. (2007). Prospection: Experiencing the future. *Science, 317*, 1351–1354.

Hamilton, K. R., Mitchell, M. R., Wing, V. C., Balodis, I. M., Bickel, W. K., Fillmore, M., Lane, S. D., Lejuez, C. W., Littlefield, A. K., Luijten, M., Mathias, C. W., Mitchell, S. H., Napier, T. C., Reynolds, B., Schütz, C. G., Setlow, B., Sher, K. J., Swann, A. C., Tedford, S. E., . . . Moeller, F. G. (2015). Choice impulsivity: Definitions, measurement issues, and clinical implications. *Personality Disorders: Theory, Research, and Treatment, 6*(2), 182–198.

Hardisty, D. J., & Weber, E. U. (2009). Discounting future green: Money versus the environment. *Journal of Experimental Psychology: General, 138*(3), 329–340.

Hassabis, D., Kumaran, D., Vann, S. D., & Maguire, E. A. (2007). Patients with hippocampal amnesia cannot imagine new experiences. *Proceedings of the National Academy of Sciences of the United States of America, 104*(5), 1726–1731.

Heatherton, T. F., & Wagner, D. D. (2011). Cognitive neuroscience of self-regulation failure. *Trends in Cognitive Sciences, 15*(3), 132–139.

Hershfield, H. E., Goldstein, D. G., Sharpe, W. F., Fox, J., Yeykelis, L., Carstensen, L. L., & Bailenson, J. N. (2011). Increasing saving behavior through age-progressed renderings of the future self. *Journal of Marketing Research, 48*, S23–S37.

Hershfield, H. E., John, E. M., & Reiff, J. S. (2018). Using vividness interventions to improve financial decision making. *Policy Insights from the Behavioral and Brain Sciences, 5*(2), 209–215.

Hill, B. (2012). Confidence in preferences. *Social Choice and Welfare, 39*(2–3), 273–302.

Hoerl, C., & McCormack, T. (2019). Thinking in and about time: A dual systems perspective on temporal cognition. *Behavioral and Brain Sciences, 42*(e244), 1–69.

Hofmann, W., Friese, M., & Strack, F. (2009). Impulse and self-control from a dual-systems perspective. *Perspectives on Psychological Science, 4*(2), 162–176.

Imas, A., Kuhn, M. A., & Mironova, V. (2018). Waiting to choose: The role of deliberation in intertemporal choice. *American Economic Journal: Microeconomics, 14*(3), 414–440.

Irish, M. (2016). Semantic memory as the essential scaffold for future-oriented mental time travel. In K. Michaelian, S. B. Klein, & K. K. Szpunar (Eds.), *Seeing the Future: Theoretical perspectives on future-oriented mental time travel* (pp. 389–408). Oxford University Press.

Irish, M., Goldberg, Z-.L., Alaeddin, S., O'Callaghan, C., & Andrews-Hanna, J. R. (2019). Age-related changes in the temporal focus and self-referential content of spontaneous cognition during periods of low cognitive demand. *Psychological Research, 83*(4), 747–760.

Irish, M., & Vatansever, D. (2020). Rethinking the episodic-semantic distinction from a gradient perspective. *Current Opinion in Behavioral Sciences, 32*, 43–49.

Jachimowicz, J. M., Chafik, S., Munrat, S., Prabhu, J. C., & Weber, E. U. (2017). Community trust reduces myopic decisions of low-income individuals. *Proceedings of the National Academy of Sciences of the United States of America, 114*(21), 5401–5406.

Jimura, K., Chushak, M. S., Westbrook, A., & Braver, T. S. (2018). Intertemporal decision-making involves prefrontal control mechanisms associated with working memory. *Cerebral Cortex (New York, N.Y.: 1991)*, 28(4), 1105–1116.

Jing, H. G., Madore, K. P., & Schacter, D. L. (2016). Worrying about the future: An episodic specificity induction impacts problem solving, reappraisal, and well-being. *Journal of Experimental Psychology: General*, 145(4), 402–418.

Jing, H. G., Madore, K. P., & Schacter, D. L. (2019). Not to worry: Episodic retrieval impacts emotion regulation in older adults. *Emotion*, 20(4), 590–604.

Kable, J. W. (2014). Valuation, intertemporal choice, and self-control. In P. W. Glimcher & E. Fehr (Eds.), *Neuroeconomics: Decision making and the brain* (2nd ed., pp. 173–192). Academic Press.

Keinan, A., & Kivetz, R. (2008). Remedying hyperopia: The effects of self-control regret on consumer behavior. *Journal of Marketing Research*, 45(6), 676–689.

Kidd, C., Palmeri, H., & Aslin, R. N. (2013). Rational snacking: Young children's decision-making on the marshmallow task is moderated by beliefs about environmental reliability. *Cognition*, 126(1), 109–114.

King, J. A., & Ehrlich, S. (2020). The elusive nature of delay discounting as a transdiagnostic process in psychiatric disorders—The devil is in the detail. *JAMA Psychiatry*, 77(3), 325–326.

King, J. A., Geisler, D., Bernardoni, F., Ritschel, F., Böhm, I., Seidel, M., Mennigen, E., Ripke, S., Smolka, M. N., Roessner, V., & Ehrlich, S. (2016). Altered neural efficiency of decision making during temporal reward discounting in anorexia nervosa. *Journal of the American Academy of Child and Adolescent Psychiatry*, 55(11), 972–979.

Kirby, K. N. (2009). One-year temporal stability of delay-discount rates. *Psychonomic Bulletin & Review*, 16(3), 457–462.

Kivetz, R., & Simonson, I. (2002). Self-control for the righteous: Toward a theory of precommitment to indulgence. *Journal of Consumer Research*, 29(2), 199–217.

Klein, S. B., Loftus, J., & Kihlstrom, J. F. (2002). Memory and temporal experience: The effects of episodic memory loss on an amnesic patient's ability to remember the past and imagine the future. *Social Cognition*, 20(5), 353–379.

Koffarnus, M. N., Jarmolowicz, D. P., Mueller, E. T., & Bickel, W. K. (2013). Changing delay discounting in the light of the competing neurobehavioral decision systems theory: A review. *Journal of the Experimental Analysis of Behavior*, 99(1), 32–57.

Konovalov, A., & Krajbich, I. (2019). Revealed strength of preference: Inference from response times. *Judgment and Decision Making*, 14(4), 381–394.

Kurth-Nelson, Z., Bickel, W., & Redish, A. D. (2012). A theoretical account of cognitive effects in delay discounting. *European Journal of Neuroscience*, 35(7), 1052–1064.

Kwan, D., Craver, C. F., Green, L., Myerson, J., Boyer, P., & Rosenbaum, R. S. (2012). Future decision-making without episodic mental time travel. *Hippocampus*, 22(6), 1215–1219.

Kwan, D., Craver, C. F., Green, L., Myerson, J., Gao, F., Black, S. E., & Rosenbaum, R. S. (2015). Cueing the personal future to reduce discounting in intertemporal choice: Is episodic prospection necessary? *Hippocampus*, 25(4), 432–443.

Kwan, D., Craver, C. F., Green, L., Myerson, J., & Rosenbaum, R. S. (2013). Dissociations in future thinking following hippocampal damage: Evidence from discounting and time perspective in episodic amnesia. *Journal of Experimental Psychology: General*, 142(4), 1355–1369.

Laibson, D. (1997). Golden eggs and hyperbolic discounting. *Quarterly Journal Of Economics*, 112(2), 443–447.

Leahy, B. P., & Carey, S. E. (2020). The acquisition of modal concepts. *Trends in Cognitive Sciences*, 24(1), 65–78.

Lee, W. S. C., & Carlson, S. M. (2015). Knowing when to be 'rational': Flexible economic decision making and executive function in preschool children. *Child Development*, 86(5), 1434–1448.

Lempert, K. M., McGuire, J. T., Hazeltine, D. B., Phelps, E. A., & Kable, J. W. (2018). The effects of acute stress on the calibration of persistence. *Neurobiology of Stress*, 8, 1–9.

Lempert, K. M., & Phelps, E. A. (2016). The malleability of intertemporal choice. *Trends in Cognitive Sciences*, 20(1), 64–74.

Lempert, K. M., Steinglass, J. E., Pinto, A., Kable, J. W., & Simpson, H. B. (2019). Can delay discounting deliver on the promise of RDoC? *Psychological Medicine, 49*(2), 190–199.

Levy, D. J., & Glimcher, P. W. (2012). The root of all value: A neural common currency for choice. *Current Opinion in Neurobiology, 22*(6), 1027–1038.

Loewenstein, G. (2018). Self-control and its discontents: A commentary on Duckworth, Milkman, and Laibson. *Psychological Science in the Public Interest, 19*(3), 95–101.

Loewenstein, G., & Elster, J. (1992). *Choice over time*. Russell Sage Foundation.

Loewenstein, G., O'Donoghue, T., & Bhatia, S. (2015). Modeling the interplay between affect and deliberation. *Decision, 2*(2), 55–81.

Loewenstein, G., & Prelec, D. (1992). Anomalies in intertemporal choice: Evidence and an interpretation. *Quarterly Journal of Economics, 107*(2), 573–597.

Logue, A. W. (1988). Research on self-control: An integrating framework. *Behavioral and Brain Sciences, 11*(04), 665–679.

Ma, F., Chen, B., Xu, F., Lee, K., & Heyman, G. D. (2018). Generalized trust predicts young children's willingness to delay gratification. *Journal of Experimental Child Psychology, 169*, 118–125.

MacLeod, A. K. (2016). Prospection, well-being and memory. *Memory Studies, 9*(3), 266–274.

Madore, K. P., Gaesser, B., & Schacter, D. L. (2014). Constructive episodic simulation: Dissociable effects of a specificity induction on remembering, imagining, and describing in young and older adults. *Journal of Experimental Psychology: Learning Memory and Cognition, 40*(3), 609–622.

Madore, K. P., & Schacter, D. L. (2014). An episodic specificity induction enhances means-end problem solving in young and older adults. *Psychology and Aging, 29*(4), 913–924.

Mahr, J. B. (2020). The dimensions of episodic simulation. *Cognition, 196*, 1–15.

Mazur, J. E. (1987). An adjusting procedure for studying delayed reinforcement. In M. L. Commons, J. E. Mazur, J. A. Nevin, & H. Rachlin (Eds.), *Quantitative analyses of behavior, Volume 5. The effect of delay and of intervening events on reinforcement value* (pp. 55–73). Lawrence Erlbaum Associates, Inc.

McClure, S. M., & Bickel, W. K. (2014). A dual-systems perspective on addiction: Contributions from neuroimaging and cognitive training. *Annals of the New York Academy of Sciences, 1327*(1), 62–78.

McClure, S. M., Laibson, D. I., Loewenstein, G. F., & Cohen, J. D. (2004). Separate neural systems value immediate and delayed monetary rewards. *Science, 306*(5695), 503–507.

McGuire, J. T., & Kable, J. W. (2013). Rational temporal predictions can underlie apparent failures to delay gratification. *Psychological Review, 120*(2), 395–410.

McGuire, J. T., & Kable, J. W. (2015). Medial prefrontal cortical activity reflects dynamic re-evaluation during voluntary persistence. *Nature Neuroscience, 18*(5), 760–766.

McGuire, J. T., & Kable, J. W. (2016). Deciding to curtail persistence. In K. D. Vohs & R. F. Baumeister (Eds.), *Handbook of self-regulation: Research, theory, and applications* (3rd ed., pp. 533–546). Guilford Press.

Mellis, A. M., Snider, S. E., Deshpande, H. U., LaConte, S. M., & Bickel, W. K. (2019). Practicing prospection promotes patience: Repeated episodic future thinking cumulatively reduces delay discounting. *Drug & Alcohol Dependence, 204*, 1–6.

Metcalfe, J., & Mischel, W. (1999). A hot/cool-system analysis of delay of gratification: Dynamics of willpower. *Psychological Review, 106*(1), 3–19.

Michaelson, L., de la Vega, A., Chatham, C., & Munakata, Y. (2013). Delaying gratification depends on social trust. *Frontiers in Psychology, 4*, 355.

Miloyan, B., Bulley, A., & Suddendorf, T. (2018). Anxiety: Here and beyond. *Emotion Review, 11*(1), 39–49.

Mischel, W. (1974). Processes in delay of gratification. *Advances in Experimental Social Psychology, 7*, 249–292.

Neroni, M. A., Gamboz, N., & Brandimonte, M. A. (2014). Does episodic future thinking improve prospective remembering? *Consciousness and Cognition, 23*(1), 53–62.

Noël, X., Jaafari, N., & Bechara, A. (2017). Addictive behaviors: Why and how impaired mental time matters? *Progress in Brain Research, 235*, 219–237.

O'Donnell, S., Oluyomi Daniel, T., & Epstein, L. H. (2017). Does goal relevant episodic future thinking amplify the effect on delay discounting? *Consciousness and Cognition, 51*, 10–16.

Odum, A. L. (2011). Delay discounting: I'm a k, you're a k. *Journal of the Experimental Analysis of Behavior, 96*(3), 427–439.

Okuda, J., Fujii, T., Ohtake, H., Tsukiura, T., Tanji, K., Suzuki, K., Kawashima, R., Fukuda, H., Itoh, M., & Yamadori, A. (2003). Thinking of the future and past: The roles of the frontal pole and the medial temporal lobes. *NeuroImage, 19*(4), 1369–1380.

O'Neill, J., Daniel, T. O., & Epstein, L. H. (2016). Episodic future thinking reduces eating in a food court. *Eating Behaviors, 20,* 9–13.

Oreopoulos, P. (2007). Do dropouts drop out too soon? Wealth, health and happiness from compulsory schooling. *Journal of Public Economics, 91*(11–12), 2213–2229.

Palombo, D. J., Keane, M. M., & Verfaellie, M. (2015). The medial temporal lobes are critical for reward-based decision making under conditions that promote episodic future thinking. *Hippocampus, 25,* 345–353.

Palombo, D. J., Keane, M. M., & Verfaellie, M. (2016). Using future thinking to reduce temporal discounting: Under what circumstances are the medial temporal lobes critical? *Neuropsychologia, 89,* 437–444.

Pepper, G. V., & Nettle, D. (2017). The behavioural constellation of deprivation: Causes and consequences. *Behavioral and Brain Sciences, 40,* 1–72.

Peters, J., & Büchel, C. (2010). Episodic future thinking reduces reward delay discounting through an enhancement of prefrontal-mediotemporal interactions. *Neuron, 66*(1), 138–148.

Peters, J., & Büchel, C. (2011). The neural mechanisms of inter-temporal decision-making: Understanding variability. *Trends in Cognitive Sciences, 15*(5), 227–239.

Peters, J., & D'Esposito, M. (2016). Effects of medial orbitofrontal cortex lesions on self-control in intertemporal choice. *Current Biology, 26*(19), 2625–2628.

Pornpattananangkul, N., Nadig, A., Heidinger, S., Walden, K., & Nusslock, R. (2017). Elevated outcome-anticipation and outcome-evaluation ERPs associated with a greater preference for larger-but-delayed rewards. *Cognitive, Affective and Behavioral Neuroscience, 17*(3), 625–641.

Pouget, A., Drugowitsch, J., & Kepecs, A. (2016). Confidence and certainty: Distinct probabilistic quantities for different goals. *Nature Neuroscience, 19*(3), 366–374.

Rangel, A., Camerer, C., & Montague, P. R. (2008). A framework for studying the neurobiology of value-based decision making. *Nature Reviews Neuroscience, 9*(7), 545–556.

Redshaw, J. (2014). Does metarepresentation make human mental time travel unique? *Wiley Interdisciplinary Reviews: Cognitive Science, 5*(5), 519–531.

Redshaw, J., & Bulley, A. (2018). Future-thinking in animals: Capacities and limits. In G. Oettingen, A. T. Sevincer, & P. M. Gollwitzer (Eds.), *The psychology of thinking about the future* (pp. 31–51). Guilford Press.

Redshaw, J., & Suddendorf, T. (2020). Temporal junctures in the mind. *Trends in Cognitive Sciences, 24*(1), 52–64.

Rösch, S. A., Stramaccia, D. F., & Benoit, R. G. (2022). Promoting farsighted decisions via episodic future thinking: A meta-analysis. *Journal of Experimental Psychology. General, 151*(7), 1606–1635.

Rung, J. M., & Madden, G. J. (2018a). Demand characteristics in episodic future thinking: Delay discounting and healthy eating. *Experimental and Clinical Psychopharmacology, 26*(1), 77–84.

Rung, J. M., & Madden, G. J. (2018b). Experimental reductions of delay discounting and impulsive choice: A systematic review and meta-analysis. *Journal of Experimental Psychology: General, 147*(9), 1349–13831.

Samuelson, P. A. (1937). A note on measurement of utility. *The Review of Economic Studies, 4*(2), 155–161.

Sasse, L. K., Peters, J., & Brassen, S. (2017). Cognitive control modulates effects of episodic simulation on delay discounting in aging. *Frontiers in Aging Neuroscience, 9,* 1–11.

Sasse, L. K., Peters, J., Büchel, C., & Brassen, S. (2015). Effects of prospective thinking on intertemporal choice: The role of familiarity. *Human Brain Mapping, 36*(10), 4210–4221.

Schacter, D. L. (2012). Adaptive constructive processes and the future of memory. *American Psychologist, 67*(8), 603–613.

Schacter, D. L. (2021). On the evolution of a functional approach to memory. *Learning & Behavior, 50*(1), 11–19.

Schacter, D. L., Addis, D. R., & Buckner, R. L. (2007). Remembering the past to imagine the future: The prospective brain. *Nature Reviews Neuroscience*, 8(9), 657–661.

Schacter, D. L., Benoit, R. G., & Szpunar, K. K. (2017). Episodic future thinking: Mechanisms and functions. *Current Opinion in Behavioral Sciences*, 17, 41–50.

Schacter, D. L., Devitt, A. L., & Addis, D. R. (2018). Episodic future thinking and cognitive aging. In B. G. Knight (Ed.), *The Oxford encyclopedia of psychology and aging* (pp. 1–22). Oxford University Press.

Schafer, J. (2021). Delayed gratification in political participation. *American Politics Research*, 49(3), 304–312.

Sellitto, M., Ciaramelli, E., & di Pellegrino, G. (2010). Myopic discounting of future rewards after medial orbitofrontal damage in humans. *Journal of Neuroscience*, 30(49), 16429–16436.

Sellitto, M., Ciaramelli, E., & Di Pellegrino, G. (2011). The neurobiology of intertemporal choice: Insight from imaging and lesion studies. *Reviews in the Neurosciences*, 22(5), 565–574.

Shaddy, F., Fishbach, A., & Simonson, I. (2021). Trade-offs in choice. *Annual Review of Psychology*, 72, 181–206.

Shadlen, M. N., & Shohamy, D. (2016). Decision making and sequential sampling from memory. *Neuron*, 90(5), 927–939.

Shamosh, N. A., DeYoung, C. G., Green, A. E., Reis, D. L., Johnson, M. R., Conway, A. R. A., Engle, R. W., Braver, T. S., & Gray, J. R. (2008). Individual differences in delay discounting: Relation to intelligence, working memory, and anterior prefrontal cortex. *Psychological Science*, 19(9), 904–911.

Shapiro, J. M. (2005). Is there a daily discount rate? Evidence from the food stamp nutrition cycle. *Journal of Public Economics*, 89(2–3), 303–325.

Shenhav, A. (2017). The perils of losing control: Why self-control is not just another value-based decision. *Psychological Inquiry*, 28(2–3), 148–152.

Shenhav, A., Rand, D. G., & Greene, J. D. (2017). The relationship between intertemporal choice and following the path of least resistance across choices, preferences, and beliefs. *Judgement and Decision Making*, 12(1), 1–18.

Snider, S. E., LaConte, S. M., & Bickel, W. K. (2016). Episodic future thinking: Expansion of the temporal window in individuals with alcohol dependence. *Alcoholism: Clinical and Experimental Research*, 40(7), 1558–1566.

Soutschek, A., Moisa, M., Ruff, C. C., & Tobler, P. N. (2021). Frontopolar theta oscillations link metacognition with prospective decision making. *Nature Communications*, 12(3943), 1–8.

Soutschek, A., & Tobler, P. N. (2020). Know your weaknesses: Sophisticated impulsiveness motivates voluntary self-restrictions. *Journal of Experimental Psychology. Learning, Memory, and Cognition*, 46(9), 1611–1623.

Soutschek, A., Ugazio, G., Crockett, M. J., Ruff, C. C., Kalenscher, T., & Tobler, P. N. (2017). Binding oneself to the mast: Stimulating frontopolar cortex enhances precommitment. *Social Cognitive and Affective Neuroscience*, 12(4), 635–642.

Spreng, R. N., Madore, K. P., & Schacter, D. L. (2018). Better imagined: Neural correlates of the episodic simulation boost to prospective memory performance. *Neuropsychologia*, 113, 22–28.

Stein, J. S., Wilson, A. G., Koffarnus, M. N., Daniel, T. O., Epstein, L. H., & Bickel, W. K. (2016). Unstuck in time: Episodic future thinking reduces delay discounting and cigarette smoking. *Psychopharmacology*, 233(21–22), 3771–3778.

Steinglass, J. E., Figner, B., Berkowitz, S., Simpson, H. B., Weber, E. U., & Walsh, B. T. (2012). Increased capacity to delay reward in anorexia nervosa. *Journal of the International Neuropsychological Society*, 18(4), 773–780.

Steinglass, J. E., Lempert, K. M., Choo, T. H., Kimeldorf, M. B., Wall, M., Walsh, B. T., Fyer, A. J., Schneier, F. R., & Simpson, H. B. (2017). Temporal discounting across three psychiatric disorders: Anorexia nervosa, obsessive compulsive disorder, and social anxiety disorder. *Depression and Anxiety*, 34(5), 463–470.

Stevens, J. R. (2011). Mechanisms for decisions about the future. In R. Menzel & J. Fischer (Eds.), *Animal thinking: Contemporary issues in comparative cognition* (pp. 95–104). MIT Press.

Stevens, J. R., & Stephens, D. (2008). Patience. *Current Biology*, 18(1), R11–R12.

Strikwerda-Brown, C., Grilli, M. D., Andrews-Hanna, J., & Irish, M. (2019). 'All is not lost'—Rethinking the nature of memory and the self in dementia. *Ageing Research Reviews*, 54(January), 100932.

Strotz, R. H. (1955). Myopia and inconsistency in dynamic utility maximization. *The Review of Economic Studies*, *23*(3), 165–180.

Suddendorf, T. (1999). The rise of the metamind. In M. C. Corballis & S. E. G. Lea (Eds.), *The descent of mind: Psychological perspectives on hominid evolution* (pp. 218–260). Oxford University Press.

Suddendorf, T. (2013). *The gap: The science of what separates us from other animals*. Basic Books.

Suddendorf, T., Brinums, M., & Imuta, K. (2016). Shaping one's future self: The development of deliberate practice. In K. Michaelian, S. B. Klein, & K. K. Szpunar (Eds.), *Seeing the future: Theoretical perspectives on future-oriented mental time travel* (pp. 343–366). Oxford University Press.

Suddendorf, T., Bulley, A., & Miloyan, B. (2018). Prospection and natural selection. *Current Opinion in Behavioral Sciences*, *24*, 26–31.

Suddendorf, T., & Corballis, M. C. (2007). The evolution of foresight: What is mental time travel, and is it unique to humans? *Behavioral and Brain Sciences*, *30*(3), 299–351.

Suddendorf, T., & Redshaw, J. (2013). The development of mental scenario building and episodic foresight. *Annals of the New York Academy of Sciences*, *1296*(1), 135–153.

Szpunar, K. K. (2010). Episodic future thought: An emerging concept. *Perspectives on Psychological Science*, *5*(2), 142–162.

Szpunar, K. K., Spreng, R. N., & Schacter, D. L. (2014). A taxonomy of prospection: Introducing an organizational framework for future-oriented cognition. *Proceedings of the National Academy of Sciences of the United States of America*, *111*(52), 18414–18421.

Szpunar, K. K., Watson, J. M., & McDermott, K. B. (2007). Neural substrates of envisioning the future. *Proceedings of the National Academy of Sciences of the United States of America*, *104*(2), 642–647.

Thakral, P. P., Madore, K. P., Addis, D. R., & Schacter, D. L. (2020). Reinstatement of event details during episodic simulation in the hippocampus. *Cerebral Cortex*, *30*, 2331–2337.

Thaler, R. H. (1981). Some empirical evidence on dynamic inconsistency. *Economics Letters*, *8*(3), 201–207.

Thaler, R. H., & Shefrin, H. (1981). An economic theory of self-control. *Journal of Political Economy*, *89*(2), 392–406.

Trope, Y., & Liberman, N. (2010). Construal-level theory of psychological distance. *Psychological Review*, *117*(2), 440.

Tulving, E. (1985). Memory and consciousness. *Canadian Psychology*, *26*(1), 1–12.

Turner, B. M., Rodriguez, C. A., Liu, Q., Molloy, M. F., Hoogendijk, M., & McClure, S. M. (2019). On the neural and mechanistic bases of self-control. *Cerebral Cortex*, *29*(2), 732–750.

van den Bos, W., & McClure, S. M. (2013). Towards a general model of temporal discounting. *Journal of the Experimental Analysis of Behavior*, *99*(1), 58–73.

Vosgerau, J., Scopelliti, I., & Huh, Y. E. (2020). Exerting self-control ≠ sacrificing pleasure. *Journal of Consumer Psychology*, *30*(1), 181–200.

Yin Mok, J. N., Kwan, D., Green, L., Myerson, J., Craver, C. F., & Rosenbaum, R. S. (2020). Is it time? Episodic imagining and the discounting of delayed and probabilistic rewards in young and older adults. *Cognition*, *199*, 1–15.

7

Working memory, intelligence, and life success

Examining relations to academic achievement, job performance, physical health, mortality, and psychological well-being

Cody A. Mashburn, Alexander P. Burgoyne, and Randall W. Engle

Some people are more successful than others when it comes to school performance, career attainment, physical health, and psychological well-being. What explains individual differences in these real-world outcomes? In this chapter, we focus on one individual difference characteristic that explains considerable variance in life success: cognitive ability. Specifically, we review evidence for the relationships between working memory capacity, attention control, and fluid intelligence, as well as their contributions to success in life. Although cognitive ability is a powerful predictor of real-world outcomes, understanding why some people excel while others do not is a complex problem that requires a complex (i.e. multivariate) solution. Thus, throughout this chapter we note just some of the other variables that may account for overlapping portions of variance in life outcomes, including socioeconomic factors such as familial income and educational attainment.

In many senses, this chapter must be reductive. Each subtopic subsumed under the umbrella term 'life success' is sufficiently broad as to define entire research programmes and careers, and the complexities of each domain are likely to exceed the grasp of non-specialists. Furthermore, there are many domains related to success in life that we do not examine, including relationship satisfaction, happiness, and creative fulfilment, to name a few. In short, a concise *and* comprehensive treatment is impossible, though we direct interested readers to Draheim et al. (2022) for a more protracted discussion of many of the topics covered in this chapter, and several we do not (e.g. cognitive training, sports, police decision-making). Rather than attempt complete coverage, our goal is to highlight primary findings in areas of ongoing research, note complications posed by extant empirical and theoretical investigations, and suggest avenues of future research.

We begin by discussing intelligence, broadly defined, and reviewing evidence for the relationships between working memory capacity, attention control, and fluid intelligence. Afterwards, we describe how these cognitive abilities relate to real-world

Cody A. Mashburn, Alexander P. Burgoyne, and Randall W. Engle, *Working memory, intelligence, and life success* In: *Memory in Science for Society*. Edited by: Robert H. Logie, Zhisheng (Edward) Wen, Susan E. Gathercole, Nelson Cowan, and Randall W. Engle, Oxford University Press. © Oxford University Press 2023. DOI: 10.1093/oso/9780192849069.003.0007 outcomes, including school performance, career attainment, physical health, and psychological well-being.

Intelligence

What is intelligence? On the face of it, this question may seem trivial for readers of this book. And yet, despite over a century of research on 'modern' intelligence tests (Binet & Simon, 1916), reasonable scientists and laypeople still disagree about what intelligence is, how it ought to be measured, and even whether it should be studied at all (e.g. Ritchie, 2015). From a conceptual standpoint, we think of intelligence as the ability to reason, solve problems, learn quickly, remember, plan, attend to what matters, and adapt to one's environment. From a technical or psychometric perspective, intelligence refers to one's level of performance on cognitive ability tests, and *general intelligence* refers to the higher-order *g* factor extracted from a battery of diverse cognitive tests (Figure 7.1). The *g* factor explains (or emerges as a result of; see Burgoyne et al., 2021; Kovacs & Conway, 2016; van der Maas et al., 2006) the positive correlations observed among broad cognitive abilities, representing what is common or shared across cognitive tests tapping different abilities (e.g. problem-solving, memory, processing speed) and content areas (e.g. verbal, numerical, visuo-spatial). *Broad* cognitive abilities refer to general classes of cognitive abilities, which, as one moves down the hierarchy from the *g* factor, can be further decomposed into more specific abilities, mechanisms, or processes measured by various cognitive assessments.

For this chapter, we focus on three highly correlated domain-general cognitive abilities and their relationships to real-world outcomes. Specifically, we consider *working memory capacity*, which refers to the ability to maintain and manipulate information; *attention control*, the ability to focus attention on task-relevant information while resisting interference and distraction by having attention captured by task-irrelevant thoughts and events; and *fluid intelligence*, the ability to reason and solve problems novel to the individual. As we will discuss, there is considerable theoretical and empirical overlap between working memory capacity and attention control (e.g. Engle, 2002, 2018). In the next section, we review evidence suggesting that attention control is important to measures of working memory capacity and plays a large role in explaining working memory capacity's relationship with other constructs, including fluid intelligence.

Working memory capacity, attention control, and fluid intelligence

Working memory refers to the cognitive system that allows us to temporarily maintain information in a readily accessible state and manipulate it to serve task goals

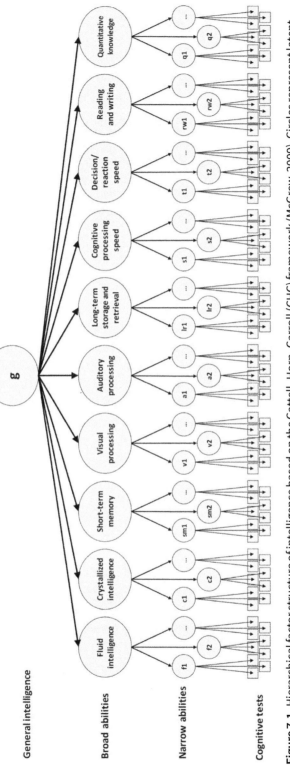

Figure 7.1 Hierarchical factor structure of intelligence based on the Cattell–Horn–Carroll (CHC) framework (McGrew, 2009). Circles represent latent factors; rectangles represent observed measures of performance on cognitive tests. Ellipses (…) indicate that there are more narrow abilities than could be depicted. The higher-order *g* factor represents general intelligence and explains the positive correlations observed among broad cognitive abilities. Adapted from Burgoyne et al. (2022).

Figure 7.2 Baddeley and Hitch's (1974) classic model of working memory.

(Baddeley, 1992). For example, solving the mental arithmetic problem '(3 × 17) + 2 =?' requires working memory because one must compute and temporarily store the product of '3 × 17' in order to add '2' to it. In classic models, the working memory system includes a controlled attention component and a short-term storage component, or components (Figure 7.2; Baddeley, 1992; Baddeley & Hitch, 1974; Cowan, 1988). Although some researchers argue that the term is antiquated (e.g. Logie, 2016), the controlled attention component is often called the *central executive* or *executive attention*, and it is responsible for coordinating the flow of information into, out of, and between the short-term storage components. Whereas the central executive is domain general, the short-term storage components are modality specific. For example, the *visuospatial sketch pad* is hypothesized to store visual information, such as mental imagery, whereas the *phonological loop* is hypothesized to store verbal and auditory information, such as speech. The phonological loop and visuospatial sketchpad are the most extensively studied of the putative storage systems, but others have been suggested (Baddeley, 2012). For the purposes of this chapter, however, we are primarily interested in working memory's interplay between controlled attention (i.e. the central executive component) and short-term memory at a more general level.

Evidence suggests that the controlled attention component of the working memory system can largely explain the correlations between measures of working memory capacity and other abilities, such as fluid intelligence (Kane et al., 2001; Kane & Engle, 2002). Early interest in the relationship between working memory capacity and fluid intelligence was spurred on by the discovery of very strong (i.e. near-perfect) correlations between the two constructs. For instance, across a series of studies of more than 2000 participants, Kyllonen and Christal (1990) found correlations ranging from $r = 0.80$ to $r = 0.90$ between working memory capacity and fluid intelligence when measured at the *latent level*.[1] This led to an exciting possibility: if working memory could explain individual differences in fluid intelligence, then, in turn, it might also explain individual differences in *g*, because *g* and fluid intelligence are often nearly perfectly correlated (e.g. Kvist & Gustafsson, 2008).

---

[1] By way of explanation, whereas observed measures capture both construct-relevant and construct-irrelevant variance, latent factors extract variance common to a set of measures and therefore come closer to approximating the theoretical constructs of interest. Latent variables also typically yield stronger correlations than observed measures because they are theoretically free of measurement error.

This excitement was tempered by subsequent findings which revealed that working memory capacity and fluid intelligence were highly correlated, yet distinct. Meta-analyses indicated that the correlation between the two constructs at the latent level was probably closer to $r = 0.70$ than 1.00, indicating that working memory capacity and fluid intelligence shared approximately 50% of their reliable variance (Ackerman et al., 2005; Kane et al., 2005; Oberauer et al., 2005). Nevertheless, this raised the question of what cognitive construct (or constructs) could account for the relationship between working memory capacity and fluid intelligence.

Our position is that most of the variance shared between working memory capacity and fluid intelligence is attributable to attention control (Burgoyne & Engle, 2020; Draheim et al., 2021; Engle et al., 1999). That is, if working memory capacity reflects the interplay between a central executive component and short-term storage components, then it is individual differences in the functioning of the central executive that largely drive the correlations between working memory capacity and other abilities and life outcomes. To test this idea, Engle et al. (1999) measured working memory capacity, short-term memory, and fluid intelligence at the latent level, using multiple tasks to measure each construct. They found that working memory capacity predicted fluid intelligence even after accounting for individual differences in short-term memory. That is, while the path from short-term memory to fluid intelligence was not significant after accounting for working memory capacity, the path from working memory capacity to fluid intelligence was substantial and significant after accounting for short-term memory (Figure 7.3). This result was corroborated and extended by Conway et al. (2002), who found that working memory capacity predicted fluid intelligence after accounting for both short-term memory and processing speed. Taken together, these findings indicated that it was not short-term storage or processing speed that explained working memory capacity's relationship with fluid intelligence, but rather, the fact that tests of working memory capacity, short-term memory, processing speed, and fluid intelligence all require controlled attention.

More recently, we examined the relationships between working memory capacity, attention control, fluid intelligence, and auditory discrimination ability, which was operationalized as one's ability to distinguish between different tones in terms of pitch, loudness, and duration (Tsukahara et al., 2020). We found that latent factors representing each broad cognitive ability were moderately to highly correlated with each other (Figure 7.4; Burgoyne et al., 2021), corroborating over a century of intelligence research (Spearman, 1904). Next, we modelled a higher-order g factor, which was specified to explain the relationships among the broad cognitive ability factors. We found that attention control had the highest loading on the g factor (i.e. a loading of 0.98), indicating that the domain-general ability to control attention is closely related to g (Figure 7.5a). Finally, we put attention control in place of the g factor, and found that once variance in attention control had been partialled out of the broad cognitive ability factors, the residual correlations between working memory capacity, fluid intelligence, and auditory discrimination

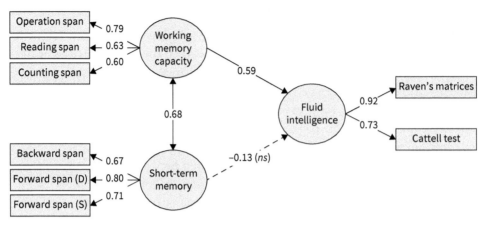

Figure 7.3 In this structural equation model adapted from Engle et al. (1999), working memory capacity predicted fluid intelligence after taking into account the independent contribution of short-term memory.

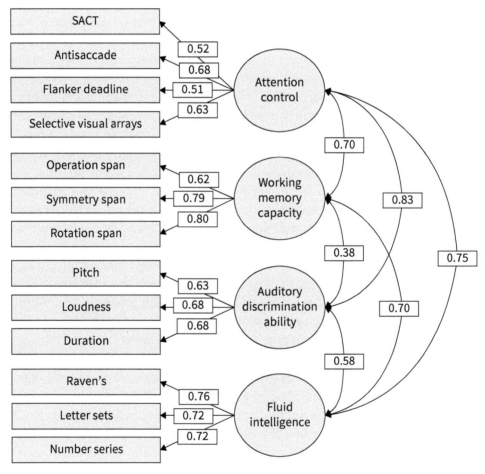

Figure 7.4 In this correlated-factors model adapted from Burgoyne et al. (2022), latent factors representing attention control, working memory capacity, auditory discrimination ability, and fluid intelligence are moderately to strongly related to each other.

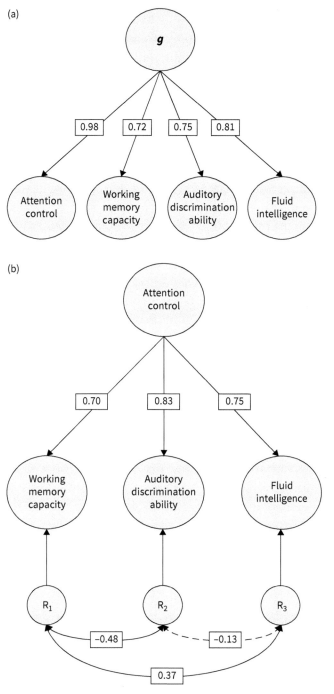

Figure 7.5 (a) This panel shows that when the latent factors depicted in Figure 7.4 are modelled under a higher-order *g* factor, all cognitive abilities load significantly on the *g* factor, with attention control having the highest loading (0.98). (b) This panel shows that when attention control is modelled as a higher-order factor that explains the covariation between broad cognitive abilities, the residual correlations between the broad cognitive abilities—representing the variance in each ability not accounted for by attention control—are significantly weaker than before, and at times reduced to non-significance. For example, compare the residual correlations depicted in Figure 7.5a to the correlations depicted in Figure 7.4. These figures were adapted from Burgoyne et al. (2022).

ability were significantly reduced, at times to non-significance (Figure 7.5b). Taken together, these results suggest that attention control is a key construct for explaining individual differences in cognitive ability, although it is almost certainly only one piece of a more complicated puzzle.

Maintenance and disengagement

If attention control is indeed key to explaining individual differences in cognitive abilities and the relationships between them, one might wonder *how*, on a mechanistic level, it supports cognitive functions such as problem-solving and remembering salient information. We have proposed the *maintenance and disengagement* theory of attention control to explain these phenomena (Burgoyne & Engle, 2020; Shipstead et al., 2016). Just as the central executive in Baddeley's working memory system is responsible for controlling what enters and exits short-term storage, we think of attention control in similar terms; it is responsible for *maintaining* information that is useful to current goals, and *disengaging* from information that is no longer useful, or was never useful to begin with. Maintenance and disengagement are brought to bear on a wide range of tasks, helping to explain why attention control is closely related to the *g*-factor, or general intelligence.

For example, maintenance and disengagement both contribute to problem-solving. As we hypothesize about potential solutions to a problem, we must keep track of the constituent ideas pertaining to each hypothesis and also remember which hypotheses we have tested and which we have not, both of which require information maintenance. Even the act of hypothesis testing requires keeping track of a prediction and comparing it against evidence, another role for maintenance. However, once a hypothesis has been tested and rejected, we must abandon it in pursuit of other hypotheses (i.e. disengagement). People who are able to flexibly attend to what matters and discard what does not are better able to solve novel problems than those who are less able to control their attention (see Krieger et al., 2019; Shipstead et al., 2016).

Maintenance and disengagement also play an important role when completing working memory tests. For instance, in the Operation Span task (Turner & Engle, 1989; Unsworth et al., 2005), participants must solve maths problems (e.g. 'Does $(4 \times 2) + 2 = 10$?') while remembering letters, which are presented in an interleaved fashion between each maths problem. The primary task is to memorize the letters, placing a clear burden on information maintenance. The challenge is that participants must attend to the maths problems to solve them, and then rapidly disengage from the arithmetic solution to attend to and remember the next letter in the sequence. This dual-tasking is a key feature of complex span tests of working memory capacity, and places a strong burden on the ability to control one's attention (Draheim et al., 2021; Kane et al., 2004).

Interim summary

Thus far, we have discussed individual differences in intelligence, with a focus on three highly correlated cognitive constructs: working memory capacity, attention control, and fluid intelligence. We presented evidence for the executive attention view, which places attention control at the centre of individual differences in working memory and helps explain why working memory capacity and other broad cognitive abilities are related to one another. In the next sections, we review evidence for how individual differences in cognitive ability are related to real-world outcomes, including school performance, career attainment, physical health, and psychological well-being.

Predicting life outcomes from cognitive abilities

Psychological science, done well, is a resource-intensive affair. Scarcity of time and money limits the quality of data generated by researchers. Unfortunately, some of the most sought-after data, such as those from nationally or internationally representative samples or from longitudinal studies, are also the most resource intensive to obtain. However, numerous longitudinal cohort studies, which often follow large, representative samples over time, have been conducted or are currently under way. While limited in their own respects (e.g. studies that began decades ago are limited by the knowledge and methods available at the start of data collection), these studies are treasure troves. Since they are often overseen by a national government, researchers have been able to track individual participants for follow-up testing for decades. This also enables researchers to link participants' data with governmental records, including health status, census data, and death records. These investigations have several obvious strengths. First, the sheer size of the samples, sometimes numbering in the tens of thousands depending upon the specific study, obviates usual concerns about power and type I (i.e. false positives) and type II error (i.e. false negatives). Second, longitudinal studies mitigate concerns about cohort effects threatening internal validity (although cohorts may still meaningfully differ from one another), while also making issues of causality easier to parse due to temporal precedence. Third, these studies often have recruitment strategies that create more representative samples than typical laboratory investigations, reducing selection effects and threats to external validity. That said, some (see Batty, Mortensen, & Osler, 2005; Calvin et al., 2011) are heavily biased by participant characteristics such as sex.

These investigations have revealed several notable findings. First, cognitive ability is quite stable over time. One study in which 101 participants tested at age 11 were retested almost *seven decades* later revealed a test–retest correlation of $r = 0.73$ after correcting for measurement unreliability, indicating astonishing consistency in rank ordering over time (Deary et al., 2000). This provides some confidence that when researchers attempt to measure intelligence, they are measuring a stable, persistent trait, or at least a facet of one. This finding was replicated with a separate set of 550 participants (316 women) at age 80: the disattenuated test–retest correlation was also $r = 0.73$. Bifurcating the sample by sex led to a disattenuated test–retest correlation of $r = 0.71$ for men and $r = 0.78$ for women, providing further evidence that intelligence tests are estimates of stable individual differences (Deary et al., 2004). Finally, Deary (2014) reviewed evidence from several other studies which administered intelligence tests to the same individuals several decades apart. In each case, the correlation was strong, with scores at each occasion correlating around $r = 0.70$.

This long-term stability provides reassurance that those studying the relationships between intelligence and life outcomes are not doing so in vain. A recent study by Brown et al. (2021) provides even more encouraging news. They analysed the relations between intelligence test scores and life outcomes (vocational/financial, social, and health outcomes) using data gleaned from four large cohort studies conducted in the UK and the US. In particular, they tested whether there were curvilinear relations which would lead to different effects of intelligence across the range of test scores. For instance, many laypersons believe that being too intelligent may actually lead to worse life outcomes, or at least that being intelligent past a certain point confers no additional advantage, a position sometimes called the threshold hypothesis (Gladwell, 2008; Robertson et al., 2010). However, the Brown et al. findings showed that intelligence tended to be a positive linear predictor of a range of outcome measures. Although intelligence test scores did not predict every outcome tested, there was little evidence that greater intelligence was ever harmful. Furthermore, where relationships with intelligence were detected, there was no evidence to support the threshold hypothesis. Even among high-intelligence subsamples, higher intelligence continued to confer benefits (Brown et al., 2021; Robertson et al., 2010).

In the following sections, we review evidence from large, longitudinal studies wherever possible. Due to the relative recency of theoretical and measurement developments in working memory capacity and attention control, many studies focus primarily or exclusively on cognitive ability as measured by intelligence tests, intelligence being the eldest of the three main constructs of interest. However, we supplement these longitudinal intelligence studies by also describing smaller laboratory-based studies of working memory capacity and attention control where possible.

Academic achievement

When it comes to academic achievement, why do some people rapidly progress through challenging courses while others struggle to master basic concepts? Individual differences in intelligence explain considerable variability in school success, with correlations typically ranging from $r = 0.50$ to $r = 0.70$ (Jensen, 1998; Rohde & Thompson, 2007). Of course, intelligence is only one piece of a complicated puzzle, and other factors such as motivation, socioeconomic status, and school quality certainly play a role (Fortier et al., 1995; Heyneman & Loxley, 1983; Sirin, 2005). Nevertheless, individual differences in cognitive ability become increasingly consequential in school as life trajectories diverge. Not only do people find their interests and begin to pursue them in earnest, but many career prospects are constrained by measures of academic achievement, including grade point averages and standardized test performance. We begin this section by reviewing evidence from longitudinal studies and meta-analyses suggesting that working memory capacity and attention control have an early influence on maths and reading achievement, followed by a discussion of potential mediating mechanisms.

Differences in cognitive abilities such as working memory and attention control are detectable early on, and they predict subsequent academic achievement. As one example, Blankenship et al. (2019) examined the development of attention control in infants and its relationship to reading achievement at age 6 in a longitudinal study of 157 children. The researchers first measured the attention control abilities of 5-month-old infants by showing them a 45-second video clip from the children's television show *Sesame Street*. The researchers estimated individual differences in attention control by measuring the longest duration the children looked at the video as well as the number of times they shifted their gaze.

Blankenship et al. (2019) found that infants who were better able to control their attention while watching *Sesame Street* performed better on a test of executive functioning called the A-not-B task 5 months later. In the A-not-B task, the infants were challenged to find a toy placed in a new location. To do so, they needed to avoid perseverating on a previously learned location. The researchers also found that individual differences in executive functioning were reliable across development; they measured executive functioning at ages 3, 4, and 6 using memory span tasks and tests of attention control and found that each measure of executive functioning was significantly related to the measure collected before it. Moreover, the relationship between the infants' ability to control their attention and their reading achievement at age 6 was mediated by executive functioning, a result which held even after controlling for verbal intelligence. This suggests that individual differences in the domain-general ability to control attention can be detected at an early age and contribute to reading achievement above and beyond domain-specific verbal abilities.

As another example, Ahmed et al. (2019) longitudinally tracked 1273 children from preschool (age 4.5) to high school (age 15) to examine the effects of working memory capacity, other executive functions, and the home environment on maths and literacy achievement. They measured working memory capacity at age 4.5 using a language recall task, in which the children were presented words, phrases, and sentences, and asked to repeat them after a delay. The other measures of executive functioning included a test of sustained attention, in which the children viewed pictures of common objects and pressed a button when a target stimulus appeared, and a test of inhibition called the Children's Stroop task, in which they were shown pictures of daytime and night-time scenes and asked to say the word 'day' in response to night-time scenes and 'night' in response to daytime scenes (Gerstadt et al., 1994). The researchers also measured the preschoolers' academic performance using a picture vocabulary test and an applied maths test designed for young children.

A number of control variables were also examined to help disentangle the effects of cognitive ability from environmental effects. For instance, Ahmed et al. (2019) measured the mothers' educational attainment, the household income-to-needs ratio, the amount of learning materials in the home, the parents' level of involvement and responsivity to the child's needs, and whether the children were ever placed in child care. They also collected demographic information such as race and sex.

Children with greater working memory capacity and other executive functions had significantly better academic performance in preschool. Furthermore, childhood working memory and executive functions predicted adolescent academic achievement at age 15 as measured by tests of reading comprehension, applied maths problems, verbal analogies, and picture vocabulary. That said, these results were qualified by including other predictors in the model, suggesting that some of the variance accounted for by cognitive ability was shared with other factors. For example, academic performance at age 4.5 accounted for meaningful variance in academic achievement at age 15, but it also reduced the effects of most of the cognitive ability predictors to non-significance. In particular, only working memory capacity remained a significant predictor of paragraph comprehension at age 15 after accounting for childhood academic achievement. Nevertheless, working memory capacity at age 4.5 remained a significant predictor even after accounting for demographic and home environment covariates, suggesting that it captured unique variance in adolescent academic performance. Thus, measures of working memory predicted children's academic achievement more than 10 years later.

Ahmed et al. (2019) report more modest findings regarding executive functions, but they should be interpreted cautiously. They found that childhood executive functions measures were no longer significant predictors of adolescent achievement after accounting for covariates. However, this *does not* indicate that executive functions are unimportant in the early stages. To the contrary, all the childhood executive function measures were significantly related to maths and vocabulary achievement at age 4.5. One interpretation of these results is that the predictive variance of the executive function measures may have been captured by childhood academic performance. Another possibility is that the childhood executive function measures were not as reliable as the working memory measure was, and, as a result, had attenuated predictive validity. Indeed, working memory capacity at age 4.5 predicted working memory capacity at age 15, whereas the other executive function measures were not significantly correlated across timepoints. This suggests that the test–retest reliability of some of the measures may have been low, a common problem in developmental studies of cognitive ability and in the executive function literature more broadly (Beck et al., 2011; Draheim et al., 2019).

Meta-analytic evidence also indicates that individual differences in working memory capacity predict reading achievement. Early evidence was provided by a meta-analysis of 77 studies (Daneman & Merikle, 1996), which revealed that complex span measures of working memory capacity predict reading comprehension to a considerable degree, with correlations ranging from $r = 0.30$ to $r = 0.52$. Daneman and Merikle (1996) also showed that complex span tasks with *verbal* stimuli yielded the strongest relationships between working memory capacity and reading comprehension, although even complex span tasks that do not use verbal stimuli for the processing subtask also predict individual differences in reading comprehension (Turner & Engle, 1989). More recently, Peng et al. (2018) meta-analysed 197 studies and found an average correlation of $r = 0.29$ between working memory capacity and reading abilities. The magnitude of the correlation was fairly similar across types of reading skills, and ranged from $r = 0.26$ for vocabulary to $r = 0.34$ for phonological coding. Like Daneman and Merikle (1996), Peng et al. (2018) also found that verbal tests of working memory had the numerically largest correlation with reading abilities, but that tests of working memory that used non-verbal stimuli also predicted individual differences in reading skill. These results suggest that tests of working memory capacity tap a domain-general ability, which we would identify as attention control, that is important for language processing.

Meta-analytic evidence also indicates that individual differences in working memory capacity predict mathematics achievement (Spiegel et al., 2021). For example, a meta-analysis of 110 studies by Peng et al. (2016) found an average correlation of $r = 0.35$ between working memory capacity and measures of maths ability. An examination of specific maths domains revealed that the largest correlation was between working memory capacity and word-problem solving ($r = 0.37$), whereas the smallest correlation, though still significant, was between working memory capacity and geometry ($r = 0.23$).

Although we have reviewed some evidence for the relationship between academic achievement and cognitive abilities such as working memory capacity and attention control, the specific mechanisms by which these cognitive abilities exert their influence has not been discussed. In light of our maintenance and disengagement framework, we now provide illustrative examples suggesting that working memory capacity and attention control contribute to school performance by supporting the maintenance of relevant information and disengagement from irrelevant and no longer relevant information.

Being able to follow instructions is critical for students to learn new skills in the classroom. Engle et al. (1991) investigated why some children are better able to follow directions than others. Their sample consisted of 120 students from first grade, third grade, and sixth grade. They asked students to follow 45 sets of directions varying in complexity, such as 'Point to the picture at the top of page three and copy it twice' (p. 256). They found that working memory capacity was significantly related to students' ability to follow directions, and it also predicted their performance on a reading comprehension test. Furthermore, Engle et al. (1991) found that as the directions became more complicated, the gap in performance between students with the highest and lowest working memory capacity increased. One interpretation of this result is that students with greater working memory capacity were better able to ignore distractions and maintain focus on task instructions. Furthermore, it seems that attentional abilities were especially important when dealing with more complex information processing demands, which would have implications for learning complicated course content. For a more comprehensive review of research on following instructions, see Allen et al., Chapter 10, this volume.

The hypothesis that individual differences in attention control explain why some students are better able to maintain focus on task-relevant information was later corroborated by a study of mind wandering, attention control, and reading comprehension (McVay & Kane, 2012). The researchers had more than 200 participants read passages of text ranging from a short article on volcanoes to five chapters of *War and Peace*. As the participants read, they were intermittently asked to report instances of task-unrelated thoughts and to describe what they were focusing on before they were interrupted. McVay and Kane (2012) found that attention control was a strong predictor of reading comprehension. Furthermore, participants with greater attention control had fewer task-unrelated thoughts, indicating that they were better able to focus on the reading material and were less susceptible to distractions. In turn, task-unrelated thoughts partially explained the relationship between attention control and reading comprehension performance, although the direct path from attention control to reading comprehension remained significant even after accounting for task-unrelated thoughts. This suggests that the ability to maintain focus and resist mind wandering captured part, but not all, of the covariation between attention control and reading comprehension.

As a final example, Tolar et al. (2009) examined the contributions of working memory capacity, computational fluency, and three-dimensional spatial visualization to algebra achievement and performance on a standardized maths test, the SAT, in a sample of over 100 undergraduate students. They found significant relationships between the measures of working memory capacity and maths achievement. Furthermore, using latent variable analyses, they found that working memory capacity had significant relationships with computational fluency and three-dimensional spatial visualization, which in turn explained the relationship between working memory capacity and maths achievement. Thus, two mechanisms by which working memory appeared to contribute to maths achievement was via computational fluency (e.g. being able to solve arithmetic problems such as $64 \div 4$), which would facilitate solving algebra problems, and three-dimensional spatial visualization (e.g. being able to mentally rotate shapes), which may be useful for solving geometry problems.

Individual differences in intelligence, and, in particular, working memory capacity and attention control, explain considerable variance in academic achievement.

Evidence suggests that differences in these cognitive abilities can be detected at an early age and that they are relatively stable across development. The ability to control attention to maintain and manipulate information in service of goals appears to influence academic achievement via mediating mechanisms such as following directions, resisting mind-wandering during task performance, computational fluency, and spatial visualization, to name a few. In the next section, we discuss how individual differences in cognitive abilities such as working memory capacity and attention control contribute to success in the occupational sector.

Job performance

It is well established that measures of cognitive ability predict job performance, training success, and career attainment (Bobko et al., 1999; Hunter, 1986; Schmidt, 2002). Indeed, companies and organizations use cognitive ability tests to select and classify personnel because it increases the productivity of the organization (Schmidt & Hunter, 1998). In this section, we review evidence for the relationship between cognitive ability and job performance, paying close attention to the roles of working memory capacity and attention control.

In now classic work, Schmidt and Hunter (1998) synthesized 85 years of research on the validity of personnel selection assessments for predicting job performance and training success. Job performance was primarily measured using supervisor ratings of worker performance, whereas training success was typically defined as the amount learned on the job. Schmidt and Hunter (1998) found that *general mental ability*, a term frequently used by industrial/organizational psychologists to refer to general intelligence, was the single best predictor of job performance ($r = 0.51$) and training success ($r = 0.56$) after correcting for the attenuating effects of criterion unreliability and restriction of range among incumbent workers. Furthermore, Schmidt and Hunter (1998) reported that the relationship between cognitive ability and job performance was moderated by job complexity; for highly complex professional/managerial jobs, the average correlation was as high as $r = 0.58$, whereas for completely unskilled jobs, the average correlation was somewhat lower, though still meaningful: $r = 0.23$.

The validity of cognitive ability for predicting job performance remains substantial even as workers gain experience on the job. McDaniel (1986) found that as the level of job experience increased, the validity of cognitive ability did not decrease. At 0–3 years of experience, the correlation was $r = 0.35$, whereas at 9–12 years of experience, the correlation was $r = 0.44$. This result runs counter to the *circumvention of limits hypothesis* (Hambrick & Meinz, 2011; Salthouse, 1991; Schmidt et al., 1988), which suggests that cognitive ability will cease to predict job performance after people have acquired domain-specific knowledge, skills, and strategies. Although domain-specific knowledge and skills may enable performers to bypass reliance on capacity-limited aspects of cognition in relatively simple, consistent tasks (Ackerman, 1988), these results do not appear to generalize to real-world jobs, which are often more complicated and change over time (see Hambrick et al., 2019). Thus, cognitive ability often remains a significant predictor of job performance even after years of training.

In addition to predicting job performance, measures of cognitive ability also predict career attainment, operationalized as occupational level or prestige within a field. For example, in a longitudinal study of over 3000 young adults (National Longitudinal Survey—Youth Cohort; Center for Human Resource Research, 1989), Wilk et al. (1995) found that general mental ability predicted subsequent job movement 2–7 years later. Higher-ability people tended to move up the job hierarchy during the intervening years, whereas lower-ability people moved down the job hierarchy. Additionally, Judge et al. (1999) found that measures of cognitive ability at age 12 predicted occupational outcomes 30–40 years later ($r = 0.51$), as well as adult income ($r = 0.53$).

Interestingly, the relationship between cognitive ability and career attainment cannot be fully explained by socioeconomic factors or other environmental variables such as school quality or differential access to opportunities. Murray (1998) controlled for these variables by sampling full biological siblings, who shared the same parents and home environment but differed in cognitive ability. Despite being raised in the same household, siblings with higher levels of cognitive ability were far more successful in the academic and occupational sectors than their lower-ability counterparts; they received more years of education, entered more prestigious professions, earned a higher income, and had more regular employment. Thus, the relationship between cognitive ability and career attainment is not merely a consequence of environmental factors such as socioeconomic status; cognitive ability apparently exerts an influence on career attainment above and beyond these factors. Taken together, the prevailing conclusion to emerge from this research is that people with greater cognitive ability learn more from job training programmes, acquire job-related skills faster, perform better, and move up the professional hierarchy more than individuals lower in cognitive ability.

Presumably, measures of specific cognitive abilities such as working memory capacity and attention control should also predict job performance to a considerable degree, given their strong relationships with general intelligence (Burgoyne et al., 2021; Conway et al., 2003). That said, while there is more than a century of evidence supporting the use of intelligence tests for personnel selection, research on the predictive validity of working memory capacity and attention control is, by comparison, still in the early stages.

Researchers have begun investigating working memory capacity and attention control tests as candidates for personnel selection assessments because they may reduce adverse impact relative to traditional tests, which often focus on acquired knowledge and demonstrate substantial differences between majority and minority groups (Burgoyne et al., 2021). *Adverse impact* refers to the disproportionate selection or promotion of members of one group over another. For an example, consider the Armed Services Vocational Aptitude Battery (ASVAB), the standardized test taken by all US military applicants. Due to mean differences in performance on the ASVAB, the selection rate for Black applicants is less than 80% of the selection rate for White applicants (ASVAB Enlistment Testing Program, 2020). Although the ASVAB predicts performance in the military to approximately the same degree for White and Black enlistees (Wise et al., 1992), its use results in the inequitable selection of Black applicants and women (see, e.g. Held et al., 2014).

One potential reason why the ASVAB results in adverse impact is because it heavily emphasizes acculturated knowledge. Although acculturated knowledge such as automotive and shop information is clearly relevant to some military jobs, Outtz and Newman (2011) found that the subtests of the ASVAB with the largest differences between White and Black applicants were those that measured technical knowledge and those that measured verbal abilities. Along similar lines, Hough et al. (2001) found that tests of acquired knowledge (e.g. verbal ability, science knowledge, and quantitative ability) tended to result in larger group differences in performance than tests of memory, processing speed, and spatial abilities. As an explanation for these differences between groups, others have argued that measures of acculturated learning may be especially sensitive to socioeconomic status and previous educational opportunities (Bosco et al., 2015; Outtz & Newman, 2011; Sternberg & Wagner, 1993). Taken together, some personnel selection tests result in less equitable outcomes than others, and so a critical goal for industrial/organizational psychology is to find tests that maintain or improve the prediction of job performance while also reducing adverse impact.

Preliminary evidence suggests that working memory capacity and attention control tests may predict job performance nearly as well as traditional tests while minimizing group differences in performance. For example, Nelson (2003) examined the relationships between working memory capacity, general intelligence, and job performance in a sample of 378 insurance agent support staff. Job performance was measured using supervisor ratings, whereas working memory capacity was measured using a reading span task and another verbal span task. Nelson (2003) found that the working memory capacity measures had good internal consistency reliability (αs = 0.85 and 0.86) and significantly predicted a cognitive job performance composite variable (average $r = 0.17$; $r = 0.24$ after correction for criterion unreliability). For comparison, the correlation between general intelligence and cognitive job performance was only slightly larger, $r = 0.22$ (or $r = 0.31$ after correction for criterion unreliability). Importantly, subgroup differences were smaller for the working memory measures than for general intelligence. Specifically, the standardized mean difference in performance on the working memory measures was $d = 0.40$ for White and Black participants and $d = 0.31$ for White and Hispanic participants, whereas the difference in performance on the general intelligence measure was $d = 1.03$ for

White and Black participants and $d = 0.73$ for White and Hispanic participants. Thus, Nelson's (2003) results suggest that tests of working memory could have less adverse impact than tests of general intelligence, while predicting job performance to a meaningful degree, perhaps due to working memory tests relying less on acquired knowledge.

As another example, Guo et al. (2020) examined the contribution of working memory capacity to job performance in a sample of 70 high-speed railway dispatchers. Working memory capacity correlated significantly with supervisor ratings of performance ($r = 0.90$) and with an objective rating of performance, train delay times in a railway simulator ($r = -0.91$). Guo et al. (2020) speculated that greater working memory capacity allowed dispatchers to maintain salient information in mind (e.g. tracking and controlling train routes) while performing their job.

As yet another example, Bosco et al. (2015) compared the validity of attention control and working memory tests to a conventional test of mental ability, the Wonderlic Personnel Test, for predicting supervisor ratings and performance in a management simulation. In a sample of 470 bank employees and undergraduate students, the attention control and working memory measures predicted performance just as well as the Wonderlic Personnel Test (compare $r = 0.35$ to $r = 0.33$) while reducing group differences between White and Black participants. Specifically, a meta-analysis of their results indicated that the attention control and working memory measures reduced group differences by around half of one standard deviation compared to the Wonderlic Personnel Test (compare $d = 0.68$ to $d = 1.09$).

As a final example, Martin et al. (2020) administered an Armed Services Vocational Aptitude Battery (ASVAB) practice test, attention control tests, fluid intelligence tests, and multitask paradigms which were used as a proxy for complex work performance to a sample of 171 young adults. They found that a latent variable representing attention control accounted for nearly one-quarter of the variance in multitasking performance above and beyond fluid intelligence and the ASVAB. Furthermore, a composite variable based on the attention control measures reduced group differences between White and Black participants by three-quarters of one standard deviation compared to the ASVAB, and two-thirds of one standard deviation compared to the Armed Forces Qualification Test (AFQT) (Burgoyne et al., 2021). That is, the group difference was $d = 1.86$ (95% confidence interval (CI) 1.47, 2.33) for the ASVAB and $d = 1.74$ (95% CI 1.32, 2.23) for the AFQT, compared to $d = 1.11$ (95% CI 0.75, 1.50) for the attention control measures. Although the absolute magnitude of these group differences is quite high, the relative difference across tests revealed much smaller group differences on the attention control measures than on the ASVAB and AFQT. This suggests that attention control tasks could reduce adverse impact while improving the prediction of work performance (Burgoyne et al., 2021).

To sum up, tests of attention control and working memory are valid predictors of job performance, with validity coefficients that are nearly equal in magnitude to those of general intelligence. While it has long been established that measures of cognitive ability predict occupational success, an exciting possibility is that tests of working memory and attention control might mitigate subgroup differences relative to traditional tests, in turn reducing adverse impact. Ultimately, this research could help attain more equitable outcomes in the world of work while continuing to benefit the productivity of organizations.

Physical health and mortality

Serious investigations of the link between cognitive ability and physical well-being are recent. In fact, protracted academic and epidemiological interest in the cognition–health association did not emerge until the 1990s (Deary et al., 2010; Gottfredson & Deary, 2004). Since then, however, several findings have been convincingly established.

First, childhood intelligence predicts later morbidity. Wraw et al. (2015) analysed data from the National Longitudinal Study of Youth 1979, a nationally representative sample of adolescents and young adults in the US who were recruited in late 1978 and periodically provided information about their health, education, and other sociodemographic details. As a measure of cognitive ability, participants completed the AFQT sometime between the ages of 14 and 21, a test which is based on mathematical and verbal ability. In a sample of over 5000 participants, adolescent intelligence was associated with fewer self-reported physical ailments, greater fitness and locomotor function, and higher estimates of health status in middle age. Greater adolescent intelligence was also associated with reduced risk for specific disorders and health conditions, including cardiovascular diseases, respiratory disease, joint discomfort, and use of movement aids.

These effects were only slightly diminished by controlling for childhood socio-economic status, but were greatly attenuated when *adult* socioeconomic status was controlled. In fact, after accounting for adult socioeconomic status, intelligence test scores no longer significantly predicted any summary health outcome, but remained a significant predictor of some specific problems, such as cardiovascular disease—including hypertension and heart attack—as well as joint discomfort. However, Wraw et al. (2015) note that the attenuating effect of adult socioeconomic status on the intelligence–health relationship should be interpreted with caution, as it is ambiguous with regard to causal direction. As Deary et al. (2010) note, some researchers argue that adult socioeconomic status can serve as a proxy for intelligence, in which case the attenuation observed by Wraw et al. (2015) would be an instance of statistical over control. For instance, Kraft et al. (2018) report that, while lower educational attainment predicts greater utilization of healthcare services even in early adulthood, the effect is partly explained by intelligence measured at age 12, which necessarily predates terminal education (see also Judge et al., 1999). This would be inconsistent with a purely socioeconomic account of the cognitive ability–health relationship.

Deary et al. (2010) review other evidence for an association between cognitive ability and health, with lower intelligence being associated with higher rates of self-harm, physical violence, cardiovascular disease (particularly coronary artery disease; see Batty, Mortensen, Andersen, & Osler, 2005; Ferrucci et al., 1993), and tentative evidence for increased rates of stomach and lung cancers, which are the cancers most associated with poor health and lifestyle decisions (e.g. smoking), and are thus arguably more preventable than other forms of cancer.

Increased rates of disease coincide with a diminished likelihood of surviving to old age. Hart et al. (2003) found children with lower intelligence at age 11 were at higher risk of dying within the succeeding 25 years. Accounting for several possible mediators such as one's occupation and neighbourhood affluence ameliorated but did not eliminate the direct effect of low intelligence on mortality rates. These results are corroborated by a meta-analysis of 16 cohort studies (Calvin et al., 2011), and a recent follow-up to the Scottish Mental Survey of 1947, which administered the Moray House Test, a test of cognitive ability, to nearly every Scottish 11-year-old born in 1936. The sample of over 70,000 individuals replicated the intelligence–health association and also revealed that it is strongest for earlier (i.e. prior to age 66), perhaps more preventable deaths (Čukić et al., 2017). Importantly, these mortality data do not differentiate by cause of death, an interesting potential moderator of the intelligence–mortality relationship. Thus, a more fine-grained analysis examining specific terminating events would likely add complexity and nuance to the relationship between cognitive ability and longevity (cf. Deary et al., 2010).

The preceding studies provide compelling evidence that cognitive ability predicts aspects of health and longevity. But why might this be the case? Several explanations have been proposed (Gottfredson & Deary, 2004). The first is that intelligence and health are both indicators of bodily/medical insults accumulated over the lifetime. By this account, the accrual of harm over time, such as by protracted physical inactivity or substance abuse, reduces intelligence while simultaneously leading to worse long-term health outcomes (Gottfredson & Deary, 2004). This account has some intuitive appeal, but the cohort studies previously described, in which intelligence is measured in childhood and used to predict adult health status, suggests that it cannot be the entire story. If it were, one would expect ability and health to track one another closely across time, but one would not necessarily expect earlier cognitive ability to predict later health outcomes. The fact that they do suggests that those who start with higher cognitive ability may behave differently in the intervening time, and the differences lead to health disparities.

This suggests a second explanation which supposes that the cause of the cognitive ability–health association lies not in common causes, but that those higher in cognitive ability are more likely to engage in health-promoting behaviours, avoid harmful behaviours, and/or engage in better self-care (Gottfredson & Deary, 2004). There is strong evidence to support this position. For instance, one study found no intelligence differences between those who began smoking cigarettes and non-smokers for Scottish people born in 1926, when the health risks of smoking were less well known. However, among smokers, those with higher intelligence test scores were more likely to quit smoking in adulthood than those with lower intelligence test scores (Taylor et al., 2003). This may be rooted in a greater ability to act on emerging knowledge about the dangers of tobacco use, such as being better at resisting the urge to smoke or prioritizing long-term health over short-term pleasure, both putative functions of attention control (cf. Shamosh et al., 2008). For example, working memory tests have been shown to predict short-term relapses in cigarette smoking, particularly when no pharmacological assistance is provided to mitigate aversive withdrawal effects (Patterson et al., 2010). Those who perform more poorly on tests of working memory are more likely to relapse because they are less able to control attention capture by their impulses to smoke, making them more likely to yield to nicotine withdrawal. For a more detailed discussion of cognition and addictive behaviours see Andrade, Chapter 14, this volume.

Working memory and executive functioning also predict aspects of healthcare management. For instance, one study found that a composite measure of working memory and executive functioning predicted medication adherence in a community sample of older adults (Insel et al., 2006). Even after accounting for variables such as age, dementia and depression symptomatology, illness severity, education, and financial well-being, the working memory/executive functioning composite measure was the sole significant predictor of treatment compliance (Insel et al., 2006; see also Stilley et al., 2010). Moreover, interventions designed to reduce reliance on working memory and executive functioning for treatment management have been linked to improved adherence rates, particularly for lower-ability individuals (Insel et al., 2016). Finally, those with low working memory capacity have been shown to struggle to comprehend, internalize, and recall health-related information, such as that pertaining to nutrition or signs of stroke, relative to more able peers (Ganzer et al., 2012; Soederberg Miller et al., 2011).

A third explanation for the relationship between cognitive ability and health is that they share common physiological mechanisms (Gottfredson & Deary, 2004). A recent notable example of this approach has been advanced by Geary (2021), who suggested that mitochondrial functioning provides a common basis for understanding an array of psychometric and epidemiological phenomenon, including the uniformly positive correlations among cognitive tests (Spearman, 1904). In particular, Geary (2021) notes that poor mitochondrial functioning limits the efficiency of important, long-range, energy-demanding neural networks underpinning complex cognitive processes such as working memory and attention control, providing a common limiting factor across many cognitive domains (cf. Detterman, 1991; Kovacs & Conway, 2016), health (i.e. poor cellular metabolism is linked to numerous adverse health outcomes and may be improved with lifestyle changes or harmed by unhealthy lifestyle choices), and ageing (i.e. mitochondrial functioning declines with age, contributing to joint declines in health and cognitive ability across adulthood; Geary, 2021).

Yet another explanation for the association between physical health and cognitive ability is rooted in resource inequities across those high and low in cognitive ability (Gottfredson & Deary, 2004). That is, those with higher ability also tend to have greater monetary resources, educational attainment, occupational prestige, and so on. These material advantages confer numerous opportunities, such as greater access to nutritious food, healthcare, and health education. This would mean that socioeconomic status may serve as a common cause for both individual differences in cognitive ability and health outcomes. While they do not investigate health outcomes, Hanscombe et al. (2012) report in a large, UK-based twin study ($N = 8716$ twin pairs) that childhood socioeconomic status does affect individual differences in intelligence, such that children from relatively more disadvantaged homes have more variable IQ estimates than their more advantaged counterparts. After partitioning variation in intelligence into genetic, shared environmental, and non-shared environmental components, the researchers conclude that shared environmental factors account for most of the observed increase in variability. That is, for low socioeconomic status children, shared family and school environments play a larger role in determining IQ than for their higher socioeconomic status peers. Importantly, the genetic effect on intelligence differences was constant across the range of socioeconomic status (Hanscombe et al., 2012; but see Turkheimer et al., 2003). Thus, individual differences in cognitive ability cannot be dismissed merely as artefacts of socioeconomic status, since low socioeconomic status affects primarily the environmental but *not* the genetic variability in intelligence. This also clarifies how controlling for childhood socioeconomic status can have little effect on the cognitive ability–health relationship: if the association between the physical well-being and cognitive ability is mainly driven by the genetic component of intelligence, then controlling for childhood socioeconomic status should leave the relationship largely intact (cf. Wraw et al., 2015).

In summary, high performance on cognitive ability tests is associated with reduced risk of disease, preventable death, and better functional outcomes. As with job performance, most of the research in this domain focuses on 'intelligence', broadly defined. As such, continued investigations into the specific relationships between health and working memory capacity and attention control will no doubt prove fruitful. In particular, we suspect that superior attention control may confer health benefits through better impulse control and maintenance of long-term goals, and future longitudinal studies should include valid measures of attention control to investigate this possibility. Despite this limitation, a major strength of the studies we reviewed is that they indicate that cognitive ability may play a causal role in health by showing that intelligence in childhood predicts health and mortality decades later. Nevertheless, the relationship between cognitive ability and health is likely bi-directional and influenced by other factors. Furthermore, future epidemiological research should examine not only the effects of cognitive ability and health, but also the relationship of health to other life outcomes such as education and career attainment.

Psychological well-being

Performance on cognitive ability tests is impaired by many psychiatric disorders as well as adverse life events. Simultaneously, high cognitive ability is a protective factor against many clinical and non-clinical psychological ailments. For instance, Batty, Mortensen, and Osler (2005) describe a study of over 7000 Danish men born in 1953 who completed an intelligence test at age 13. Participants were tracked via a government health database for incidents of psychiatric disorders occurring between 1969 and 2002. Children with lower test scores were at elevated risk of developing a psychiatric disorder in the following years compared to children with higher intelligence test scores, a pattern which held even after controlling for participants' birthweight and their father's occupation.

Ohi et al. (2022) took a different approach by assessing whether developing major depression, bipolar disorder, or schizophrenia lowered participants' intelligence test scores. Ohi et al. (2022) estimated premorbid intelligence using the Japanese translation of the National Adult Reading Test, a measure of crystallized intelligence. Crystallized intelligence is thought to be relatively stable throughout the lifespan and is often spared in psychological disorders (Horn & Cattell, 1967; Ohi et al., 2022; Russell, 1980; Wang & Kaufman, 1993). Importantly, theorists (Cattell, 1987; Kvist & Gustafsson, 2008) have proposed a causal link between fluid and crystallized intelligence, such that higher fluid intelligence in youth coupled with more opportunities for learning and investment by the learner lead to higher crystallized intelligence: an individual's level of knowledge is a function of their fluid intelligence at the time they learned that knowledge. Thus, the use of a crystallized intelligence test to estimate premorbid intelligence appears sensible. For comparison, the researchers measured current intelligence using the Wechsler Adult Intelligence Scale, which incorporates aspects of both crystallized and fluid intelligence. Notably, fluid intelligence is more sensitive to effects of ageing, injury, and disease (Horn & Cattell, 1967; Ohi et al., 2022; Russell, 1980; Wang & Kaufman, 1993). Consistent with Batty, Mortensen, Andersen, and Osler (2005), participants with a clinical diagnosis of major depressive disorder, bipolar disorder, or schizophrenia had lower premorbid intelligence than healthy controls (Ohi et al., 2022). More strikingly, they also demonstrated marked declines from premorbid to current intelligence in participants with a clinical diagnosis, whereas intelligence estimates for healthy controls were stable across the two time points. These results held after controlling for age and sex (Ohi et al., 2022). Taken together, these results suggest that while those lower in cognitive ability are at elevated risk for psychiatric disorder, disorders themselves may also lead to further impairment.

Further evidence for an association between cognitive ability and psychological well-being comes from the post-traumatic stress disorder (PTSD) literature. Macklin et al. (1998) found that individuals with lower cognitive ability may be at higher risk of developing PTSD. Superficially, they found that US Vietnam War veterans who scored lower on pre-combat intelligence tests were more likely to develop PTSD post combat; they also reported more severe PTSD symptomology. Importantly, these results held even after accounting for combat exposure and post-combat intelligence. This last point is crucial, since lower scores on *pre-combat* aptitude tests increase the likelihood that one will see active combat (Macklin et al., 1998). Breslau et al. (2013) also reported increased risks of PTSD symptomatology among adolescents who scored poorly on childhood intelligence tests. In their study, 6-year-olds were recruited from hospitals and administered the Wechsler Intelligence Test for Children—Revised. They were contacted again at age 17 and asked to report instances of traumatic events and completed PTSD diagnostic screenings. Of the 713 participants, approximately 75% reported some traumatizing event. Of these, 45% met criteria for a PTSD diagnosis according to the fourth edition of the *Diagnostic and Statistical Manual of Mental Disorders*. Within this subgroup, those with lower childhood intelligence scores were at higher risk, with a one standard deviation decrease in childhood intelligence being associated with approximately a 50% increased risk of PTSD diagnosis following trauma exposure (Breslau et al., 2013).

As was the case for physical well-being, there are several non-mutually exclusive explanations for why cognitive ability predicts aspects of psychological well-being. Here, we focus on one aspect of psychological well-being in particular: *emotion regulation*. Emotion regulation is a critical life skill, and studies show that effective emotion regulation may be an important component of long-term psychological and physical well-being. For instance, a nationally representative sample of over 1100 adults in the US answered questions about daily stressors (e.g. having an argument) and negative affect (Leger et al., 2018) for 8 consecutive days. Participants' stressors and negative affect were then used to predict health outcomes 10 years later. Importantly, researchers omitted days on which participants reported experiencing a stressor from their analyses, focusing instead on lingering negative affect on the day *following* a stressor. Participants with greater lingering negative affect likely suffer from poor emotion regulation and may be at higher risk for adverse health events, perhaps due to prolonged physiological stress responses damaging biological systems (Carlson et al., 2012; Geary, 2021). Indeed, lingering negative affect predicted incidents of chronic disease and difficulties completing daily activities without assistance. Moreover, these results held even after accounting for health 10 years prior, sex, age, and educational attainment (Leger et al., 2018).

Evidence suggests that those higher in cognitive ability may be more effective at coping with stress and regulating negative emotions. Garrison and Schmeichel

(2022) probed participants to answer questions about their stress and affect as they went about their day. Participants also completed two versions of the operation span. The researchers found that, while participants generally experienced increased negative affect following stressful events, those who performed well on the operation span tasks experienced less negative affect following negative events than participants who scored lower (Garrison & Schmeichel, 2022; see also Coifman et al., 2021). Moreover, a meta-analysis conducted by Moran (2016) which included over 22,000 data points found that working memory capacity moderately predicted anxiety ($r = -0.33$). In particular, domain-general, attentionally demanding working memory tasks showed the strongest, most consistent relationship and was associated with both facets of anxiety: cognitive (i.e. worry) and affective (i.e. arousal; Moran, 2016). This suggests that attention control likely has a particularly important part to play in explaining anxiety.

Indeed, attention control has been central to theories of emotion regulation more generally (Ochsner et al., 2012). For instance, Engen and Anderson (2018) contend that proficient control of memory is the cognitive basis of emotion regulation. They note that many affective disorders coincide with memory difficulties, including rumination (i.e. the tendency to spontaneously elaborate on or reexperience emotional thoughts and memories) and intrusive thoughts (see Buckley et al., 2000; Pe et al., 2013; Yoon et al., 2014). Moreover, they propose several mechanisms by which this may occur. The first is direct suppression, and involves preventing unwanted negative memories from entering active memory, perhaps through cognitive inhibition (Cohen et al., 2014) or memory updating (Yoon et al., 2014). A second is thought substitution, in which alternative memories to the problematic ones are retrieved. Over time, this strengthens retrieval structures for the preferable memories, increasing the likelihood that they will be retrieved in contexts which previously elicited negative memories.

While Engen and Anderson (2018) prefer the term 'memory control', the concepts of direct suppression and thought substitution have clear conceptual overlap with working memory, intelligence, and attention control. Rosen and Engle (1998), for instance, found evidence that high-working memory individuals were superior at suppressing unwanted items from working memory. Across two experiments, participants learned three lists of paired associates. For some participants, all three lists contained unique associates (i.e. the non-interference list). For other participants, the first and third lists were duplicates, while the second list paired words seen in the first list with new associates (i.e. the interference lists; see Table 7.1 for example list items). This should have created competition for activation between the two associates paired with the same word. After learning the lists, participants attempted to recall the correct paired associate for a given word in a given list. The main question of interest was what effect these interference manipulations would have on list recall by participants determined to be high and low in working memory capacity.

Table 7.1 Examples of paired associates used by Rosen and Engle (1998)

	List 1	List 2	List 3
Non-interference			
Participant #1	dust–pan	bird–dawn	eye–glass
Participant #2	dust–pan	eye–tear	bird–dawn
Interference			
Participant #3	bird–bath	bird–dawn	bird–bath
Participant #4	eye–glass	eye–tear	eye–glass

Table reproduced from Rosen and Engle (1998).

When participants were instructed to rapidly recall items from the interference lists, high working memory capacity participants had fewer between-list intrusion errors in the second list than did low working memory participants. That is, high working memory capacity participants did a better job of *not* recalling the word that had been paired with the test item in the *previous* list. In a second experiment where participants were asked to emphasize accuracy, high working memory individuals in the interference condition were slower to retrieve items from the third list than items from the first list. Retrieval times for low working memory individuals, meanwhile, did not differ. This pattern would be expected if high working memory individuals suppressed paired associates from the first list in order to minimize interference with the new associates in the second list. Together, these experiments suggest that high working memory capacity individuals are better at suppressing unwanted items in memory.

Rosen and Engle's (1998) results are similar to those reported by Brewin and Beaton (2002), who employed the 'white bear' paradigm. Participants were first trained on a think-aloud protocol in which they continually verbalized their stream of consciousness. Next, participants were asked to avoid thinking about a white bear. They were asked to ring a bell any time they thought about or mentioned a white bear. Participants also completed measures of working memory capacity, fluid intelligence, and crystallized intelligence, to determine whether any of these cognitive abilities predicted success at avoiding intrusive thoughts about white bears. Working memory capacity and fluid intelligence were significant independent predictors of participants' ability to avoid thinking of a white bear, with high-ability participants having fewer intrusive thoughts. Together, the results from Rosen and Engle 91998) and Brewin and Beaton (2002) establish a clear connection between memory suppression and cognitive ability as indexed by working memory capacity and fluid intelligence tasks.

We have argued that controlling the contents of working memory according to current goals is the key function of attention control, and that working memory capacity tasks emphasize maintenance of immediately relevant information in working memory whereas fluid intelligence tasks place greater emphasis on disengaging from or removing no longer relevant information (Shipstead et al., 2016). In our own framework, Engen and Anderson's (2018) notion of direct suppression of accords nicely with attentional disengagement, whereas thought substitution likely incorporates aspects of attentional maintenance (e.g. allocating attention to a preferable memory/interpretation rather than a psychologically distressing one) with attention control ability important to both. Engen and Anderson (2018) also nominate direct suppression and thought substitution as conjointly undergirding *cognitive reappraisal of memories*, one of the most well-studied emotion regulation strategies, in which new, more positive/adaptive meanings, evaluations, or interpretations are retrospectively made of negative experiences. This, of course, entails that effective reappraisal is related to cognitive ability as indexed by measures of attention control, working memory capacity, and fluid intelligence. This appears substantiated (Andreotii et al., 2013; Cohen et al., 2014; Gan et al., 2017; McRae et al., 2012; Ochsner et al., 2012; Zaehringer et al., 2018). For example, Pe et al. (2013) found that participants who reported engaging in more attempts at reappraising negative memories also reported lower levels of negative affect. However, this effect only held for participants who performed well on an *n*-back task that used emotionally valenced stimuli, which was used to measure working memory updating ability. Furthermore, updating ability moderated the association between negative affect and self-reported rumination, such that habitual rumination led to less negative affect for participants high in working memory updating ability (see also Cohen et al., 2014; Joorman & Gotlib, 2008; Yoon et al., 2014).

Better emotion regulation by way of superior attention control is unlikely to be the sole explanation for why those high in cognitive ability seem to cope better with stress and have fewer instances of psychological disorder. However, it is a promising avenue of research that warrants further study.

Conclusion and future directions

In this chapter, we described the relationships between individual differences in working memory capacity, attention control, fluid intelligence, and general intelligence, broadly defined. We also reviewed evidence that these cognitive abilities predict four important classes of life outcomes, including academic achievement, job performance, physical health and longevity, and psychological well-being. We attempted to mitigate some common concerns about psychological research by focusing on large, longitudinal studies where possible, and relying on smaller laboratory studies to test theoretical explanations for the relationships between cognitive ability and life outcomes. Overall, the evidence suggests an association between cognitive ability and numerous life outcomes, with high cognitive ability generally conferring benefits and low cognitive ability generally entailing greater risk.

While researchers have made efforts to control for plausible covariates, mediating factors, and alternative explanations for the cognitive ability–life success relationship, it is fair to ask how successful these attempts have been and how much we should be persuaded by these findings. While much attention is given to adequately measuring and/or manipulating independent and predictor variables, adequate control is equally important for sound interpretations of research findings. Of particular concern to many readers, no doubt, are socioeconomic factors, and the risk of perpetuating systematic disadvantages through the use of 'cognitive ability tests'. After all, some studies report that cognitive ability tests add no incremental validity for predicting some life outcomes beyond measures of adult socioeconomic status (e.g. Wraw et al., 2015). In other cases, while cognitive ability tests technically provide incremental validity over socioeconomic metrics, it is dubious whether the indicators included by researchers adequately represent the construct 'socioeconomic status', which is typically operationally defined as familial income, educational attainment, and/or occupational status. This is a clear weakness of many of the studies we reviewed, and one that future researchers ought to keep in mind. It is notable, however, that measures of adult socioeconomic status partially or completely account for the relationships between cognitive ability and life outcomes in several of the studies we reviewed, whereas controlling for measures of childhood socioeconomic status generally leaves these association intact. This is especially noteworthy in light of evidence that higher cognitive ability may lead to greater educational attainment and job success, both of which help comprise adult socioeconomic status (cf. Deary et al., 2010; Wraw et al., 2015). This is not to say that material disadvantage has no adverse effects on cognitive ability nor that poor socioeconomic conditions are merely a by-product of low cognitive ability. Indeed, low socioeconomic status can be quite limiting (e.g. Batty et al., 2006; Hanscombe et al., 2012). It does mean, however, that even within materially disadvantaged populations, greater cognitive ability is generally associated with better life outcomes.

We would like to close this chapter with a cautionary note for well-intentioned researchers and practitioners about applying the knowledge in this chapter, particularly as it pertains to developing interventions to increase the health and success of participants who are low in cognitive ability. We sympathize with the sentiments that motivate such attempts, but warn that, if not done carefully, interventions intended to aid those low in cognitive ability may inadvertently *increase* the disparities between them and those higher in cognitive ability, a pattern referred to as the *Matthew effect* or the 'rich get richer effect' (Ceci & Papierno, 2005). While those low in cognitive ability may benefit from interventions designed to help them, they often benefit less than their higher ability counterparts if resources are allocated equally to all groups. From one perspective, this is unproblematic. After all, in this scenario, everyone benefits, even if they do not benefit equally. From another perspective, however, this pattern can be frustrating, as interventions wind up benefitting least the very people they were designed to help (Ceci & Papierno, 2005). We make no claim as to which perspective is more correct, but believe it is a complication of which everyone interested in applying psychological theory to solve human problems should be aware.

Acknowledgements

This work was supported by Office of Naval Research Grants N00173-20-2-C003 and N00173-20-P-0135 to Randall W. Engle.

References

Ackerman, P. L. (1988). Determinants of individual differences during skill acquisition: Cognitive abilities and information processing. *Journal of Experimental Psychology: General, 117*(3), 288–318.

Ackerman, P. L., Beier, M. E., & Boyle, M. O. (2005). Working memory and intelligence: The same or different constructs? *Psychological Bulletin, 131*(1), 30–60.

Ahmed, S. F., Tang, S., Waters, N. E., & Davis-Kean, P. (2019). Executive function and academic achievement: Longitudinal relations from early childhood to adolescence. *Journal of Educational Psychology, 111*(3), 446–458.

Andreotti, C., Thigpen, J. E., Dunn, M. J., Watson, K., Potts, J., Reising, M. M., Robinson, K. E., Rodriguez, E. M., Roubinov, D., Luecken, L., & Compas, B. E., (2013). Cognitive reappraisal and secondary control coping: Associations with working memory, positive and negative affect, and symptoms of anxiety and depression. *Anxiety, Stress & Coping, 26*(1), 20–35.

ASVAB Enlistment Testing Program. (2020, 13 August). *Fairness information.* https://www.officialasvab.com/researchers/fairness-information/

Baddeley, A. D. (1992). Working memory. *Science, 255*(5044), 556–559.

Baddeley, A. (2012). Working memory: Theories, models, and controversies. *Annual Review of Psychology, 63*, 1–29.

Baddeley, A. D., & Hitch, G. (1974). Working memory. In G. H. Bower (Ed.), *Psychology of learning and motivation* (Vol. 8, pp. 47–89). Academic Press.

Batty, G. D., Der, G., Macintyre, S., & Deary, I. J. (2006). Does IQ explain socioeconomic inequalities in health? Evidence from a population based cohort study in the west of Scotland. *BMJ (Clinical Research Ed.), 332*(7541), 580–584.

Batty, G. D., Mortensen, E. L., Andersen, A.-M. N., & Osler, M. (2005). Childhood intelligence in relation to adult coronary artery disease and stroke risk: Evidence from a Danish birth cohort study. *Paediatric and Perinatal Epidemiology, 19*(6), 452–459.

Batty, G. D., Mortensen, E. L., & Osler, M. (2005). Childhood IQ in relation to later psychiatric disorder: Evidence from a Danish birth cohort study. *British Journal of Psychiatry, 187*, 180–181.

Beck, D. M., Schaefer, C., Pang, K., & Carlson, S. M. (2011). Executive function in preschool children: Test–retest reliability. *Journal of Cognition and Development, 12*(2), 169–193.

Binet, A., & Simon, T. (1916). New methods for the diagnosis of the intellectual level of subnormals. (*L'Année Psychologique*, 1905, pp. 191–244). In A. Binet & T. Simon, *The development of intelligence in children (The Binet-Simon Scale)* (E. S. Kite, Trans.) (pp. 37–90). Williams & Wilkins Co.

Blankenship, T. L., Slough, M. A., Calkins, S. D., Deater-Deckard, K., Kim-Spoon, J., & Bell, M. A. (2019). Attention and executive functioning in infancy: Links to childhood executive function and reading achievement. *Developmental Science, 22*(6), e12824.

Bobko, P., Roth, P. L., & Potosky, D. (1999). Derivation and implications of a meta-analytic matrix incorporating cognitive ability, alternative predictors, and job performance. *Personnel Psychology*, *52*(3), 561–589.

Bosco, F., Allen, D. G., & Singh, K. (2015). Executive attention: An alternative perspective on general mental ability, performance, and subgroup differences. *Personnel Psychology*, *68*(4), 859–898.

Breslau, N., Chen, Q., & Luo, Z. (2013). The role of intelligence in posttraumatic stress disorder: Does it vary by trauma severity? *PLoS One*, *8*(6), e65391.

Brewin, C. R., & Beaton, B. (2002). Thought suppression, intelligence, and working memory capacity. *Behavior Research and Therapy*, *40*(8), 923–930.

Brown, M. I., Wai, J., & Chabris, C. F. (2021). Can you ever be too smart for your own good? Comparing linear and nonlinear effects of cognitive ability on life outcomes. *Perspectives on Psychological Science*, *16*(6), 1337–1359.

Buckley, T. C., Blanchard, E. B., & Neill, W. T., (2000). Information processing and PTSD: A review of the empirical literature. *Clinical Psychology Review*, *20*(8), 1041–1065.

Burgoyne, A., & Engle, R. (2020). Attention control: A cornerstone of higher-order cognition. *Current Directions in Psychological Science*, *29*(6), 624–630.

Burgoyne, A. P., Mashburn, C. A., & Engle, R. W. (2021). Reducing adverse impact in high- stakes testing. *Intelligence*, *87*, 101561.

Burgoyne, A. P., Mashburn, C. A., Tsukahara, J. S., & Engle, R. W. (2022). Attention control and process overlap theory: Searching for cognitive processes underpinning the positive manifold. *Intelligence*, *91*, 101629.

Calvin, C. M., Deary, I. J., Fenton, C., Roberts, B. A., Der, G., Leckenby, N., & Batty, G. D. (2011). Intelligence in youth and all-cause-mortality: Systematic review with meta-analysis. *International Journal of Epidemiology*, *40*(3), 626–644.

Carlson, J. M., Dikecligil, G. N., Greenberg, T., & Mujica-Parodi, L. R. (2012). Trait reappraisal is associated with resilience to acute psychological stress. *Journal of Research in Personality*, *46*(5), 609–613.

Cattell, R. B. (1987). *Intelligence: Its structure, growth, and action*. Elsevier Science Publishers B.V.

Ceci, S. J., & Papierno, P. B. (2005). The rhetoric and reality of gap closing: When the 'have- nots' gain, but the 'haves' gain even more. *American Psychologist*, *60*(2), 149–160.

Center for Human Resource Research. (1989). *NLS codebook*. Ohio State University.

Cohen, N., Daches, S., Mor, N., & Henik, A. (2014). Inhibition of negative content—A shared process in rumination and reappraisal. *Frontiers in Psychology*, *5*, 662.

Coifman, K. G., Kane, M. J., Bishop, M., Matt, L. M., Nylocks, K. M., & Aurora, P. (2021). Predicting negative affect variability and spontaneous emotion regulation: Can working memory span tasks estimate emotion regulatory capacity? *Emotion*, *21*(2), 297–314.

Conway, A. R., Cowan, N., Bunting, M. F., Therriault, D. J., & Minkoff, S. R. (2002). A latent variable analysis of working memory capacity, short-term memory capacity, processing speed, and general fluid intelligence. *Intelligence*, *30*(2), 163–183.

Conway, A. R., Kane, M. J., & Engle, R. W. (2003). Working memory capacity and its relation to general intelligence. *Trends in Cognitive Sciences*, *7*(12), 547–552.

Cowan, N. (1988). Evolving conceptions of memory storage, selective attention, and their mutual constraints within the human information-processing system. *Psychological Bulletin*, *104*(2), 163–191.

Čukić, I., Brett, C. E., Calvin, C. M., Batty, G. D., & Deary, I. J. (2017). Childhood IQ and survival to 79: A follow-up of 94% of the Scottish Mental Survey 1947. *Intelligence*, *63*, 45–50.

Daneman, M., & Merikle, P. M. (1996). Working memory and language comprehension: A meta-analysis. *Psychonomic Bulletin & Review*, *3*(4), 422–433.

Deary, I. J. (2014). The stability of intelligence from childhood to old age. *Current Directions in Psychological Science*, *23*(4), 239–245.

Deary, I. J., Weiss, A., & Batty, G. D. (2010). Intelligence and personality as predictors of illness and death: How researchers in differential psychology and chronic disease epidemiology are collaborating to understand and address health inequalities. *Psychological Science in the Public Interest*, *11*(2), 53–79.

Deary, I. J., Whalley, L. J., Lemmon, H., Crawford, J. R., & Starr, J. M. (2000). The stability of individual differences in mental ability from childhood to old age: Follow-up of the 1932 Scottish Mental Survey. *Intelligence*, *28*, 49–55.

Deary, I. J., Whiteman, M. C., Starr, J. M., Whalley, L. J., & Fox, H. C. (2004). The impact of childhood intelligence on later life: Following up the Scottish Mental Surveys of 1932 and 1947. *Journal of Personality and Social Psychology*, *86*(1), 130–147.

Detterman, D. K. (1991). A reply to Deary and Pagliari: Is *g* intelligence or stupidity? *Intelligence*, *15*(2), 251–255.

Draheim, C., Mashburn, C. A., Martin, J. D., & Engle, R. W. (2019). Reaction time in differential and developmental research: A review of problems and alternatives. *Psychological Bulletin*, *145*(5), 508–535.

Draheim, C., Pak, R., Draheim, A. A., & Engle, R. W. (2022). The role of attention control in complex real-world tasks. *Psychonomic Bulletin & Review*, *29*(4), 1143–1197.

Draheim, C., Tsukahara, J. S., Martin, J. D., Mashburn, C. A., & Engle, R. W. (2021). A toolbox approach to improving the measurement of attention control. *Journal of Experimental Psychology: General*, *150*(2), 242–275.

Engen, H. G., & Anderson. M. C. (2018). Memory control: A fundamental mechanism of emotion regulation. *Trends in Cognitive Sciences*, *22*(11), 982–995.

Engle, R. W. (2002). Working memory capacity as executive attention. *Current Directions in Psychological Science*, *11*(1), 19–23.

Engle, R. W. (2018). Working memory and executive attention: A revisit. *Perspectives on Psychological Science*, *13*(2), 190–193.

Engle, R. W., Carullo, J. J., & Collins, K. W. (1991). Individual differences in working memory for comprehension and following directions. *Journal of Educational Research*, *84*(5), 253–262.

Engle, R. W., Tuholski, S. W., Laughlin, J. E., & Conway, A. R. (1999). Working memory, short-term memory, and general fluid intelligence: A latent-variable approach. *Journal of Experimental Psychology: General*, *128*(3), 309–331.

Ferrucci, L., Guralnik, J. M., Marchionni, N., Costanzo, S., Lamponi, M., & Baroni, A. (1993). Relationship between health status, fluid intelligence, and disability in a non demented elderly population. *Aging Clinical and Experimental Research*, *5*(6), 435–443.

Fortier, M. S., Vallerand, R. J., & Guay, F. (1995). Academic motivation and school performance: Toward a structural model. *Contemporary Educational Psychology*, *20*(3), 257–274.

Gan, S., Yang, J., Chen, X., Zhang, X., & Yang, Y. (2017). High working memory load impairs the effect of cognitive reappraisal on emotional response: Evidence from an event-related potential study. *Neuroscience Letters*, *639*, 126–131.

Ganzer, C. A., Insel, K. C., & Ritter, L. S. (2012). Associations between working memory, health literacy, and recall of stroke symptoms among older adults. *Journal of Neuroscience Nursing*, *44*(5), 236–243.

Garrison, K. E., & Schmeichel, B. J. (2022). Getting over it: Working memory capacity and affective responses to stressful events in daily life. *Emotion*, *22*(3), 418–429.

Geary D. C. (2021). Mitochondrial functioning and the relations among health, cognition, and aging: Where cell biology meets cognitive science. *International Journal of Molecular Sciences*, *22*(7), 3562.

Gerstadt, C. L., Hong, Y. J., & Diamond, A. (1994). The relationship between cognition and action: Performance of children 3 1/2–7 years old on a Stroop-like day-night test. *Cognition*, *53*(2), 129–153.

Gladwell, M. (2008). *Outliers: The story of success*. Little, Brown and Company.

Gottfredson, L. S., & Deary, I. J. (2004). Intelligence predicts health and longevity, but why? *Current Directions in Psychological Science*, *13*(1), 1–4.

Guo, Z., Zou, J., He, C., Tan, X., Chen, C., & Feng, G. (2020). The importance of cognitive and mental factors on prediction of job performance in Chinese high-speed railway dispatchers. *Journal of Advanced Transportation*, *2020*, 1–13.

Hambrick, D. Z., Burgoyne, A. P., & Oswald, F. L. (2019). Domain-general models of expertise: The role of cognitive ability. In P. Ward, J. Maarten Schraagen, J. Gore, & E. Roth (Eds.), *The Oxford handbook of expertise* (pp. 56–84). Oxford University Press.

Hambrick, D. Z., & Meinz, E. J. (2011). Limits on the predictive power of domain-specific experience and knowledge in skilled performance. *Current Directions in Psychological Science, 20*(5), 275–279.

Hanscombe, K. B, Trzaskowski M., Haworth C. M. A., Davis O. S. P., Dale P. S., & Ploimin, R. (2012). Socioeconomic status (SES) and children's intelligence (IQ): In a UK-representative sample SES moderates the environmental, not genetic, effect on IQ. *PLoS One, 7*(2), e30320.

Hart, C. L., Taylor, M. D., Smith, G. D., Wahlley, L. J., Starr, J. M., Hole, D. J., Wilson, V., & Deary, I. J. (2003). Childhood IQ, social class, deprivation, and their relationships with mortality and later morbidity risk in later life: Prospective observational study linking the Scottish Mental Survey 1932 and the Midspan Studies. *Psychosomatic Medicine, 65*(5), 877–883.

Held, J. D., Carretta, T. R., & Rumsey, M. G. (2014). Evaluation of tests of perceptualspeed/accuracy and spatial ability for use in military occupational classification. *Military Psychology, 26*(3), 199–220.

Heyneman, S. P., & Loxley, W. A. (1983). The effect of primary-school quality on academic achievement across twenty-nine high-and low-income countries. *American Journal of Sociology, 88*(6), 1162–1194.

Horn, J. L., & Cattell, R. B. (1967). Age differences in fluid and crystallized intelligence. *Acta Psychologica, 26*(2), 107–129.

Hough, L. M., Oswald, F. L., & Ployhart, R. E. (2001). Determinants, detection and amelioration of adverse impact in personnel selection procedures: Issues, evidence and lessons learned. *International Journal of Selection and Assessment, 9*(1–2), 152–194.

Hunter, J. E. (1986). Cognitive ability, cognitive aptitudes, job knowledge, and job performance. *Journal of Vocational Behavior, 29*(3), 340–362.

Insel, K. C., Einstein, G. O., Morrow, D. G., Koerner, K. M., & Hepworth, J. T. (2016). A multifaceted prospective memory intervention to improve medication adherence. *Journal of the American Geriatrics Society, 64*(3), 561–568.

Insel, K. C., Morrow, D., Brewer, B., & Figueredo, A. (2006). Executive function, working memory, and medication adherence of older adults. *Journal of Gerontology: Psychological Sciences, 61*(2), 102–107.

Jensen, A. R. (1998). *The g factor: The science of mental ability*. Prager.

Joorman, J., & Gotlib, I. H. (2008). Updating the contents of working memory in depression: Interference from irrelevant negative material. *Journal of Abnormal Psychology, 117*(1), 182–192.

Judge, T. A., Higgins, C. A., Thoresen, C. J., & Barrick, M. R. (1999). The big five personality traits, general mental ability, and career success across the life span. *Personnel Psychology, 52*(3), 621–652.

Kane, M. J., Bleckley, M. K., Conway, A. R., & Engle, R. W. (2001). A controlled-attention view of working-memory capacity. *Journal of Experimental Psychology: General, 130*(2), 169–183.

Kane, M. J., & Engle, R. W. (2002). The role of prefrontal cortex in working-memory capacity, executive attention, and general fluid intelligence: An individual-differences perspective. *Psychonomic Bulletin & Review, 9*(4), 637–671.

Kane, M. J., Hambrick, D. Z., & Conway, A. R. (2005). Working memory capacity and fluid intelligence are strongly related constructs: Comment on Ackerman, Beier, and Boyle (2005). *Psychological Bulletin, 131*(1), 66–71.

Kane, M. J., Hambrick, D. Z., Tuholski, S. W., Wilhelm, O., Payne, T. W., & Engle, R. W. (2004). The generality of working memory capacity: A latent-variable approach to verbal and visuospatial memory span and reasoning. *Journal of Experimental Psychology: General, 133*(2), 189–217.

Kovacs, K., & Conway, A. R. A. (2016). Process overlap theory: A unified account of the general factor of intelligence. *Psychological Inquiry, 27*(3), 151–177.

Kraft, M., Arts, K., Traag, T., Otten, F., & Bosma, H. (2018). The contribution of intellectual abilities to young adults' educational differences in healthcare use: A prospective cohort study. *Intelligence, 68*, 1–5.

Krieger, F., Zimmer, H. D., Greiff, S., Spinath, F. M., & Becker, N. (2019). Why are difficult figural matrices hard to solve? The role of selective encoding and working memory capacity. *Intelligence, 72*, 35–48.

Kvist, A. V., & Gustafsson, J. E. (2008). The relation between fluid intelligence and the general factor as a function of cultural background: A test of Cattell's investment theory. *Intelligence, 36*(5), 422–436.

Kyllonen, P. C., & Christal, R. E. (1990). Reasoning ability is (little more than) working- memory capacity?! *Intelligence, 14*(4), 389–433.

Leger, K. A., Charles, S. T., & Almeida, D. M. (2018). Let it go: Lingering negative affect in response to daily stressors is associated with physical health years later. *Psychological Science*, *29*(8), 1283–1290.

Logie, R. H. (2016). Retiring the central executive. *Quarterly Journal of Experimental Psychology*, *69*(10), 2093–2109.

Macklin, M. L., Metzger, L. J., Litz, B. T., McNally, R. J., Lasko, N. B., Orr, S. P., & Pitman, R. K. (1998). Lower precombat intelligence is a risk factor for post-traumatic stress disorder. *Journal of Consulting and Clinical Psychology*, *66*(2), 323–326.

Martin, J., Mashburn, C. A., & Engle, R. W. (2020). Improving the validity of the Armed Service Vocational Aptitude Battery with measures of attention control. *Journal of Applied Research in Memory and Cognition*, *9*(3) 323–335.

McDaniel, M. A. (1986). The evaluation of a causal model of job performance: The interrelationships of general mental ability, job experience, and job performance (Unpublished doctoral dissertation). George Washington University, Washington, DC.

McGrew, K. S. (2009). CHC theory and the human cognitive abilities project: Standing on the shoulders of the giants of psychometric intelligence research. *Intelligence*, *37*(1), 1–10.

McRae, K., Jacobs, S. E., Ray, R. D., John, O. P., & Gross, J. J. (2012). Individual differences in reappraisal ability: Links to reappraisal frequency, well-being, and cognitive control. *Journal of Research in Personality*, *46*(1), 2–7.

McVay, J. C., & Kane, M. J. (2012). Why does working memory capacity predict variation in reading comprehension? On the influence of mind wandering and executive attention. *Journal of Experimental Psychology: General*, *141*(2), 302–320.

Moran, T. P. (2016). Anxiety and working memory capacity: A meta-analysis and narrative review. *Psychological Bulletin*, *142*(8), 831–864.

Murray, C. (1998). *Income and inequality*. AEI Press.

Nelson, L. C. (2003). *Working memory, general intelligence, and job performance* (Publication No. 305322924) [Doctoral dissertation, University of Minnesota]. ProQuest Dissertations Publishing.

Oberauer, K., Schulze, R., Wilhelm, O., & Süß, H. M. (2005). Working memory and intelligence— Their correlation and their relation: Comment on Ackerman, Beier, and Boyle (2005). *Psychological Bulletin*, *131*(1), 61–65.

Ohi, K., Takai, K., Sugiyama, S., Kitagawa, H., Katoaka, Y., Soda, M., Kitaich, K., Kawasaki, Y., Ito, M., & Shiori, T. (2022). Intelligence declines across major depressive disorder, bipolar disorder, and schizophrenia. *CNS Spectrums*, *27*(4), 468–474.

Ochsner, K. N., Silvers, J. A., & Buhle, J. T. (2012). Functional imaging studies of emotion regulation: A synthetic review and evolving model of the cognitive control of emotion. *Annals of the New York Academy of Sciences*, *1251*(1), E1–E24.

Outtz, J. L., & Newman, D. A. (2011). A theory of adverse impact. In J. L. Outtz (Ed.), *Adverse impact: Implications for organizational staffing and high stakes selection* (pp. 53–94). Routledge.

Patterson, F., Jepson, C., Loughead, J., Perkins, K., Strasser, A. A., Siegel, S., Frey, J., Gur, R., & Lerman, C. (2010). Working memory deficits predict short-term smoking resumption following brief abstinence. *Drug and Alcohol Dependence*, *106*, 61–64.

Pe, M. L., Raes, F., Koval, P., Brans, K., Verduyn, P., & Kuppens, P. (2013). Interference resolution moderates the impact of rumination and reappraisal on affective experiences in daily life. *Cognition & Emotion*, *27*(3), 492–501.

Peng, P., Barnes, M., Wang, C., Wang, W., Li, S., Swanson, H. L., Dardick, W., & Tao, S. (2018). A meta-analysis on the relation between reading and working memory. *Psychological Bulletin*, *144*(1), 48–76.

Peng, P., Namkung, J., Barnes, M., & Sun, C. (2016). A meta-analysis of mathematics and working memory: Moderating effects of working memory domain, type of mathematics skill, and sample characteristics. *Journal of Educational Psychology*, *108*(4), 455–473.

Ritchie, S. J. (2015). *Intelligence: All that matters*. Teach Yourself.

Robertson, K. F., Smeets, S, Lubinski, D., & Benbow, C. P. (2010). Beyond the threshold hypothesis: Even among the gifted and top math/science graduate students, cognitive abilities, vocational interests, and lifestyle preferences matter for career choice, performance, and persistence. *Current Directions in Psychological Science*, *19*(6), 346–351.

Rohde, T. E., & Thompson, L. A. (2007). Predicting academic achievement with cognitive ability. *Intelligence, 35*(1), 83–92.

Rosen, V. M., & Engle, R. W. (1998). Working memory capacity and suppression. *Journal of Memory and Language, 39*(3), 418–436.

Russell, E. W. (1980). Fluid and crystallized intelligence: Effects of diffuse brain damage on the WAIS. *Perceptual and Motor Skills, 51*(1), 121–122.

Salthouse, T. A. (1991). Expertise as the circumvention of human processing limitations. In K. A. Ericsson & J. Smith (Eds.), *Toward a general theory of expertise: Prospects and limits* (pp. 286–300). Cambridge University Press.

Schmidt, F. L. (2002). The role of general cognitive ability and job performance: Why there cannot be a debate. *Human Performance, 15*(1–2), 187–210.

Schmidt, F. L., & Hunter, J. E. (1998). The validity and utility of selection methods in personnel psychology: Practical and theoretical implications of 85 years of research findings. *Psychological Bulletin, 124*(2), 262–274.

Schmidt, F. L., Hunter, J. E., Outerbridge, A. N., & Goff, S. (1988). Joint relation of experience and ability with job performance: Test of three hypotheses. *Journal of Applied Psychology, 73*(1), 46–57.

Shamosh, N. A., DeYoung, C. G., Green, A. E., Reis, D. L., Johnson, M. R., Conway, A. R. A., Engle, R. W., Braver, T. S., & Gray, J. R. (2008). Individual differences in delay discounting: Relation to intelligence, working memory, and anterior prefrontal cortex. *Psychological Science, 19*(9), 904–911.

Shipstead, Z., Harrison, T. L., & Engle, R. W. (2016). Working memory capacity and fluid intelligence: Maintenance and disengagement. *Perspectives on Psychological Science, 11*(6), 771–799.

Sirin, S. R. (2005). Socioeconomic status and academic achievement: A meta-analytic review of research. *Review of Educational Research, 75*(3), 417–453.

Soederberg Miller, L. M., Gibson, T. N., Applegate, E. A., & de Dios, J. (2011). Mechanisms underlying comprehension of health information in adulthood: The roles of prior knowledge and working memory capacity. *Journal of Health Psychology, 16*(5), 794–806.

Spearman, C. (1904). 'General intelligence,' objectively determined and measured. *American Journal of Psychology, 15*, 201–293.

Spiegel, J. A., Goodrich, J. M., Morris, B. M., Osborne, C. M., & Lonigan, C. J. (2021). Relations between executive functions and academic outcomes in elementary school children: A meta-analysis. *Psychological Bulletin, 147*(4), 329–351.

Sternberg, R. J., & Wagner, R. K. (1993). The g-centric view of intelligence and job performance is wrong. *Current Directions in Psychological Science, 2*, 1–4.

Stilley, C. S., Bender, C. M., Dunbar-Jacob, J., Sereika S., & Ryan, C. M. (2010). The impact of cognitive function on medication management: Three studies. *Health Psychology, 29*(1), 50–55.

Taylor, M. D., Hart, C. L., Davey Smith, G., Starr, J. M., Hole, D. J., Whalley, L. J., Wilson, V., & Deary, I. J. (2003). Childhood mental ability and smoking cessation in adulthood: Prospective observational study linking the Scottish Mental Survey 1932 and the Midspan studies. *Journal of Epidemiology and Community Health, 57*(6), 464–465.

Tolar, T. D., Lederberg, A. R., & Fletcher, J. M. (2009). A structural model of algebra achievement: Computational fluency and spatial visualisation as mediators of the effect of working memory on algebra achievement. *Educational Psychology, 29*(2), 239–266.

Tsukahara, J. S., Harrison, T. L., Draheim, C., Martin, J. D., & Engle, R. W. (2020). Attention control: The missing link between sensory discrimination and intelligence. *Attention, Perception, & Psychophysics, 82*(7), 3445–3478.

Turkheimer, E., Haley, A., Waldron, M., d'Onofrio, B., & Gottesman, I. I. (2003). Socioeconomic status modifies heritability of IQ in young children. *Psychological Science, 14*(6), 623–628.

Turner, M. L., & Engle, R. W. (1989). Is working memory capacity task dependent? *Journal of Memory and Language, 28*(2), 127–154.

Unsworth, N., Heitz, R. P., Schrock, J. C., & Engle, R. W. (2005). An automated version of the operation span task. *Behavior Research Methods, 37*(3), 498–505.

van der Maas, H. L., Dolan, C. V., Grasman, R. P., Wicherts, J. M., Huizenga, H. M., & Raijmakers, M. E. (2006). A dynamical model of general intelligence: The positive manifold of intelligence by mutualism. *Psychological Review, 113*(4), 842–861.

Wang, J.-J., & Kaufman, A. S. (1993). Changes in fluid and crystallized intelligence across the 20- to 90-year age range on the K-Bit. *Journal of Psychoeducational Assessment, 11*(1), 29–37.

Wilk, S. L., Desmarais, L. B., & Sackett, P. R. (1995). Gravitation to jobs commensurate with ability: Longitudinal and cross-sectional tests. *Journal of Applied Psychology, 80*(1), 79–85.

Wise, L., Welsh, J., Grafton, F., Foley, P., Earles, J., Sawin, L., & Divgi, D. R. (1992). *Sensitivity and fairness of the Armed Services Vocational Aptitude Battery (ASVAB) technical composites.* Personnel Testing Division, Defense Manpower Data Center, Department of Defense. https://www.officialasvab.com/wp-content/uploads/2019/08/AS92009_Sensitivity_Fairness_of_ASVAB_Tech_Composites.pdf

Wraw, C., Deary, I. J., Gale, C. R., & Der, G. (2015). Intelligence in youth and health at age 50. *Intelligence, 53*, 23–32.

Yoon, K. L., LeMoult, J., & Joorman, J. (2014). Updating emotional content in working memory: A depression-specific deficit? *Journal of Behavior Therapy and Experimental Psychiatry, 45*(3), 368–374.

Zaehringer, J., Falquez, R., Schubert, A-L., Nees, F., & Barnow, S. (2018). Neural correlates of reappraisal considering working memory capacity and cognitive flexibility. *Brain Imaging and Behavior, 12*(6), 1529–1543.

8

The phonological loop as a neural network

From specific models to general principles

Graham J. Hitch

Introduction

One of the classic ways science makes progress is by subjecting explanations to challenge and where necessary modifying them. This is readily illustrated in the case of the phonological loop, the component of working memory capable of maintaining verbal sequences for just a few seconds (Baddeley & Hitch, 1974) and its assumption that information is lost through time-based decay (see, e.g. Jalbert et al., 2011; Lewandowsky et al., 2009; Service, 1998). However, the broader claim that information in the loop undergoes rapid forgetting still stands. More significant challenges arise from the deliberately simplistic way the loop was originally described, with no mention of links with long-term memory or any realistic mechanism for maintaining serial order. In subsequent years, there have been several attempts to fill these gaps by means of computational modelling, an approach that has become increasingly common thanks to advances in technology and techniques. Computational models have several desirable features. For example, their assumptions have to be specified explicitly, they can be objectively assessed by their capacity to simulate human performance, and they readily generate novel, testable predictions. However, such models also have a number of less desirable aspects. One is the need to make numerous arbitrary assumptions and adjustments simply in order to get a computer program to run coherently. Another is the sheer technical detail and complexity involved, making it sometimes hard to understand which of the many assumptions are crucial to a model's behaviour. Yet another is the problem of communicating what has been learned to non-specialist audiences. In this chapter, I will give a short history of attempts to model the phonological loop aimed at non-specialists. I have concentrated mainly on my own collaborations as this makes it relatively straightforward for me to tell the story of gradual progress over time. I have not been involved in the technical aspects of modelling and I hope my qualitative feel for the way the models work will help meet the challenge of communication.

It is important to acknowledge from the outset the many competing models and their developers without whose work the story I tell here would not have been possible. To provide some sense of this broader context I have included a short overview

Graham J. Hitch, *The phonological loop as a neural network* In: *Memory in Science for Society*. Edited by: Robert H. Logie, Zhisheng (Edward) Wen, Susan E. Gathercole, Nelson Cowan, and Randall W. Engle, Oxford University Press. © Oxford University Press 2023.
DOI: 10.1093/oso/9780192849069.003.0008

of the various alternative types of models that have been proposed, but I am conscious it does not do sufficient justice to the individual models themselves. It does, however, give an opportunity to underline the broad lessons that have been learned. The focus throughout is mainly on the problem of serial order and somewhat less on links with long-term memory where the problems have turned out to be more challenging and progress has, so far, been more modest.

Working memory as a multicomponent system

As has been described many times, the idea of working memory as a multicomponent system stemmed from a research project in the early 1970s in which Alan Baddeley and I set out to explore the claim that the limited-capacity short-term store in Atkinson and Shiffrin's (1968) influential model of human memory serves as a general-purpose working memory in cognition. This was based on the idea that many aspects of cognition involve maintaining temporary information. However, a number of observations raised our doubts, one of which was the absence of obvious difficulties in learning or language comprehension in a neuropsychological patient whose auditory–verbal short-term memory capacity was selectively impaired (Shallice & Warrington, 1970; Warrington & Shallice, 1969). Given the rarity of such patients, we decided to explore further by simulating the neuropsychological deficit in healthy adults. We did so by loading their short-term store with extraneous information while they attempted to perform a cognitive task such as verbal reasoning, language comprehension, or free recall learning. We assumed that if Atkinson and Shiffrin's claim is correct, occupying the short-term store with irrelevant information should make these tasks more difficult to perform. We investigated the effect of systematically varying the load by altering the number of irrelevant items to be maintained. The result was a striking common pattern whereby loads approaching the capacity of the short-term store did disrupt reasoning, comprehension, and long-term learning, but the interference was unexpectedly moderate and by no means catastrophic, with small loads causing no discernible interference.

We interpreted this outcome as suggesting that the short-term store does not by itself serve as a working memory but, rather, forms one component of a working memory system that also includes other capacities. The multicomponent model was our attempt to capture this interpretation in the simplest terms (Baddeley & Hitch, 1974). It viewed working memory as consisting of a limited-capacity attentional workspace (the central executive) controlling a phonemic buffer store that can be refreshed by subvocal rehearsal. The latter was broadly equivalent to Atkinson and Shiffrin's short-term store and is nowadays commonly referred to as the phonological loop. We also suggested the possibility of a similar buffer store for visual information based on evidence that short-term retention draws on qualitatively different resources when it involves visuospatial imagery (Brooks, 1967). For example,

Baddeley, Grant, et al. (1975) showed that tracking a moving visual target disrupted a short-term memory task that involved visual imagery but had no effect on a corresponding task that involved verbal coding. We assumed that the resources of the central executive can be allocated flexibly over attention-demanding control processes and different forms of temporary storage according to task demands. Thus, in terms of the model, our results suggested that verbal reasoning, comprehension, and free recall place minimal demands on the phonological loop and draw most heavily on the central executive.

Since it was first proposed nearly 50 years ago, the multicomponent model has undergone significant elaboration in the light of further research. However, its core has nevertheless remained substantially the same. The first elaboration described in more detail how the visual buffer acts as a 'visuospatial sketchpad' capable of maintaining and manipulating images and shifted to a view of the resources of the central executive as purely attentional (Baddeley, 1986). A second major elaboration revived the idea of temporary storage at the centre of the system by adding a multimodal episodic buffer, closely linked to but separate from the central executive (Baddeley, 2000). This fourth component holds cross-linked bindings of information from the domain-specific buffers and interfaces with episodic long-term memory. It is regarded as crucial for the coherent operation of the system as a whole. Attempts have also been made to say more about the central executive from time to time (Baddeley, 1996, 2002) and we have recently made some progress in distinguishing the roles of internal and external (perceptual) attentional processes (Hitch et al., 2020). However, given our present focus on the phonological loop I will only touch on aspects of the broader model when they become relevant.

Original concept of the phonological loop

Previous research on short-term memory had shown that immediate recall of a sequence of digits, letters, or words involves storing the items in a speech-based code. Thus, immediate recall is impaired when the items are phonologically similar, even when presented visually (Conrad & Hull, 1964). Evidence that maintaining information in short-term memory involves subvocal rehearsal came from the observation that repeating an irrelevant word such as 'the' impairs recall and, with visual presentation, removes the effect of phonological similarity (Murray, 1968). These effects can be explained by assuming a limited-capacity phonological store in which information is rapidly forgotten but can be refreshed by subvocal rehearsal.

Given the sequential nature of verbal rehearsal, the number of items that can be maintained should depend on the rate of refreshing relative to that of forgetting. Initial support for this came from the word length effect in immediate recall, whereby the longer words take to say, the fewer can be recalled (Baddeley, Thomson, & Buchanan, 1975). There was a linear relationship between the number of items that could be recalled and the rate at which they could be articulated, consistent with trace decay offset by rehearsal. Other evidence showed that the word length effect could be removed by articulatory suppression, in accordance with its dependence on subvocal rehearsal. Indeed, the existence of a cluster of conceptually related variables that interacted systematically provided strong support for the idea of a phonological loop. It turned out that these interactions depend subtly on whether items are presented auditorily or visually, and these effects helped elaborate the model. Specifically, suppression removes the phonological similarity effect when items are presented visually but not when they are presented auditorily, while removing the word length effect regardless of presentation modality (Baddeley et al., 1984). This pattern of results was interpreted as suggesting that speech input accesses the phonological buffer directly whereas printed words do so indirectly via subvocalization.

Other research has shown that articulatory suppression disrupts the ability to decide whether printed words rhyme but not whether they have the same sound (Baddeley & Lewis, 1981; Besner et al., 1981). This suggests a non-articulatory code which Baddeley and Lewis (1981) labelled the 'inner ear' that allows simple homophone judgements but does not support manipulation. They distinguished this from the 'inner voice' of the articulatory code, which allows items to be held and manipulated, as in stripping away the initial phoneme of a word in order to make a rhyme judgement. However, even with these elaborations, including the evidence that forgetting is not solely due to trace decay mentioned earlier, the phonological loop remains a relatively simple concept. We will illustrate some of the ways it turned out to be unexpectedly fruitful before going on to discuss its shortcomings and attempts to address them through computational modelling.

The phonological loop in practice
Cross-linguistic differences in digit span

One of the most striking early successes of the phonological loop was to offer a simple explanation for cross-linguistic differences in digit span. For example, Ellis and Hennelly (1980) found that Welsh–English bilinguals have lower spans in Welsh than in English, and concluded that this reflected different rates of rehearsal as the digits take longer to articulate in Welsh than in English. Subsequent studies went on to show that variations in digit span in Arabic, English, Finnish, Greek, Hebrew, Japanese, Malay, and Swedish can also be accounted for in terms of the rate at which the digits can be articulated in each language (Chan & Elliot, 2011, 2020; Chincotta & Underwood, 1997; Hoosain & Salili, 1987; Naveh Benjamin & Ayres, 1986; Stigler et al., 1986). Further evidence showed that cross-linguistic differences in digit span are reduced by articulatory suppression, consistent with an interpretation in terms of the phonological loop (Chincotta & Underwood, 1997). Digit span is of special interest psychometrically as a component of intelligence tests and it is clearly important that effects of language are taken into account when interpreting performance, while of course recognizing that many other factors also contribute to cross-cultural differences (Ostrosky-Solis & Lozano, 2006).

Mental calculation

It is a common experience that mental calculation becomes more difficult when it puts more demands on short-term memory. In most adults, multidigit mental calculations are carried out through a series of simpler steps with the consequence that any interim results have to be briefly held in store until they can be utilized. A simple model that assumes temporary information is rapidly forgotten during delays forced by the calculation strategy predicts patterns of error in mental calculation remarkably well (Hitch, 1978). However, evidence for rapid forgetting is not by itself sufficient to implicate the phonological loop. If the loop plays an important role, mental arithmetic should be disrupted by articulatory suppression, and indeed a number of studies have shown this is the case (Fürst & Hitch, 2000; Heathcote, 1994; Logie et al., 1994). These studies used different forms of mental calculation but that had in common the requirement to maintain temporary information over short intervals. In contrast, articulatory suppression has little or no effect on the ability to add a pair of integers, where in adults the answer is typically retrieved directly from long-term memory rather than calculated (De Rammelaere et al., 2001; Hecht, 2002). More generally, reviews of dual-task studies show that arithmetic places different patterns of demand on the components of working memory depending on the task and context (Chen & Bailey, 2021; De Stefano & LeFevre, 2004). Thus, while the phonological loop can clearly make an important contribution to mental arithmetic performance, it is by no means the whole story. This conclusion is underlined by the observation of unimpaired at mental arithmetic ability in a neuropsychological patient with a marked deficit in auditory–verbal short-term memory (Butterworth et al., 1996).

Vocabulary acquisition

During the 1980s, a colleague once famously dismissed the phonological loop as a 'pimple on the face of cognition', implying it served no useful function in everyday life. This was an arguable position given some of the results we have just discussed. However, subsequent evidence pointed to one aspect of cognition where the phonological loop does seem to play a vital role, namely learning new words, as in vocabulary acquisition (Baddeley et al., 1988).

Baddeley et al. (1988) described another patient with a selective deficit in auditory–verbal short-term memory who displayed a severe learning difficulty specific to new words. This was shown by the patient's contrasting performance in two versions of an auditory–verbal paired-associate learning task. In one, familiar words in her native vocabulary were paired with words in an unfamiliar, foreign vocabulary. In the other, stimuli and responses were both familiar native vocabulary words. The patient was unable to learn a single pair involving a foreign word, performing much worse than healthy controls, but learned pairs of words in her native vocabulary at just the same rate as controls. The crucial difference would appear to be that learning an unfamiliar word involves forming a representation of a novel phonological sequence whereas learning a pair of familiar words involves associating pre-existing lexical representations. The patient's difficulty suggests that a fully functioning phonological loop is important for acquiring vocabulary in a second language. Indeed, on reflection it is hard to see how it would be possible to acquire native vocabulary without being able to maintain a short-term representation of novel phonological sequences. Speech is distinctive in that its meaningful content is never simultaneously present in the stimulus, and in that learning new forms involves not only recognition but also production. These characteristics seem to demand short-term storage in order for new long-term representations to be rapidly established.

A series of follow-up experiments on healthy adults provided complementary evidence for the role of the phonological loop in learning new words. Thus, articulatory suppression and phonological similarity impaired the learning of word–non-word pairs but had relatively little effect on learning word–word pairs (Papagno et al., 1991; Papagno & Vallar, 1992). Other evidence confirmed that the phonological loop is important in children's vocabulary acquisition. One indication was studies of individual differences showing a correlation between the ability to repeat a novel word on first hearing and scores on standard vocabulary tests. Cross-lagged longitudinal analyses suggested that early in development the ability to repeat non-words drives vocabulary acquisition whereas later on causality is reversed and non-word repetition benefits from the possession of a larger vocabulary (Gathercole & Baddeley, 1989). In other evidence, children diagnosed with specific language impairment and poor vocabulary have been found markedly impaired in non-word repetition (Gathercole & Baddeley, 1990). These and other findings led Baddeley et al. (1998) to propose that the evolutionary function of the phonological loop is to support the learning of new words. This stands in marked contrast to its less critical role in language comprehension and verbal reasoning according to our original investigation (Baddeley & Hitch, 1974), and in cognitive skills such as mental arithmetic (e.g. Butterworth et al., 1996).

A neural network model of the phonological loop

So far, we have seen how the idea of a phonological loop came into being to account for a set of inter-related effects in verbal short-term memory and stimulated research on its role in various aspects of cognition, most notably vocabulary learning. However, these developments draw attention to the limited explanatory power of the initial concept, which likened the loop to a fading but refreshable audio recording of limited duration. This offers a simple, clear picture, but cannot begin to explain how a novel phonological sequence becomes a familiar word. Nor can it begin to explain why the most common error in immediate serial recall is to remember items in the wrong order (Bjork & Healy, 1974), why such errors tend to involve small movements of position within the sequence (Healy, 1974), or why they are highly sensitive to rhythm (Ryan, 1969a, 1969b). The problem of how the phonological loop deals with serial order was the first to be tackled through computational modelling.

In an influential paper, Lashley (1951) discussed the general problem of serial order in behaviour and presented persuasive arguments against the then-dominant view that it involves reactivating a chain of associative links between successive elements. He proposed instead that the elements of a forthcoming sequence are simultaneously available in a plan for action and selected one by one for its realization in behaviour. In this account it is natural to think of order errors as resulting from mistakes in the selection process. However, at first sight it seems far from obvious how such a process might operate. Implementations of Lashley's idea had to await developments in computational modelling in terms of networks of units, sometimes described as artificial neurons. These transmit activation to one another through excitatory and inhibitory connections that can vary in strength.

Neural network models came to prominence in the 1980s through demonstrations of their applicability to several aspects of cognition (e.g. McClelland et al., 1986). Alan Baddeley was quick to spot their potential in the case of working memory and encouraged me and a number of others to explore further. However, an immediate obstacle was that constructing and analysing such models requires expertise I lacked, having unfortunately forgotten much of what I had learned from a first degree in natural sciences. Luckily, the opportunity arose to collaborate with Neil Burgess who was working on models with the relevant mathematics as part of his PhD in theoretical physics in Manchester and was interested in their applications in cognitive science. Together we set out to develop a network model of the phonological loop taking inspiration from previous work that implemented Lashley's ideas in a neural network (Grossberg, 1978a, 1978b). The mechanism involved representing a sequence of items as a set of simultaneously active units and then repeatedly selecting the currently most active unit in a winner-takes-all competition. Perseveration on a single response was prevented by strongly suppressing a unit's activation immediately after its selection. We were greatly encouraged by the development of a model demonstrating that such a mechanism could be used to select the constituents of a word in the correct order when saying it (Houghton, 1990). This led us to consider the possibility that a similar 'competitive queuing' (CQ) mechanism might operate at the level of items (i.e. words, digits, letters) in verbal short-term memory (Burgess & Hitch, 1992).

The architecture of our model consists of layers of artificial neurons or units capable of being activated via interconnections of variable strength (Figure 8.1). Individual units in different layers represent words, input phonemes, or output phonemes while units in a fourth layer form a distributed 'context signal', a pattern of activation that changes progressively during exposure to a series of items. CQ is performed within the word layer by means of a competitive inhibitory filter. To put this

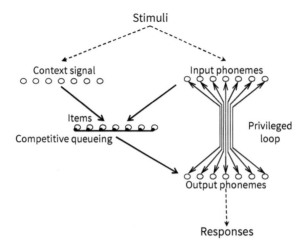

Figure 8.1 Basic architecture of the Burgess and Hitch (1992, 1999, 2006) model of the phonological loop as an artificial neural network. Circles denote units whose activation can vary. Units in the item layer represent individual words, digits, or letters, and the horizontal line with filled circles permits CQ, a winner-take-all inhibitory interaction with subsequent decaying inhibition of the winning unit. Units in the input and output phoneme layers represent individual phonemes. The activation of units in the context layer forms a pattern that changes progressively during stimulus presentation. Arrows between layers denote excitatory connections. Thick arrows indicate modifiable connections from all units in one layer to all units in the next. Thin double-ended arrows are fixed one-to-one connections between corresponding input and output phonemes. In later versions of the model, visual stimuli activate the item layer directly whereas auditory stimuli activate it indirectly via the input phoneme layer, as illustrated. See text for further explanation.

arrangement in perspective, the word and phoneme layers and their interconnections implement the original idea of the phonological loop with the input and output phoneme layers representing the 'inner ear' and 'inner voice' respectively. The context signal and competitive filter are new elements introduced specifically to handle serial ordering. The strengths of connections between the word and phoneme layers are preset to reflect the system's knowledge of the phonemic composition of words. During exposure to a series of items, connections linking the current word to both its context (an approximate representation of its position) and the immediately preceding word (through a chaining association not shown in Figure 8.1) are strengthened, the balance between the two being adjustable via a free parameter. These changes in connection strengths decay rapidly over time. Recalling the sequence begins by resetting the context signal to its initial state, allowing activations to propagate around the loop and selecting the winning item for response. The recall process continues by advancing the context signal and then repeating the cycle until the end of the sequence is reached.

The initial model proved capable of generating immediate serial recall with a limited memory span sensitive to phonemic similarity, word length, and articulatory

suppression, the trio of effects forming the core empirical signature of the phonological loop. The limit on memory span arose from activation levels being inherently
noisy, the phonemic similarity effect arose from crosstalk in connections to and from
the phonemic layers, and the word length effect arose from the rapid forgetting of
modified connections. Articulatory suppression was modelled by disconnecting the
input and output phoneme layers which resulted in increased errors and reduced effects of phonemic similarity and word length, much as in human performance. The
model also generated serial position curves and distributions of order errors broadly
similar to human performance and was observed to work best when set up without
chaining. These serial ordering phenomena were of course the ones the original description of the phonological loop did not address. However, despite these successes,
the model clearly left much room for improvement. Most obviously, the effects of
phonemic similarity, word length and suppression, and the primacy gradient in the
serial position curve were less marked than in humans, and differences between
auditory and visual presentation were not addressed. Perhaps most strikingly to us at
the time, the model failed to generate the characteristic zig-zag serial position curves
observed in the recall of lists in which alternate items are either phonemically similar
or dissimilar (Baddeley, 1968; see also Henson et al., 1996).

A revised model of the phonological loop

Our initial model succeeded in demonstrating the viability of implementing the
phonological loop as an artificial neural network, but clearly fell some way short of
an adequate account. In subsequent work, we revised the model to address its shortcomings, principally by dropping the chaining component entirely and separating
the cueing of items by the context signal from phonological interference (Burgess
& Hitch, 1999). At the same time, we extended the scope of the model to encompass differences between auditory and visual presentation, effects of presenting
items rhythmically, in a temporally grouped fashion (Ryan, 1969a, 1969b), and effects of item familiarity and long-term sequence learning. All this was achieved
through relatively minor adjustments. Over and above the removal of chaining, the
main changes were (1) the addition of an acoustic input buffer, with visual inputs
activating the item layer directly; (2) an elaboration of the context signal to incorporate elements that respond to rhythmic timing in a series of stimuli; and (3) a provision for modifiable connection weights to decay at two rates, one rapid as in the
initial model the other much slower. The slowly decaying weights were intended to
provide a mechanism for item familiarity effects and long-term learning.

The revised model continued to simulate the core phenomena concerning phonemic similarity, word length, and articulatory suppression. However, unlike the initial version, suppression removed the phonological similarity effect for visual but
not auditory sequences while removing the word length effect for each modality,
as in human performance. In addition, auditory presentation resulted in greater

recency due to the auditory input buffer, and the separation between cueing items by the context signal and phonological interference resulted in zig-zag serial position curves in the recall of alternating sequences, as in human data.

Temporal grouping effects were modelled by adding an extra dimension to the context signal. To illustrate, for a sequence presented rhythmically by inserting a short pause after every third item, one dimension of the context signal changes progressively with position within each group of three items while the other changes in an analogous way between groups. The effect of the extra dimension was to increase the discriminability of the context signal for temporally grouped lists, with simulations showing enhanced recall, scalloped serial position curves, and a small increase in interposition errors (where an item from one group is recalled as an error at the corresponding position in a different group), as is typically observed in human recall (Ryan, 1969b). The model also generated the novel predictions that temporal grouping effects should be independent of articulatory suppression, phonemic similarity, and word length, and these were confirmed experimentally (Hitch et al., 1996).

As already mentioned, the inclusion of long-term connection weights was intended as a starting place for capturing effects of item familiarity and long-term learning. For example, the assumption of a long-term component in item–phoneme connections provides a simple way of accounting for the observation that memory span is greater for familiar than unfamiliar words (see, e.g. Gregg et al., 1989; Hulme et al., 1991), as this component is stronger for more familiar items. This assumption can also account for the enhanced recall of non-words that are similar to known words (Roodenrys & Hinton, 2002), as they benefit from long-term item–phoneme connections for words they resemble. However, we regarded this account as only a beginning, as the tendency for a non-word to activate a highly similar word would in the long run lead to a problem of distinguishing the new word from the known word.

The revised model also attempted to explain how merely holding information in phonological short-term memory can lead to long-term learning. The classic experimental demonstration involves an immediate serial recall task in which one list is surreptitiously repeated on different trials. Recall of the repeated list gradually improves over trials, a phenomenon known as the Hebb effect (Hebb, 1961). In these experiments, the lists typically contain highly familiar items such as digits, so what is learned is principally the serial order of the items in the repeated list. In the revised model, this is achieved by strengthening long-term context–item connection weights. The slow decay of these connections accounts for human data showing that the Hebb effect disappears when the interval between repetitions is too great (Melton, 1963). The sensitivity of the context signal to rhythm accounts for the finding that the Hebb effect is greatly reduced when the temporal grouping pattern is changed each time a repeated list is presented (Bower & Winzenz, 1969). We replicated this observation in experiments of our own and went on to test the model's counterintuitive prediction that disrupting the operation of the phonological layers would have no influence on Hebb repetition learning. We confirmed this, finding that articulatory suppression and phonological similarity had no effect on the learning of a repeated sequence despite having their usual effects of substantially impairing its immediate recall (Hitch et al., 2009). The marked contrast between these phonological effects and that of temporal grouping encouraged us to regard the architecture of the model as broadly along the right lines.

Further revisions

Two close colleagues, Mike Page and Dennis Norris, soon drew our attention to a problem with our revised model. They had their own competing model (Page & Norris, 1998), about which more later, and we enjoyed lively exchanges in meetings of our working memory discussion group. Mike and Dennis pointed out that if repeating a list strengthens context–item connections in the way we proposed, learning should transfer to a test list in which alternate items continue to occupy the same serial position but the remaining items are shuffled randomly around. However, when they investigated experimentally they did not find the predicted effect (Cumming et al., 2003). To further emphasize the problem, we already had data of our own in which we observed simultaneous Hebb effects when two different lists were repeated in parallel (Hitch et al., 2009). This would be impossible if repeated presentation of a list strengthens associations to states of the same context signal. Fortunately, a clue as to how to solve both problems came from a series of experiments exploring the effects of repeating parts of a sequence rather than an entire list (Hitch et al., 2005). The interesting finding was that repeating a subsequence of items at the start of a list results in much faster learning than repeating a subsequence of equivalent length at the end of a list. This asymmetry suggests that learning a new sequence involves some sort of cumulative matching process during presentation that links the current input to relevant past inputs. With these observations in mind, we embarked on a modest though significant further revision to the model (Burgess & Hitch, 2006).

The revision was inspired by the solution to a similar problem of cue overload in word pronunciation whereby each word is assumed to have its own dedicated context signal (Houghton, 1990). We applied this idea in our model by having many context sets rather than a single set, keeping the overall architecture the same. The strengths of connections to different context sets reflect the model's previous learning through exposure to different sequences. At each step during exposure to a sequence, the cumulative match between these connections and the current input is computed and context sets that do not match are discarded. In this way the cohort of candidate context sets becomes progressively smaller, resulting in either one set winning or the recruitment of an uncommitted context set. In the first case, the sequence is recognized as familiar and undergoes further learning. In the second case, the sequence is treated as entirely novel. Simulations showed that these changes were successful in replicating our experimental findings on the Hebb effect (Hitch et al.,

2009), the absence of position-specific transfer effects (Cumming et al., 2003), and the particular importance of the start of the list (e.g. Hitch et al., 2005).

At this point it will be useful to summarize briefly. Overall, the network model retains the original idea of the phonological loop but captures a wider range of phenomena by adding mechanisms for serial ordering and links with long-term memory. At the same time, it has some obvious limitations. One is that the model has become more complicated with each revision, a potential drawback in using and applying it. Another is that it is mainly concerned with serial order at the level of lexical items. This is a useful starting point for modelling classic findings on verbal short-term memory but is less immediately useful when it comes to simulating vocabulary acquisition, where the challenge of serial ordering is primarily at the phonological level. Gupta and McWhinney (1997) outlined an integrative model for verbal short-term memory and vocabulary acquisition that adopts a similar approach and combines separate CQ processes at the phonological and lexical levels (see also Gupta & Tisdale, 2009). However, it is not straightforward to achieve an appropriate balance between ordering mechanisms operating at different levels when such a model is implemented (see also Glasspool et al., 1995).

An alternative approach is to assume a single CQ mechanism operating at multiple levels simultaneously. Our revised model made a start in this direction with its two-dimensional context signal where one component changes with the presentation of individual items and the other changes more slowly with the presentation of groups. This approach can in principle be extended by adding a further dimension consisting of a more rapidly changing component corresponding to phonemes. However, once again it is not straightforward to specify a mechanism that would achieve this. For example, even in the relatively simple case of encoding a temporally grouped sequence of digits, the context signal has to be specified by the modeller to reflect the way the sequence is grouped (Burgess & Hitch, 1999). In humans, however, grouping effects are observed whether or not the grouping pattern is known in advance (Ryan, 1969a), suggesting the context signal is automatically generated on-line during exposure to a sequence. A similar computational challenge arises in modelling the repetition of a non-word, in that errors suggest the serial ordering mechanism is sensitive to syllable structure (Trieman & Danis, 1988). As syllable structure varies unpredictably between items, any context signal must somehow capture the structure as it is perceived. In a significant piece of work, Hartley and Houghton (1996) identified a mechanism capable of achieving this based on the observation that local changes in sonority, a property of the speech signal, track syllable structure. They developed a model for sequencing at the phonological level in which a cyclical context signal is driven by changes in the sonority of the auditory input. This approach was successful in generating the characteristic error patterns seen in non-word repetition, where phonemes tend to migrate to corresponding positions in adjacent syllables rather than adjacent positions within the same syllable.

A model for deriving a multilevel context signal from auditory input

In another stroke of fortune, Tom Hartley moved to my department at the University of York where he developed his approach further by showing how sonority could be detected using a bank of filters tuned to different frequencies that respond to local changes in the amplitude and phase of a spoken input (Hartley, 2002). At about the same time, Mark Hurlstone and I were looking at temporal grouping effects in the immediate recall of spoken nine-digit lists presented in three groups of various sizes. We found substantial differences in the accuracy of recall for the 28 grouping arrangements studied, replicating previous work by Ryan (1969a) together with no effect of informing participants of the grouping pattern in advance of list presentation, a new finding. The latter is exactly what would be expected if the context signal is driven directly by spoken input in the manner Tom was proposing and the three of us realized that the recall data offered an excellent opportunity to assess the plausibility of the mechanism. We devised a minimalist computational model that derives a context signal from the auditory input stream by means of detectors tuned to different frequencies and uses this to drive a CQ process. The model was remarkably successful in reproducing human performance when presented with each of the 28 grouping arrangements, including not only effects on the overall accuracy of recall but also the finer details of serial position curves and frequencies of different types of error (Hartley et al., 2016). We were also able to show that these were emergent properties of the mechanism and not a result of fitting parameters post hoc. The model, which we refer to as 'BUMP' (Bottom-Up Multi-oscillator Population) provides an interesting insight into the empirical observation that the overall accuracy of recall increases broadly in line with the product of the group sizes, being poorest for the various 1–7–1 groupings and best for 3–3–3. In terms of BUMP, regular groups generate a strong component of the context signal at the group presentation frequency and this tends to increase the discriminability of the context signal for different items, thereby reducing transposition errors and enhancing recall. However, there is also a cost as items occupying corresponding positions within groups of the same size have a component of the context signal with the same phase, resulting in a small increase in interposition errors, as in human data.

It is important to emphasize that BUMP is a minimalist model of only the 'front-end' of the phonological loop driven by spoken input and omits entirely its phonological and lexical components and interconnections with long-term memory. Thus, by itself it does not address effects of variables such as phonological similarity and articulatory suppression, or recall following visual presentation, or effects of phonotactic, syntactic, and semantic knowledge. However, despite these limitations, it seems a useful step towards linking serial ordering in list recall and nonword repetition, and, potentially, vocabulary acquisition. It also serves to illustrate the inherent complexity of the loop with regard to its links to speech perception and production and, more broadly, the implications of such complexity for developing the multicomponent model.

Is it necessary to assume a context signal?

So far, discussion has focused on my own collaborations in developing computational ideas about the phonological loop and has made no attempt to discuss the many other investigators developing competing models of their own at the same time. One of these, the primacy model, was particularly influential. This was the model I referred to earlier developed by Mike Page and Dennis Norris at the Cognition and Brain Sciences Unit in Cambridge, UK. They took a broadly similar approach except that they rejected any idea of a context signal, invoking in its place a 'primacy gradient' of activation levels over units representing items (Page & Norris, 1998). The basic idea of a primacy gradient is that when exposure to an item activates its internal representation, the boost is smaller for each successive item. Other aspects of the primacy model were similar to ours in that changes triggered by sequence presentation decay with time, and recalling an item involves CQ among simultaneously active items followed by retrieval of the winner's phonological composition followed by inhibition of its lexical representation. The primacy model gave rise to a steeper, more human-like primacy effect in serial position curves than our own model and did a better job of accounting for 'fill-in' errors in serial recall, where recalling an item one position early is typically followed by the error of recalling the item it displaced. Fill-in errors follow naturally in the primacy model as the displaced item will tend to be the most active at the point in question, whereas they are not so frequent when a context signal is used to select items.

In a subsequent development, the primacy model has been extended to address the Hebb repetition effect and, in principle, the learning of phonological word forms (Page & Norris, 2009). It achieves this by adding a chunking mechanism whereby exposure to a novel sequence of items results in a previously uncommitted input unit starting to engage in representing the entire sequence. When the same sequence is encountered again, the effect of the engaged unit is to steepen the primacy gradient over individual items, reducing errors in the CQ process and increasing the accuracy of recall. Eventually, after a sufficient number of exposures to the repeated list, the engaged input unit becomes fully committed and outcompetes the units representing individual items. In this way, the network acquires a new chunk (Miller, 1956) that represents the repeated list as a whole and this takes over from its representation as a series of individual items. Simulations show that the model correctly reproduces characteristic features of the Hebb effect such as its disappearance when list repetitions are separated by a large number of filler lists (Melton, 1963), simultaneous learning of more than one list (Hitch et al., 2009), and no transfer of learning to a list where alternate items occupy the same positions (Cumming et al., 2003).

Chunk formation is an extremely important aspect of long-term learning and the main contribution of the extended model is in suggesting a possible mechanism that makes some correct predictions. Page and Norris (2009) make much of the fact that their model achieves this without appealing to positional context units of the type assumed in our own model (Burgess & Hitch, 2006). However, they did not attempt to simulate aspects of the Hebb effect that are more difficult for their model to explain, but which ours handles well such as the effect of varying the temporal grouping pattern of the repeated list, which greatly reduces the rate of long-term learning as compared with keeping the grouping pattern the same (Bower & Winzenz, 1969; Hitch et al., 2009). Indeed, a general limitation of the primacy model is that it does not address effects of temporal grouping and rhythm. As we saw earlier, these include effects of different patterns of grouping on recall and systematic errors such as interpositions (see above) that clearly reflect the selective loss of positional information at different levels for individual items. Explaining these phenomena is one of the strongest arguments for assuming a context signal of some sort. Another seeming problem for the primacy model is that articulatory suppression would appear to disrupt the part of the network that steepens the primacy gradient and would therefore slow the rate of learning seen in the Hebb effect, whereas this is not the case in human experiments (Hitch et al., 2009).

It is important to emphasize that the point of these criticisms is not to weigh on whether one of the two modelling approaches is better than the other, but rather to draw attention to the fact that each has a different combination of strengths and weaknesses. Indeed, a weakness common to both is the assumption that forgetting is due to decay, which is at the very least an oversimplification given the empirical evidence we noted earlier. Another is that they do not address the formidable challenge of scale set by the size of human vocabulary, which in adults typically comprises tens of thousands of words. As for strengths, my best guess at the time of writing is that future models will combine a mechanism for chunk formation with some form of context signal, while continuing to retain the important distinction between the processing and representation of order and item information.

Computational principles versus computational models

The disagreement about whether or not to assume a context signal raises the general problem of how to choose between competing models, a problem amplified by the development of several alternative models of immediate serial recall that do not base themselves on the phonological loop (e.g. Botvinick & Plaut, 2006; Farrell & Lewandowsky, 2002; Grossberg, 1978a, 1978b; Grossberg & Pearson, 2008; Lewandowsky & Murdock, 1989; Murdock, 1993, 1995; Solway et al., 2012). One source of difficulty is that each model embodies several assumptions, some of which are regarded as key to its operation while others are seen as details of the way it is implemented. The challenge here is that as noted earlier it can be difficult to understand how a complex computational model works. A good example is the model proposed by Botvinick and Plaut (2006). They disagreed with the idea that short-term memory for serial order depends on transient context–item associations and presented a recurrent neural network model that reproduced core features of human performance without explicitly making this assumption. However, the model had to be given extensive training and appears to achieve its success by learning to form such associations, with the result that they were implicit in its operation. Another problem is that modellers have made different choices of core phenomena to simulate when assessing their models (as we saw above in the case of the Hebb effect). This makes it difficult to compare models using published papers alone. One way of addressing these challenges is to classify models in terms of the underlying computational principles they embody and evaluate these principles against a common set of core empirical phenomena. This approach has the advantage of avoiding becoming bogged down by the details of how individual models are implemented and differences in the way they have been assessed. Mark Hurlstone adopted it as part of his PhD, having been inspired by earlier work along these lines by Farrell and Lewandowsky (2004).

Mark's review of computational principles focused on serial order, concentrating mainly on verbal short-term memory, where human performance has been studied most extensively, but including visual and spatial short-term memory too (Hurlstone et al., 2014). The core phenomena chosen for consideration were performance and error patterns associated with serial position, sequence length, temporal grouping, item similarity, item repetition, and list (Hebb) repetition. This is a more comprehensive set than is typical in evaluations of individual models. However, it might seem surprising that it excludes word length and articulatory suppression, key features of the phonological loop. This was partly because the chosen phenomena were ones that could be considered across all three domains—verbal, visual, and spatial—and partly because effects of word length and articulatory suppression were judged not directly relevant to the problem of serial order. Turning to principles, models were broadly classified as falling into two types according to whether the mechanism for serial ordering is chaining (e.g. Lewandowsky & Murdock, 1989; Murdock, 1993, 1995; Solway et al., 2012) or CQ (e.g. Burgess & Hitch, 1992, 1999, 2006; Farrell & Lewandowsky, 2002; Grossberg, 1978a, 1978b; Grossberg & Pearson, 2008; Henson, 1998; Page & Norris, 1998, 2009). In the case of verbal short-term memory, chaining was rejected because of its difficulty explaining zig-zag serial position curves in the recall of sequences of alternating phonologically similar and dissimilar items (Baddeley, 1968; Henson et al., 1996). Chaining also has difficulty accounting for the high frequency of fill-in errors after the error of recalling an item one position too soon in a sequence. This is because chaining associations should tend to cue the item that originally followed the item recalled too soon, at variance with human data (Farrell et al., 2013; Suprenant et al., 2005). The main focus of the review was therefore on models employing CQ. However, it would be premature to write off chaining completely and indeed there are models that continue to employ this approach (Logan, 2021).

CQ models can be broadly classified as to whether the mechanism driving the activations of competing items is context free (Farrell & Lewandowsky, 2002; Grossberg, 1978a, 1978b; Grossberg & Pearson, 2008; Page & Norris, 1998) or context dependent (Brown et al., 2000, 2007; Burgess & Hitch, 1992, 1999, 2006; Farrell, 2006; Henson, 1998; Lewandowsky & Farrell, 2008). In context-free models, each successive item is less strongly activated than its predecessor, as in the primacy model. In context-dependent models, each item is associated with the current state of a context signal that changes progressively during sequence presentation. Serial recall is achieved by returning the context signal to its original state and allowing it to evolve as before, with context–item associations altering the activation gradient over items at each step. As a further distinction, context-dependent models differ as to whether the context signal is event based, that is, changing with each item (Burgess & Hitch, 1992; Farrell, 2006; Henson, 1998; Lewandowsky & Farrell, 2008); time based, that is, changing as a function of absolute time (Brown et al., 2000); or a combination of the two (Burgess & Hitch, 1999, 2006; Hartley, et al., 2016).

CQ provides a parsimonious account of several aspects of verbal immediate serial recall, most notably the tendency for order errors to move to adjacent serial positions which arises naturally from the effect of noise on the selection process. Context-dependent CQ models give a good account of the recall of temporally grouped lists whereas context-free CQ models do less well on this and other phenomena that suggest an internal representation of serial position such as 'protrusion errors' where an item that occupied a position in a preceding list appears as an error at the corresponding position in the current list (Conrad, 1960; Henson, 1999). Context-dependent CQ models also give a ready account of the effects of repeating an item within a sequence where the typical finding is that the second occurrence is poorly recalled, a phenomenon known as the Ranschburg effect (Jahnke, 1969). This observation can be explained in terms of response suppression after recalling the repeated item's first occurrence. Context-free CQ models have more difficulty handling item repetition effects, as extra assumptions have to be made to prevent the repeated item from being recalled first. However, as we have already noted, context-free CQ models provide a better account of the high frequency of fill-in errors in serial recall. Taken together, these observations suggest no single approach has all the answers and indeed when evidence from the verbal domain was reviewed, it favoured hybrid models that combine context-free and context-dependent CQ (Hurlstone et al., 2014). Thus, as suggested earlier when discussing the primacy model as a competitor to our own, a more adequate model will include desirable features of each.

Serial ordering in non-verbal domains

Such evidence as there was on the immediate recall of spatial and visual sequences pointed to the same conclusion as for verbal sequences (Hurlstone et al., 2014). However, several core phenomena in these domains had not been investigated and could not be considered. This prompted experiments designed to look more closely at serial order in short-term memory for spatial and visual sequences (Hurlstone & Hitch, 2015, 2018). The results provided further support for the conclusion that broadly the same principles apply in each of these very different domains. This has also been shown in situations where the immediate recall of verbal sequences involves a combination of visual and phonological coding as, for example, when Japanese speakers recall of a series of visually presented kanji characters. Recall is not only poorer when the characters are either visually or phonemically similar (Saito et al., 2008), but furthermore shows a zig-zag pattern when the visual similarity of items alternates in mixed lists (Logie et al., 2016), analogous to the pattern that proved so diagnostic for accounts of serial order in the verbal domain (Henson et al., 1996).

Further research suggests that the same underlying principles for serial ordering may also apply to short-term memory for musical sequences (Gorin et al., 2018). This in turn raises the question of whether a common multimodal CQ mechanism controls serial behaviour across domains. Something like this would seem intuitively necessary to account for cross-domain coordination and synchronicity in everyday serial behaviour such as gesturing while speaking or singing while dancing. The obvious locus for such a mechanism in the multicomponent model of working memory is the multimodal episodic buffer (Baddeley, 2000). If so, we would expect to see cross-domain interference in the control of serial order under appropriate conditions, and there is some evidence that this does occur. For example, manual tapping disrupts verbal short-term memory for order more than memory for items (Henson et al., 2003). Similarly, tapping to an irregular rhythm disrupts the retention of order information in short-term memory for musical sequences (Gorin et al., 2016). However, other studies emphasize the domain specificity of ordering processes (e.g. Soemer & Saito, 2016) suggesting that serial order may be handled differently depending on task and level of representation. There is clearly much more to find out about how this is achieved.

Neuroscience

The insights gained through modelling can be applied in numerous ways, a good example being the search for neural mechanisms for serial order. In a pioneering study, Averbeck et al. (2002) looked for neural evidence of CQ when non-human primates prepare to carry out a sequence of movements. They taught rhesus macaques to draw simple geometric shapes, each involving the execution of a unique sequence of movement segments. Recordings from individual neurons in prefrontal cortex showed that each segment was associated with a distinct pattern of neural activity that waxed and waned in turn as the drawing was carried out. Crucially, these distinctive patterns were also simultaneously present when preparing to draw, their relative strengths reflecting the order in which the segments were to be executed, the first being strongest. This is in line with Lashley's (1951) original proposal and emphasizes once more the wider generality of the CQ mechanism identified in human verbal short-term memory. Similar results for motor sequencing have been obtained in humans (Kornysheva et al., 2019). Participants learned sequences of finger presses that varied in the order and timing of responses and subsequently reproduced them in ~an~ a magnetoencephalography scanner. Patterns of neural activity were classified for each finger press and dynamic changes in their strengths were analysed during both planning and task execution. As in non-human primates, neural activation patterns for forthcoming responses were simultaneously active during the planning phase, with their relative strengths reflecting their serial positions in the forthcoming sequence. Further analyses revealed that these activations generalize across differences in the responses and their timings in the forthcoming sequence and therefore correspond to an ordinal template. The overall picture to emerge is one in which the brain combines this high-level ordinal template with lower-level representations of individual actions and their timings to generate behaviour. It is interesting to note that the very same question of how serial order is handled at multiple levels arose when discussing CQ in the context of the phonological loop, and in working memory more generally, encouraging the view that investigations of the control of serial order in different domains have mutual implications, just as in Lashley's (1951) far-sighted proposal.

Other neuroimaging studies have focused on the neural basis of serial order in auditory–verbal short-term memory. For example, Kalm and Norris (2014) used functional magnetic resonance imaging to detect neural activity in a task in which participants heard three bisyllabic non-words for immediate serial vocal recall. By using a small set of items, they were able to dissociate patterns of activity representing items from those representing their order. The results showed that brain areas associated with auditory processing store associations between items and their serial positions. In subsequent work, Kalm and Norris (2017) used functional magnetic resonance imaging to investigate the representation of temporal order information in the presentation, retention, and response phases of a visual short-term memory task involving serial recognition. They found that voxels in the anterior temporal lobe represented serial position consistently across task phases, providing further support for positional accounts of context. They speculated that the shared code might reflect verbal recoding, but whether this is so and whether the conclusions generalize to longer sequences and different types of items remain to be explored.

Theoretical questions arising from computational modelling have also been used to guide investigations of the neural basis of sequence learning. Thus, in another study, Kalm and Norris (2021) collected functional magnetic resonance imaging data in a visual memory task that involved the immediate serial recall of sequences of four Gabor patches by means of a four-button box where each button mapped onto a single item. The same items were presented on each trial and only their order varied such that on some trials the sequence was one that had been learned previously, and on others it was entirely novel, as in studies of the Hebb effect. By comparing patterns of neural activity for the novel and learned sequences it was possible to compare the predictions of models that assume learning reflects the strengthening of position–item associations against the alternative that a familiar sequence is recoded as a novel chunk. The results were inconsistent with the strengthening of position–item associations and supported instead a chunking account. However, the use of very short sequences would favour chunking and the data do not speak to learning longer sequences, where sequential matching processes and positional associations might be expected to play a role (Burgess & Hitch, 2006). Nevertheless, this study and the others we have described briefly here clearly illustrate the value of computational models for guiding research into the neural processes underpinning serial order and sequence learning not only in the verbal domain but others too.

Conclusion

I have described how the phonological loop was the first component of working memory to be studied in any depth, building on prior research on verbal short-term memory. The initial concept was relatively simple and easy to investigate empirically, being associated with converging evidence from a small set of interrelated experimental manipulations. I went on to describe how the phonological loop can account for cross-linguistic variation in digit span, how it serves the useful function of storing partial results in mental arithmetic tasks, and its crucial role in supporting vocabulary acquisition. As a broad theoretical concept, therefore, it has held up remarkably well over time. It continues to feature prominently in many aspects of cognitive psychology (Baddeley & Hitch, 2019), in cognitive neuropsychology (Shallice & Papagno, 2019), and neuroscience (Buchsbaum & D'Esposito, 2019) and has even entered areas such as anthropology (Aboitiz et al., 2010).

The initial description of the phonological loop was, however, lacking in two respects that are important for understanding how it operates and its role in cognition. Specifically, it failed to deal with the maintenance of serial order information and how short-term storage interfaces with long-term memory. My aim here has been to explain how modelling the loop as an artificial neural network has helped develop our theoretical understanding on these two fronts. The story that has emerged through my collaborations with expert modellers resonates with the progress made by the many other contributors to the field. The main conclusion is that computational modelling has greatly extended our view of the phonological loop so as to account for more aspects of verbal short-term memory, including sequencing errors and their distributions, effects of rhythm, and long-term sequence learning, as in the Hebb repetition effect. A second conclusion is that serial ordering involves simultaneous activation of the elements of a sequence and selection of each in turn by means of CQ, just as Lashley (1951) foresaw. The activations that drive the CQ process are partly determined by internal, context-free dynamics and partly by associations to an externally driven context signal, the latter being particularly important in accounting for effects of rhythm. In the case of a spoken sequence, the context signal appears to depend on perceptually driven amplitude and phase detectors that track rhythm at multiple frequencies (corresponding to groups, words, syllables, etc.) simultaneously. Progress regarding long-term learning and interactions with long-term memory has been made but is more limited, with attempts to model the Hebb repetition effect and vocabulary acquisition as yet only moderately successful (see also Gupta & Tisdale, 2009; Jones et al., 2007, for approaches not discussed here).

Finally, it is interesting to reflect briefly on some of the wider benefits arising from attempts to model the phonological loop computationally. An early goal was to provide a more detailed account of the selective neuropsychological impairment of auditory–verbal short-term memory that was influential in developing the original multicomponent model of working memory (Baddeley & Hitch, 1974). In the Burgess and Hitch (1999) model, we showed how this deficit could be straightforwardly simulated by lesioning phonological input units and their lexical connections. However, we also noted a potentially important limitation of modelling in that a patient's residual abilities may involve compensation strategies using subsystems that fall outside the scope of the model. It came as no surprise, therefore, that a comprehensive review of cases of the selective neuropsychological impairment of auditory–verbal short-term memory concluded that the broader multicomponent model of working memory offers a more useful framework for theoretical interpretation than do detailed models of the subsystem itself (Shallice & Papagno, 2019).

Shallice and Papagno's (2019) conclusion raises the general question whether computational models are too narrow and specific to have broader applications. However, one of the strongest arguments against this is the evidence we have described implicating CQ in the control of serial order in a range of domains besides auditory–verbal short-term memory. The point here is that a computational mechanism identified in modelling one set of phenomena may have general application even if the model itself has limited scope. Another outcome of attempts to model the phonological loop has been to boost the search for separate neural mechanisms underpinning order and item information. This has started to pay off in the verbal domain with evidence for a double dissociation in brain-damaged adults (Attout et al., 2012) and neuroimaging data showing the progressive specialization of neural correlates of serial order in normal development (Attout et al., 2019). The order/

item distinction is also becoming increasingly important in specifying the role of the phonological loop in cognitive skills such as arithmetic. One example is longitudinal evidence that serial order abilities at age 5 predict later mental calculation performance whereas item abilities do not (Attout et al., 2014). Another is evidence that developmental dyscalculia is associated with a deficit specific to serial ordering (e.g. Attout & Majerus, 2014).

In conclusion, developing computational models of the phonological loop has improved our understanding of the loop and its role in working memory and cognition and established links with mechanisms for serial order in other quite different domains. The story that has emerged illustrates something of what stands to be gained by focusing on a specific problem while not forgetting the broader picture and sets the scene for further questions.

Acknowledgements

I am grateful to all my collaborators without whom none of the work described here would have been possible. Special thanks go to Alan Baddeley, Neil Burgess, Tom Hartley, and Robert Logie for comments that led to substantial improvements in the final version.

References

Aboitiz, F., Aboitiz, S., & Garcia, R. R. (2010). The phonological loop: A key innovation in human evolution. *Current Anthropology*, *51*(Suppl 1), S55–S64.

Atkinson, R. C., & Shiffrin, R. M. (1968). Human memory: A proposed system and its control processes. In K. W. Spence (Ed.), *The psychology of learning and motivation: Advances in research and theory* (Vol. 2, pp. 89–195). Academic Press.

Attout, L., Magro, L. O., Szmalec, A., & Majerus, S. (2019). The developmental neural substrates of item and serial order components of verbal working memory. *Human Brain Mapping*, *40*(5), 1541–1553.

Attout, L., & Majerus, S. (2014). Working memory deficits in developmental dyscalculia: The importance of serial order. *Child Neuropsychology*, *21*(4), 432–450.

Attout, L., Noel, M. P., & Majerus, S. (2014). The relationship between working memory for serial order and numerical development: A longitudinal study. *Developmental Psychology*, *50*(6), 1667–1679.

Attout, L., Van der Kaa, M. A., George, M., & Majerus, S. (2012). Dissociating short-term memory and language impairment: The importance of item and serial order information. *Aphasiology*, *26*(3–4), 355–382.

Averbeck, B. B., Chafee, M. V., Crowe, D. A., & Georgopoulos, A. P. (2002). Parallel processing of serial movements in prefrontal cortex. Proceedings of the New York Academy of Sciences, *99*(20), 13172–13177.

Baddeley, A. D. (1968). How does acoustic similarity influence short-term memory? *Quarterly Journal of Experimental Psychology*, *20*(3), 249–264.

Baddeley, A. D. (1986). *Working memory*. Oxford University Press.

Baddeley, A. D. (1996). Exploring the central executive. *Quarterly Journal of Experimental Psychology*, *49*(1), 5–28.

Baddeley, A. D. (2000). The episodic buffer: A new component of working memory? *Trends in Cognitive Sciences*, *4*(11), 417–423.

Baddeley, A. D. (2002). Is working memory still working? *European Psychologist, 7*(2), 85–97.

Baddeley, A. D., Gathercole, S., & Papagno, C. (1998). The phonological loop as a language learning device. *Psychological Review, 105*(1), 158–173.

Baddeley, A. D., Grant, S., Wight, E., & Thomson, N. (1975). Imagery and visual working memory. In P. M. A. Rabbitt & S. Dornic (Eds.), *Attention and performance V* (pp. 205–217). Academic Press.

Baddeley, A. D., & Hitch, G. J. (1974). Working memory. In G. H. Bower (Ed.), *The psychology of learning and motivation: Advances in research and theory* (Vol. VIII, pp. 47–90). Academic Press.

Baddeley, A. D., & Hitch, G. J. (2019). The phonological loop as a buffer store: An update. *Cortex, 112*, 91–106.

Baddeley, A. D., & Lewis, V. J. (1981). Inner active processes in reading: The inner voice, the inner ear and the inner eye. In A. M. Lesgold & C. A. Perfetti (Eds.), *Interactive processes in reading* (pp. 107–129). Lawrence Erlbaum.

Baddeley, A. D., Lewis, V., & Vallar, G. (1984). Exploring the articulatory loop. *Quarterly Journal of Experimental Psychology, 36*(2), 233–252.

Baddeley, A. D., Papagno, C., & Vallar, G. (1988). When long-term learning depends on short-term storage. *Journal of Memory and Language, 27*(5), 586–595.

Baddeley, A. D., Thomson, N., & Buchanan, M. (1975). Word length and structure of short-term memory. *Journal of Verbal Learning and Verbal Behavior, 14*(6), 575–589.

Besner, D., Davies, J., & Daniels, S. (1981). Reading for meaning: The effects of concurrent articulation. *Quarterly Journal of Experimental Psychology, 33*(4), 415–438.

Bjork, E. L., & Healy, A. F. (1974). Short-term order and item retention. *Journal of Verbal Learning and Verbal Behavior, 13*(1), 80–97.

Botvinick, M. M., & Plaut, D. C. (2006). Short-term memory for serial order: A recurrent neural network model. *Psychological Review, 113*(2), 201–233.

Bower, G., & Winzenz, D. (1969). Group structure, coding, and memory for digit series. *Journal of Experimental Psychology, 80*(2, Pt. 2), 1–17.

Brooks, L. R. (1967). The suppression of visualization by reading. *Quarterly Journal of Experimental Psychology, 19*(4), 289–299.

Brown, G. D. A., Neath, I., & Chater, N. (2007). A temporal ratio model of memory. *Psychological Review, 114*(3), 539–576.

Brown, G. D. A., Preece, T., & Hulme, C. (2000). Oscillator-based memory for serial order. *Psychological Review, 107*(1), 127–181.

Buchsbaum, B. R., & D'Esposito, M. (2019). A sensorimotor view of verbal working memory. *Cortex, 112*, 134–148.

Burgess, N., & Hitch, G. J. (1992). Toward a network model of the articulatory loop. *Journal of Memory and Language, 31*(4), 429–460.

Burgess, N., & Hitch, G. J. (1999). Memory for serial order: A network model of the phonological loop and its timing. *Psychological Review, 106*(3), 551–581.

Burgess, N., & Hitch, G. J. (2006). A revised model of short-term memory and long-term learning of verbal sequences. *Journal of Memory and Language, 55*(4), 627–652.

Butterworth, B., Cipolotti, L., & Warrington, E. K. (1996). Short-term memory impairment and arithmetic ability. *Quarterly Journal of Experimental Psychology, 49*(1), 251–262.

Chan, M. E., & Elliot, J. M. (2011). Cross-linguistic differences in digit memory span. *Australian Psychologist, 46*(1), 23–50.

Chen, E. H., & Bailey, D. H. (2021). Dual-task studies of working memory and arithmetic performance: A meta-analysis. *Journal of Experimental Psychology: Learning, Memory, and Cognition, 47*(2), 220–233.

Chincotta, D., & Underwood, G. (1997). Digit span and articulatory suppression: A cross-linguistic comparison. *European Journal of Cognitive Psychology, 9*(1), 89–96.

Conrad, R. (1960). Serial order intrusions in immediate memory. *British Journal of Psychology, 51*, 45–48.

Conrad, R., & Hull, A. J. (1964). Information, acoustic confusion and memory span. *British Journal of Psychology, 55*, 429–432.

Cumming, N., Page, M., & Norris, D. (2003). Testing a positional model of the Hebb effect. *Memory, 11*(1), 43–63.

De Rammelaere, S., Stuyven, E., & Vandierendonck, A. (2001). Verifying simple arithmetic sums and products: Are the phonological loop and the central executive involved? *Memory & Cognition, 29*(2), 267–273.

DeStefano, D., & LeFevre, J. A. (2004). The role of working memory in mental arithmetic. *European Journal of Cognitive Psychology, 16*(3), 353–386.

Ellis, N. C., & Hennelly, R. A. (1980). Bilingual word-length effect: Implications for intelligence-testing and the relative ease of mental calculation in Welsh and English. *British Journal of Psychology, 71*(1), 43–51.

Farrell, S. (2006). Mixed-list phonological similarity effects in delayed serial recall. *Journal of Memory and Language, 55*(4), 587–600.

Farrell, S., Hurlstone, M. J., & Lewandowsky, S. (2013). Sequential dependencies in recall of sequences: Filling in the blanks. *Memory & Cognition, 41*(6), 938–952.

Farrell, S., & Lewandowsky, S. (2002). An endogenous distributed model of ordering in serial recall. *Psychonomic Bulletin & Review, 9*(1), 59–79.

Farrell, S., & Lewandowsky, S. (2004). Modelling transposition latencies: Constraints for theories of serial order memory. *Journal of Memory and Language, 51*(1), 115–135.

Fürst, A., & Hitch, G. J. (2000). Separate roles for executive and phonological components of working memory in mental arithmetic. *Memory & Cognition, 28*(5), 774–782.

Gathercole, S., & Baddeley, A. (1989). Evaluation of the role of phonological STM in the development of vocabulary in children: A longitudinal study. *Journal of Memory and Language, 28*(2), 200–213.

Gathercole, S. E., & Baddeley, A. D. (1990). Phonological memory deficits in language-disordered children: Is there a causal connection? *Journal of Memory and Language, 29*(3), 336–360.

Glasspool, D. W., Houghton, G., & Shallice, T. (1995). Interactions between knowledge sources in a dual-route connectionist model of spelling. In: L. S. Smith & P. J. B. Hancock (Eds.), *Neural computation and psychology: Workshops in computing* (pp. 209–226). Springer.

Gorin, S., Kowialiewski, B., & Majerus, S. (2016). Domain-generality of timing-based serial order processes in short-term memory: New insights from musical and verbal domains. *PLoS One, 11*(12), 1–25.

Gorin, S., Mengal, P., & Majerus, S. (2018). A comparison of serial order short-term memory across verbal and musical domains. *Memory & Cognition, 46*(3), 464–481.

Gregg, V. H., Freedman, C. M., & Smith, D. K. (1989). Word frequency, articulatory suppression and memory span. *British Journal of Psychology, 80*(3), 363–374.

Grossberg, S. (1978a). A theory of human memory: Self-organization and performance of sensory-motor codes, maps, and plans. In R. Rosen & F. Snell (Eds.), *Progress in theoretical biology* (Vol. 5, pp. 233–374). Academic Press.

Grossberg, S. (1978b). Behavioral contrast in short-term memory: Serial binary memory models or parallel continuous memory models? *Journal of Mathematical Psychology, 17*(3), 199–219.

Grossberg, S., & Pearson, L. R. (2008). Laminar cortical dynamics of cognitive and motor working memory, sequence learning and performance: Towards a unified theory of how the cerebral cortex works. *Psychological Review, 115*(3), 677–732.

Gupta, P., & MacWhinney, B. (1997). Vocabulary acquisition and verbal short-term memory: Computational and neural bases. *Brain and Language, 59*(2), 267–333.

Gupta, P., & Tisdale, J. (2009). Does phonological short-term memory causally determine vocabulary learning? Toward a computational resolution of the debate. *Journal of Memory and Language, 61*(4), 481–502.

Hartley, T. (2002). Syllabic phase: A bottom-up representation of the temporal structure of speech. In J. Bullinaria & W. Lowe (Eds.), *Connectionist models of cognition and perception* (pp. 277–288). World Scientific Publishing Co.

Hartley, T., & Houghton, G. (1996). A linguistically constrained model of short-term memory for nonwords. *Journal of Memory and Language, 35*(1), 1–31.

Hartley, T., Hurlstone, M. J., & Hitch, G. J. (2016). Effects of rhythm on memory for spoken sequences: A model and tests of its stimulus-driven mechanism. *Cognitive Psychology, 87*, 135–178.

Healy, A. F. (1974). Separating item from order information in short-term memory. *Journal of Verbal Learning and Verbal Behavior, 13*(6), 644–655.

Heathcote, D. (1994). The role of visuo-spatial working memory in the mental addition of multi digit addends. *Cahiers de Psychologie Cognitive, 13*, 207–245.

Hebb, D. (1961). *The organization of behavior*. John Wiley.

Hecht, S. A. (2002). Counting on working memory in simple arithmetic when counting is used for problem solving. *Memory & Cognition, 30*(3), 447–455.

Henson, R. N. A. (1998). Short-term memory for serial order: The start-end model. *Cognitive Psychology, 36*(2), 73–137.

Henson, R. N. A. (1999). Positional information in short-term memory: Relative or absolute? *Memory & Cognition, 27*(5), 915–927.

Henson, R., Hartley, T., Burgess, N., Hitch, G., & Flude, B. (2003). Selective interference with verbal short-term memory for serial order information: A new paradigm and tests of a timing signal hypothesis. *Quarterly Journal of Experimental Psychology, 56*(8), 1307–1334.

Henson, R. N. A., Norris, D. G., Page, M. P. A., & Baddeley, A. D. (1996). Unchained memory: Error patterns rule out chaining models of immediate serial recall. *Quarterly Journal of Experimental Psychology, 49*(1), 80–115.

Hitch, G. J. (1978). The role of short-term working memory in mental arithmetic. *Cognitive Psychology, 10*(3), 302–323.

Hitch, G. J., Allen, R. J., & Baddeley, A. D. (2021). Attention and binding in visual working memory: Two forms of attention and two kinds of buffer storage. *Attention, Perception and Psychophysics, 82*(1), 280–293.

Hitch, G. J., Burgess, N., Towse, J. N., & Culpin, V. (1996). Temporal grouping effects in immediate recall: A working memory analysis. *Quarterly Journal of Experimental Psychology, 49*(1), 140–158.

Hitch, G. J., Fastame, M. C., & Flude, B. (2005). How is the serial order of a verbal sequence coded? Some comparisons between models. *Memory, 13*(3–4), 247–258.

Hitch, G. J., Flude, B., & Burgess, N. (2009). Slave to the rhythm: Experimental tests of a model for verbal short-term memory and long-term sequence learning. *Journal of Memory and Language, 61*(1), 97–111.

Hoosain, R., & Salili, F. (1987). Language differences in pronunciation speed for numbers, digit span and mathematical ability. *Psychologia, 30*(1), 34–38.

Houghton, G. (1990). The problem of serial order: A neural network model of sequence learning and recall. In R. Dale, C. Mellish, & M. Zock, (Eds.), *Current research in natural language generation* (pp. 287–319). Academic Press.

Hulme, C., Maughan, S., & Brown, G. D. A. (1991). Memory for familiar and unfamiliar words: Evidence for a long-term memory contribution to short-term memory span. *Journal of Memory and Language, 30*(6), 685–701.

Hurlstone, M. J., & Hitch, G. J. (2015). How is the serial order of a spatial sequence represented? Insights from transposition latencies. *Journal of Experimental Psychology: Learning, Memory, and Cognition, 41*(2), 295–324.

Hurlstone, M. J., & Hitch, G. J. (2018). How is the serial order of a visual sequence represented? Insights from transposition latencies. *Journal of Experimental Psychology: Learning, Memory, and Cognition, 44*(2), 167–192.

Hurlstone, M. J., Hitch, G. J., & Baddeley, A. D. (2014). Memory for serial order across domains: An overview of the literature and directions for future research. *Psychological Bulletin, 140*(2), 339–373.

Jahnke, J. C. (1969). The Ranschburg effect. *Psychological Review, 76*(6), 592–605.

Jalbert, A., Neath, I., Bireta, T. J., & Surprenant, A. M. (2011). When does length cause the word length effect? *Journal of Experimental Psychology: Learning, Memory, and Cognition, 37*(2), 338–353.

Jones, G., Gobet, F., & Pine, J. M. (2007). Linking working memory and long-term memory: A computational model of the learning of new words. *Developmental Science, 10*(6), 853–873.

Kalm, K., & Norris, D. (2014). The representation of order information in auditory-verbal short-term memory. *Journal of Neuroscience, 34*(20), 6879–6879.

Kalm, K., & Norris, D. (2017). A shared representation of order between encoding and recognition in visual short-term memory. *Neuroimage, 155*, 138–146.

Kalm, K., & Norris, D. (2021). Sequence learning recodes cortical representations instead of strengthening initial ones. *PLoS Computational Biology, 17*(5), e1008969.

Kornysheva, K., Bush, D., Meyer, S. S., Sadnicka, A., Barnes, G., & Burgess, N. (2019). Neural CQ or ordinal structure underlies skilled sequential action. *Neuron*, *101*(6), 1–15.

Lashley, K. S. (1951). The problem of serial order in behavior. In L. A. Jeffress (Ed.), *Cerebral mechanisms in behavior: The Hixon symposium* (pp. 112–146). John Wiley.

Lewandowsky, S., & Farrell, S. (2008). Short-term memory: New data and a model. In B. H. Ross (Ed.), The *psychology of learning and motivation: Advances in research and theory* (Vol. 49, pp. 1–48). Elsevier Academic Press.

Lewandowsky, S., & Murdock, B. B. (1989). Memory for serial order. *Psychological Review*, *96*(1), 25–57.

Lewandowsky, S., Oberauer, K., & Brown, G. D. A. (2009). No temporal decay in verbal short-term memory. *Trends in Cognitive Sciences*, *13*(3), 120–126.

Logan, G. D. (2021). Serial order in perception, memory, and action. *Psychological Review*, *128*(1), 1–44.

Logie, R. H., Gilhooly, K., & Wynn, V. (1994). Counting on working memory in arithmetic problem-solving. *Memory & Cognition*, *22*(4), 395–410.

Logie, R. H., Saito, S., Morita, A., Varma, S., & Norris, D. (2016). Recalling visual serial order for verbal sequences. *Memory & Cognition*, *44*(4), 590–607.

McClelland, J. C., Rumelhart, D. E., & Hinton, G. E. (1986). The appeal of parallel distributed processing. In D. E. Rumelhart, J. L. McClelland, & the PDP Research Group (Eds.), *Parallel distributed processing* (Vol. 1, pp. 3–44). MIT Press.

Melton, A. W. (1963). Implications of short-term memory for a general theory of memory. *Journal of Verbal Learning and Verbal Behavior*, *2*(1), 1–21.

Miller, G. A. (1956). The magical number seven, plus or minus two: Some limits on our capacity to process information. *Psychological Review*, *63*(2), 81–97.

Murdock, B. B. (1993). TODAM2: A model for the storage and retrieval of item, associative, and serial-order information. *Psychological Review*, *100*(2), 183–203.

Murdock, B. B. (1995). Developing TODAM: Three models for serial order information. *Memory & Cognition*, *23*(5), 631–645.

Murray, D. J. (1968). Articulation and acoustic confusability in short-term memory. *Journal of Experimental Psychology*, *78*(4, Pt. 1), 679–684.

Naveh-Benjamin, M., & Ayres, T. J. (1986). Digit span, reading rate, and linguistic relativity. *Quarterly Journal of Experimental Psychology*, *38*(4), 739–751.

Ostrosky-Solis, F., & Lozano, A. (2006). Digit span: Effect of education and culture. *International Journal of Psychology*, *41*(5), 333–341.

Page, M. P. A., & Norris, D. (1998). The primacy model: A new model of immediate serial recall. *Psychological Review*, *105*(4), 761–781.

Page, M. P. A., & Norris, D. (2009). A model linking immediate serial recall, the Hebb repetition effect and the learning of phonological word forms. *Philosophical Transactions of the Royal Society of London. Series B, Biological Sciences*, *364*(1536), 3737–3753.

Papagno, C., Valentine, T., & Baddeley, A. D. (1991). Phonological short-term memory and foreign language vocabulary learning. *Journal of Memory and Language*, *30*(3), 331–347.

Papagno, C., & Vallar, G. (1992). Phonological short-term memory and the learning of novel words: The effect of phonological similarity and item length. *Quarterly Journal of Experimental Psychology*, *44*(1), 47–67.

Roodenrys, S., & Hinton, M. (2002). Sublexical or lexical effects on serial recall of nonwords? *Journal of Experimental Psychology: Learning, Memory, and Cognition*, *28*(1), 29–33.

Ryan, J. (1969a). Grouping and short-term memory: Different means and patterns of grouping. *Quarterly Journal of Experimental Psychology*, *21*(2), 137–147.

Ryan, J. (1969b). Temporal grouping rehearsal and short-term memory. *Quarterly Journal of Experimental Psychology*, *21*(2), 148–155.

Saito, S., Logie, R. H., Morita, A., & Law, A. (2008). Visual and phonological similarity effects in verbal immediate serial recall: A test with kanji materials. *Journal of Memory and Language*, *59*(1), 1–17.

Service, E. (1998). The effect of word length on immediate serial recall depends on phonological complexity, not articulatory duration. *Quarterly Journal of Experimental Psychology Section A*, *51*(2), 283–304.

Shallice, T., & Papagno, C. (2019). Impairments of auditory-verbal short-term memory: Do selective deficits of the input phonological buffer exist? *Cortex*, *11*, 107–121.

Shallice, T., & Warrington, E. K. (1970). Independent functioning of verbal memory stores: A neuropsychological study. *Quarterly Journal of Experimental Psychology*, *22*(2), 261–273.

Soemer, A., & Saito, S. (2016). Domain-specific processing in short-term serial order memory. *Journal of Memory and Language*, *88*, 1–17.

Solway, A., Murdock, B. B., & Kahana, M. J. (2012). Positional and temporal clustering in serial order memory. *Memory & Cognition*, *40*(2), 177–190.

Stigler, J. W., Lee, S. Y., & Stevenson, H. W. (1986). Digit memory in Chinese and English: Evidence for a temporally limited store. *Cognition*, *23*(1), 1–20.

Surprenant, A. M., Kelley, M. R., Farley, L. A., & Neath, I. (2005). Fill-in and infill errors in order memory. *Memory*, *13*(3–4), 267–273.

Treiman, R., & Danis, C. (1988). Short-term memory errors for spoken syllables are affected by the linguistic structure of the syllables. *Journal of Experimental Psychology: Learning, Memory, and Cognition*, *14*(1), 145–152.

Warrington, E. K., & Shallice, T. (1969). The selective impairment of auditory-verbal short-term memory. *Brain*, *92*(4), 885–896.

PART 2

MEMORY DEVELOPMENT

9

Working memory and child development with its windfalls and pitfalls

Nelson Cowan

This chapter includes sections that deal with the embedded-processes theoretical model of working memory and information processing, typical child development of working memory, and practical issues emerging in childhood, within the context of the lifespan, leading finally to some brief concluding comments integrating the cognitive and developmental theory of working memory with practical implications.

Link between theory and practice

My research on working memory has been motivated by wanting to understand consciousness, but with the fervent hope that educational and medical needs would also be served somehow by this work. Among various influences on my interests, when I was an adolescent I saw a children's science show explaining how difficult it is for humans to control their own attention to generate random numbers without including unintended runs (e.g. 4, 5, 6) and later understood that this demonstration was most likely based on research by Baddeley (1966). It is noteworthy also that I began graduate school in 1974, the year that the seminal chapter on working memory by Baddeley and Hitch (1974) came out, and it is clear that their approach in that paper and subsequent ones greatly influenced the course of my career.

My writing has mostly covered the theoretical part, including adult processes, and a typical child development component funded by the National Institute of Child Health and Human Development since 1984. Along the way, however, I have had intensive collaborations with investigators of applied topics including language impairment, dyslexia, bilingualism, autism, schizophrenia, amnesia, Parkinson's disease, alcoholic intoxication, and adult ageing. I am glad to be asked about practical applications of theory; in this chapter, I will focus on child development and how my collaborations in practical topics have interfaced with theory. I have sometimes been disappointed at how slowly the ideas seep from research into practice, and I hope this concentrated review can help. In the remainder of this section, I will discuss the link between my theoretical work and practical implications for child development,

Nelson Cowan, *Working memory and child development with its windfalls and pitfalls* In: *Memory in Science for Society*. Edited by: Robert H. Logie, Zhisheng (Edward) Wen, Susan E. Gathercole, Nelson Cowan, and Randall W. Engle, Oxford University Press.
© Oxford University Press 2023. DOI: 10.1093/oso/9780192849069.003.0009

followed by a main section on the embedded-processes approach and a section on practical developmental applications.

The thrust of the embedded-processes theory (Cowan, 1988, 1995, 1999, 2005/ 2016, 2019) has been to emphasize the link between attention and memory, a move away from the focus on the automatic nature of storage (e.g. Baddeley, 1986) and one that may have helped to bring about similar changes in other theories (e.g. Baddeley, 2001). It divides working memory storage into some information held in the focus of attention and other information held outside of that focus. Mnemonic strategies are cast as activities to retain information in the focus of attention or, more often, boost or reorganize information held outside of the focus, in the activated portion of long-term memory, facilitating rapid learning of that information. Before the embedded-processes view, the field emphasized loss of information from working memory through temporal decay (e.g. Baddeley, 1986; Hulme & Tordoff, 1989; Schweickert & Boruff, 1986). Illustrating that there are limitations to the explanatory power of decay and combatting it through speedier responding, Cowan, Elliott et al. (2006) found that children who were taught to repeat lists as quickly as adults usually do were able to gain this speed but found that it did not improve their digit spans at all. The embedded-processes model moved away from the primacy of timing factors and placed a core limit on the number of chunks that could be held at once in the focus of attention independent of one another when the information could not be maintained using strategies such as grouping or rehearsal (Cowan, 2001; Chen & Cowan, 2009; Cowan et al., 2012). This approach assisted a movement in the field towards an emphasis on chunk capacity limits that had mostly been in abeyance since Miller (1956).

Interference effects in the embedded-processes model (Cowan, 1988) were recast as due to similar features of many types (acoustic, phonological, orthographic, spatial, haptic, semantic, etc.) in multiple stimuli, a similarity principle more general than the notion of stimuli entering the same or different domain-specific buffers, phonological or visuospatial (e.g. Baddeley, 1986). Attentional filtering was recast as habituation of the attentional orienting response (Sokolov, 1963), with dishabituation of changed stimuli attracting attention in competition with voluntarily controlled central executive processes. This provided a specific mechanism for the attenuation theory of Treisman (1964).

The implications of these theoretical changes for practical issues are that there is more expectation of developmental changes and pathologies that are general across domains, as opposed to reflecting domain-specific issues, and more based on capacity rather than speed. In keeping with this expectation, we have the example of developmental changes in working memory capacity attributed to better detection of patterns that can allow information to be off-loaded out of the focus of attention (in child development, Cowan et al., 2018; in adult ageing, Greene et al., 2020). In pathology, we have the example of memory in dyslexic children suffering from poor serial order memory not only for spoken digits, but also for spatial locations in a running span task (Cowan et al., 2017).

There are, nevertheless, cases in which speed seems to be important for recall (Cowan et al., 1998; Gaillard et al., 2011), but it is not clear if fundamental differences in speed control the apparent capacity, or vice versa (Lemaire et al., 2018), a question for future research. This question has implications for interpreting specific deficits, for example, differential deficits in Down and Williams syndromes (Jarrold et al., 2004), to be discussed later.

Some of my early interest in working memory was stimulated by graduate school reading of Alan Baddeley's work on that emerging topic (e.g. Baddeley et al. 1975) and at some point Miller (1956), along with early works on attention (e.g. Kahneman, 1973; Posner, 1980; Sokolov, 1963; Treisman, 1964), information processing (e.g. Atkinson & Shiffrin, 1968), neuropsychology (e.g. a book by Alexander Luria on neuropsychological symptoms following bullet wounds to the head), and the consciousness-related perceptual moment (e.g. Allport, 1968). My 1980 PhD was in the developmental area at the University of Wisconsin, and my early published work covered echoic memory and speech-related memory in infants (Cowan et al., 1982; Goodsitt et al., 1984) and adults (e.g. Cowan, 1984, 1991; Cowan & Morse, 1986).

Early in my current job at the University of Missouri, I published a *Psychological Bulletin* paper, my second after the 1984 paper on acoustic memory. This new paper was aimed at reconceptualizing the information processing system (Cowan, 1988). It led to Oxford University Press offering a book contract to expand upon the ideas, culminating in *Attention and Memory: An Integrated Framework* (Cowan, 1995). In the process of doing research for that book, in the autumn of 1990 I spent several weeks in the UK, one stop being to visit Donald Broadbent at Oxford. He was greatly responsible for helping to kick-start the modern field of cognitive psychology, especially with his 1958 book, *Perception and Communication*. The book included a footnote with a roughly sketched processing diagram that was the precursor to the model of Atkinson and Shiffrin (1968). Anyway, Broadbent looked at the list of researchers I planned to visit all over the UK and remarked that they all had worked with him (including Alan Baddeley, Graham Hitch, Susan Gathercole, Charles Hulme, Dylan Jones, Robert Logie, and Clive Frankish among others).

In earlier days, Broadbent had directed the Applied Psychology Unit in Cambridge, UK, and it was remarkable that this focus on applied psychology yielded so much prime theoretical information. Broadbent described to me how he had been focused on solutions to practical problems encountered in World War II. He told me a story about how pilots landing planes concurrently could not receive their radio instructions without interference from a soup of instructions intended for other pilots. He invented an ingenious method in which the left and right ears would receive instructions on different carrying frequencies, so that only the signal would be identical in the left and right ears, whereas the noise would be uncorrelated, making the intended signal pop out from the background. The method apparently was never used because the invention of FM radio made it unnecessary, but practical problems like this led to an amazing amount of basic new information about immediate memory and selective attention.

When I visited Alan Baddeley during my 1990 trip, he was Director of the Applied Psychology Unit. We went for a walk as he told me about the patient who could remember no more than one or two digits and updated me with the patient's changes upon partial adaptation to her injury. During a hiatus from meeting people at the Applied Psychology Unit, I visited the library and found a copy of Baddeley's doctoral dissertation from the mid 1960s. I was amazed at how closely the issues and methods from the dissertation mapped onto work carried out decades later, and I think that this indicates a kind of persistence in pursuing and finishing important initiatives, which paid off quite well.

In what follows, I will first explain a little about how and why, from the aforementioned experiences, my embedded-processes view of information processing emerged. Next, I will discuss typical lifespan development. Following this, I will describe various applied outcomes that are the products of collaborations with investigators of these applied topics. In each case, I will explain the relation between theory and practice, which will be foreshadowed by a theoretical account explained in a way that is meant to emphasize the processes that will be of practical use.

Embedded-processes theoretical view of working memory

Working memory, as the assembly of mental faculties used to keep a limited amount of information in an accessible state temporarily (Cowan, 2017a), is an important part of the embedded-processes approach to information processing. I have given more extensive accounts of the embedded-processes view elsewhere (Cowan, 1988, 1995, 1999, 2019; Cowan, Morey, & Naveh-Benjamin, 2021). Figure 9.1 shows a

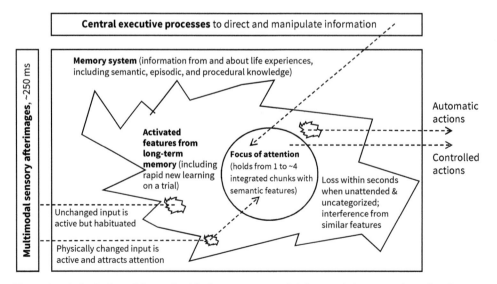

Figure 9.1 A depiction of the embedded-processes model that explains several mechanisms.

graphic representation of the theoretical model. Here I am attempting to reconstruct the spirit of the theoretical approach and its potential relevance for practical issues. During my graduate training and early work in the field, I carried out some research closely aligned with the initiatives of researchers tying together the limited capacity of immediate memory and selective attention (e.g. Broadbent, 1958; Norman, 1968; Treisman, 1964). The explanations and visual representations of information processing models from the 1960s and 1970s did not satisfy me, for several reasons, and I handled those situations in my 1988 paper (with several subsequent refinements and clarifications) as follows.

Immediate memory storage and attention

In the traditional literature on the persistence of information in mind, there are two different traditions. First, according to James (1890), there is a primary memory for the recently experienced present that appears to be intimately related to attention. What is attended can be retained in that manner for a short while. Second, according to Hebb (1949), cell assemblies are active patterns of nerve cell firing that represent an active idea, and this nerve cell firing might not depend on attention. It might occur for stimuli entering the nervous system as part of the panorama of the world that we experience, but outside of our attention. Or it might still occur for a while for information that was attended recently but is no longer attended. In the embedded-processes model of Cowan (1988), both of these mechanisms were said to occur. With or without attention, features of the current environment and of recent thoughts remain still in an activated form, as active cell assemblies representing these features. Embedded in this activated portion of long-term memory is a more strictly limited amount of information (Cowan, 2001) that gains access to more coherent, complete processing, in the focus of attention, as Figure 9.1 shows. The activated long-term memory is supposed to be limited by rapid decay over some seconds, though this decay may be less rapid when the information has been categorized (Ricker & Cowan, 2010, 2014; Ricker et al., 2014, 2020). The focus of attention is said to be limited to only a few coherent ideas or chunks (Miller, 1956) of information at once. A chunk can be a single presented item (e.g. the word *candy*) or a series of items that can be associated to make a larger pattern (e.g. presentation of the letters U, S, and A in that order, associated to make the single, known chunk USA). Across both verbal and spatial domains, the number of separate chunks held concurrently seems to be limited to three or four, except when it is possible to apply some special mnemonic processes to group or rehearse the information (Cowan, 2001). The act of paying attention allows most of the semantic processing of the stimuli, and sometimes allows a reorganization of the stimuli when a meaningful or consistent pattern can be perceived or established.

Immediate storage of information in the focus of attention presumably results in rapid, new learning that may still be active in long-term memory and usable during the immediate-memory trial (Cowan, 2019). Forsberg, Guitard, and Cowan (2021) found that after brief, single exposures to arrays of two familiar items, about half could be remembered later and after brief exposures to larger arrays, about 0.2 of those that made it into working memory could be remembered later—regardless of the array size of four, six, or eight items. This working memory to long-term memory ratio was also shown to be fixed across child development starting in the early elementary school years (Forsberg et al., 2022). Jiang and Cowan (2020) found that memory for which items appeared together in lists was stored in long-term memory for items towards the end of the list, even when the list task did not include instructions to remember the items.

Temporary storage of information from different domains

In modular models of working memory (Baddeley, 1986; Baddeley & Hitch, 1974; Logie, 2016; Vandierendonck, 2016) there are said to be separate stores to hold verbal and visual information. The evidence comes from dual tasks in which two types of information are to be stored at once (e.g. Cocchini et al., 2002) or in which something is to be saved while something else is to be processed (e.g. Baddeley, 1986). Typically, two verbal tasks interfere with one another, and two visual tasks interfere with one another, more than a verbal task and a visual task interfere with each other. In the embedded process view, there is just one focus of attention for all types of information and, surrounding it, a complex medium of activated long-term memory.

Cowan (1988) noted that modular stores were not the only way to account for this specificity of interference. In activated long-term memory, representations might include features that are sensory, phonological, orthographic, spatial, tactile, and semantic, among others. In principle, any kind of similarity of features between two items held in activated long-term memory might cause mutual interference. Therefore, finding less interference between a visual and a verbal set may be only confirmatory information for the modular type of view. If there are also, say, effects of other kinds of similarity (e.g. memory interference between two sets of line orientations or two sets of coloured spots, with less interference when there is one set of each) then it becomes more difficult to use a verbal–visual storage separation to account for all of the results. In keeping with the view that feature similarity gives rise to interference is the finding that in addition to the large detrimental effect of phonological similarity effects between words in a list to be recalled (Conrad, 1964), orthographic similarity also matters, not only for visually presented words (Logie et al., 2000) but even for acoustically presented words (Guitard & Cowan, 2020). There are differences between the amount of interference from different features within a modality (e.g. an array of line orientations and an array of coloured spots interfering with one another less than two arrays of the same kind) and there are differences between modalities in the amount of cross-modality interference (with more interference on visual memory from acoustic or verbal memory as compared to the converse; Li & Cowan, 2021; Morey et al., 2013).

A common criticism of this feature-similarity view is that, in some immediate-memory contexts, semantic similarity between words in a list does not seem to affect immediate recall (e.g. Baddeley, 1972). This absence of an effect, however, might occur because the phonological features are better suited to recall of a random set of words, with the phonological sequence rapidly learned (Baddeley et al., 1998; Cowan, 2019) whereas the semantic features cannot easily be organized. In contrast, when the semantic features can be organized, as is the case in a coherent sentence, their semantic similarity may make a big difference to recall. One might have more trouble retaining the sentence 'The blue glove has yellow stitching and a purple border' while carrying out a silent colour-matching task as opposed to some non-colour-related task.

There is also considerable evidence that even very different tasks, such as memory for tones and colours, still show non-negligible interference between tasks, both in the situation of dual-set storage (Cowan et al., 2014; Uittenhove et al., 2019) and in the situation of storage and processing (Cowan et al., 2018; Doherty et al., 2019; Morey & Bieler, 2013; Vergauwe et al., 2010). This kind of interference could come not from a shared set of features between the tasks but from a conflict regarding how the focus of attention can be used to assist storage and carry out processing; it may not have the capability to do its best work on both sets of information at once. The focus of attention may have a relation to versions of modular theories that include an episodic buffer (Baddeley, 2000, 2001). Possibly, the focus of attention and episodic buffer handle the same sorts of situations that cannot be comfortably discussed as resulting from separate phonological and visuospatial buffers, in the modular view, and cannot be comfortably discussed as resulting from activated long-term memory, in the embedded-processes view.

Reconceptualization of attentional filtering

Since Broadbent (1958), researchers have used the concept of an attentional filter to explain some incredibly interesting aspects of information processing. If one is in an environment with several meaningful streams of information, generally only one stream can be processed at a time. The participant has some control over which information gets processed, up to but not beyond the limit. Yet, there is a lot of residual information about unattended streams for some seconds afterwards, during which time attention can still be usefully switched to these sensory traces. A real-world example would be if someone asks you what time it is while you are reading, and you can use auditory sensory memory to figure out what was asked of you.

In Broadbent's (1958) theoretical construct, this attention limit was represented by an attention filter. Vast amounts of sensory information impinge on the filter, but only very limited information makes it through the filter to be further remembered and processed. One issue that puzzled me about this theoretical construct is that changes in sensory information are noticed. For example, in a selective listening task, one repeats a speech message from one ear while a different message presented to the other ear is ignored (and cannot be processed at the same time). A change in the voice of that unattended channel is quickly noticed, and a more subtle change in the properties of the speech, from forward to backward speech, can be noticed after some seconds (Cherry, 1953; replicated by Wood & Cowan, 1995a). A problem for the filter theory is that, if the sensory information does not make it through the filter because it is unselected, it is unclear how it can be noticed when it changes.

Cowan (1988) reconceptualized selective attention based on other information, about the habituation of the orienting response (see Elliott & Cowan, 2001; Sokolov, 1963). The notion is that all information from the environment stimulates the brain in some ways and activates long-term memory representations of past stimulation. The individual cannot process all types of features of all stimuli but whatever can be processed contributes to a neural model of the environment. When the processed information is unchanged, then no change is needed in the neural model. However, an abrupt physical change will be processed as discrepant from the neural model and will recruit orienting of attention, which allows the changed stimulus to be assessed in greater detail. This might happen, for example, if you are reading this chapter and, suddenly, there is a loud noise or the room lighting flickers off and on. If the attentional system judges that the change is important, then further changes towards those stimuli will occur. For example, the loud noise could be judged to be either thunder that one would then forget about, or an explosion that one might worry about.

There was evidence suggesting that some semantic information might be processed without attention, also. For example, Moray (1959, replicated by Wood & Cowan, 1995b) found that presentations of the participant's name in the unattended channel was sometimes noticed. Treisman (1964) accounted for that situation with the proposal that the attention filter does not completely eliminate information that is filtered out, but only attenuates it. The orienting response in the embedded-processes model can accomplish the same thing.

An unanswered question was just how much semantic information is processed without attention. Only about a third of participants notice their names in an ignored channel. It is possible that those participants notice their names because they have enough attention to monitor the allegedly unattended channel while repeating (shadowing) their names, but an alternative possibility is that these participants notice their names because their attention wandered over to the channel that was supposed to be ignored. Conway et al. (2001) found that the latter is more apt, inasmuch as 65% of the bottom quartile noticed their names, compared to only 20% of the top quartile. The basic findings have been replicated twice (Naveh-Benjamin et al., 2014; Röer & Cowan, 2021). Therefore, there may be no processing of semantic information without attention. For another demonstration of that point showing that one must use a sufficiently demanding shadowing task to show that semantic information does not easily come through the unattended channel, see Wood et al. (1997).

One relatively uninvestigated topic is the supposed use of the focus of attention to encode serial order information in working memory. Guitard et al. (2021) presented two lists, varying the type of forewarning of the type of recall test to follow (definitely an item test for which order was irrelevant, definitely an order task with the items provided, or no forewarning of which kind of task). The finding was that the expectation of two order tasks interfered with retention of order compared to expectation of an order task for one list and an item task for the other. In contrast, expectation of two item tasks had much less effect on the retention of items. Order retention seems to have a greater demand for attention.

Working memory in typical child development

Working memory is extremely important for ordinary cognition and for education (Cowan, 2014). When one listens to a sentence, the beginning must be held in working memory until it can be integrated with the later part. Multiple instructions must be remembered while carrying out a task; and when solving a problem, various types of relevant information have to be held in mind and used together. The key limit can occur in how much information can be considered or, in understanding a new concept, how many elements can be associated (Halford et al., 2007). A large, striped cat is a tiger; a large, striped horse is a zebra; a small, striped cat is just a striped cat; and so on. In English grammar, *he* can be followed by *goes*, whereas *they* must be followed by *go* instead; and so on. Holding the relevant information in working memory at the same time may be the first step to learning associations between them (Baddeley, 2000; Cowan, 2019; Halford et al., 2007; Jiang & Cowan, 2020; Oberauer, 2021). Let's examine a few of the things that are known about working memory at different points in ordinary child development, exemplified here primarily in my own work.

Infant development

Understanding early cognitive development, including immediate memory, may be important for assessing whether a particular infant is developing without special issues. It is still not clear what changes with early development, but there are some interesting leads.

Cowan et al. (1982) examined infants' acoustic memory for vowels and learned that it seems to have less temporal speed or precision than is found in adults. Pairs of vowels were presented and 6-week-old infants were to suck on a pacifier that rewarded them with more changes in the vowels. Instead of a very monotonous series, sucking on the pacifier resulted in a change in the first vowel in each pair, from 'ah-ah' to 'eh-ah' and back again, or the second vowel in each pair. When the second vowel in each pair changed, this was easily detected and it increase the rate of sucking. For changes in the first vowel in a pair, the changes were detected by the infants when there were 400 ms between vowels, but hardly at all when there were 250 ms between vowels in a pair, even though this shorter interval would be long enough for adults to notice. At 250 ms, the second vowel in a pair may backward-mask the first vowel, cutting off its sensory afterimage before its perceptual processing is complete. This suggests that infants already have acoustic sensory memory, but that it may persist longer or be the basis of a longer-lasting perceptual process than what happens in older participants. The new discrimination procedure used in this study yielded quite robust results and would be helpful for other studies with young infants.

Goodsitt et al. (1984) examined speech discrimination at 6 months old using a head-turning procedure and found that infants' ability to detect a change (turning their head to the change, which activated a motorized animal as a reward) depended on the complexity of the working memory context. Infants could detect a change, for example, from the sequence 'tee-bah-tee' to 'tee-doo-tee' or from 'ko-bah-ko' to 'ko-doo-ko'. They had more difficulty detecting changes when the context contained more exemplars, such as the change between 'tee-bah-ko' to 'tee-doo-ko'. This result is a precursor of the effects of grouping and redundancy on memory in adults. It seems clear that many of the rules of perception and working memory in adults get a robust start early in life.

Cowan (2016) noted a paradox in the literature on the development of working memory that led to a proposal to resolve this (Cowan, 2017b). The paradox is that the infant discrimination literature seems to suggest that infants can retain about three items, which is no different from adults when they are unable to engage in rehearsal or grouping processes (e.g. Cowan, 2001). Cowan (2017b) proposed that the difference is an increase with age in the amount of detail that can be included in each of several items. Zosh and Feigenson (2012) placed several toys into an opaque container, in view of a 1.5-year-old, and studied the tendency of the infant to search for toys in the box. In some conditions, the toys were covertly replaced with other ones before the infant was allowed to search. With one or two toys hidden in the box, infants kept searching longer after the new toys were removed, presumably in search of toys they originally saw hidden. When three toys were hidden, they searched no longer when the toys were covertly replaced than when they were able to remove the toys they had seen hidden. Presumably, the infants could retain three items, but not with the individual identity information. With fewer items to retain, more individual identity information about each one was retained. This research contrasts with adult research suggesting that three simple objects seen in a brief array with their identities can easily be retained (e.g. Luck & Vogel, 1997). We are currently investigating whether the increase in the amount of information per object develops further in middle childhood.

There is a period of development from about 2–3 years of age, toddlerhood, during which little is known because the children tend to be too mobile for the infant procedures, yet not disciplined enough for procedures used with older children. The Wechsler Preschool and Primary Scale of Intelligence (WPPSI), Version IV, has a couple of non-verbal working memory tests for children 2–7 years of age normed on a large sample, with some interesting properties. Cowan (2021) showed that one of the tests of working memory, 'Picture Memory', is closely tied to the other intelligence measures in the test because it depends on real-world knowledge, whereas the other test, 'Zoo Locations', is not dependent on knowledge but is a purer measure of working memory per se, making it better for assessing working memory but not as good for assessing overall intelligence.

Child development

It is clear that during the childhood years, working memory increases dramatically (e.g. Gathercole et al., 2004). If we could determine what aspects of working memory are responsible for this increase, it might be possible to adjust educational materials to be optimal for the child's current learning ability. The problem is that many qualities develop concurrently and it is not easy to disentangle these factors and determine which developmental growth is most basic. Cowan (2016, 2017b) summarized a programme of research that helps to show that certain hypotheses favoured by the field are not sufficient. Working memory development is not completely caused by developmental increases in knowledge, in the ability to allocate attention to the assigned task while excluding distracting information, or to the ability to improve in encoding speed, precision, or rehearsal. With each of those factors controlled, increases in working memory between 7 years and adulthood persist (Clark et al., 2018; Cowan et al., 1999, 2010, 2011, 2015; Gilchrist et al., 2009).

One research strategy to determine the nature of working memory development is to determine whether there is a process that accounts for the developmental change, such that if one is able to control for that factor, developmental change is erased. One such factor is processing speed. Case et al. (1982) showed that the speed at which children can identify items is related to how many of those items they can recall. By using nonsense items for adults, low in familiarity for them, both the time of identification and the number of items recalled were regressed to what younger children were able to do with known items. In a digit span test, Cowan et al. (1998) showed that two speeds—the time between items in recall and the time it took to recite a known series such as 1–10—were independent of one another and, taken together, accounted for the increase with age in digit span during the elementary school years. Gaillard et al. (2011) showed that by manipulating stimuli to equalize all processing and refreshing times across age groups, complex span scores no longer included a developmental trend. In all of these studies, however, the causal path is unclear. Processing times may be viewed as a cause of behaviour but, alternatively, they may be viewed as an effect of something else. For example, it is possible that memory is searched multiple items at a time up to a capacity limit and that participants with smaller working memory capacities therefore search at a slower rate, as a consequence of the smaller capacity (a possibility supported by a modelling investigation by Lemaire et al., 2018).

Another research strategy is to divide the developmental increase in capacity into components that change with development and others that do not, narrowing down the candidates for what the fundamental change could be. Cowan et al. (2018) were able to use this strategy. Participants were to remember a visual display (an array of coloured squares) and an acoustic display (a series of spoken digits or tones) with one probe to be judged present from a stimulus set or absent from both sets. In some trial blocks, there were instructions indicating in advance that one of these stimulus sets would be tested on every trial. Using performance from both single-set and dual-set attention, it was possible to divide performance into several components. Two components were the number of items remembered in each modality regardless of attention, called peripheral memory, and an additional number of items in working memory that could be allocated to the visual or acoustic stimuli, or split between them, depending on the attention, called central memory. Peripheral visual and acoustic memory increased markedly with age from the early elementary school years to adulthood, whereas central memory did not increase. According to the interpretation by Cowan et al., we can rule out a theory in which children grow to retain more and more information in the focus of attention as they get older. Instead, they learn to use attention to encode items in a way that allows them to be retained better throughout the trial without constant attention, such as encoding patterns formed by the items and perhaps rapidly learning those patterns.

A final research strategy is to find qualitatively different patterns of memory, which can indicate developmental changes in encoding methods. For example, Camos and Barrouillet (2011) showed that 6-year-olds did not refresh or rehearse items in a complex span task, and therefore lost information as a function of elapsed time for forgetting, whereas slightly older children did use these mnemonic strategies, and therefore lost information not as a function of elapsed time, but as a function of the cognitive load that prevented refreshing. Cowan, AuBuchon et al. (2021) carried out a dual task in which a variable number of coloured squares was to be remembered and, in some trials, a rapid key press (to the same or opposite side as a signal) was to be carried out during the retention period during which colours were to be maintained for a subsequent test. Younger children in the early elementary school years (6–9-year-olds) exhibited a reactive style, tending to drop the colour information in order to carry out the key press, whereas older age groups became progressively more proactive, trying to maintain the colours during the key press task, resulting in a moderate dual-task cost on both tasks, not just on the colours, but with better performance than the younger children overall.

Our recent work (Adams & Cowan, 2021) suggests that working memory can be used in various ways in children's learning and processing, for example, to ensure the quality of their responses. Preschool children were taught to carry out a delayed imitations of passive-voice descriptions of actions. The finding was that, when performing under a working memory load, children surprisingly became more, not less, prone to using the passive voice as instructed. The interpretation was that their imitations of the passive voice included errors and that children with free working memory capacity tried to avoid those errors by translating the descriptions into the more familiar, active voice. Under a memory load, the passive voice was used more often but in a relatively unanalysed manner, with more errors, some of them silly-sounding such as *The girl was watered by the flowers*.

In sum, then, the most fundamental basis of developmental increase in working memory overall is unknown, although we have clues. Children increase in a fundamental capacity, in processing speed, and in a proactive stance that involves mnemonic activities or maintenance strategies, though it is unclear if one of these changes is most fundamental and allows the others to happen. This is an important area for future research.

Individual differences within an age group

A lot of work has shown that multiple aspects of working memory can differ among individuals. These can include aspects of attention (e.g. Cowan, Elliott et al., 2005; Doherty et al., 2019; Kane et al., 2004), but also phonological storage and processing (e.g. Gray et al., 2017). According to some work, the individual differences in the focus of attention used for storage and central executive processing are separable (e.g. Cowan, Fristoe et al., 2006; Gray et al., 2017). Individual differences provide one way to understand the structure of the working memory system, on the assumption that mechanisms that are independent can be independently better functioning or worse functioning. A stronger version of this logic is that neural damage can independently impair different working memory functions, though there are disagreements about what these neural deficits show (Buchsbaum & D'Esposito, 2019; Hanley & Young, 2019; Logie, 2019; Morey, 2019; Morey et al., 2019, 2020; Shallice & Papagno, 2019).

Some have argued that individuals with better working memory are better able to evade proactive interference. This interference might come, for example, in a list-search task when a previous trial contained items closely related to the items in the present list, which might make it harder to retain items in the present list if it is too large to be completely saved in working memory (Cowan, Johnson, & Saults, 2005). Cowan and Saults (2013), however, found that the advantage of high-span individuals comes instead from them being better able to take advantage of situations in which they can count on there being not much proactive interference. This might occur because of the better use of attention for memory generally in high-working-memory individuals. Rosen and Engle (1997) found that high-span individuals recalling items from long-term memory are less likely to repeat themselves, so they appear to have a better episodic memory for their own responses. This might be because long-term memory retrieval is attention demanding (Craik et al., 1996; Unsworth & Engle, 2007).

Some implications for later development

Although this chapter focuses primarily on child development, it is worth pointing out that similar factors should be important throughout the lifespan. Most working memory skills decline with age (e.g. Jenkins et al., 1999). Just as the ability to re-tain associations between elements may be key to concept formation in childhood (Halford et al., 2007), increasing difficulty with associations (e.g. Old & Naveh-Benjamin, 2008) may lead to a decline in reasoning and problem-solving ability with age in adulthood (e.g. Oberauer et al., 2007). Older adults may find it difficult to exclude irrelevant information from working memory (e.g. Logie & Morris, 2015; Weeks et al., 2020).

For a general theory of development across the lifespan, it is especially interesting to consider the similarities and differences between the suboptimal performance of children and older adults, compared to younger adults. In both cases, there may be a suboptimal state of neural processing. In the case of children, however, there is also less knowledge than in adults, whereas older adults presumably retain their know-ledge. Therefore, a lifespan comparison can yield interesting observations about the basis of developmental change. In an examination of memory for coloured squares and their locations within briefly presented arrays, Cowan, Naveh-Benjamin, et al. (2006) found that third-grade children (8–9 years old) and older adults had com-parable levels of performance in some tasks, lower than young adults. The pattern of change in peripheral as opposed to central components of working memory are also comparable in children and older adults (Greene et al., 2020). However, the change across the lifespan in bias was monotonic, tending with age more towards under-standing that a change might have been missed.

The core portion of working memory, the number of items that can be held without rehearsal, does seem to have a lifespan course that is described by an inverted-U shape, with the relevant knowledge held constant. In particular, Gilchrist et al. (2008, 2009) studied memory for lists of unconnected, semantically unrelated spoken sentences, easy enough to be easily understood by second-grade children (7–8 years old), yet long enough so that verbal rehearsal of the list would not be possible. In recall of these lists from second grade through to old age, the sentences proved comprehensible enough that participants of all ages were able to recall most (about 80%) of the words within a sentence, if they remembered that sentence at all; relevant linguistic knowledge was in that way controlled across age groups. Yet, the number of sentences remembered on a trial increased as children matured (to about three items) and then decreased again in old age.

It is still not exactly clear why working memory increases in childhood and why it decreases in older adults. According to one prominent view, it has to do with the ability to suppress irrelevant information so it will not clutter working memory (Vogel et al., 2005; Weeks et al., 2020) or the similar ability to disengage from no-longer-relevant information (Martin et al., 2020). Rhodes et al. (2019) carried out a dual task with instructions to place more emphasis on the storage task or the processing task and found that compared to younger adults, older adults had no difficulty allocating attention to one task or the other, but they did have trouble co-ordinating a dual task. All of these findings and others (such as the child developmental findings of Cowan et al., 2005, and the adult age effects in selective listening found by Naveh-Benjamin et al., 2014) seem consistent with a view that the scope of attention changes with age in an inverted-U-shaped manner. A smaller focus of attention could make a participant more vulnerable to clutter in working memory because there is less storage or processing space to spare. Yet, the source of individual differences within an age group might be primarily in the control of attention rather than its scope. There appear to be separable individual differences in both scope and control of attention in young adults (Cowan, Fristoe et al., 2006).

Working memory and individual influences on cognitive development

Now that we have discussed factors influencing working memory in typical development, I would like to review a variety of practical issues that affect working memory. I have had the honour of working with colleagues who examine a number of issues in pathology that are related to working memory. These studies help to shed some light on the nature of the pathologies, and on the nature of working memory. Figure 9.2 indicates the approximate points during the lifespan development of working memory at which these pathological issues typically tend to occur, although some of the issues can spread across many years. The theoretical question asked here is which component or components of the working memory system can be compromised by various pathologies, what implications they have for the pathologies, and what implications they have for the nature of working memory. The life course is shown in Figure 9.2 but this chapter will focus on issues that most often emerge in childhood.

Cognitive deficits

By distinguishing between patterns of working memory deficits in two different groups of children with cognitive deficits, it is possible to provide information relevant to both the groups and the deficits. Jarrold et al. (2004) examined the role of speech timing during recall in working memory within responses in Down syndrome and Williams syndrome, and in groups of participants matched to these two

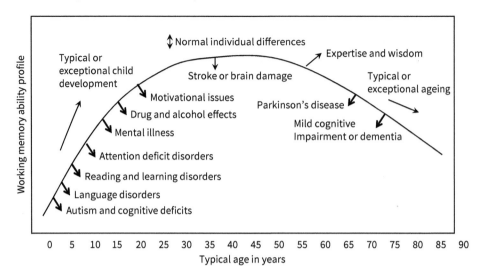

Figure 9.2 A schematic diagram of the trajectory of working memory ability during the lifespan, showing some potential windfalls and pitfalls that produce individual differences in ability at each point in development.

groups on receptive vocabulary. The duration of pauses between words has been taken as an indication of how quickly working memory can be searched for the next item to say; for a given list length, the pauses between words increase linearly with the list length and the slope of this increase diminishes with age in childhood (Cowan, Elliott et al., 2002). Jarrold et al. (2004) found that, in Williams syndrome, the inter-word pauses in recall were longer than in matched controls, and that this additional pause time was closely correlated with the poorer memory for word lists in these individuals. In contrast, although individuals with Down syndrome performed more poorly than their matched controls, they were not commensurately slow in their recall. It is not clear how these two patterns are to be explained but it seems like an important clue for further research.

Autism spectrum disorders

Although autism tends to be diagnosed early in childhood, some of the symptoms tend to remain throughout life. Bodner et al. (2019) examined the tendency of young adults with autism spectrum disorder to allocate attention appropriately. Individuals with severe cognitive impairment were excluded. Participants saw arrays of two, three, or four highly relevant items (e.g. squares whose colours are tested on 75% of trials) and, on each trial, an equal number of less relevant items (e.g. circles whose colours are tested only on 25% of trials). Attention allocation can be expressed as the proportion of items retained in working memory from the highly relevant set out of all items in working memory. In control participants, as the number of array items increased, the allocation of attention veered more towards the highly relevant set, from under 60% with two items per shape up to almost 80% with four items per shape. The participants with autistic spectrum disorder, however, looked like the control participants for two or three array items per shape, but with an attention allocation ratio that then decreased slightly rather than increasing for four items per shape. Apparently, these individuals have normal working memory capacities but become overwhelmed when the capacity is reached and do not focus attention as reliably as control participants. This kind of result suggests that avoiding an overstimulating or overtaxing environment could allow the individuals with autism spectrum disorders to pay attention and use working memory in a manner comparable to other people.

Language disorders and dyslexia

According to the embedded-processes approach to working memory, I have proposed that serial position information about list items is learned with items starting in the focus of attention and resulting in rapid new learning of the associations between items, and between items and their serial positions (Cowan, 2019). If this is the case, then deficits in serial order information might be independent of the modality of the information. On the other hand, material that is phonological may be better suited for serial order coherence than other kinds of items, in which case there might be a special component of serial ordering that is disturbed specifically for verbal information.

It turns out that serial order information is impaired in language development. Gillam et al. (1995) examined memory for spoken verbal lists that sometimes were followed by an interfering item in the same voice or 'suffix', which greatly impairs memory for the last few list items. The suffix effect on item recall was similar in language-impaired and typically developing 9-year-old children. However, when recall was scored for serial order correctness, the effect of the suffix was far worse for language-impaired children than for typically developing control children. We don't know how they would fare on non-verbal items.

Children as old as 9 years with language impairment also tend to have dyslexia. Cowan et al. (2017) examined memory for items and order across children with dyslexia, with or without concomitant spoken language impairment. Serial order impairment was found within not only digit span, but also within span for locations, in difficult situations (running memory span). The serial order deficit was obtained even when the children with and without dyslexia were matched for non-verbal intelligence and language. However, children with concomitant spoken language disorder had an additional deficit in phonological memory when the stimuli included consonant clusters. These results suggest that there are both general serial ordering deficits that could be attention related, and language-specific deficits in serial ordering based on phonological coding factors.

Implications for deficits later in life

The embedded-processes approach and related attention- and capacity-based approaches to working memory have been useful to examine a number of challenges in later development, and will be briefly sampled here. Gold et al. (2006) showed that individuals with schizophrenia can select relevant items as well as control participants. Unlike the individuals with autism spectrum disorder tested by Bodner et al. (2019), the working memory capacities for visual arrays demonstrated by participants with schizophrenia were smaller than the capacities of control individuals. It may prove to be quite useful in the future to confirm a double dissociation between the working memory and attention characteristics of those with autism spectrum disorder versus schizophrenia. Attention-related deficits in schizophrenia may account for why encoding is impaired but time-based forgetting is not impaired by schizophrenia (Javitt et al., 1997), as in young children compared to older children and adults (Keller & Cowan, 1994). Attention- and capacity-related deficits also come into play in Parkinson's disease (Lee et al., 2010) and drug and alcohol effects (Bartholow et al., 2018; Saults et al., 2007).

Summary: the relation between theoretical and practical issues of working memory development

In sum, the relation between theoretical aspects of working memory and its childhood and later development is a two-way relation. Theoretical issues provide a context with which to interpret practical issues and, in turn, the practical issues carry lessons that sometimes help in the refinement of theory. One additional case will illustrate this point, having to do with motivated thinking in working memory tasks and their development: the case of metamemory.

Metamemory means knowing whether you remember, or will remember, certain information or not. It makes a big difference for problem-solving and communication because if you believe that you have forgotten some essential information, you will try to have that information repeated. If, on the other hand, you believe that your memory is working flawlessly, you will proceed to make decisions on those grounds even though you may be incorrect.

Demonstrating the role of metamemory in adults, Cowan et al. (2016) showed a brief array of coloured squares on every trial followed by a test array with items in the same locations, the task being to indicate how many of the squares had changed colour. From zero to all of the squares could change. In some trials, the retention interval included a question about how many items the participant thought they remembered. On average, these adult participants overestimated and this point altered their responding. If they missed a change, they were too confident that the change had not occurred, which led participants to adopt a bias in which they were further off the mark when many items had changed and closer to the mark when few items had changed.

Younger children overestimate the number of items they hold in working memory more than older children or adults do (Forsberg, Blume, & Cowan, 2021). Further research could be helpful to understand how that metamemory deficit affects their communication and problem-solving. Take, for example, the role of working memory in following instructions (e.g. Jaroslawska et al., 2016). One possibility is that missing information from working memory is filled in with previously learned information even when that previous learning is not entirely applicable. So if children receive novel instructions, they might disregard part of those instructions without realizing it, having filled in the missing parts with things they think they already know. Of course, it is merely human to do this and some humans are more prone to thinking they know than are others (Cowan et al., 2019) but metamemory provides one good example in which theoretical aspects of working memory matter within practical contexts.

Conclusion

The developmental trends and individual differences, both typical and abnormal ones, have been presented here in a manner that incorporates mechanisms in the embedded-processes model. There are some recurrent themes that reinforce the mechanisms in the model.

1. Throughout development, there are issues of the control of attention that change with development, and attention control deficiencies that accompany atypical processing deficiencies such as language and reading impairment in children and attentional filtering in autistic young adults.
2. In infants, younger children, and older adults, there can be non-optimal use of attention to reformulate stimuli in a way that allows them to be off-loaded to free up the focus of attention for subsequent processing; the ability to do so may involve pattern perception and attention-based refreshing. This pattern perception and off-loading capability results in increased working memory capacity. At least children in the elementary school years have further room to improve in the coordination of two tasks, in that they tend to be reactive instead of proactive, that is, they tend to drop a memory load when a second demand arises. Older adults tend to lose the ability to coordinate two tasks gracefully, and this may be an issue for many developmental practical issues such as attention deficit disorder.

3. There are deficiencies in serial order processing (e.g. in language and reading impairments) that are not limited to one domain, though these problems do seem somewhat worse for verbal materials compared to visual and spatial materials. Serial encoding may depend on attentional involvement in serial order encoding and rapid learning and, as a backup method, verbal rehearsal.

4. Younger children, as well as some individuals with disabilities such as schizophrenia, appear to encode information (e.g. tones) non-optimally. There are some indications of the importance of individual differences in sensory decay, though these require further research.

These developmental findings may help to increase the specificity of the embedded-processes model. If the capacity of the focus of attention does not increase with development, perhaps its basis is somewhat biologically determined and fundamental, unlike the control of attention. The theoretical model is not the only one that can provide an explanation for the collection of developmental and practical trends observed, but it seems basically favourable to that approach in comparison to a more modular approach. The theoretical and practical investigation of working memory should proceed hand in hand, and the present review should help establish the case for that approach.

Acknowledgement

This work was completed with support from NIH Grant R01 HD021338.

References

Adams, E. J., & Cowan, N. (2021). The girl was watered by the flower: Effects of working memory loads on syntactic production in young children. *Journal of Cognition and Development*, *22*(1), 125–148.

Allport, D. A. (1968). Phenomenal simultaneity and the perceptual moment hypothesis. *British Journal of Psychology*, *59*(4), 395–406.

Atkinson, R. C., & Shiffrin, R. M. (1968). Human memory: A proposed system and its control processes. In K. W. Spence & J. T. Spence (Eds.), *The psychology of learning and motivation: Advances in research and theory* (Vol. 2, pp. 89–195). Academic Press.

Baddeley, A. D. (1966). The capacity for generating information by randomization. *Quarterly Journal of Experimental Psychology*, *18*(2), 119–129.

Baddeley, A. D. (1972). Retrieval rules and semantic coding in short-term memory. *Psychological Bulletin*, *78*(5), 379–385.

Baddeley, A. D. (1986). *Working memory*. Clarendon Press.

Baddeley, A. (2000). The episodic buffer: A new component of working memory? *Trends in Cognitive Sciences*, *4*(11), 417–423.

Baddeley, A. (2001). The magic number and the episodic buffer. *Behavioral and Brain Sciences*, *24*(1) 117–118.

Baddeley, A. D., Gathercole, S. E., & Papagno, C. (1998). The phonological loop as a language learning device. *Psychological Review*, *105*(1), 158–173.

Baddeley, A. D., & Hitch, G. (1974). Working memory. In G. H. Bower (Ed.), *The psychology of learning and motivation* (Vol. 8, pp. 47–89). Academic Press.

Baddeley, A. D., Thomson, N., & Buchanan, M. (1975). Word length and the structure of short term memory. *Journal of Verbal Learning and Verbal Behavior, 14*(6), 575–589.

Bartholow, B. D., Fleming, K. A., Wood, P. K., Cowan, N., Saults, J. S., Altamirano, L., Miyake, A., Martins, J., & Sher, K. J. (2018). Alcohol effects on response inhibition: Variability across tasks and individuals. *Experimental and Clinical Psychopharmacology, 26*(3), 251–267.

Bodner, K. E., Cowan, N., & Christ, S. E. (2019). Contributions of filtering and attentional allocation to working memory performance in individuals with autism spectrum disorder. *Journal of Abnormal Psychology, 128*(8), 881–891.

Broadbent, D. E. (1958). *Perception and communication.* Pergamon Press.

Buchsbaum, B. R., & D'Esposito, M. (2019). A sensorimotor view of verbal working memory. *Cortex, 112*, 134–148.

Camos, V., & Barrouillet, P. (2011). Developmental change in working memory strategies: From passive maintenance to active refreshing. *Developmental Psychology, 47*(3), 898–904.

Case, R., Kurland, D. M., & Goldberg, J. (1982). Operational efficiency and the growth of short term memory span. *Journal of Experimental Child Psychology, 33*(3), 386–404.

Chen, Z., & Cowan, N. (2009). Core verbal working memory capacity: The limit in words retained without covert articulation. *Quarterly Journal of Experimental Psychology, 62*(7), 1420–1429.

Cherry, E. C. (1953). Some experiments on the recognition of speech, with one and with two ears. *Journal of the Acoustical Society of America, 25*(5), 975–979.

Clark, K. M., Hardman, K., Schachtman, T. R., Saults, J. S., Glass, B. A., & Cowan, N. (2018). Tone series and the nature of working memory capacity development. *Developmental Psychology, 54*(4), 663–676.

Cocchini, G., Logie, R. H., Della Sala, S., MacPherson, S. E., & Baddeley, A. D. (2002). Concurrent performance of two memory tasks: Evidence for domain-specific working memory systems. *Memory & Cognition, 30*(7), 1086–1095.

Conrad, R. (1964). Acoustic confusion in immediate memory. *British Journal of Psychology, 55*(1), 75–84.

Conway, A. R. A., Cowan, N., & Bunting, M. F. (2001). The cocktail party phenomenon revisited: The importance of working memory capacity. *Psychonomic Bulletin & Review, 8*(2), 331–335.

Cowan, N. (1984). On short and long auditory stores. *Psychological Bulletin, 96*(2), 341–370.

Cowan, N. (1988). Evolving conceptions of memory storage, selective attention, and their mutual constraints within the human information processing system. *Psychological Bulletin, 104*(2), 163–191.

Cowan, N. (1991). Recurrent speech patterns as cues to the segmentation of multisyllabic sequences. *Acta Psychologica, 77*(2), 121–135.

Cowan, N. (1995). *Attention and memory: An integrated framework* (Oxford Psychology Series, No. 26). Oxford University Press.

Cowan, N. (1999). An embedded-processes model of working memory. In A. Miyake & P. Shah (Eds.), *Models of working memory: Mechanisms of active maintenance and executive control* (pp. 62–101). Cambridge University Press.

Cowan, N. (2001). The magical number 4 in short-term memory: A reconsideration of mental storage capacity. *Behavioral and Brain Sciences, 24*(1), 87–185.

Cowan, N. (2014). Working memory underpins cognitive development, learning, and education. *Educational Psychology Review, 26*(2), 197–223.

Cowan, N. (2016). *Working memory capacity: Classic edition* (with new Foreword). Routledge. (Original work published 2005).

Cowan, N. (2016). Working memory maturation: Can we get at the essence of cognitive growth? *Perspectives on Psychological Science, 11*(2), 239–264.

Cowan, N. (2017a). The many faces of working memory and short-term storage. *Psychonomic Bulletin & Review, 24*(4), 1158–1170.

Cowan, N. (2017b). Mental objects in working memory: Development of basic capacity or of cognitive completion? *Advances in Child Development and Behavior, 52*, 81–104.

Cowan, N. (2019). Short-term memory based on activated long-term memory: A review in response to Norris (2017). *Psychological Bulletin, 145*(8), 822–847.

Cowan, N. (2021). Differentiation of two working memory tasks normed on a large U.S. sample of children 2–7 years old. *Child Development, 92*(6), 2268–2283.

Cowan, N., Adams, E. J., Bhangal, S., Corcoran, M., Decker, R., Dockter, C. E., Eubank, A. T., Gann, C. L., Greene, N. R., Helle, A. C., Lee, N., Nguyen, A. T., Ripley, K. R., Scofield, J. E., Tapia, M. A., Threlkeld, K. L., & Watts, A. L. (2019). Foundations of arrogance: A broad survey and framework for research. *Review of General Psychology, 23*(4), 425–443.

Cowan, N., AuBuchon, A. M., Gilchrist, A. L., Blume, C. L., Boone, A. P., & Saults, J. S. (2021). Developmental change in the nature of attention allocation in a dual task. *Developmental Psychology, 57*(1), 33–46.

Cowan, N., AuBuchon, A. M., Gilchrist, A. L., Ricker, T. J., & Saults, J. S. (2011). Age differences in visual working memory capacity: Not based on encoding limitations. *Developmental Science, 14*(5), 1066–1074.

Cowan, N., Elliott, E. M., & Saults, J. S. (2002). The search for what is fundamental in the development of working memory. In R. Kail & H. Reese (Eds.), *Advances in child development and behavior* (Vol. 29, pp. 1–49). Academic Press.

Cowan, N., Elliott, E. M., Saults, J. S., Morey, C. C., Mattox, S., Hismjatullina, A., & Conway, A. R. A. (2005). On the capacity of attention: Its estimation and its role in working memory and cognitive aptitudes. *Cognitive Psychology, 51*(1), 42–100.

Cowan, N., Elliott, E. M., Saults, J. S., Nugent, L. D., Bomb, P., & Hismjatullina, A. (2006). Rethinking speed theories of cognitive development: Increasing the rate of recall without affecting accuracy. *Psychological Science, 17*(1), 67–73.

Cowan, N., Fristoe, N. M., Elliott, E. M., Brunner, R. P., & Saults, J. S. (2006). Scope of attention, control of attention, and intelligence in children and adults. *Memory & Cognition, 34*(8), 1754–1768.

Cowan, N., Hardman, K., Saults, J. S., Blume, C. L., Clark, K. M., & Sunday, M. A. (2016). Detection of the number of changes in a display in working memory. *Journal of Experimental Psychology: Learning, Memory, and Cognition, 42*(2), 169–185.

Cowan, N., Hogan, T. P., Alt, M., Green, S., Cabbage, K. L., Brinkley, S., & Gray, S. (2017). Short-term memory in childhood dyslexia: Deficient serial order in multiple modalities. *Dyslexia, 23*(3), 209–233.

Cowan, N., Johnson, T. D., & Saults, J. S. (2005). Capacity limits in list item recognition: Evidence from proactive interference. *Memory, 13*(3–4), 293–299.

Cowan, N., Li, Y., Glass, B., & Saults, J. S. (2018). Development of the ability to combine visual and acoustic information in working memory. *Developmental Science, 21*(5), e12635.

Cowan, N., Morey, C. C., AuBuchon, A. M., Zwilling, C. E., & Gilchrist, A. L. (2010). Seven-year-olds allocate attention like adults unless working memory is overloaded. *Developmental Science, 13*(1), 120–133.

Cowan, N., Morey, C. C., & Naveh-Benjamin, M. (2021). An embedded-processes approach to working memory: How is it distinct from other approaches, and to what ends? In R. H. Logie, V. Camos, & N. Cowan (Eds.), *Working memory: State of the science* (pp. 44–84). Oxford University Press.

Cowan, N., & Morse, P. A. (1986). The use of auditory and phonetic memory in vowel discrimination. *Journal of the Acoustical Society of America, 79*(2), 500–507.

Cowan, N., Naveh-Benjamin, M., Kilb, A., & Saults, J. S. (2006). Life-span development of visual working memory: When is feature binding difficult? *Developmental Psychology, 42*(6), 1089–1102.

Cowan, N., Nugent, L. D., Elliott, E. M., Ponomarev, I., & Saults, J. S. (1999). The role of attention in the development of short-term memory: Age differences in the verbal span of apprehension. *Child Development, 70*(5), 1082–1097.

Cowan, N., Ricker, T. J., Clark, K. M., Hinrichs, G. A., & Glass, B. A. (2015). Knowledge cannot explain the developmental growth of working memory capacity. *Developmental Science, 18*(1), 132–145.

Cowan, N., Rouder, J. N., Blume, C. L., & Saults, J. S. (2012). Models of verbal working memory capacity: What does it take to make them work? *Psychological Review, 119*(3), 480–499.

Cowan, N., & Saults, J. S. (2013). When does a good working memory counteract proactive interference? Surprising evidence from a probe recognition task. *Journal of Experimental Psychology: General, 142*(1), 12–17.

Cowan, N., Saults, J. S., & Blume, C. L. (2014). Central and peripheral components of working memory storage. *Journal of Experimental Psychology: General, 143*(5), 1806–1836.

Cowan, N., Suomi, K., & Morse, P. A. (1982). Echoic storage in infant perception. *Child Development, 53*(4), 984–990.

Cowan, N., Wood, N. L., Wood, P. K., Keller, T. A., Nugent, L. D., & Keller, C. V. (1998). Two separate verbal processing rates contributing to short-term memory span. *Journal of Experimental Psychology: General, 127*(2), 141–160.

Craik, F. I., Govoni, R., Naveh-Benjamin, M., & Anderson, N. D. (1996). The effects of divided attention on encoding and retrieval processes in human memory. *Journal of Experimental Psychology: General, 125*(2), 159–180.

Doherty, J. M., Belletier, C., Rhodes, S., Jaroslawska, A. J., Barrouillet, P., Camos, V., Cowan, N., Naveh-Benjamin, M., & Logie, R. H. (2019). Dual-task costs in working memory: An adversarial collaboration. *Journal of Experimental Psychology: Learning, Memory, and Cognition, 45*(9), 1529–1551.

Elliott, E. M., & Cowan, N. (2001). Habituation to auditory distractors in a cross-modal, color-word interference task. *Journal of Experimental Psychology: Learning, Memory, and Cognition, 27*(3), 654–667.

Forsberg, A., Blume, C., & Cowan, N. (2021). The development of metacognitive accuracy in working memory across childhood. *Developmental Psychology, 57*(8), 1297–1317.

Forsberg, A., Guitard, D., Adams, E. J., Pattanakul, D., & Cowan, N. (2022). Children's long-term retention is directly constrained by their working memory capacity limitations. *Developmental Science, 25*(2), e13164.

Forsberg, A., Guitard, D., & Cowan, N. (2021). Working memory limits severely constrain long-term retention. *Psychonomic Bulletin & Review, 28*(2), 537–547.

Gaillard, V., Barrouillet, P., Jarrold, C., & Camos, V. (2011). Developmental differences in working memory: Where do they come from? *Journal of Experimental Child Psychology, 110*(3), 469–479.

Gathercole, S. E., Pickering, S. J., Ambridge, B., & Wearing, H. (2004). The structure of working memory from 4 to 15 years of age. *Developmental Psychology, 40*(2), 177–190.

Gilchrist, A. L., Cowan, N., & Naveh-Benjamin, M. (2008). Working memory capacity for spoken sentences decreases with adult ageing: Recall of fewer, but not smaller chunks in older adults. *Memory, 16*(7), 773–787.

Gilchrist, A. L., Cowan, N., & Naveh-Benjamin, M. (2009). Investigating the childhood development of working memory using sentences: New evidence for the growth of chunk capacity. *Journal of Experimental Child Psychology, 104*(2), 252–265.

Gillam, R. B., Cowan, N., & Day, L. S. (1995). Sequential memory in children with and without language impairment. *Journal of Speech & Hearing Research, 38*(2), 393–402.

Gold, J. M., Fuller, R. L., Robinson, B. M., McMahon, R. P., Braun, E. L., & Luck, S. J. (2006). Intact attentional control of working memory encoding in schizophrenia. *Journal of Abnormal Psychology, 115*(4), 658–673.

Goodsitt, J., Morse, P., Ver Hoeve, J., & Cowan, N. (1984). Infant speech recognition in multisyllabic contexts. *Child Development, 55*(3), 903–910.

Gray, S., Green, S., Alt, M., Hogan, T., Kuo, T., Brinkley, S., & Cowan, N. (2017). The structure of working memory in young school-age children and its relation to intelligence. *Journal of Memory and Language, 92*, 183–201.

Greene, N. R., Naveh-Benjamin, M., & Cowan, N. (2020). Adult age differences in working memory capacity: Spared central storage but deficits in ability to maximize peripheral storage. *Psychology and Aging, 35*(6), 866–880.

Guitard, D., & Cowan, N. (2020). Do we use visual codes when information is not presented visually? *Memory & Cognition, 48*(8), 1522–153.

Guitard, D., Saint-Aubin, J., & Cowan, N. (2021). Asymmetrical interference between item and order information in short-term memory. *Journal of Experimental Psychology: Learning, Memory, and Cognition, 47*(2), 243–263.

Halford, G. S., Cowan, N., & Andrews, G. (2007). Separating cognitive capacity from knowledge: A new hypothesis. *Trends in Cognitive Sciences, 11*(6), 236–242.

Hanley, J. R., & Young, A. W. (2019). ELD revisited: A second look at a neuropsychological impairment of working memory affecting retention of visuo-spatial material. *Cortex, 112*, 172–179.

Hebb, D. O. (1949). *Organization of behavior*. Wiley.

Hulme, C., & Tordoff, V. (1989). Working memory development: The effects of speech rate, word length, and acoustic similarity on serial recall. *Journal of Experimental Child Psychology*, *47*(1), 72–87.

James, W. (1890). *The principles of psychology*. Henry Holt.

Jaroslawska, A. J., Gathercole, S. E., Logie, M. R., & Holmes, J. (2016). Following instructions in a virtual school: Does working memory play a role? *Memory & Cognition*, *44*(4), 580–589.

Jarrold, C., Cowan, N., Hewes, A. K., & Riby, D. M. (2004). Speech timing and verbal short-term memory: Evidence for contrasting deficits in Down syndrome and Williams syndrome. *Journal of Memory and Language*, *51*(3), 365–380.

Javitt, D. C., Strous, R., Grochowski, S., Ritter, W., & Cowan, N. (1997). Impaired precision, but normal retention, of auditory sensory ('echoic') memory information in schizophrenia. *Journal of Abnormal Psychology*, *106*(2), 315–324.

Jenkins, L., Myerson, J., Hale, S., & Fry, A. F. (1999). Individual and developmental differences in working memory across the life span. *Psychonomic Bulletin & Review*, *6*(1), 28–40.

Jiang, Q., & Cowan, N. (2020). Incidental learning of list membership is affected by serial position in the list. *Memory*, *28*(5), 669–676.

Kahneman, D. (1973). *Attention and effort*. Prentice Hall.

Kane, M. J., Hambrick, D. Z., Tuholski, S. W., Wilhelm, O., Payne, T. W., & Engle, R. W. (2004). The generality of working memory capacity: A latent variable approach to verbal and visuospatial memory span and reasoning. *Journal of Experimental Psychology: General*, *133*(2), 189–217.

Keller, T. A., & Cowan, N. (1994). Developmental increase in the duration of memory for tone pitch. *Developmental Psychology*, *30*(6), 855–863.

Lee, E., Cowan, N., Vogel, E. K., Rolan, T., Valle-Inclán, F., & Hackley, S. A. (2010). Visual working memory deficits in Parkinson's patients are due to both reduced storage capacity and impaired ability to filter out irrelevant information. *Brain*, *133*(9), 2677–2689.

Lemaire, B., Pageot, A., Plancher, G., & Portrat, S. (2018). What is the time course of working memory attentional refreshing? *Psychonomic Bulletin & Review*, *25*(1), 370–385.

Li, Y., & Cowan, N. (2021). Attention effects in working memory that are asymmetric across sensory modalities. *Memory & Cognition*, *49*(5), 1050–1065.

Logie, R. H. (2016). Retiring the central executive. *Quarterly Journal of Experimental Psychology*, *69*(10), 2093–2109.

Logie, R. H. (2019). Converging sources of evidence and theory integration in working memory: A commentary on Morey, Rhodes, and Cowan (2019). *Cortex*, *112*, 162–171.

Logie, R. H., Della Sala, S., & Wynn, V., & Baddeley, A. D. (2000). Visual similarity effects in immediate verbal serial recall. *Quarterly Journal of Experimental Psychology*, *53*(3), 626–646.

Logie, R. H., & Morris, R. G. (Eds.) (2015). *Working memory and ageing*. Psychology Press.

Luck, S. J., & Vogel, E. K. (1997). The capacity of visual working memory for features and conjunctions. *Nature*, *390*(6657), 279–281.

Martin, J. D., Shipstead, Z., Harrison, T. L., Redick, T. S., Bunting, M., & Engle, R. W. (2020). The role of maintenance and disengagement in predicting reading comprehension and vocabulary learning. *Journal of Experimental Psychology: Learning, Memory, and Cognition*, *46*(1), 140–154.

Miller, G. A. (1956). The magical number seven, plus or minus two: Some limits on our capacity for processing information. *Psychological Review*, *63*(2), 81–97.

Moray, N. (1959). Attention in dichotic listening: Affective cues and the influence of instructions. *Quarterly Journal of Experimental Psychology*, *11*(1), 56–60.

Morey, C. C. (2019). Working memory theory remains stuck: Reply to Hanley and Young. *Cortex*, *112*, 180–181.

Morey, C. C., & Bieler, M. (2013). Visual short-term memory always requires attention. *Psychonomic Bulletin & Review*, *20*(1), 163–170.

Morey, C. C., Morey, R. D., van der Reijden, M., & Holweg, M. (2013). Asymmetric cross-domain interference between two working memory tasks: Implications for models of working memory. *Journal of Memory and Language*, *69*(3), 324–348.

Morey, C. C., Rhodes, S., & Cowan, N. (2019). Sensory-motor integration and brain lesions: Progress toward explaining domain-specific phenomena within domain-general working memory. *Cortex*, *112*, 149–161.

Morey, C. C., Rhodes, S., & Cowan, N. (2020). Co-existing, contradictory working memory models are ready for progressive refinement: Reply to Logie. *Cortex*, *123*, 200–202.

Naveh-Benjamin, M., Kilb, A., Maddox, G., Thomas, J., Fine, H., Chen, T., & Cowan, N. (2014). Older adults don't notice their names: A new twist to a classic attention task. *Journal of Experimental Psychology: Learning, Memory, and Cognition*, *40*(6), 1540–1550.

Norman, D. A. (1968). Toward a theory of memory and attention. *Psychological Review*, *75*(6), 522–536.

Oberauer, K. (2021). Towards a theory of working memory: From metaphors to mechanisms. In R. H. Logie, V. Camos, & N. Cowan (eds.), *Working memory: State of the science* (pp. 116–149). Oxford University Press.

Oberauer, K., Süß, H. M., Wilhelm, O., & Sander, N. (2007). Individual differences in working memory capacity and reasoning ability. In A. Miyake, A. Conway, C. Jarrold, M. J. Kane, & J. N. Towse (Eds.), *Variation in working memory* (pp. 49–75). Oxford University Press.

Old, S. R., & Naveh-Benjamin, M. (2008). Differential effects of age on item and associative measures of memory: A meta-analysis. *Psychology and Aging*, *23*(1), 104–118.

Posner, M. I. (1980). Orienting of attention. *Quarterly Journal of Experimental Psychology*, *32*(1), 3–25.

Rhodes, S., Jaroslawska, A. J., Doherty, J. M., Belletier, C., Naveh-Benjamin, M., Cowan, N., Camos, V., Barrouillet, P., & Logie, R. H. (2019). Storage and processing in working memory: Assessing dual task performance and task prioritization across the adult lifespan. *Journal of Experimental Psychology: General*, *148*(7), 1204–1227.

Ricker, T. J., & Cowan, N. (2010). Loss of visual working memory within seconds: The combined use of refreshable and non-refreshable features. *Journal of Experimental Psychology: Learning, Memory, and Cognition*, *36*(6), 1355–1368.

Ricker, T. J., & Cowan, N. (2014). Differences between presentation methods in working memory procedures: A matter of working memory consolidation. *Journal of Experimental Psychology: Learning, Memory, and Cognition*, *40*(2), 417–428.

Ricker, T. J., Sandry, J., Vergauwe, E., & Cowan, N. (2020). Do familiar memory items decay? *Journal of Experimental Psychology: Learning, Memory, and Cognition*, *46*(1), 60–76.

Ricker, T. J., Spiegel, L. R., & Cowan, N. (2014). Time-based loss in visual short-term memory is from trace decay, not temporal distinctiveness. *Journal of Experimental Psychology: Learning, Memory, and Cognition*, *40*(6), 1510–1523.

Röer, J. P., & Cowan, N. (2021). A preregistered replication and extension of the cocktail party phenomenon: One's name captures attention, unexpected words do not. *Journal of Experimental Psychology: Learning, Memory, and Cognition*, *47*(2), 234–242.

Rosen, V. M., & Engle, R. W. (1997). The role of working memory capacity in retrieval. *Journal of Experimental Psychology: General*, *126*(3), 211–227.

Saults, J., Cowan, N., Sher, K. J., & Moreno, M. V. (2007). Differential effects of alcohol on working memory: Distinguishing multiple processes. *Experimental and Clinical Psychopharmacology*, *15*(6), 576–587.

Schweickert, R., & Boruff, B. (1986). Short term memory capacity: Magic number or magic spell? *Journal of Experimental Psychology: Learning, Memory, and Cognition*, *12*(3), 419–425.

Shallice, T., & Papagno, C. (2019). Impairments of auditory-verbal short-term memory: Do selective deficits of the input phonological buffer exist? *Cortex*, *112*, 102–121.

Sokolov, E. N. (1963). *Perception and the conditioned reflex*. Pergamon Press.

Treisman, A. M. (1964). Selective attention in man. *British Medical Bulletin*, 20, 12–16.

Uittenhove, K., Chaabi, L., Camos, V., & Barrouillet, P. (2019). Is working memory storage intrinsically domain-specific? *Journal of Experimental Psychology: General*, *148*(11), 2027–2057.

Unsworth, N., & Engle, R. W. (2007). The nature of individual differences in working memory capacity: Active maintenance in primary memory and controlled search from secondary memory. *Psychological Review*, *114*(1), 104–132.

Vandierendonck, A. (2016). A working memory system with distributed executive control. *Perspectives on Psychological Science*, *11*(1), 74–100.

Vergauwe, E., Barrouillet, P., & Camos, V. (2010). Do mental processes share a domain general resource? *Psychological Science*, *21*(3), 384–390.

Vogel, E. K., McCollough, A. W., & Machizawa, M. G. (2005). Neural measures reveal individual differences in controlling access to working memory. *Nature*, *438*(7067), 500–503.

Weeks, J. C., Grady, C. L., Hasher, L., & Buchsbaum, B. R. (2020). Holding on to the past: Older adults show lingering neural activation of no-longer-relevant items in working memory. *Journal of Cognitive Neuroscience*, *32*(10), 1946–1962.

Wood, N., & Cowan, N. (1995a). The cocktail party phenomenon revisited: Attention and memory in the classic selective listening procedure of Cherry (1953). *Journal of Experimental Psychology: General*, *124*(3), 243–262.

Wood, N., & Cowan, N. (1995b). The cocktail party phenomenon revisited: How frequent are attention shifts to one's name in an irrelevant auditory channel? *Journal of Experimental Psychology: Learning, Memory, & Cognition*, *21*(1), 255–260.

Wood, N. L., Stadler, M. A., & Cowan, N. (1997). Is there implicit memory without attention? A re-examination of task demands in Eich's (1984) procedure. *Memory & Cognition*, *25*(6), 772–779.

Zosh, J. M., & Feigenson, L. (2012). Memory load affects object individuation in 18-month-old infants. *Journal of Experimental Child Psychology*, *113*(3), 322–336.

10
Working memory in action

Remembering and following instructions

Richard J. Allen, Amanda H. Waterman, Tian-xiao Yang, and Agnieszka J. Graham

Background

Working memory is often identified as being the keystone (Barrouillet & Camos, 2021) or hub (Haberlandt, 1997) of cognition, central to many different abilities and activities of daily life. It is also frequently defined in this way (Cowan, 2017), with theoretical approaches to working memory (see Logie, Camos, & Cowan, 2021) often explicitly incorporating reference to the service of complex cognition (Baddeley et al., 2021), ongoing information processing (Cowan et al., 2021), and the control of thoughts and actions (Oberauer, 2021), and behaviour (Postle, 2021), that are goal-directed in nature (Barrouillet & Camos, 2021; Mashburn et al., 2021; Vandierendonck, 2021). In keeping with this central position in cognition, working memory may operate at the interface between perception, internal control, and stored knowledge; between externally driven attention and internal control (Chun et al., 2011; Hitch et al., 2020); and between input, thought, and action (Baddeley, 2007, 2012). Indeed, working memory as a term has been used not only to denote specific models and theoretical frameworks, but also at a practical level as a way of describing a range of important cognitive abilities that underlie performance in laboratory-based and everyday tasks (Logie, Belletier, & Doherty, 2021).

In this context, empirical explorations of working memory, employing either experimental or individual difference approaches, typically focus on tasks requiring recall (verbal, typed, or via computer interface) or recognition of sequences or displays of information in tasks designed to measure memory for verbal or visuospatial information. This is understandable, given the dominant and information-rich role these forms of information have in our interactions with the world. Assessing verbal or visuospatial memory also allows for relatively straightforward and replicable methods of stimulus presentation and response mode. However, our everyday experience clearly extends beyond the verbal and the visuospatial, and our behaviour is not limited to the simple recall or recognition of such stimulus domains. Given the widely accepted characterization of working memory as an interface between perception, long-term memory, and subsequent goal-directed thought and action, it is important to explore working memory in action, and in particular how perceived,

Richard J. Allen, Amanda H. Waterman, Tian-xiao Yang, and Agnieszka J. Graham, *Working memory in action* In: *Memory in Science for Society*. Edited by: Robert H. Logie, Zhisheng (Edward) Wen, Susan E. Gathercole, Nelson Cowan, and Randall W. Engle, Oxford University Press. © Oxford University Press 2023. DOI: 10.1093/oso/9780192849069.003.0010 instructed, anticipated, and planned action might be generated, temporarily retained, and produced as output (e.g. Raw et al., 2019; Rosenbaum, 2005; Rosenbaum & Feghhi, 2019; Tomasino & Gremese, 2016).

Following instructions

One broad key question, but until recently relatively underexplored, is how instructions are encoded and retained over brief periods of time, for the purposes of recall and implementation. This may represent a direct example of the involvement of working memory in supporting everyday action and behaviour. We encounter instructional sequences across many different real-world contexts. Illustrative examples include following a recipe, building flatpack furniture, airline safety advice, medication use, following directions, craft and home improvement activities (e.g. using YouTube videos), or taking on board and implementing guidance from a driving, skiing, dance, fitness, or tennis instructor. It even extends to space exploration, with astronauts on the International Space Station often required to read and implement lengthy sets of instructions when carrying out tasks (Goemare et al., 2018; Krikalev et al., 2010). It is also worth noting that instruction memory does not have to centre on actual or anticipated motor movement; it could refer to learning how to use a new software platform, for example. Thus, being able to hold on to and act on verbal or demonstrated instructions is 'a pervasive instance of everyday cognition' (Engle et al., 1991, p. 255).

One context for which following instructions (sometimes abbreviated as FI) is particularly relevant is education. It is now well established that working memory ability develops through childhood (e.g. Gathercole, Pickering, Ambridge, & Wearing, 2004; Hill et al., 2021). Working memory shows a close relationship with academic progress in primary and secondary education across areas such as literacy, mathematics, and science (e.g. Gathercole et al., 2003, Gathercole, Pickering, Knight, & Stegmann, 2004; Wiklund-Hornqvist et al., 2016), with children identified as having poor working memory typically exhibiting slower academic progress (Gathercole & Pickering, 2000; Holmes & Adams, 2006). This is likely to reflect both domain-specific and general impacts of working memory on the likelihood of successfully completing learning activities (Gathercole et al., 2006). Gathercole and Alloway (2008; Gathercole et al., 2006) identified the working memory demands inherent in many structured learning activities and suggested that children with poor working memory are unable to meet these demands due to memory overload and/or distraction, resulting in incomplete or inaccurate task performance and missed learning opportunities that accumulate over time and lead to learning delays.

One important example of this is in the ability to remember and follow instructions provided by the teacher. In their observational study of three 5–6-year-old children identified as having poor working memory, Gathercole et al. (2006) noted that the most consistent difficulty was in following classroom instructions, particularly when these instructions were longer and fell outside typical routines (e.g. 'Put your sheets on the green table, put your arrow cards in the packet, put your pencil away, and come and sit on the carpet'). The children with poor working memory appeared to forget what they were required to do in such circumstances, and either left the activity incomplete or had to ask for a reminder. The findings from this important small-scale observational study appear to fit with those emerging from a larger ($N = 1425$) questionnaire survey of teachers in primary and secondary education (Atkinson et al., 2021). When asked about behavioural signs associated with poor working memory, as well as reporting general problems with information retention and attentional control, nearly half the respondents spontaneously identified difficulty with classroom/learning activities, and with following instructions. Many teachers also noted an increased dependence on others and/or classroom resources. Thus, not least from an educational perspective, there is a clear need to understand the cognitive underpinnings of memory for instructions and how this key ability might be supported and improved.

Linking working memory and following instructions

Early investigations of instruction-guided behaviour were focused on its relationship to intelligence. For example, Thorndike (1912) included following-instructions assessments in his test of intelligence. A few years later, Brener (1940) studied the ability to implement written commands on letter displays (e.g. 'Put a comma below B, put a circle around A') as a proxy for memory span. He used a span-type procedure and reported positive correlations between the ability to follow instructions and simple span measures (e.g. digit span). This was the first study to suggest that working memory capacity may be a source of individual variation in instruction following. Further evidence of the link between working memory skills and the ability to follow instructions was obtained using the Token Test developed by De Renzi and colleagues (De Renzi & Faglioni, 1978; De Renzi & Vignolo, 1962) for use in neuropsychological assessment. In this task, participants are asked to carry out sequences of commands on objects in a display (e.g. 'After picking up the green rectangle, touch the white circle'). Task difficulty is manipulated by increasing the length and grammatical complexity of the instruction sequences of commands across blocks. Performance on this measure has been shown to correlate with verbal and visual aspects of short-term memory (Lesser, 1976).

Adopting a more applied approach, Kaplan and White (1980) investigated instruction following in the context of the classroom by recording and analysing the types of instructions children were required to complete over the course of a typical school day. They identified that instructions typically differ in terms of grammatical complexity and the number of steps involved. Kaplan and White then produced their own set of instructions which were described in terms of the number of behaviours and the number of qualifiers that a command contained. For instance, a command such as 'Open your books on page three and solve the first four problems' consists of two behaviours (i.e. 'open your books' and 'solve the problems') and two qualifiers (i.e. 'on page three' and 'first four'). Kaplan and White's materials included varying numbers of qualifiers and behaviours, and the difficulty of a single sequence was indexed by the number of these components. The instructions were administered to 215 children (aged between 5–11 years) who were required to execute the sequences immediately. The ability to follow instructions increased steadily with age and increasing the sentence complexity (by adding qualifiers) had a detrimental effect on performance, especially for younger children (up to around age 7). Although Kaplan and White did not administer any assessments of memory ability, it is plausible to suggest that the increasing sentence complexity imposed higher demands on working memory and, therefore, that superior performance in older children was reflective of their increased working memory ability.

More direct evidence of the relationship between working memory and following instructions comes from a study conducted by Engle et al. (1991). The authors modified Kaplan and White's (1980) instruction task and included both pencil-and-paper commands (e.g. 'Point to the picture at the top of page three and copy it twice') and action-oriented commands (e.g. 'Sit down and open your books'). Consistent with previous findings, Engle and colleagues reported that instruction following improved with age (in a sample of 7–12-year-olds). In addition, working memory, as measured by forward word span and sentence span tasks, was closely associated with the ability to follow practical multistep instructions (an association that increased with age).

Building on this earlier work, Gathercole et al. (2008) developed a laboratory analogue of a following-instructions task, using verbally presented sequences of action directions motivated in part by their classroom observational work (Gathercole et al., 2006). In this study, children aged 5–6 years were presented with sequences such as 'Touch the red ruler then pick up the yellow pencil and put it in the black box'. These sequences varied in length (i.e. the number of to-be-performed or to-be-repeated steps) but not in linguistic complexity or grammatical construction. Only simple active constructions concatenated with the adverb *then* were used (e.g. 'Touch the white bag, then pick up the yellow ruler, then put it in the blue folder'). In addition, a span-type procedure was employed in which the number of steps in the instructions increased progressively up to the point at which the child was no longer able to perform accurately. Children were required to follow the instructions given and either manipulate three-dimensional objects placed in front of them, or verbally repeat the instructions back to the experimenter. Gathercole et al. (2008) found that working memory (as indexed by forwards and backwards digit recall) was significantly and positively correlated with enacted instruction performance, but not with verbal repetition. The former relationship was also stronger for backwards digit recall, typically assumed to capture more complex working memory ability in children (e.g. Alloway et al., 2006; Gathercole, Pickering, Ambridge, & Wearing, 2004; Hill et al., 2021), relative to forwards recall. This influential study is important in extending the findings of Engle et al. (1991) linking working memory and instruction performance while also offering an experimental paradigm that can be used and adapted to systematically explore such links.

Nevertheless, a possible limitation of the experimental paradigms described thus far is that they somewhat simplify the practical demands of real-life situations in which following instructions is required. In a classroom environment, for example, a teacher's instructions are often more complex, varied, and unpredictable, and the entire sequence may take more time to complete. To explore the role of working memory in situations that vary the amount of time over which instructions must be retained, Jaroslawska, Gathercole, Logie, and Holmes (2016) developed two-dimensional virtual versions of Gathercole et al.'s (2008) following-instructions paradigm (Figure 10.1d). These tasks required 7–11-year-old participants to navigate through a computerized school environment to perform a sequence of spoken instructions. For example, in one version of the task, participants moved between

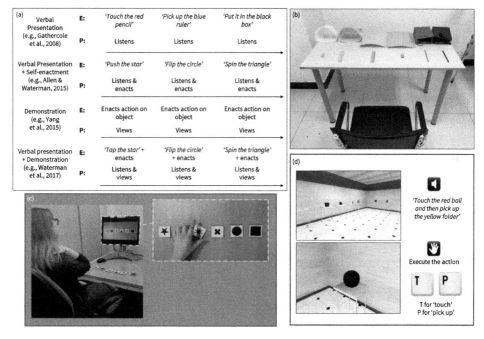

Figure 10.1 Illustration of methods in following instructions tasks. (a) Schematic illustration of a selection of the encoding conditions explored to date using variants of the following instructions task, displaying how the experimenter (E) and participant (P) engages in each case. Demonstration can be via video or in-person. Following sequence presentation, P attempts to reproduce the whole sequence through verbal or enacted recall. (b) Real-world object set-up (e.g. Yang et al., 2019). (c) Geometric shape set-up (e.g. Allen et al., 2020), showing a video demonstration condition. (d) Virtual MySchool paradigm (Jaroslawska et al., 2016).

multiple rooms in a virtual school building to implement instructions (e.g. 'Go to the IT suite and touch the red ball and then go to Mrs Bolton's room and pick up the green box'). The additional navigation demands of these activities aligned the tasks more closely with the everyday practical demands imposed on children in their school life. Jaroslawska, Gathercole, Logie, and Holmes (2016) found that children's abilities to carry out sequences of actions following spoken instructions were closely related to their verbal working memory skills, further emphasizing that verbal aspects of working memory involved in both simple and complex span tasks play a specific role in following instructions over extended periods of activity.

The importance of action

Instructions guide us to complete procedural tasks and are most often organized as a series of successive steps that need to be carried out. For example, instructions on how to scan and send a document via email might include (1) placing the document on the glass surface, text facing down; (2) entering an email address in the field provided; (3) pressing start to scan and send the email; and (4) checking the display to confirm whether the transmission was successful. Such procedural instructions can take various forms, from text and images describing how to install new software to graphical, interactive maps explaining how to find one's way in an unfamiliar environment. Whatever their form, it takes work to transform instructions into concrete actions: from efficient sensory perception to language comprehension, memory storage, and, ultimately, action execution. As a result, a range of factors determine whether instructions will be followed through to successful completion. The focus of this section of the chapter is the importance of physical movement in the context of instruction following. In principle, there are two points in time at which physical movement might influence whether the instruction is later implemented: performing the instruction when it is first *encoded* or performing the instruction when it is *recalled*. We will begin with the latter.

Enacted recall

Research suggests that information encoded for the purpose of future action is stored or organized differently from information retained for future verbal recall. Indeed, anticipating physical performance during encoding has been shown to improve subsequent recall (e.g. Engelkamp, 1997; Koriat et al., 1990), even when the participant does *not* execute the instruction at recall (e.g. Engelkamp et al., 1994; Koriat & Pearlman-Avnion, 2003; Kormi-Nouri et al., 1994). For example, Koriat et al. (1990) presented short sequences of action phrases relating to real objects (e.g. 'move the eraser, lift the cup, flip the coin') and manipulated whether participants enacted or verbally repeated the instructions at test. The key finding—replicated many times since—was that enacted recall was markedly more accurate than verbal recall. This advantage for phrases that participants expected to recall through performance at test occurred even when recall was, unexpectedly, tested verbally. Subsequently, merely the intention to perform the described actions in the future has been shown to improve memory recall (Jahn & Engelkamp, 2003; Maylor et al., 2001), recognition accuracy (Jahn & Engelkamp, 2003), recognition latencies (Goschke & Kuhl, 1993), and lexical decision latencies (Marsh et al., 1998) for content associated with instructions participants intended to perform and not for content associated with instructions they did not intend to perform.

To explain the enacted-recall advantage, Koriat et al. (1990) suggested that participants did not simply verbally rehearse the instructions and then translate them into actions during retrieval. Instead, it was proposed that action commands were encoded in a motoric form to take advantage of the richness of the visual and kinaesthetic representations that underlie action performance. The authors further hypothesized that the benefit of action recall lies in the encoding stage and argued that planning for actions facilitates the formation of an integrated multimodal representation involving phonological, visual, and motor codes. This multimodal representation (which is not present when verbal repetition is expected) integrates elements from various channels into a coherent representation. In other words, they assumed that the positive memory effect of to-be-performed actions is also due to non-verbal encoding or, more specifically, to sensorimotor encoding, and that there is little difference between the intention to perform at test and performing actions at study. Later studies carried out by Engelkamp (1997) allowed assessment of the independent contribution to memory of the expectation during encoding to perform the instructions at recall and of performing the instructions at recall. The results indicated that the critical factor is action planning and were taken by Engelkamp (1997) to suggest that both the enactment-at-encoding effect and the enactment-at-recall effect are due to encoding rather than to retrieval differences. These findings were further supported by Watanabe (2003) in a series of studies probing the effects of encoding style, expectation of retrieval mode, and retrieval style on memory for action phrases. Watanabe (2003) instructed participants to memorize action phrases with a variety of encoding styles (i.e. verbal, performance, imaging), expectation of retrieval modes (i.e. to-be-remembered, to-be-performed, to-be-forgotten), and retrieval styles (verbal, enacted). Results consistently showed that long-term memory for action phrases was affected by the mode of encoding (i.e. type of encoding and expectation of retrieval at encoding), rather than retrieval.

Extending the work of Koriat and others, Gathercole and colleagues' examination of children's memory for instructions using physical enactment or verbal repetition was the study to explicitly compare these response modes in a working memory paradigm (Gathercole et al., 2008). They found a substantial span advantage for enacted over verbal recall, with performance being more than twice as accurate for the former response type. Following this initial finding, the enacted-recall advantage has emerged as an extraordinarily reliable and robust effect, and has been replicated in multiple studies across a variety of populations including typically developing children (e.g. Jaroslawska et al., 2016; Waterman et al., 2017; Yang et al., 2021), children with attention deficit hyperactivity disorder (ADHD; Yang et al., 2017), adults (e.g. Allen & Waterman, 2015; Gathercole et al., 2008; Jaroslawska et al., 2018; Li et al., 2022; Lui et al., 2018; Koriat et al., 1990; Makri & Jarrold, 2021; Yang et al., 2014, 2017), individuals with schizophrenia (Lui et al., 2018), and healthy older adults (Jaroslawska et al., 2021; Yang, Su, et al., 2022). This is illustrated in Figure 10.2, which displays a summary of the enacted recall advantage in comparable conditions

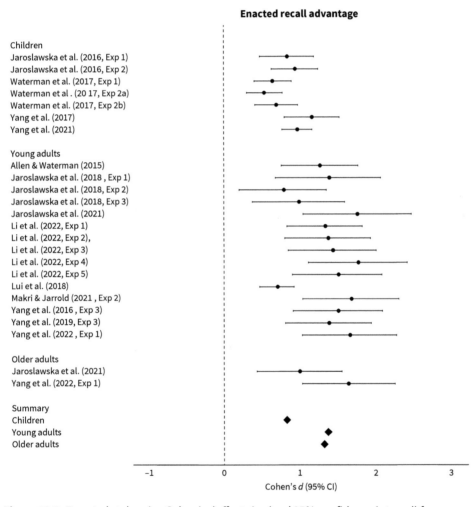

Figure 10.2 Forest plot showing Cohen's *d* effect size (and 95% confidence interval) for the enacted recall effect (i.e. the difference in recall accuracy between enacted and verbal recall) across studies, where this was implemented as a within-subjects manipulation. Only conditions involving auditory–verbal presentation and no other required activity during encoding were included.

across multiple studies featuring children, young adults, or older adults. The effect has also been reported in bilingual speakers when instructions were presented in both the participants' first and second language (Shekari, 2020). These outcomes are typically measured in terms of recall accuracy, though analysis of response timings in young adults also indicates that participants show faster response preparation and implementation times for enacted compared to verbal recall (Yang et al., 2019).

To pinpoint the cognitive source of the action advantage, Yang et al. (2014, 2016) instructed (either through visual text or auditory presentation) young adults to perform a series of actions on objects (using a version of the Gathercole et al., 2008 paradigm; see Figure 10.1b) and asked them to either physically perform or verbally repeat the sequence. Participants engaged in concurrent articulatory suppression, backwards counting, or spatial tapping during the encoding phase. These tasks were selected with the aim of disrupting subcomponents of working memory as described by the multicomponent model (Baddeley, 1986, 2012; Baddeley & Hitch, 1974; Baddeley et al., 2021), namely the phonological loop, central executive, and visuospatial sketchpad components of working memory, respectively. Recall was disrupted by all three concurrent activities, indicating that encoding and retention of verbal instructions depend on multiple aspects of working memory, with the phonological loop retaining verbal information and the visuospatial sketchpad storing visual and spatial information (see also Jaroslawska, Gathercole, Logie, & Holmes, 2016). Crucially, performance was better when participants had to physically perform rather than verbally repeat the sequences, and this action advantage remained intact under dual-task conditions. Thus, whatever underlies the enacted recall advantage in a working memory context appears to be distinct from the 'classic' components targeted by Yang et al. (2016).

This raises the question of what contributes to such a reliable and substantial effect, if not verbal, visuospatial, or executive attentional resources. Research is still required on this issue, but some insights can be drawn from work applying concurrent movement as dual-task manipulations. Jaroslawska et al. (2018) found that asking participants to perform certain types of motor activity (namely, repeated arm movement) during encoding reduced and in fact eliminated the enacted recall advantage. This might indicate that this effect is driven by a specific component of the working memory system that is separable from those involved in encoding verbal sequences more generally and may be associated with motor coding. These results were recently explored further in a series of five experiments by Li et al. (2022) that systematically manipulated the scale (gross vs fine), complexity, and familiarity of movement that participants were required to make during encoding. Rather than using the Gathercole et al. (2008) method, Li et al. (2022) employed a slightly different paradigm developed by Allen and Waterman (2015) involving sequences of simple action–object pairs (e.g. 'tap the circle, spin the triangle, flip the cross'; see next section for more discussion). Li et al. again replicated the enacted recall advantage in all experiments. However, this effect was variously reduced in size or abolished by different forms of concurrent motor movement, especially when these movements were more complex or unfamiliar. Extending the initial findings reported by Jaroslawska et al. (2018), this may represent evidence for a previously underexplored motor component in working memory that supports anticipated action planning.

The enacted recall advantage was also recently explored by Makri and Jarrold (2021) using a different paradigm, involving actions and physical (foam) objects depicting numbers. They found evidence that enactment planning may critically impact action–object binding, as the effect was only present when participants recalled action–object pairs (Exp 2), and not in a first experiment in which only individual features had to be recalled (which required the 'unbinding' of action and object). This also fits with additional analysis of feature and binding memory reported by Yang et al. (2016); here, planning for enacted recall seemed to particularly enhance action–object binding. Thus, anticipating an enacted response at recall may result not only in activation of motor coding but also the binding of this with representations of the relevant objects in the environment.

Enactment during encoding

Given the highly reliable superiority of physically enacted recall over verbal repetition, and the possibility that this may in part represent motor-related planning during encoding, this raises the question of how action-based manipulations targeted at the encoding phase might also impact performance. Such a consideration represents a possible route to better understanding the enacted recall advantage and memory for instruction, and as a way of improving performance in a practical sense.

There have been many studies in the long-term memory literature demonstrating that instructions that are acted out by the participant during encoding are more likely to be subsequently recalled than instructions that are only verbally encoded into memory (e.g. Cohen, 1981; Engelkamp & Krumnacker, 1980; Saltz & Donnenwerth-Nolan, 1981). The standard procedure is that participants are asked to remember a series of short action commands, such as 'open the book' or 'lift the pen', with a single encoding phase often consisting of around 30–40 such commands, followed by a response phase. The crucial aspect of these studies is that participants are not simply instructed to remember the information but are, additionally, required to execute the actions during encoding (e.g. to lift the pen). In a typical control condition, the participants are presented with the same verbal action phrases as in the subject-performed tasks condition, but without any instruction to perform the actions. Consistent findings show that the recall of action phrases is significantly better when participants are instructed to execute the actions during the acquisition phase, relative to simply passively listening to the commands. This so-called subject-performed task or enactment effect has been observed on tests of long-term memory in a wide variety of experimental settings (see Cohen, 1989; Engelkamp, 1998; Engelkamp & Zimmer, 1989, 1994; Zimmer et al., 2001; for reviews).

More recently, this encoding-based self-enactment procedure has been adapted and integrated into research examining the immediate recall of instructions. In this approach, the participant performs each action command immediately after its presentation, during the encoding phase. Allen and Waterman (2015) developed this procedure by orthogonally manipulating enactment at both encoding and recall. This experiment (with young adults) examined the impact of enactment during both encoding and recall. Thus, participants either simply listened to the sequence, or physically performed each action on the object set laid out in front of them, as the sequence progressed (Figure 10.1a). They then attempted to recall the whole sequence either verbally or via enactment. The impacts of these manipulations were measured using a task requiring memory for sequences of simple action and object pairs (e.g. 'flip the triangle, spin the cross, touch the square'). The paradigm used in this study is notable for two reasons. Firstly, a larger range of actions was used than in the original Gathercole et al. (2008) task, thus increasing the difficulty and motoric complexity of the task. Secondly, rather than using real objects, participants interacted with card-based depictions of simple geometric shapes (Figure 10.1c). This was influenced by research in the visual working memory domain (e.g. Allen et al., 2006, 2012, 2014), and was implemented to minimize any contribution of object affordances or long-term memory and offer a task that particularly emphasized working memory. In reality, of course, objects often vary in affordance, kinaesthetic, and tactile information, as well as being associated with prior experience and associations with certain types of movement. It remains an open question for future research to explore exactly how these dimensions might impinge on performance and interact with the factors established to date.

Nevertheless, Allen and Waterman (2015) not only replicated the enacted recall advantage, but also found that the enactment of actions during the encoding phase significantly facilitated subsequent recall performance. Furthermore, this effect was marked for verbal repetition but attenuated for enacted recall. In keeping with the more recent work on motor dual-tasking in this context (Jaroslawska et al., 2018; Li et al., 2022), these findings were interpreted as reflecting a beneficial role of spatial motoric coding in working memory, that can be engaged either through action planning or through physical performance. There is no 'double boost' though; enactment during encoding recruits this additional coding and therefore potentially facilitates subsequent verbal recall, but this is relatively redundant in the case of enacted recall where such coding is already engaged during the process of action planning.

These findings were replicated and extended by Jaroslawska, Gathercole, Allen, & Holmes (2016), again demonstrating that encoding- and recall-based enactment effects may have a common source. Using a version of the Gathercole et al. (2008) paradigm (i.e. a reduced number of actions, and real objects), children aged 7–9 years were instructed to recall sequences of spoken instructions under presentation and recall conditions that either did or did not involve their physical performance. Memory accuracy was enhanced by carrying out the instructions as they were presented and when they were performed at recall. Crucially, the benefits of action-based recall were reduced following enactment during encoding, suggesting that the positive effects of action at encoding and recall may have a common origin. This basic finding has since been replicated in children, and younger and older adults (Jaroslawska et al., 2021; Lui et al., 2018; Waterman et al., 2017). Jaroslawska, Gathercole, Allen, and Holmes (2016) proposed that when performing physical actions during encoding, or planning for action recall, participants may actively construct action plans that incorporate spatial–motoric information and representations of intended movements (Koriat et al., 1990; Wolpert & Ghahramani, 2000), which are held in a specialized motor store in working memory (see also Jaroslawska et al., 2018; Smyth & Pendleton, 1989, 1990). By this account, planning to execute action sequences at recall did not enhance children's performance in the enactment condition because the action sequence was already represented in the hypothesized motor store.

Demonstration

Many procedural tasks, such as learning how to use computer software, knit a sweater, or bake a pie, are increasingly learned through video demonstrations uploaded to platforms such as YouTube (Torrey et al., 2009). Experimental explorations of the impact of visual demonstration (sometimes referred to as observation, or experimenter performed task) on long-term memory performance have indeed consistently shown it to facilitate performance (e.g. Cohen, 1981; Engelkamp & Dehn, 2000; Schult et al., 2014; Steffens, 2007), often to the same extent as self-enactment (see Steffens et al., 2015).

Different forms of demonstration (including video demonstration) have also recently been explored in the context of working memory. In a series of three experiments, Waterman et al. (2017) investigated how planning for action, self-enactment, and researcher demonstration influenced 6–10-year-old's abilities to follow instructions (Figure 10.1). They found that the benefits of self-enactment at encoding were contingent on sequence complexity as determined by the sequence length and number of possible actions and objects in the experimental set. More specifically, self-enactment benefits were only evident when sequences were shorter, with less variation in the possible actions or objects involved. Demonstration, on the other hand, enhanced recall across the board, regardless of the level of complexity. Yang et al. (2015) extended this result to the use of video demonstration (see also Yang et al., 2019), finding that this facilitated adults' recall performance, compared to auditory presentation of verbal instruction sequences. This was further observed both in children with ADHD and in age-matched typical controls (Yang et al., 2017). Finally, Allen et al. (2020) again showed that demonstration via video led to substantially better verbal recall, compared to spoken presentation, while self-enactment only had relatively small effects. They also found that performance improvements were limited to what was demonstrated; video clips showing each action–object pair facilitated pair recall, whereas videos only showing action or object produced recall improvements for those features in isolation. Taken together, these findings clearly illustrate that visual demonstration, either in isolation or alongside verbal instruction, can benefit performance in working memory.

Of interest in the context of the findings mentioned above is the apparent difference in efficacy between enactment during encoding and demonstration during encoding (Allen et al., 2020; Coats et al., 2021; Waterman et al., 2017; but see Lui et al., 2018 for a different finding). Waterman et al. (2017) suggested that this disparity might reflect the increased cognitive cost associated with self-enactment compared to demonstration. By this account, self-enactment is a more resource-demanding means of acquiring information compared to passive observation. Any benefits from self-enactment are, therefore, offset by the competing executive costs associated with self-generating the visuospatial–motoric information—a cost that does not emerge to the same extent when the participant is simply observing the actions. Similar claims have been made in the context of novel activity learning. A study by von Stülpnagel et al. (2016) found that self-enactment of instruction steps in two different learning tasks (namely, paper frog origami and knot tying) led to higher concurrent task costs, compared to an observation encoding condition. This objective evidence emerged despite the participants reporting lower subjective costs of the enactment condition.

In line with this interpretation, it has been shown that although incorporating physical engagement within instructions may have the potential to accelerate performance among young adults, this may be less useful for older adults who demonstrate performance decrements on tasks requiring the concurrent storage and processing of information (e.g. Bier et al., 2017; Forsberg et al., 2020; Holtzer et al., 2004; Jaroslawska & Rhodes, 2019; Rhodes et al., 2019). For example, Coats et al. (2021) examined the effects of self-enactment and demonstration at encoding on verbal recall of instructions in older and younger adults and found that relative to the spoken-only condition, demonstration significantly improved younger and older adults' serial recall performance, but self-enactment only enhanced performance in the younger group, and this boost was smaller than the one gained through demonstration. Thus, older adults gained from observing demonstration of instructions rather than enacting these instructions themselves during encoding, though it is worth noting that the demonstration gains were also smaller in the older adult group, relative to younger adults. In a similar study, Jaroslawska et al. (2021) instructed younger and older adults to recall spoken action commands under presentation and recall conditions that either did or did not involve their physical performance. When action was involved at encoding or recall, the difference in performance between the two age groups became more distinct. Enactment-based encoding significantly improved younger but not older adults' ability to follow spoken instructions, and

though the older group exhibited an enacted recall advantage, this was relatively reduced compared to that seen in young adults. Extending these findings, Yang, Su,
et al. (2022) has also recently reported demonstration benefits, and a somewhat
smaller enacted recall advantage, in older adults.

Taken together, such work indicates that individuals with smaller working
memory capacities (such as children and older adults) are more likely to benefit from
planning for action (i.e. in anticipation of action-based recall), rather than from action per se. Consequently, the use of demonstration or motor imagery rather than
actual physical movement, may be more effective at improving memory performance in these populations.

Finally, it is worth considering whether demonstration during encoding mirrors
enactment during encoding in showing larger impacts on verbal relative to enacted
recall. Unlike enacted encoding, the evidence here is rather more mixed. The first
study to examine demonstration effects (Yang et al., 2015) found no clear evidence
for such a pattern in young adults, though a cross-experiment analysis tentatively
indicated that the enacted recall effect was only apparent with spoken and not demonstrated presentation. Subsequent to this, evidence for an interaction between
presentation and recall modality was found by Yang et al. (2017) in children, Lui
et al. (2018) in adults, and Yang, Su, et al. (2022) with younger and older adults, but
not in the young adult study reported by Yang et al. (2019). Finally, Waterman et al.
(2017) found that demonstration facilitated verbal and enacted recall to the same
extent. This mixed picture may reflect in part the considerable variation in methodology and population samples across these studies; for example, in contrast to the
Waterman et al. study, the work by Yang and colleagues typically uses real objects
and video demonstration that is often instead of (rather than in addition to) spoken
presentation (Figure 10.1). However, it might also suggest a reduced mechanistic
overlap with action planning, relative to the common basis for enacted and planned
enactment discussed earlier. It will be of value to explore the commonalities and differences between self- and other-performance at the encoding phase, and at recall.

Imagination

Encoding-based manipulations have so far focused primarily on actual observable
activity, either in the form of disruptive dual-tasking, or the (typically) beneficial effects of self-enactment or observed demonstration. As it is assumed that the enacted
recall effect at least partly reflects the generation of an imagined action plan during
encoding, it is of interest to attempt to harness this directly by explicitly asking participants to imagine performing the instructions during presentation. Previously in
the long-term memory literature, researchers have investigated the benefit of imagining actions during encoding in healthy adults and in patients with obsessive–
compulsive disorders (Ecker & Engelkamp, 1995). Participants remembered lists of

action phrases under four conditions—motor encoding condition (physically act out with imaginary objects), motor-imaginal condition (imagine yourself doing it), visual imaginal condition (imagine someone else doing it), and subvocal rehearsal condition—and then free recall the lists by writing down the action phrases. Memory performance was most accurate in the motor encoding condition, followed by the visual imagery and motor imagery conditions, with accuracy lowest following subvocal rehearsal. This finding indicates that both enacted and imagined actions during encoding can facilitate long-term memory, and the benefit of actual enactment (albeit with imagined objects) seems larger than imaging actions.

Extending from long-term memory, recent studies examined have the potential benefit of imagination on following instructions in working memory (Waterman & Allen, 2019; Yang et al., 2021). In the study by Yang et al. (2021), typically developing children aged 7–12 years encoded spoken instructions using either a motor imagery strategy (i.e. imagine performing the actions) or a verbal rehearsal strategy, followed by either verbal or enacted recall. Recall performance was superior when children imagined the actions compared to verbal rehearsal. Moreover, this motor imagery benefit was similar in size for verbal and enacted recall. There was also a trend for a larger motor imagery benefit in older than younger children. These findings provide direct evidence for the superiority in forming an action-based representation via motor imagery when children are encoding spoken instructions, compared to generation of a verbal-based representation via rehearsal.

Following the same design as Yang et al. (2021), the potential benefit of motor imagery strategy over verbal rehearsal strategy was further examined in young adults (Yang, Allen, et al., 2022). The patterns in young adults mirrored the findings in children, with superior recall performance after motor imagery compared with a verbal rehearsal strategy, for both verbal and enacted recall. Comparing the effect sizes of the motor imagery benefit in the two studies indicates a somewhat larger motor imagery benefit in young adults ($h_p^2 = 0.43$) than children ($h_p^2 = 0.29$); it would be useful for future work to systematically map out how imagery effects emerge across the lifespan. Another study by Waterman and Allen (2019) also investigated the benefit of motor imagery in following spoken instructions in young adults, but they compared the motor imagery condition to a control condition where participants were instructed not to engage in active imagination at encoding. Again, a benefit of motor imagery emerged (relative to the no motor imagery condition), which was similar in size for verbal and enacted recall, replicating the findings above.

These findings emphasize the important role of motor imagery in helping children and adults to remember spoken instructions. One pattern to emerge from each of these initial explorations is the apparent additive effects of instructed motor imagery and enacted recall. We assume that the instruction to use motor imagery encourages the participant to generate visuospatial information during encoding. It is therefore intriguing that motor imagery appears to benefit verbal and enacted recall to the same extent, as this does not mirror the interactive patterns that emerge with actual enactment during encoding and recall, or (at least in some studies) between demonstration and enacted recall. Explicit instruction to imagine during encoding appears to provide benefits that go beyond those that are derived from planning enactment at recall. Alternatively, directing participants to use either a verbal rehearsal strategy or not to imagine the actions may represent a negative manipulation by instead reducing the opportunity for motoric coding and forcing a more verbal-based representation. While promising, research on how imagination influences working memory for instructions is in its early stages, with more work required to dissect the mechanisms involved and explore their possible practical utility.

Following instructions in atypical populations

The ability to follow instructions is essential for atypical and clinical populations, particularly when remembering medical practitioners' guidance. While researchers have investigated the deficits of populations using more traditional working memory tasks (Forbes et al., 2009; Huntley & Howard, 2010; Kasper et al., 2012; Kercood et al., 2014), less is known about action-based working memory (including following instructions). Here, we summarize findings of four relatively small-scale studies that have examined the ability to follow instructions in people with neurodevelopmental disorders or mental illness (Charlesworth et al., 2014; Lui et al., 2018; Wojcik et al., 2011; Yang et al., 2017).

Wojcik et al. (2011) found that children with autism performed worse in following spoken instructions than typically developing children. This difference between the two groups was limited to the baseline condition (only listening to the instructions) and the self-enactment condition (performing actions in addition to listening to the instructions), but not in the condition involving additional demonstrations of actions by an experimenter (although performance levels were very high in this study and ceiling effects should be considered). Moreover, children with autism showed better performance in the self-enactment and demonstration conditions than in the baseline conditions. These findings suggest that while children with autism benefited from both demonstration and self-enactment, viewing others' actions during encoding can help them reach a similar level of performance as typically developing children.

Also in a developmental context, Yang et al. (2017) investigated the ability of children with ADHD to follow spoken or silently demonstrated instructions. Compared with typically developing children, children with ADHD had worse performance in all conditions. However, both groups showed similar action-based advantages, including the enacted-recall advantage and a boost to verbal recall of instructions from demonstration at encoding. Thus, action-based manipulations at the encoding or retrieval stage can enhance instruction-following ability in children with ADHD, though it should be noted that they still failed to reach the level of typically developing children.

Turning to an adult population, Lui et al. (2018) examined the ability to follow spoken instructions in patients with schizophrenia. They found that while patients performed worse in following spoken instructions, their performance was improved by self-enactment, or by observing demonstration during encoding, to a similar extent as controls. This improvement was larger with self-enactment than with observation, similar to the findings in the long-term memory literature (Engelkamp, 2001). Patients also showed an enacted-recall advantage that was similar to controls, reflecting their intact ability to plan for actions during encoding.

Finally, Charlesworth et al. (2014) examined whether self-enactment during encoding can improve ability to follow instructions in older adults with mild Alzheimer's disease. While patients with mild Alzheimer's disease were worse in recalling instructions than older adult controls, their performance was improved by self-enacting the actions during encoding to a similar extent as controls. However, it should be noted that, unlike most other research on following instructions, Charlesworth et al. (2014) used a free recall scoring method. Whether the same enactment benefit would exist for serial recall of spoken instructions in patients with mild Alzheimer's disease remains to be investigated, and indeed, there is some evidence from the long-term memory literature that enactment during encoding benefits memory for individual objects and actions, but not relational or order memory (e.g. Engelkamp & Dehn, 2000).

In summary, the ability to follow instructions has to date been examined in four clinical populations (autism, ADHD, mild Alzheimer's disease, and schizophrenia), with two broad and consistent findings emerging. First, each of these groups was impaired in following spoken instructions compared with controls, in line with the working memory deficits commonly seen in such groups (e.g. Belleville et al., 2007; Bennetto et al., 1996; Forbes et al., 2009; Martinussen et al., 2005). Second, action-based manipulations at the encoding stage (in autism, ADHD, Alzheimer's disease, and schizophrenia) or the retrieval stage (in ADHD and schizophrenia) can improve memory of instructions in these groups. These findings are intriguing in showing that despite general low working memory, and poorer performance in following instructions, clinical populations are still able to benefit from the types of action-based advantage also exhibited by typical population samples. This might suggest that such effects may be relatively automatic or non-strategic and arise without substantially taxing the limited working memory resources in such groups. However, the relatively small sample sizes and varying methodology used across these studies mean that further work is required to establish the extent to which such action-based manipulations might offer practical ways of supporting atypical working memory functioning in the real world.

Applying working memory theory

This area is interesting but challenging to interpret from a theoretical perspective because the underlying processes have the potential to be complex and multidimensional in nature, possibly drawing in verbal, semantic, visuospatial, and motoric information (as a non-exclusive list). What is more, the possible types of representation and process that are involved in memory for instruction tasks are likely to depend to a large extent on the nature of the encoding context and response task. We assume that the starting 'baseline' mode of representation changes with input mode. Thus, spoken presentation of instructions will result in a baseline phonological representation, as will presentation via written text provided verbal recoding is possible, while visual demonstration generates a visuospatial representation. Semantic representations are likely to be activated in each case given that actions and objects from within instructions sequences are typically meaningful, even if the links between these are often arbitrary by design in the experimental paradigms described here. It is also possible that motor coding is activated automatically when hearing action-related language (Pulvermüller, 2005) or viewing actions performed by someone else (Rizzolatti & Craighero, 2004), though we do not have clear evidence for how this may contribute to instruction following in working memory at present. These 'baseline' representations can then be supplemented with resulting performance improvements, in a range of ways. This might be through engagement activity during encoding, for example, if the participant enacts each instruction component or imagines enacting them. This will activate additional forms of coding, likely motoric and visuospatial. Anticipation of performance (i.e. planning for enacted recall) will also increase the likelihood of engaging with environmental cues and activating motoric representations. Each of these is much more likely to meaningfully facilitate performance when it adds something extra to the representational profile, while redundancy only appears to bring limited additional benefits.

It is useful to consider these insights in the context of the leading frameworks of working memory theory, while also acknowledging that more carefully specified models capturing the details of following instructions across different task contexts will be required to develop a more precise understanding. This follows the process described by Baddeley (2012) of using broad frameworks such as the multicomponent model (Baddeley, 2012; Baddeley et al., 2021) to identify and conceptualize problems at a broad level, before attempting to drill down into specific components and tasks in a more detailed way, as achieved, for example, in empirical and computational work on the phonological loop (e.g. Burgess & Hitch, 1999; Hurlstone et al., 2014; Page & Norris, 1998). Theoretical accounts need to incorporate the varying representational codes that may be involved in following-instructions tasks, along with a way of pulling together these disparate forms of representation into a unified or at least connected event/object file. Most current theories do offer such a framework, at least in broad terms, though they differ in how this is described. It is important to note at this point that work to date has typically assumed a multicomponent perspective for the purposes of conceptualization, design, and interpretation. For example, dual-task approaches (e.g. Yang et al., 2014, 2016) have used tasks designed to load on to the subcomponents of the Baddeley and Hitch (1974; Baddeley, 1986) tripartite framework. This reflects a strength of this framework as developed by Alan

Baddeley, Graham Hitch, and colleagues in generating tractable questions that can lead to expanded scope and new findings (Baddeley et al., 2019, 2021). However, the research in the area has typically not been explicitly developed with the aim of contrasting one model against another, and so it is not straightforward at least with the present evidence base to do so. Instead, it is useful to note the extent to which different theoretical views might engage with the evidence to date, and possibly guide future exploration.

Starting at a broad level, Laird et al. (2017) synthesized different cognitive architectures (ACT-R, Sigma, and SOAR) designed to model 'the humanlike mind', proposing a 'standard model' constituting declarative and procedural long-term memory, and working memory, with the latter connected directly to perception and motor control modules. Laird et al. suggest that these modules may connect to perceptual and motor buffers operating within working memory. This is useful in placing working memory within a broader cognitive context and identifying links with perception and action. How these connections might be characterized is increasingly a focus of debate for working memory theory.

The influential multicomponent framework of Baddeley and colleagues (Baddeley, 1986, 2012; Baddeley et al., 2011, 2021) incorporates a set of central executive control resources and specialized subcomponents for phonological and visuospatial information, along with a way of bringing together these different types of information in combination with knowledge drawn from long-term memory (i.e. the episodic buffer). This approach would be successful in capturing, at least at a broad level, the encoding and storage of spoken or demonstrated instructions (via the phonological loop and visuospatial sketchpad), and the amalgamation of different sources of information and components of the instruction sequence within the episodic buffer. However, this framework, as currently described, does not explicitly incorporate a role for motor coding and action planning. This may be represented within working memory via a distinct, specialized motor buffer, as suggested by Jaroslawska et al. (2016, 2018; see also Smyth & Pendleton, 1989, 1990; Wood, 2007). Such a component could be added to the Baddeley framework, albeit with a possible cost to model parsimony. Alternatively, Li et al. (2022) have speculated that motor coding could be attributed to the visuospatial component of the framework (see also Allen & Waterman, 2015), in line with a more complex conceptualization of the existing working memory subsystems as ways of receiving and compressing different dimensions of information. These would then be made available to conscious awareness and focused attention via the episodic buffer. This view is broadly illustrated in Figure 10.3, which adapts and combines the different depictions of the model as set out in Baddeley et al. (2021) and adds motor coding to the visuospatial component.

Representation of space and movement has already been incorporated into multicomponent approaches. Logie (1995) described an 'inner scribe' that operates alongside a visual cache, analogous to the storage and articulatory components of the phonological loop. In a review of the literature on (predominantly long-term) action

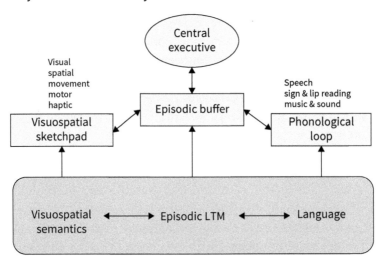

Figure 10.3 A multicomponent model of working memory, illustrating how certain forms of information might be incorporated into the system, and links to long-term memory. LTM, long-term memory (adapted from Baddeley et al., 2021).

memory and possible links with working memory, Logie et al. (2001) suggested a role for the inner scribe as a spatial, movement-based system in encoding motor or kinaesthetic information associated with enactment. Logie et al. also emphasized how a range of different codes (verbal and visual, but also semantic, motoric, kinaesthetic, or tactile, among others) might potentially be recruited into working memory depending on the nature of the task in hand. Building on this, Logie and colleagues (e.g. Logie, 2011, 2018; Logie, Belletier, & Doherty, 2021) outline an alternative way of characterizing a multicomponent framework, with a range of domain-specific systems operating to deal with input from the environment and long-term memory. This 'cognitive toolbox' combines in different ways for different tasks, feeding into what emerges as working memory capacity and eventual task performance, while being consciously monitored within a focus of awareness. This approach allows flexibility in leaving open the nature of the domain-specific components, meaning that phonological, visuospatial, and motoric information could each be incorporated into the framework. Logie's approach is also pertinent to the research on working memory for instruction and action in emphasizing how the contribution of different subcomponents might change depending on the individual's strategic approach and the task demands they encounter (e.g. materials, encoding context, and response requirements). Finally, another leading model of working memory, the time-based resource-sharing framework (Barrouillet et al., 2004; Barrouillet & Camos, 2015, 2021) was originally focused on capturing the relationship between storage and processing but has more recently incorporated this into a multicomponent architecture. This includes the working memory representation itself, characterized as a mental model analogous to Baddeley's episodic buffer, along with peripheral input buffers for phonological, visuospatial, and output buffers for speech and motor information (with additional buffers also accepted as being likely).

Embedded processes approaches are often considered as an alternative to multicomponent perspectives. For example, Oberauer (2009, 2021; Oberauer et al., 2013) describes a focus of attention that holds one (or possibly more than one) object and their spatial or temporal position, within a region of direct access emerging from activated long-term memory. This theoretical development also incorporates both item representations and the action representations associated with them, as part of procedural working memory. Oberauer suggests that the critical procedural component holds a small set of condition–action bindings that make up the task set, with a one-item response focus (analogous to the focus of attention) selecting the next action to be implemented. These ideas are useful in tackling the key question of how representations of stimuli are connected to, and converted into, action responses. However, the procedural aspect of this approach is perhaps less directly relevant for the examination of instruction memory that has been carried out to date (see earlier sections). In this work, there is typically no dynamic switching of task sets within condition blocks, and the constituent actions are presented as to-be-remembered items within instruction sequences. As such, they would be categorized as part of declarative, rather than procedural, working memory within Oberauer's approach.

The embedded processes approach set out by Cowan and colleagues (Cowan, 1988, 1999; Cowan et al., 2021; see Cowan, Chapter 9, this volume) characterizes working memory as activated long-term memory, with a limit of around three or four chunks of information held in the focus of attention, along with a specialized phonological storage/rehearsal component. Cowan et al. note that multiple different forms of representation may be activated in a working memory task, presumably including visual, spatial, and motoric information. If this information is useful and goal relevant, it might then be held within working memory as part of the modality-general focus of attention. Again, this would provide a broad way of understanding how enactment and action planning might feed into working memory, with the focus of attention capable of temporarily holding information drawn from multiple domains. While there are several similarities between such an approach and multicomponent frameworks, embedded processes models have also been linked (e.g. Morey et al., 2019) to dynamic sensorimotor approaches that view the functions of working memory as emergent properties reflecting links between perception and action (Buchsbaum & D'Esposito, 2019; Jones & Macken, 2018; Postle, 2021). For example, Jones and Macken (2018) describe an object-oriented action system in which perceptual-motor processes are opportunistically co-opted for the purposes of producing motor responses following sensory input in short-term memory tasks. Such accounts may offer routes to capturing how we encode sequences of instructions or actions, and plan for enacted recall, by focusing on the systems responsible for perception and action, although Li et al. (2022) have questioned the capability of such an approach to explain key findings from the following-instructions literature such as the enacted recall advantage.

More broadly, the idea of contributions from perception and action feeding into working memory is in fact not controversial. Logie, Belletier, and Doherty (2021) note that there is no major contemporary model of typical working memory functioning that sees this as operating independently and separately from the processes supporting perception and action. Indeed, multicomponent perspectives generally assume that systems responsible for perception and action can be co-opted in the service of working memory performance (Li et al., 2022). However, we would ultimately agree with the argument made by Logie, Belletier, and Doherty (2021) that to a certain extent it does not matter if a memory buffer can also be described as a sensory or motor buffer; this primarily represents a difference in labelling. If we are interested in temporary representations, then we call these 'working memory'.

It is apparent that a range of theoretical approaches can potentially contribute to our understanding of how working memory might incorporate complex multidimensional coding that responds dynamically to a strategic approach and task context, as is emerging in the following-instructions literature. In each case, appropriate consideration of this literature is likely to require adaption or extension to the assumptions and structures built into these models. The frameworks that are plausible, intuitive, generalizable, and capable of generating testable questions, while offering both structure and flexibility, are likely to be successful in making this step.

Taking it into the real world

The chapter so far has discussed experimental work investigating performance on following-instructions tasks including a summary of work from within the long-term memory paradigm as well as descriptions of the studies that have explored this from a working memory perspective. This has included how key manipulations have affected the ability to remember instructions, both at the encoding and recall stage, and how this might differ across different populations and across different tasks. Considerations around how this relates back to different models and theories suggest that no single framework fully captures all the complexities within the following-instructions paradigm, but that each model contributes useful insights into the processes underlying this ability. But how might we apply these findings in the real world? What recommendations could we make to practitioners, from across a range of contexts including education, medicine, and social care?

Most of the studies that have been conducted within this field have been laboratory based (e.g. Allen et al., 2020; Allen & Waterman, 2015; Waterman et al., 2017). Some of those studies have used real-world objects coupled with more familiar actions (e.g. 'pick up the yellow folder'), but they have still been run under relatively controlled conditions (Gathercole et al., 2008; Jaroslawska, Gathercole, Allen, &

Holmes, 2016b; Yang et al., 2019). This raises two issues. First, will the key findings to emerge from the research to date be replicated within a more ecologically valid setting and can we eventually achieve successful implementation science (Soicher et al., 2020)? In a typical classroom, instructions are delivered within an environment that contains several distractions (e.g. display boards, other pupils, additional items on desks). Further, real-world instructions often involve multistep sequences that build towards a desired end state. This includes the need to link together procedural steps, and build a coherent mental model involving subgoals, as well as the overarching end goal (Irrazabal et al., 2016). To date, following-instructions research has tended to adopt the more traditional working memory approach of studying arbitrarily constructed, relatively unconnected multi-item sequences. Second, even if the findings are replicated, some of them may not easily translate across to real-world settings. Again, if we consider a typical classroom, it may not be possible, or even desirable, for pupils to self-enact instructions as they are being read out.

Therefore, any recommendations regarding how to apply these findings to real-world contexts need to be made with these potential limitations in mind. Nevertheless, there are some clear themes to emerge from the literature that might be useful for practitioners to know. The most replicated finding from across different studies and different laboratories is that enactment at recall boosts performance compared to verbal recall (Allen & Waterman, 2015; Coats et al., 2021; Engelkamp, 1997; Gathercole et al., 2008; Hill et al., 2021; Jaroslawska et al., 2018; Koriat et al., 1990; Yang et al., 2017). When considering the education setting, there may be certain subjects, such as music, art, design and technology, physical education, or practical classes within the sciences, where instructions could be suitably reframed to require enacted responses to engage learning, rather than verbal or written responses. This general principle might also be useful within the medical setting. For example, if a patient needs to take home some equipment to monitor or support their recovery, while the medical professional is providing instructions on how to use a piece of equipment, rather than the individual taking notes or simply trying to hold the instructions in mind, they might be encouraged to listen to the instructions and then physically act out those instructions straight afterwards. Therefore, using action at recall may provide a relatively easy way to improve recall of instructions within working memory, which then would facilitate effective transfer into long-term memory.

The role of action at encoding is more complex. Self-enactment appears to boost performance for young adults (Allen & Waterman, 2015; Cohen, 1981; Engelkamp, 1998; Jaroslawska et al., 2018), but does not always provide such a boost either for children (Waterman et al., 2017) or for older adults (Coats et al., 2021). In contrast, demonstration at encoding has been shown to provide a more robust boost to recall across different experiments and different populations (Allen et al., 2020; Coats et al., 2021; Waterman et al., 2017; Yang et al., 2015, 2017; Yang, Su, et al., 2022). Using demonstration to aid recall in real-world contexts is something that could be applied across different tasks and situations. When considering the earlier example relating to medical equipment, patients might also benefit from watching the medical professional demonstrate the use of the equipment in tandem with the spoken instructions. Some studies would also suggest that watching a demonstration during encoding, and then being asked physically to enact those instructions immediately afterward, might provide a 'double boost' (Waterman et al., 2017), although more research is needed to establish the additivity or interactivity of such effects. When thinking about the education setting, it is clear to see how teachers could make use of demonstration while providing verbal instructions within certain subjects, for example, demonstrating the steps for a practical that students have then to carry out within a science lesson.

Beyond the educational realm, there are many situations where memory for, and understanding of, instructions are required. In such situations, incorporating action components such as demonstration may be much more effective than verbal description. A good example is giving pre-flight safety information on a plane by demonstration versus reading a set of instructions. Indeed, there is some evidence that tablet-based video demonstration of safety instructions may be more effective than card presentation of instructions, with further enhancements in understanding and performance added by requiring the individual to carry out interactive gestures that map on to the real-world actions (Chittaro, 2017). Another is the teaching of sports skills such as learning to ski where verbally describing the posture necessary for turning is likely to be much less effective than demonstrating it. The problem becomes acute, however, when direct demonstration is not possible, as anyone who has had to follow purely visual instructions to assemble an unfamiliar piece of domestic equipment or furniture is likely to have experienced.

Another, more recent, line of research within the following-instructions paradigm has been the exploration of strategies at encoding to support recall. The only known published study to date that has explicitly examined strategy use for encoding instructions with a working memory paradigm is by Yang et al. (2021). It was found that 7–12-year-olds showed improved recall in a following-instructions task following mental imagery (compared to verbal rehearsal) during encoding, a finding that is supported by unpublished data on the use of mental imagery with adults (Waterman & Allen, 2019). This supplements the data on demonstration at encoding and suggests that actively imagining yourself acting out instructions at encoding could provide a similar boost to watching someone demonstrate the actions at encoding.

All these recommendations would benefit from further studies that utilize a more ecologically valid paradigm, perhaps by taking the experiment into real-world settings such as the classroom or the hospital and working alongside practitioners. These types of experiments can be challenging; there are many confounds that occur within natural settings that are often outside the control of the researcher. One study by Jaroslawska, Gathercole, Logie, and Holmes (2016) used a virtual school environment to allow investigation of a following-instructions task within a situation that was more applicable to everyday instructions, while still enabling control over experimental variables. The researchers compared children's (aged 7–11 years) performance on three different types of following instruction tasks, a standard manual task, a task that took place within a single room within the virtual school, and a task that required navigating between different rooms within the virtual school building. They found that performance on the manual task correlated with performance on the two virtual tasks, and that children with lower verbal working memory ability performed less well on all the tasks. The authors argue that this demonstrates the potential usefulness of virtual environments to enable exploration of following instruction tasks within a more ecologically valid context, but with more control over extraneous variables.

When considering recommendations for application to real-world practice on following instructions, we can also use what we know about minimizing demands on working memory in general (Berry et al., 2019; Cowan, 2014; Gathercole & Alloway, 2008). If the aim is to improve people's ability to remember instructions, then we need to be mindful of the demands that instructions can place on a limited capacity system. If instruction sequences are too long, or too complex, then people may struggle to retain them in working memory. By reducing these demands, and therefore reducing the cognitive load, we create the opportunity for optimal engagement with the instruction sequence. Examples could include breaking down the sequence into a series of shorter sequences, chunking similar aspects of the instructions, repeating the instructions, minimizing distractions while delivering instructions, and potentially using visual or physical aids to allow cognitive offloading of certain elements of the sequence.

Conclusion

The question of how people remember and follow instructions has broad and far-reaching applications across many different real-world contexts. This has historically been a relatively underexplored area in the working memory literature, possibly reflecting the complexity involved in measuring the ability to follow instructions, the variability in likely mechanisms across different contexts, and the comparative lack of focus on movement, action, and multidomain processing in general. Research has started to explore this question, developing candidate paradigms and manipulations designed to enhance or disrupt performance, with resulting insights for both theory and practical application. This remains a relatively young area of research, however, and much more work is required to further our understanding along multiple lines of enquiry.

We have seen to varying extents how planning for enactment, imagined enactment, observed enactment, and actual self-enactment can all benefit recall, with some indication of representational overlap, but further work is needed to map this out in full. From a theoretical perspective, it will be important for models to incorporate the various ways in which action can enhance performance, but also disrupt performance when it is introduced as a dual-task manipulation. Recent work (including the research discussed in this chapter) has started to consider the ways in which multicomponent, embedded processes, and other models might capture encoding, maintenance, and implementation of verbal, visual, spatial, and motoric information. It should also be useful to explore the extent to which following instructions in working memory maps on to the established literature in the long-term memory domain. Do the same principles apply, and does this indicate overlapping, analogous, or distinct mechanisms and representations? One possibility is that working memory may operate during initial encoding to enhance long-term episodic and procedural memory by pulling together the different forms of coding that are elicited by a given task context (Logie et al., 2001).

From a methodological perspective, while there are useful methodological commonalities across studies, somewhat inevitably though given both the youth and the complexity of the research area, there is no 'gold standard' paradigm for examining following instructions in working memory, and it remains to be seen to what extent variation between methodologies across studies matters, and what this might tell us. Similarly, work is needed to identify the optimal ways to develop these paradigms in a way that maps on to more varied and realistic real-world challenges. Building on this, and perhaps most importantly, the aim should be for insights regarding following instructions that continue to be derived from cognitive psychology to eventually be transferred to the real world. This would continue the long tradition of productively linking theory, research, and application as exemplified by the work of Alan Baddeley.

References

Allen, R. J., Baddeley, A. D., & Hitch, G. J. (2006). Is the binding of visual features in working memory resource-demanding? *Journal of Experimental Psychology: General, 135*(2), 298–313.

Allen, R. J., Baddeley, A. D., & Hitch, G. J. (2014). Evidence for two attentional components in visual working memory. *Journal of Experimental Psychology: Learning, Memory, and Cognition, 40*(6), 1499–1509.

Allen, R. J., Hill, L. J., Eddy, L. H., & Waterman, A. H. (2020). Exploring the effects of demonstration and enactment in facilitating recall of instructions in working memory. *Memory & Cognition, 48*(3), 400–410.

Allen, R. J., Hitch, G. J., Mate, J., & Baddeley, A. D. (2012). Feature binding and attention in working memory: A resolution of previous contradictory findings. *Quarterly Journal of Experimental Psychology, 65*(12), 2369–2383.

Allen, R. J., & Waterman, A. H. (2015). How does enactment affect the ability to follow instructions in working memory? *Memory & Cognition, 43*(3), 555–561.

Alloway, T. P., Gathercole, S. E., & Pickering, S. J. (2006). Verbal and visuospatial short-term and working memory in children: Are they separable? *Child Development, 77*(6), 1698–1716.

Atkinson, A. L., Allen, R. J., & Waterman, A. H. (2021). Exploring the understanding and experience of working memory in teaching professionals: A large-sample questionnaire study. *Teaching and Teacher Education, 103*, 103343.

Baddeley, A. D. (1986). *Working memory.* Oxford University Press.

Baddeley, A. (2007). *Working memory, thought, and action.* Oxford University Press.

Baddeley, A. (2012). Working memory: Theories, models, and controversies. *Annual Review of Psychology, 63*, 1–29.

Baddeley, A. D., Allen, R. J., & Hitch, G. J. (2011). Binding in visual working memory: The role of the episodic buffer. *Neuropsychologia, 49*(6), 1393–1400.

Baddeley, A. D., & Hitch, G. (1974). Working memory. *Psychology of Learning and Motivation, 8*, 47–89.

Baddeley, A. D., Hitch, G. J., & Allen, R. J. (2019). From short-term store to multicomponent working memory: The role of the modal model. *Memory & Cognition, 47*(4), 575–588.

Baddeley, A. D., & Hitch, G, & Allen, R. (2021). A multicomponent model of working memory. In R. H. Logie, V. Camos, & N. Cowan (Eds.), *Working memory: State of the science* (pp. 10–43). Oxford University Press.

Barrouillet, P., Bernardin, S., & Camos, V. (2004). Time constraints and resource sharing in adults' working memory spans. *Journal of Experimental Psychology: General, 133*(1), 83–100.

Barrouillet, P., & Camos, V. (2015). *Working memory: Loss and reconstruction.* Psychology Press.

Barrouillet, P., & Camos, V. (2021). The time-based resource sharing model of working memory. In R. H. Logie, V. Camos, & N. Cowan (Eds.), *Working memory: State of the science* (pp. 85–115). Oxford University Press.

Belleville, S., Chertkow, H., & Gauthier, S. (2007). Working memory and control of attention in persons with Alzheimer's disease and mild cognitive impairment. *Neuropsychology, 21*(4), 458–469.

Bennetto, L., Pennington, B. F., & Rogers, S. J. (1996). Intact and impaired memory functions in autism. *Child Development, 67*(4), 1816–1835.

Berry, E. D., Allen, R. J., Mon-Williams, M., & Waterman, A. H. (2019). Cognitive offloading: Structuring the environment to improve children's working memory task performance. *Cognitive Science, 43*(8), e12770.

Bier, B., Lecavalier, N. C., Malenfant, D., Peretz, I., & Belleville, S. (2017). Effect of age on attentional control in dual-tasking. *Experimental Aging Research, 43*(2), 161–177.

Brener, R. (1940). An experimental investigation of memory span. *Journal of Experimental Psychology, 26*, 467–482.

Buchsbaum, B. R., & D'Esposito, M. (2019). A sensorimotor view of verbal working memory. *Cortex, 112*, 134–148.

Burgess, N., & Hitch, G. J. (1999). Memory for serial order: A network model of the phonological loop and its timing. *Psychological Review, 106*(3), 551–581.

Charlesworth, L. A., Allen, R. J., Morson, S., Burn, W. K., & Souchay, C. (2014). Working memory and the enactment effect in early Alzheimer's disease. *ISRN Neurology, 2014*, 694761.

Chittaro, L. (2017). A comparative study of aviation safety briefing media: Card, video, and video with interactive controls. *Transportation Research Part C: Emerging Technologies, 85*, 415–428.

Chun, M. M., Golomb, J. D., & Turk-Browne, N. B. (2011). A taxonomy of external and internal attention. *Annual Review of Psychology, 62*, 73–101.

Coats, R. O., Waterman, A. H., Ryder, F., Atkinson, A. L., & Allen, R. J. (2021). Following instructions in working memory: Do older adults show the enactment advantage? *Journals of Gerontology: Series B, 76*(4), 703–710.

Cohen, R. L. (1981). On the generality of some memory laws. *Scandinavian Journal of Psychology, 22*(1), 267–281.

Cohen, R. L. (1989). Memory for action events: The power of enactment. *Educational Psychology Review, 1*(1), 57–80.

Cowan, N. (1988). Evolving conceptions of memory storage, selective attention, and their mutual constraints within the human information-processing system. *Psychological Bulletin, 104*(2), 163–191.

Cowan, N. (1999). An embedded-processes model of working memory. In A. Miyake & P. Shah (Eds.), *Models of working memory: Mechanisms of active maintenance and executive control* (pp. 62–101). Cambridge University Press.

Cowan, N. (2014). Working memory underpins cognitive development, learning, and education. *Educational Psychology Review, 26*(2), 197–223.

Cowan, N. (2017). The many faces of working memory and short-term storage. *Psychonomic Bulletin & Review, 24*(4), 1158–1170.

Cowan, N., Morey, C. C., Naveh-Benjamin, M. (2021). An embedded-processes approach to working memory: How is it distinct from other approaches and to what ends? In R. H. Logie, V. Camos, & N. Cowan (Eds.), *Working memory: State of the science* (pp. 44–84). Oxford University Press.

De Renzi, E., & Faglioni, P. (1978). Normative data and screening power of a shortened version of the Token Test. *Cortex, 14*(1), 41–49.

De Renzi, E., & Vignolo, L. A. (1962). The Token Test: A sensitive test to detect receptive disturbances in aphasics. *Brain, 85*(4), 665–678.

Ecker, W., & Engelkamp, J. (1995). Memory for actions in obsessive-compulsive disorder. *Behavioural and Cognitive Psychotherapy, 23*(4), 349–371.

Engelkamp, J. (1997). Memory for to-be-performed tasks versus memory for performed tasks. *Memory & Cognition, 25*(1), 117–124.

Engelkamp, J. (1998). *Memory for actions.* Psychology Press/Taylor & Francis.

Engelkamp, J. (2001). Action memory: A system-oriented approach. In H. D. Zimmer, R. Cohen, M. Guynn, J. Engelkamp, R. Kormi-Nouri, & M. Foley (Eds.), *Memory for action: A distinct form of episodic memory* (pp. 49–96). Oxford University Press.

Engelkamp, J., & Dehn, D. M. (2000). Item and order information in subject-performed tasks and experimenter-performed tasks. *Journal of Experimental Psychology: Learning, Memory, and Cognition, 26*(3), 671–682.

Engelkamp, J., & Krumnacker, H. (1980). Image-and motor-processes in the retention of verbal materials. *Zeitschrift für Experimentelle und Angewandte Psychologie, 27*(4), 511–533.

Engelkamp, J., & Zimmer, H. D. (1989). Memory for action events: A new field of research. *Psychological Research, 51*(4), 153–157.

Engelkamp, J., & Zimmer, H. D. (1994). Motor similarity in subject-performed tasks. *Psychological Research, 57*(1), 47–53.

Engelkamp, J., Zimmer, H. D., Mohr, G., & Sellen, O. (1994). Memory of self-performed tasks: Self-performing during recognition. *Memory & Cognition, 22*(1), 34–39.

Engle, R. W., Carullo, J. J., & Collins, K. W. (1991). Individual differences in working memory for comprehension and following directions. *Journal of Educational Research, 84*(5), 253–262.

Forbes, N. F., Carrick, L. A., McIntosh, A. M., & Lawrie, S. M. (2009). Working memory in schizophrenia: A meta-analysis. *Psychological Medicine, 39*(6), 889–905.

Forsberg, A., Fellman, D., Laine, M., Johnson, W., & Logie, R. H. (2020). Strategy mediation in working memory training in younger and older adults. *Quarterly Journal of Experimental Psychology, 73*(8), 1206–1226.

Gathercole, S. E., & Alloway, T. P. (2008). *Working memory and learning: A practical guide for teachers.* Sage.

Gathercole, S. E., Brown, L., & Pickering, S. J. (2003). Working memory assessments at school entry as longitudinal predictors of National Curriculum attainment levels. *Educational and Child Psychology, 20*(3), 109–122.

Gathercole, S. E., Durling, E., Evans, M., Jeffcock, S., & Stone, S. (2008). Working memory abilities and children's performance in laboratory analogues of classroom activities. *Applied Cognitive Psychology, 22*(8), 1019–1037.

Gathercole, S. E., Lamont, E., & Alloway, T. P. (2006). Working memory in the classroom. In S. J. Pickering (Ed.), *Working memory and education* (pp. 219–240). Academic Press.

Gathercole, S. E., & Pickering, S. J. (2000). Working memory deficits in children with low achievements in the national curriculum at 7 years of age. *British Journal of Educational Psychology, 70*(2), 177–194.

Gathercole, S. E., Pickering, S. J., Ambridge, B., & Wearing, H. (2004). The structure of working memory from 4 to 15 years of age. *Developmental Psychology, 40*(2), 177–190.

Gathercole, S. E., Pickering, S. J., Knight, C., & Stegmann, Z. (2004). Working memory skills and educational attainment: Evidence from national curriculum assessments at 7 and 14 years of age. *Applied Cognitive Psychology, 18*(1), 1–16.

Goemaere, S., Beyers, W., De Muynck, G. J., & Vansteenkiste, M. (2018). The paradoxical effect of long instructions on negative affect and performance: When, for whom and why do they backfire? *Acta Astronautica*, *147*, 421–430.

Goschke, T., & Kuhl, J. (1993). Representation of intentions: Persisting activation in memory. *Journal of Experimental Psychology: Learning, Memory, and Cognition*, *19*(5), 1211–1226.

Haberlandt, K. (1997). *Cognitive psychology*. Allyn & Bacon.

Hill, L. J., Shire, K. A., Allen, R. J., Crossley, K., Wood, M. L., Mason, D., & Waterman, A. H. (2021). Large-scale assessment of 7–11-year-olds' cognitive and sensorimotor function within the Born in Bradford longitudinal birth cohort study. *Wellcome Open Research*, *6*, 53.

Hitch, G. J., Allen, R. J., & Baddeley, A. D. (2020). Attention and binding in visual working memory: Two forms of attention and two kinds of buffer storage. *Attention, Perception, & Psychophysics*, *82*(1), 280–293.

Holmes, J., & Adams, J. W. (2006). Working memory and children's mathematical skills: Implications for mathematical development and mathematics curricula. *Educational Psychology*, *26*(3), 339–366.

Holtzer, R., Stern, Y., & Rakitin, B. C. (2004). Age-related differences in executive control of working memory. *Memory & Cognition*, *32*(8), 1333–1345.

Huntley, J. D., & Howard, R. J. (2010). Working memory in early Alzheimer's disease: A neuropsychological review. *International Journal of Geriatric Psychiatry*, *25*(2), 121–132.

Hurlstone, M. J., Hitch, G. J., & Baddeley, A. D. (2014). Memory for serial order across domains: An overview of the literature and directions for future research. *Psychological Bulletin*, *140*(2), 339–373.

Irrazabal, N., Saux, G., & Burin, D. (2016). Procedural multimedia presentations: The effects of working memory and task complexity on instruction time and assembly accuracy. *Applied Cognitive Psychology*, *30*(6), 1052–1060.

Jahn, P., & Engelkamp, J. (2003). Design-effects in prospective and retrospective memory for actions. *Experimental Psychology*, *50*(1), 4–15.

Jaroslawska, A. J., Bartup, G., Forsberg, A., & Holmes, J. (2021). Age-related differences in adults' ability to follow spoken instructions. *Memory*, *29*(1), 117–128.

Jaroslawska, A. J., Gathercole, S. E., Allen, R. J., & Holmes, J. (2016). Following instructions from working memory: Why does action at encoding and recall help? *Memory & Cognition*, *44*(8), 1183–1191.

Jaroslawska, A. J., Gathercole, S. E., & Holmes, J. (2018). Following instructions in a dual-task paradigm: Evidence for a temporary motor store in working memory. *Quarterly Journal of Experimental Psychology*, *71*(11), 2439–2449.

Jaroslawska, A. J., Gathercole, S. E., Logie, M. R., & Holmes, J. (2016). Following instructions in a virtual school: Does working memory play a role? *Memory & Cognition*, *44*(4), 580–589.

Jaroslawska, A. J., & Rhodes, S. (2019). Adult age differences in the effects of processing on storage in working memory: A meta-analysis. *Psychology and Aging*, *34*(4), 512–530.

Jones, D. M., & Macken, B. (2018). In the beginning was the deed: Verbal short-term memory as object-oriented action. *Current Directions in Psychological Science*, *27*(5), 351–356.

Kaplan, C. H., & White, M. A. (1980). Children's direction-following behavior in grades K-S. *Journal of Educational Research*, *74*(1), 43–48.

Kasper, L. J., Alderson, R. M., & Hudec, K. L. (2012). Moderators of working memory deficits in children with attention-deficit/hyperactivity disorder (ADHD): A meta-analytic review. *Clinical Psychology Review*, *32*(7), 605–617.

Kercood, S., Grskovic, J. A., Banda, D., & Begeske, J. (2014). Working memory and autism: A review of literature. *Research in Autism Spectrum Disorders*, *8*(10), 1316–1332.

Koriat, A., Ben-Zur, H., & Nussbaum, A. (1990). Encoding information for future action: Memory for to-be-performed tasks versus memory for to-be-recalled tasks. *Memory & Cognition*, *18*(6), 568–578.

Koriat, A., & Pearlman-Avnion, S. (2003). Memory organization of action events and its relationship to memory performance. *Journal of Experimental Psychology: General*, *132*(3), 435–454.

Kormi-Nouri, R., Nyberg, L., & Nilsson, L. G. (1994). The effect of retrieval enactment on recall of subject-performed tasks and verbal tasks. *Memory & Cognition*, *22*(6), 723–728.

Krikalev, S. K., Kalery, A. Y., & Sorokin, I. V. (2010). Crew on the ISS: Creativity or determinism? *Acta Astronautica*, *66*(1–2), 70–73.

Laird, J. E., Lebiere, C., & Rosenbloom, P. S. (2017). A standard model of the mind: Toward a common computational framework across artificial intelligence, cognitive science, neuroscience, and robotics. *AI Magazine, 38*(4), 13–26.

Lesser, R. (1976). Verbal and non-verbal memory components in the Token Test. *Neuropsychologia, 14*(1), 79–85.

Li, G., Allen, R. J., Hitch, G. J., & Baddeley, A. D. (2022). Translating words into actions in working memory: The role of spatial-motoric coding. *Quarterly Journal of Experimental Psychology, 75*(10), 1959–1975.

Logie, R. H. (1995). *Visuo-spatial working memory*. Psychology Press.

Logie, R. H. (2011). The functional organization and capacity limits of working memory. *Current Directions in Psychological Science, 20*(4), 240–245.

Logie, R. (2018). Human cognition: Common principles and individual variation. *Journal of Applied Research in Memory and Cognition, 7*(4), 471–486.

Logie, R. H., Belletier, C., & Doherty, J. M. (2021). Integrating theories of working memory. In R. H. Logie, V. Camos, & N. Cowan (Eds.), *Working memory: State of the science* (pp. 389–429). Oxford University Press.

Logie, R. H., Camos, V., & Cowan, N. (2021). *Working memory: State of the science*. Oxford University Press.

Logie, R. H., Engelkamp, J., Dehn, D., & Rudkin, S. (2001). Actions, mental actions, and working memory. In M. Denis, R. H. Logie, C. Cornoldi, J. Engelkamp, & M. De Vega (Eds.), *Imagery, language and visuo-spatial thinking* (pp. 161–184). Psychology Press.

Lui, S. S., Yang, T. X., Ng, C. L., Wong, P. T., Wong, J. O., Ettinger, U., & Chan, R. C. (2018). Following instructions in patients with schizophrenia: The benefits of actions at encoding and recall. *Schizophrenia Bulletin, 44*(1), 137–146.

Makri, A., & Jarrold, C. (2021). Investigating the underlying mechanisms of the enactment effect: The role of action-object bindings in aiding immediate memory performance. *Quarterly Journal of Experimental Psychology, 74*(12), 2084–2096.

Martinussen, R., Hayden, J., Hogg-Johnson, S., & Tannock, R. (2005). A meta-analysis of working memory impairments in children with attention-deficit/hyperactivity disorder. *Journal of the American Academy of Child & Adolescent Psychiatry, 44*(4), 377–384.

Mashburn, C. A., Tsukahara, J. S., & Engle, R. W. (2021). Individual differences in attention control: Implications for the relationship between working memory capacity and fluid intelligence. In R. H. Logie, V. Camos, & N. Cowan (Eds.), *Working memory: State of the science* (pp. 175–211). Oxford University Press.

Marsh, R. L., Hicks, J. L., & Bink, M. L. (1998). Activation of completed, uncompleted, and partially completed intentions. *Journal of Experimental Psychology: Learning, Memory, and Cognition, 24*(2), 350–361.

Maylor, E. A., Chater, N., & Brown, G. D. (2001). Scale invariance in the retrieval of retrospective and prospective memories. *Psychonomic Bulletin & Review, 8*(1), 162–167.

Morey, C. C., Rhodes, S., & Cowan, N. (2019). Sensory-motor integration and brain lesions: Progress toward explaining domain-specific phenomena within domain-general working memory. *Cortex, 112*, 149–161.

Oberauer, K. (2009). Design for a working memory. *Psychology of Learning and Motivation, 51*, 45–100.

Oberauer, K. (2021). Towards a theory of working memory: From metaphors to mechanisms. In R. H. Logie, V. Camos, & N. Cowan (Eds.), *Working memory: State of the science* (pp. 116–149). Oxford University Press.

Oberauer, K., Souza, A. S., Druey, M. D., & Gade, M. (2013). Analogous mechanisms of selection and updating in declarative and procedural working memory: Experiments and a computational model. *Cognitive Psychology, 66*(2), 157–211.

Page, M., & Norris, D. (1998). The primacy model: A new model of immediate serial recall. *Psychological Review, 105*(4), 761–781.

Postle, B. R. (2021). Cognitive neuroscience of visual working memory. In R. H. Logie, V. Camos, & N. Cowan (Eds.), *Working memory: State of the science* (pp. 333–357). Oxford University Press.

Pulvermüller, F. (2005). Brain mechanisms linking language and action. *Nature Reviews Neuroscience*, 6(7), 576–582.

Raw, R. K., Wilkie, R. M., Allen, R. J., Warburton, M., Leonetti, M., Williams, J. H., & Mon-Williams, M. (2019). Skill acquisition as a function of age, hand and task difficulty: Interactions between cognition and action. *PloS One*, 14(2), e0211706.

Rhodes, S., Jaroslawska, A. J., Doherty, J. M., Belletier, C., Naveh-Benjamin, M., Cowan, N., Camos, V., Barrouillet, P., & Logie, R. H. (2019). Storage and processing in working memory: Assessing dual task performance and task prioritization across the adult lifespan. *Journal of Experimental Psychology: General*, 148(7), 1204–1227.

Rizzolatti, G., & Craighero, L. (2004). The mirror-neuron system. *Annual Reviews of Neuroscience*, 27, 169–192.

Rosenbaum, D. A. (2005). The Cinderella of psychology: The neglect of motor control in the science of mental life and behavior. *American Psychologist*, 60, 308–317.

Rosenbaum, D. A., & Feghhi, I. (2019). The time for action is at hand. *Attention, Perception, & Psychophysics*, 81(7), 2123–2138.

Saltz, E., & Donnenwerth-Nolan, S. (1981). Does motoric imagery facilitate memory for sentences? A selective interference test. *Journal of Verbal Learning and Verbal Behavior*, 20(3), 322–332.

Schult, J., von Stülpnagel, R., & Steffens, M. C. (2014). Enactment versus observation: Item-specific and relational processing in goal-directed action sequences (and lists of single actions). *PloS One*, 9(6), e99985.

Shekari, E. (2020). *Following spoken instructions in L1 and L2: The effects of language dominance, working memory, cognitive complexity, and language acquisition background* [Unpublished doctoral dissertation]. McMaster University.

Smyth, M. M., & Pendleton, L. R. (1989). Working memory for movements. *Quarterly Journal of Experimental Psychology*, 41(2), 235–250.

Smyth, M. M., & Pendleton, L. R. (1990). Space and movement in working memory. *Quarterly Journal of Experimental Psychology Section A*, 42(2), 291–304.

Soicher, R. N., Becker-Blease, K. A., & Bostwick, K. C. (2020). Adapting implementation science for higher education research: The systematic study of implementing evidence-based practices in college classrooms. *Cognitive Research: Principles and Implications*, 5(1), 1–15.

Steffens, M. C. (2007). Memory for goal-directed sequences of actions: Is doing better than seeing? *Psychonomic Bulletin & Review*, 14(6), 1194–1198.

Steffens, M. C., von Stülpnagel, R., & Schult, J. C. (2015). Memory recall after 'learning by doing' and 'learning by viewing': Boundary conditions of an enactment benefit. *Frontiers in Psychology*, 6, 1907.

Thorndike, E. L. (1912). The measurement of educational products. *Journal of Secondary Education*, 20(5), 289–299.

Tomasino, B., & Gremese, M. (2016). Effects of stimulus type and strategy on mental rotation network: An activation likelihood estimation meta-analysis. *Frontiers in Human Neuroscience*, 9, 693.

Torrey, C., Churchill, E. F., & McDonald, D. W. (2009, April). Learning how: The search for craft knowledge on the internet. In *Proceedings of the SIGCHI Conference on Human Factors in Computing Systems* (pp. 1371–1380). Association for Computing Machinery.

Vandierendonck, A. (2021). Multicomponent working memory system with distributed executive control. In R. H. Logie, V. Camos, & N. Cowan (Eds.), *Working memory: State of the science* (pp. 150–174). Oxford University Press.

von Stülpnagel, R., Schult, J. C., Richter, C., & Steffens, M. C. (2016). Cognitive costs of encoding novel natural activities: Can 'learning by doing' be distracting and deceptive? *Quarterly Journal of Experimental Psychology*, 69(8), 1545–1563.

Watanabe, H. (2003). Effects of encoding style, expectation of retrieval mode, and retrieval style on memory for action phrases. *Perceptual and Motor Skills*, 96(3), 707–727.

Waterman, A. H., & Allen, R. J. (2019, June). Following instructions: The effect of imagining on recall [Paper presentation]. Working Memory Meeting Conference, Yorkshire, UK.

Waterman, A. H., Atkinson, A. L., Aslam, S. S., Holmes, J., Jaroslawska, A. J., & Allen, R. J. (2017). Do actions speak louder than words? Examining children's ability to follow instructions. *Memory & Cognition*, 45(6), 877–890.

Wiklund-Hörnqvist, C., Jonsson, B., Korhonen, J., Eklöf, H., & Nyroos, M. (2016). Untangling the contribution of the subcomponents of working memory to mathematical proficiency as measured by the national tests: A study among Swedish third graders. *Frontiers in Psychology*, *7*, 1062.

Wojcik, D., Allen, R., Brown, C., & Souchay, C. (2011). Memory for actions in autism spectrum disorder. *Memory*, *19*(6), 549–558.

Wolpert, D. M., & Ghahramani, Z. (2000). Computational principles of movement neuroscience. *Nature Neuroscience*, *3*(11), 1212–1217.

Wood, J. N. (2007). Visual working memory for observed actions. *Journal of Experimental Psychology: General*, *136*(4), 639–652.

Yang, T. X., Allen, R. J., & Gathercole, S. E. (2016). Examining the role of working memory resources in following spoken instructions. *Journal of Cognitive Psychology*, *28*(2), 186–198.

Yang, T. X., Allen, R. J., Holmes, J., & Chan, R. C. (2017). Impaired memory for instructions in children with attention-deficit hyperactivity disorder is improved by action at presentation and recall. *Frontiers in Psychology*, *8*, 39.

Yang, T. X., Allen, R. J., Waterman, A. H., Jaroslawska, A. J., Su, X. M., & Gao, Y. (2022). *Exploring strategies for encoding spoken instructions: Comparing rehearsal, motor imagery, enactment and observation*. Manuscript in preparation.

Yang, T. X., Allen, R. J., Waterman, A. H., Zhang, S. Y., Su, X. M., & Chan, R. C. (2021). Comparing motor imagery and verbal rehearsal strategies in children's ability to follow spoken instructions. *Journal of Experimental Child Psychology*, *203*, 105033.

Yang, T. X., Allen, R. J., Yu, Q. J., & Chan, R. C. (2015). The influence of input and output modality on following instructions in working memory. *Scientific Reports*, *5*(1), 1–8.

Yang, T. X., Jia, L. X., Zheng, Q., Allen, R. J., & Ye, Z. (2019). Forward and backward recall of serial actions: Exploring the temporal dynamics of working memory for instruction. *Memory & Cognition*, *47*(2), 279–291.

Yang, T. X., Gathercole, S. E., & Allen, R. J. (2014). Benefit of enactment over oral repetition of verbal instruction does not require additional working memory during encoding. *Psychonomic Bulletin & Review*, *21*(1), 186–192.

Yang, T. X., Su, X., Allen, R. J., Ye, Z., & Jia, L. X. (2022). Improving older adults' ability to follow instructions: Benefits of actions at encoding and retrieval in working memory. *Memory*, *30*(5), 610–620.

Zimmer, H. D., Cohen, R. L., Foley, M. A., Guynn, M. J., Engelkamp, J., & Kormi-Nouri, R. (2001). *Memory for action: A distinct form of episodic memory?* Oxford University Press.

11

Parent–child autobiographical reminiscing as a foundation for literacy, memory, and science education

Robyn Fivush, Catherine A. Haden, and Elaine Reese

Research on autobiographical memory has blossomed since 1988, when Baddeley famously asked, 'But what the hell is it for?' Research from a cognitive perspective has focused on the accuracy, organization, and duration of personal memories (e.g. Conway & Pleydell-Pearce, 2000; Rubin 2006), and research from a personality perspective has focused on the ways in which personal memories, and how we express them through narrative, define identity, relationships, values, and meaning and purpose in life (Fivush, 2011; McAdams, 1992). In this chapter, we focus on yet another critical function of our autobiographical memories—as a foundation for the growth of cognitive skills. We demonstrate that coherent and expressive autobiographical narratives help children build critical skills in literacy, deliberate memory, and scientific understanding. More specifically, we present a sociocultural developmental model of how children are scaffolded into coherent and expressive narratives of their own personal experiences that is foundational for the development of more abstract cognitive skills (Fivush, 2019; Rogoff et al., 2018).

In the first section, we lay out our theoretical framework, defining autobiographical memory and explicating the sociocultural developmental model of autobiographical memory development. We then turn to the ways in which parents, and especially mothers, structure, or scaffold, reminiscing conversations with their preschool children, helping their children learn the socioculturally mediated forms for expressing their lived experience, and how these resulting individual and sociocultural forms of narrative reminiscing eventuate in individual differences in children's developing autobiographical memories. With this as a base, we discuss relations between early mother–child reminiscing styles and the development of literacy. We then turn to relations between reminiscing and deliberate memory skills, and, in the following section, we elucidate relations between reminiscing and understanding scientific concepts. Our argument throughout is that the ways in which parents help children understand and organize their personal experience is foundational to how children come to use their personal experiences to develop more abstract, academic knowledge. Thus, in the last section, we draw the threads across the domains of

Robyn Fivush, Catherine A. Haden, and Elaine Reese, *Parent–child autobiographical reminiscing as a foundation for literacy, memory, and science education* In: *Memory in Science for Society*. Edited by: Robert H. Logie, Zhisheng (Edward) Wen, Susan E. Gathercole, Nelson Cowan, and Randall W. Engle, Oxford University Press. © Oxford University Press 2023. DOI: 10.1093/oso/9780192849069.003.0011 literacy, deliberate memory, and scientific understanding as emerging from mother–child autobiographical reminiscing across childhood.

The sociocultural developmental model of autobiographical memory

Defining autobiographical memory

Autobiographical memories are consciously accessible accounts of lived experience. They are a type of explicit declarative memory and can be either episodic or semantic in form (Baddeley, 2013). In the most widely accepted model of autobiographical memory, Conway and Pleydall-Pearce (2000) lay out a hierarchical organization of autobiographical memories with the overall life narrative at the top of the hierarchy, and life periods (when I was in high school; when I lived in Minneapolis) and life themes (romance; career) at lower levels, which further organize more specific event experiences that include episodic and sensory detail. These various levels of autobiography interact; specific episodes, recurring events, and extended events are dynamically interrelated, and, importantly, are verbalized in a narrative form.

To be clear, it is not that autobiographical memories are represented as narratives in the mind/brain; we know that autobiographical memories are multimodal and multisensory (Rubin, 2006). Rather, narratives are the sociocultural tools that allow us to construct coherent accounts of what happened and, perhaps more importantly, what it means for the self, the essential quality of autobiography. Narratives carve the flow of experience into discrete units with beginnings, middles, and ends (Ricoeur, 1991) to allow individuals to organize distinct sequences and life periods that connect across time, creating a sense of continuity and coherence of self, intentions, and goals (Conway et al., 2004; Fivush, 2011; McAdams, 1992). Narratives move from recounting what happened to integrating internal reactions, thoughts, emotions, motivations, and intentions, in ways that create a meaningful human experience of self and other. Thus, when fully developed, the autobiographical memory system is a complicated tapestry of episodic, extended and recurring events and life periods, organized within themes, that tell the story of me (Fivush & Waters, 2019).

Developmental considerations

Nelson and Fivush (2004, 2020) argue that the development of such a complicated autobiographical system must depend on, and contribute to, multiple neurological, socioemotional, and cognitive achievements. Whereas it is now quite clear that virtually all other animals must have some form of episodic memory ability, that is, the ability to recall specific events bounded in time and space (see Roberts, 2002, for a review), what is uniquely human is the ability to weave these experiences into a coherent narrative of self, the story of how I became the person I am and the person I wish to become (Fivush, 2011; McAdams, 1992). This type of memory goes beyond simple episodic memory in at least three ways. First, it references self. It is not simply that something happened at some point in time and space but that this happened to *me*; thus autobiographical memory involves memory of self in the past. Second, this past self must be linked in some way to the current self who is remembering (Fivush, 2011; Nelson & Fivush, 2020). Thus, there must be some continuity of self/autobiographical consciousness across time. Third, there must be some awareness that my memory is unique to me. Even if you experienced the event with me, you will have different thoughts and emotions at the time and each of us will have different thoughts and emotions evolving over time, interweaving with different unique individual memories, weaving a different life story (Fivush & Nelson, 2006). This kind of complex cognitive meta-reflection on experience, memory, self, and other, emerges gradually across the preschool years, and continues to develop across middle childhood and adolescence (Fivush et al., 2011), and, indeed, throughout the lifespan (Bluck, 2003). Some of this meta-understanding is linked to theory of mind, the idea that each individual has desires, emotions, and beliefs that are unique to the self, and that these mental states emerge from direct and indirect experiences in the world, and evolve over time (Wellman, 2018).

Of critical importance, Nelson and Fivush (2004, 2020) argue that the developing understanding of the continuity and uniqueness of individual autobiographical consciousness over time, a core aspect of theory of mind, is only possible in the context of being able to share experiences of the past with others, and this is most easily accomplished through language. An example best illustrates this point, as in this co-constructed narrative between 8-year-old Rachel and her mother, reminiscing about a bike trip during a family camping vacation:

MOTHER: And we were all goin' on a bike and you didn't wanna go on a bike and so you were just going to jog but you got so tired—
REBECCA: NOT TIRED! [Very loud voice.]
MOTHER: [Laughing.] You didn't get tired. OK. You didn't get tired—
REBECCA: [Giggles.]
MOTHER: —but you wanted to sit on the bike seat I was peddling. What do you remember about that?
REBECCA: Wanting you to go really really slow. My legs were hurting.
MOTHER: [Laughs] Why were your legs hurting?
REBECCA: Cuz I was like this [spreads legs wide to show how she was riding on the mother's handlebars] all the time.
MOTHER: Cuz your legs were spread apart like that.
REBECCA: Yeah, but if you went slowly I could relax.
MOTHER: Uh huh.

REBECCA: And you went too fast.

MOTHER: But you had fun, though, didn't you?

REBECCA: It was great!

MOTHER: What was that, a half mile or something?

REBECCA: I was afraid I might, uh, you might go flying off the edge [both laughing], edge of the bridge and, umm, I just wanted to jog.

MOTHER: And you were afraid of riding on the bike with me across the bridge, huh?

REBECCA: Uh huh uh huh uh huh.

Several aspects of this conversation are noteworthy. First, the fact that Rebecca and her mother are quite clearly referring to the same past experience which exists only as a shared mental experience with no physical cues in the present environment already suggests the complexity of remembering as a shared activity. Humans do this virtually all the time with ease, 'Remember when?' indicating the role of language in creating a shared representation (Sutton et al., 2010). But, even though clearly focused on the same shared mental experience, both Rebecca and her mother inflect the conversation with their own perspectives, both agreeing on some points and disagreeing on others, understanding that each had a unique lens both as the event unfolded and as they recollect it. Parenthetically, both are highly engaged and enjoying savouring the shared past together, highlighting the emotional bonding aspect of reminiscing. Again, outside of language it is hard to imagine how this kind of complex interplay could be communicated. And it is through this complex interplay of reminiscing with others, agreeing, disagreeing, negotiating, and constructing shared representations of individual pasts that we each create our own sense of a unique autobiographical consciousness (Nelson & Fivush, 2020).

The sociocultural context

If autobiographical memory emerges from social interactions, then it is clear that we must examine the sociocultural contexts in which the past is recollected in order to understand the development, forms, and functions of autobiographical reminiscing. The cultural value of autobiographical memories is both universal and culturally variable (Nelson, 2003). Universally, humans understand lived experience through stories (Bruner, 1991) and all cultures, including our ancestral forbearers (Donald, 2001), communicate through stories. However, cultures also vary in the forms and functions of these personal narratives. In modern, industrialized, urban environments, being able to tell an autonomous, achievement-oriented narrative is related to a cultural focus on mobility and self-actualization, whereas in more rural, eastern, traditional cultures, narratives of community and relatedness are deemed more appropriate (Tweed & Lehman, 2002; Wang, 2016). Note that this very broad theoretical dichotomy is actually quite nuanced in reality; cultures vary along multiple dimensions, and within each culture, specific contexts call for different emphases (Mistry & Dutta, 2015). Thus, development becomes a complicated process of differentiating and practising skills that may be particular to specific contexts over others. In thinking about the forms and functions of autobiographical narratives, the question must always be asked: what narrative is being told to whom in what context for what purpose?

Early in development, during what is often called the preschool years in industrialized cultures, children engage in reminiscing conversations mostly within the family, and mostly with mothers (Fivush & Zaman, 2013), and much of this interaction focuses on helping young children to create coherent narratives of recently experienced events, such as going to the park earlier that afternoon or the supermarket yesterday. Mothers report engaging in these kinds of reminiscing conversations in order to help their children to learn about themselves and to construct emotional bonds through a shared history over time (Kulkofsky et al., 2009). As we will see, although this may be the explicit reason that mothers report for engaging in reminiscing, these early reminiscing conversations also lay the foundation for cognitive and academic skills that emerge as children enter more formal educational settings. Thus, both for the development of autobiography and for these emerging cognitive skills, it becomes important to examine how mothers structure these reminiscing conversations in ways that facilitate these various developmental outcomes. Further, these local mother–child reminiscing conversations must be understood within the larger cultural contexts in which they are embedded.

Maternal reminiscing style

Individual and cultural differences

Substantial research over the past three decades has confirmed that mothers in multiple cultures vary along a dimension of elaboration when reminiscing with their young children (see Fivush, 2019, and Fivush et al., 2006, for a review). When their children are quite young, from about 16 months to 3 years of age, highly elaborative mothers provide richly detailed questions about past experiences, adding new information with each new question, essentially providing a coherent narrative of what occurred through their scaffolding, as the child participates mainly through yes/no and one-word responses. Highly elaborative mothers also use more internal state language in their reminiscing, providing a causal structure linking internal states to external actions though motivations, volitions, and emotions. In this way, highly elaborative mothers provide both a model for narrating the past and communicate the positive values of sharing oneself through sharing the past with others. As children begin to engage in reminiscing to a greater extent, at ages 4 and 5, highly elaborative mothers continue to provide more detailed information, asking more open-ended questions and following in on children's responses, weaving them into an ongoing narrative. Initial research, mostly with middle-class mothers in the US, discovered that mothers were consistent across time, with highly elaborative mothers reminiscing in more elaborate ways across the preschool years, and consistently across siblings. Importantly, elaborative mothers are not simply more talkative; mothers who elaborate more during reminiscing are not necessarily more talkative during play or routine caregiving activities. These results, now confirmed in dozens of studies, indicate that reminiscing is a specific context in which highly elaborative mothers draw children into co-constructing coherent detailed narratives of their past experiences.

Whereas the dimension of elaboration is remarkably robust across cultures, there are also broad cultural differences both in how much mothers elaborate and on what information they elaborate. Drawing broad distinctions, Wang (2013, 2016) has shown that mothers in Western cultures that value high levels of autonomy and independence are more elaborative overall, and especially so about the child's individual emotions and motivations, than mothers in Asian cultures that value community and connection to a greater extent. Mothers from Asian cultures, in contrast, reminisce more about moral behaviours and social commitments than do mothers from Western cultures. Indigenous Māori mothers from New Zealand, who hail from a strong oral tradition, reminisce in greater detail about events of significance to the whole family, such as the birth of a new child, compared to European New Zealand mothers (Reese et al., 2008). Moreover, Māori mothers with a stronger affiliation to cultural traditions report reminiscing more frequently about a wide range of past events (Reese & Neha, 2015). Delving deeper into these distinctions, Keller and her colleagues (Tougu et al., 2012; Tulviste et al., 2016) have examined cultures that vary along multiple dimensions, including urban/rural, independent/interdependent, and relational/autonomous. Across the cultures studied, they also find that mothers vary along a dimension of elaboration in predictable ways, with more urban, independent, and autonomous-oriented cultures linked to more elaborative reminiscing than more rural, interdependent, relational cultures. Critically, as all these cultural theorists argue, reminiscing itself is not a monolithic conversational context. There is reliable variability in maternal elaborative reminiscing style both within and across cultures.

Developmental effects on autobiographical narratives

The sociocultural developmental model of autobiographical memory development asserts that children learn the forms and functions of autobiographical reminiscing by participating in scaffolded co-constructed narratives of their personal past. Indeed, correlational, longitudinal, and experimental research concurs that children of mothers who are highly elaborative early in development come to tell detailed and coherent personal narratives over time (see Fivush, 2019, and Fivush et al., 2006 for reviews). More specifically, in the most extensive longitudinal study examining the role of maternal reminiscing style from 19 months of age through adolescence, Reese and colleagues (Farrant & Reese, 2000; Reese, Jack, & White, 2010) confirmed previous research demonstrating that highly elaborative maternal reminiscing early in the preschool years predicted children's abilities to participate in both joint reminiscing and independent autobiographical narrative recall in more detailed and coherent ways by the end of the preschool years, and this study extended previous findings in that these effects were still apparent into adolescence. Adolescents whose mothers were highly elaborative had an earlier age of first memory, and provided more memories from early childhood than adolescents of less elaborative mothers (Reese & Robertson, 2019).

Whereas this kind of longitudinal research is compelling, it is, of course, still correlational. Several studies in which mothers were trained to be more elaborative have shown more direct causal effects. Mothers are both easily trained to be more elaborative during reminiscing, by instructing them to use more open-ended questions and by expanding on information provided by the child, and children of mothers who are trained to be more elaborative show more detailed and coherent autobiographical memories years later as compared to children of mothers in a control group (Boland et al., 2003; Peterson et al., 1999; Reese et al., 2020; Reese & Newcombe, 2007).

The overall patterns lead to several important conclusions. First, maternal reminiscing style is individually consistent and predicts important outcomes for children's autobiographical memory development. Second, maternal reminiscing style is a critical factor in this development as experimental studies show causal connections; and third, although maternal elaborative reminiscing is variable across cultures, the long-term effects of maternal reminiscing style seem to be quite consistent across cultures, with children of highly elaborative mothers developing elaborated and coherent independent autobiographical narratives. Whereas the bulk of research on maternal reminiscing style has focused on the longitudinal outcomes for children's autobiographical memory development and related socioemotional understanding of self and others, a smaller, but still compelling, body of research has examined the ways in which maternal reminiscing style sets the foundations for the emergence and development of more cognitive and academically oriented skills, including literacy, strategic memory, and scientific concepts. It is to these outcomes that we now turn.

Reminiscing style and literacy

Literacy is perhaps the most essential tool in modern industrialized cultures, opening children up to a world beyond direct experiences and personal conversations (Britto, 2012; Harris et al., 2018), as well as providing myriad professional opportunities. Research on early literacy and improving literacy skills abounds, and children's early literacy in the preschool years comprises a complex set of skills, including vocabulary knowledge, phonological awareness (of speech sounds), story comprehension and production, and print skills (recognizing and writing letters and words) (Shanahan & Lonigan, 2010). All of these abilities in the preschool years are necessary for successful reading and writing, as well as overall academic achievement (National Early Literacy Panel, 2008; Suggate et al., 2018).

Many types of adult–child conversations foster children's early literacy in the preschool years, including, not surprisingly, shared book reading. Both the frequency of book reading and the quality of the resulting conversations lead to growth especially in children's vocabulary (Dowdall et al., 2020), but also their phonological awareness, story comprehension and production, and print skills (see Reese, 2018 for a review). Yet parent–child reminiscing also contributes to many of these same early literacy skills (see Reese, 2018; Salmon & Reese, 2016, for reviews). In 1995, Reese first identified maternal reminiscing as a longitudinal predictor of children's independently assessed vocabulary, storybook comprehension, and print concepts at school entry. Subsequent correlational studies with larger and culturally diverse samples replicated and extended these findings to highlight maternal reminiscing as a unique contributor to preschool children's early literacy skills (Leyva, Reese, & Wiser, 2012; Leyva, Sparks, & Reese, 2012; Neha et al., 2020; Rowe, 2012; Sparks & Reese, 2013). Experimental evidence, in which one group of mothers was randomly assigned to a reminiscing coaching condition compared to both book-reading and no-treatment control conditions, underscored a causal role for elaborative reminiscing on children's vocabulary and story comprehension (Peterson et al., 1999; Reese, Leyva, et al., 2010).

The mechanism for the effects on children's print skills is not immediately obvious. The content of parent–child reminiscing is about the dyad's personal past experiences and does not include any obvious literacy focus. Why would these memory discussions lead to growth in children's early literacy, and specifically to children's ability to recognize letters and words? Reese (1995) proposed that these conversations could boost children's symbolic capacity through connecting children's mental representations of the past event with the words used to represent the event in conversation—a highly abstract skill. Reminiscing conversations are, by their very nature, a decontextualized form of talk that stretches children beyond the here and now. To date, however, this hypothesis has not been tested directly through assessing the effects of elaborative reminiscing on children's symbolic capacity, and the two experimental studies conducted so far did not show direct effects of elaborative reminiscing training on children's print skills.

The mechanisms for the effects of maternal reminiscing on children's vocabulary and story comprehension are more understandable. As noted above, the elaborative reminiscing style involves building a coherent and complete story about a past event. The skills children gain in creating these personal narratives are likely to transfer to fictional narratives. Studies that delve into the aspects of reminiscing that are important for children's vocabulary and story comprehension are particularly revealing. For instance, Reese (1995) identified mothers' 'metamemory' comments in the conversations, not their elaborations, as the strongest correlate of children's early literacy skills. Metamemory comments encourage the child to become aware of the act of remembering (e.g. 'I don't remember that part because I was looking the other way'). Likewise, Tompkins and colleagues (2019) isolated mothers' elaborations that are inferential in nature (e.g. 'Who was Abraham Lincoln anyway?') as a stronger predictor of children's later story comprehension skills than were mothers' literal elaborations (e.g. 'Who went with us to the Lincoln Memorial?'). In the same sample, Tompkins and colleagues (2019) showed that mothers' rare words during reminiscing (like 'reality' and 'fortune') also predicted children's later story comprehension. Notably, mothers issue more rare words during reminiscing than when sharing a wordless storybook or during free play (Tompkins, 2015). There are even hints that reminiscing supports children's numeracy skills (Neha et al., 2020), perhaps through discussions of sequence and time (Reese et al., 2008), but this link needs to be tested more rigorously through in-depth analyses of the reminiscing conversations.

Reminiscing is truly a rich context for children's literacy learning. Indeed, multiple mechanisms could be at play. The abstract nature of the reminiscing conversation as a whole could directly help children's ability to engage in symbolic thinking, which in turn could support their reading and writing skills. An elaborative reminiscing style produces a well-rounded narrative model for children that could transfer to their more general story comprehension skills. Within the conversations, some utterance types have specific links to aspects of children's complex language. Elaborations that are more inferential (called 'associative talk' in Reese, 1995) could help children's story comprehension by increasing their vocabularies, general knowledge about the world, and ability to draw causal connections between events. These skills in turn are all known to support children's reading comprehension later in development (Kendeou et al., 2009; Suggate et al., 2018). Cultures in which literacy is prized often focus on shared book reading, but research on maternal reminiscing style suggests that there might be additional links between oral cultures and literate cultures (Gutierrez & Rogoff, 2003).

Reminiscing style and deliberate memory

As with literacy, memory as a deliberate skill emerges in industrialized cultures that value the ability to commit information to memory, to organize information in more taxonomic rather than thematic forms, and to build hierarchically organized conceptual information in the service of decontextualized knowledge (Gauvain, 2001; Rogoff, 1990). Deliberate memory involves using mnemonic strategies—such as naming or grouping—to study to-be-remembered materials (e.g. objects, words, pictures) to prepare in an intentional way for a future assessment of remembering

(Bjorklund et al., 2009). Even young preschoolers can demonstrate 'strategic' behaviour under certain circumstances. For example, when asked to remember the location of a familiar stuffed toy that was hidden in a room, 18-month-olds utilized a number of rudimentary strategies (pointing, peeking, and naming) so that the toy could be retrieved after a delay (DeLoache et al., 1985). Older preschoolers have a firmer understanding of the need to do something in order to prepare for an assessment of memory, although young children's initial strategic behaviours are not unambiguously related to their memory performance. For example, when 4-, 5-, and 6-year-olds are asked to simply play with a set of objects for 2 minutes (either free play or specific instructions to play) versus to 'try to remember' those objects, children in the 'try to remember' group engaged in more labelling or naming of objects and less playing than those in the play conditions. The memory instructions seemed to engender a more studious approach to the task for the 4-, 5-, and 6-year-olds alike. However, only among the 6-year-olds were the strategic behaviours exhibited associated with higher levels of recall (Baker-Ward et al., 1984) The literature on deliberate memory contains many demonstrations of what Miller (1990) dubbed *utilization deficiencies*: whereas young children spontaneously produce appropriate strategies, these intentional mnemonic behaviours do not seem to initially correspond to improvements in the amount recalled.

Indeed, a comparison of children's performance on tasks that assess memory for personally experienced events and those that required deliberate remembering reveals the relatively late emergence of deliberate memory skills. By 8 or 9 years of age, children are very adept at providing rich reports about their experiences, but at the same time, their skills in deploying mnemonic strategies in situations that call for them remain quite limited. To illustrate the relatively late emergence of these deliberate memory skills, consider for example, when children between the ages of 9 and 14 are asked to remember a list of words and are prompted to talk aloud as each word is presented. In this type of overt rehearsal task, 9-year-olds' rehearsal is passive, repeating each word in isolation from the others (Ornstein et al., 1975). A 9-year-old might say *table, table, table* when table is presented first, *car, car, car* when car is presented second, and *flower, flower, flower* when flower is presented third, and so forth. By contrast, 14-year-olds rehearse more cumulatively or actively, adding new words as each is presented: *table, table* when table is presented first; *table, car, table, car* when car is presented; and *table, car, flower* when flower is presented. These changes in rehearsal styles for memorization are paralleled by comparable developments in organizational strategies to remember arbitrary materials. For example, when presented with a set of low-associated pictures or words and asked to 'form groups that will help you remember', 9-year-olds tend not to form groups based on semantic relations among the items. On the other hand, 12-year-olds routinely do create groups that are semantically constrained (Ornstein et al., 1975). It is not that the younger children lack the understanding of the semantic linkages among items, as they can readily sort even low-associated items on the basis of meaning when instructed to do so (Bjorklund et al., 1977). Rather, the age differences in performance seem to reflect age differences in understanding how underlying knowledge can be applied strategically in the service of a memory goal.

A developmental perspective would therefore suggest that children's abilities to remember their previous experiences and to talk about past events precede their later skills in preparing deliberately for future assessment of remembering. But autobiographical and deliberate memory are routinely treated in distinct literatures, and only recently has longitudinal research considered whether and how developments in skills for reminiscing about past events may set the stage for the emergence and refinement of deliberate memory abilities. Nevertheless, Ornstein et al. (2006) argue we should expect to see such developmental linkages, because event memory and deliberate memory share the same underlying processes of encoding, storage, retrieval, and reporting of information. Event memory involves encoding information during an event without the intent to remember, but efforts to search memory for details of an experience after the fact are under the deliberate control of the child. Likewise, the deployment of strategies to remember is primarily deliberate, albeit with automatic contributions to remembering being possible as the result of knowledge-based associations between to-be-remembered items. Ornstein and colleagues stress the idea that the same memory processes might operate across memory contexts, and that reminiscing about personal experiences might provide a foundation not only for the development of event and autobiographical memory skills, but also for the development of memory more broadly. For example, by responding to parents' wh-questions (who, what, where, when) about events during reminiscing, children can gain practice and learn skills for searching and retrieving information from memory and organizing this information into coherent narrative reports. In this way, because children need to engage in such deliberate efforts to retrieve and report memories of personal experiences to participate in reminiscing, reminiscing conversations may support competencies in deliberate remembering (Ornstein et al., 2006).

Linkages between autobiographical and deliberate memory have found empirical support. For example, Haden et al. (2001) reported associations between 2½-year-old children's memory for features of an event experienced with their mothers and their recall of objects in a deliberate memory task a year later. With the same sample, Rudek and Haden (2005) found that mothers who used more mental state language (e.g. remember, think, forget) during reminiscing when their children were 2.5 years of age had children who engaged in more strategic behaviours during the object memory task at 3.5 years old. Moreover, Rudek (2004) studied children at 3.5, 4.5, and 5 years of age and found concurrent and longitudinal associations between the extent of maternal elaboration during reminiscing and children's deliberate remembering of objects. Specifically, the children whose mothers were classified as high elaborative during reminiscing engaged in more strategic behaviours—particularly naming—during a study period, and recalled more of the to-be-remembered objects than did children of mothers who used a less elaborative reminiscing style. These findings suggest that early maternal reminiscing is important for the development of children's deliberate memory skills.

Results of a recent large-scale longitudinal study reported by Langley et al. (2017) also point to shared social-conversational mechanisms in the development of autobiographical and deliberate memory. In this work, a socioeconomically diverse sample of children was observed reminiscing with their mothers and engaging in a deliberate memory task when they were 3, 5, and 6 years old. In addition to the associations between maternal elaborative reminiscing and children's autobiographical memory performance, there were also associations between maternal reminiscing style and children's deliberate memory recall. Children whose mothers used a high elaborative reminiscing style at age 3 started off with better skills for deliberate remembering than children whose mothers used a low elaborative reminiscing style. But contrary to what was expected, children of mothers with a low elaborative reminiscing style at 3 years of age displayed a faster rate of growth in recall on the object memory task from ages 3 to 6, such that there were no differences between the groups by age 6. The researchers suggest that children with high elaborative mothers may get a leg up on their peers and advance in their deliberate memory skills at an earlier age. Children of low elaborative mothers catch up, potentially as a function of increases in processing speed, knowledge, and exposure to formal schooling (e.g. Coffman et al., 2008). In particular, the impact of the classroom might be especially important for children who have not had early exposure to a high elaborative reminiscing style (Ornstein et al., 2011). This interpretation highlights the essential aspects of culture more broadly. Cultures in which children are exposed to many reminiscing partners, as in cultures that live in more extended families and communities, may facilitate children's more elaborative and more nuanced autobiographical skills. However, cultures that value formal schooling and instruction may build stronger bridges between autobiographical reminiscing and more deliberate memory skills that are more highly valued in these cultures (Misty & Dutta, 2015).

Reminiscing style and science learning

Success in formal education, and especially in the science, technology, engineering, and mathematics (STEM) fields, is highly valued in industrialized cultures, where technical information is necessary for professional and financial success. Academic success clearly depends on the development of deliberate memory skills to master complex knowledge domains, but also relies on learning particular ways of thinking and analysing problems that are more abstract and decontextualized (National Academies of Sciences, Engineering, and Medicine, 2018). Intriguingly, this kind of abstract scientific thinking, while clearly taught in formal educational settings, is also supported by everyday parentally scaffolded conversations. Haden and colleagues (2021) have conducted a systematic research programme that examines how conversations between parents and children during and after experiences in informal educational settings (e.g. museums) can support learning about science and engineering. Museum exhibits for young children are designed to encourage hands-on activities, reflecting the notion that children learn best through direct engagement with objects (e.g. Piaget, 1970). Extending this argument, and drawing on sociocultural theory and research on autobiographical memory, Haden and her colleagues argue that parent–child conversations can be crucial for children to represent knowledge gained from exhibits in ways that enhance memory and transfer of science learning beyond the museum experience (Haden et al., 2016).

The conversations children and their caregivers have about experiences after visiting a museum exhibit can provide a vantage point for observing learning outcomes as well as the process of learning. For example, in some studies, parents are asked to record reminiscing conversations with their children shortly after experiences in museums, and days and weeks following museum visits, and these conversations can be very revealing of what information children understood and retained (e.g. Benjamin et al., 2010; Haden et al., 2014; Jant et al., 2014). Additionally, reminiscing conversations shortly after an exhibit experience can be viewed as part of the extended encoding of the event and part of the learning process (Haden, 2010). These conversations can help support children's initial learning from hands-on activities, help build more organized and coherent representations of the experience, supplement knowledge, and help consolidate and transfer specific information into more abstract understandings (Haden et al., 2016). Most of the work involving observations of parent–child conversations in museums has focused on the scaffolding of science learning during the encoding of the exhibit experience as it unfolds. Nonetheless, reminiscing shortly after an experience can support consolidation, the step in the learning process in which labile and fleeting patterns of experience are strengthened and transformed into long-lasting memory representations (Pagano et al., 2019).

Reflection is foundational in modern science and engineering education (e.g. *Next Generation Science Standards*, National Research Council, 2013; *A Framework for K-12 Science Education*, National Research Council, 2012; *Strands of Informal Science Learning*, National Research Council, 2009). Moreover, reflection in the form of parent–child reminiscing may be especially important for the consolidation of learning from hands-on science and engineering activities (Pagano et al., 2019). This is because reminiscing can facilitate what Sigel (1993) called *distancing* and Goldstone and Sakamoto (2003) called *concreteness fading*—learning to focus less on specific objects and more on the general knowledge and concepts that can be learned from object manipulation. Reminiscing shortly after a science-related experience in a museum can provide mechanisms for filling in the blanks in children's understandings, connecting new and prior knowledge, and elaborating what was initially encoded (Haden et al., 2016). At the same time, conversational reflection can be part of the learning process that can enable the storage of information in long-term memory, and enhance retrieval of learning for later use (Haden, 2010). Reminiscing may be most influential in the process of abstracting science-related semantic knowledge and facilitating lasting and coherent representations of science and engineering experiences when it involves an elaborative conversational style.

Pagano et al. (2019) showed that reminiscing conversations recorded shortly after science-related experiences could provide important diagnostic information about whether and how different programmes in the exhibit encourage science learning. Using a special multimedia exhibit component at a children's museum called *Story Hub: The Mini Movie Memory Maker*, families recorded their own reminiscing conversations following one of two programmes in a tinkering exhibit. Some families made recordings after participating in a tool-focused programme (Woodshop Plus) whereas others recorded reminiscing conversations following an engineering-focused programme (Make it Roll). As it happened, some families had their tinkering creations with them when reminiscing in *Story Hub* (opting to bring their creations home with them) and others did not. Overall, the most detailed and elaborated reminiscing conversations were observed among families who participated in the engineering-focused programme compared to the tool-focused programme. There was also more talk about engineering during reminiscing among families who participated in the engineering-focused programme compared to the tool-focused programme. These effects were amplified when families had their creations with them. Given prior work showing that the frequency of specific language inputs, such as spatial and relational language (Pruden et al., 2011) and number words (Gunderson & Levine, 2011), can predict children's skills in STEM domains, the increased talk about engineering during reminiscing was encouraging of the notion that engineering-focused tinkering programmes can advance STEM learning opportunities for children. Further, the presence of the creation was associated with discussions beyond what was perceptually available: talk about scientific and engineering practices, such as planning, testing, redesigning, as well as metacognitions and evaluations, such as figuring things out, making mistakes, and being un/successful.

In subsequent work, Haden and her colleagues (Pagano et al., 2020) observed a different sample of families both when families reminisced about their tinkering experiences and during a tinkering activity itself. In this study, families with 6–8-year-olds participated in one of two engineering-focused programmes that identified what families' creations should do and provided exhibit spaces for testing (Make it Roll, Make it Fly) or one of two programmes that did not (Make a Robot, Make Something that Does Something). As in the first study, there was more engineering talk in the reminiscing conversations of families who participated in the engineering-focused programme, as illustrated by this example from a family reminiscing about planning, testing, and redesigning their creation in a Make it Fly programme. In this excerpt, 7-year-old David (pseudonym) discusses his project with his mother:

MOTHER: Alright what was your mission?

DAVID: My mission was to make a plane with a propeller. The propeller kept holding it down. And I tried all kinds of different ways. Aluminium foil, um paper . . .

MOTHER: Cardboard.

DAVID: Cardboard and paper. Paper worked best because it was the lightest. I also used this paper bag to keep it in shape. The propeller kept holding it down because it had metal and really hard paper.

MOTHER: So how did you know if it worked or not? Talk about the testing.

DAVID: The testing was a little bit hard because I had to figure out what was keeping it down, what was keeping it from going all the way up out of the tube.

MOTHER: So what happened when you first tested it?

DAVID: When I first tested it, it did not go very far. It stayed down and its propeller was sticking down to the ground.

MOTHER: And how did you want it?

DAVID: I wanted it to go all the way up in the air flying out of the tube.

MOTHER: And you picked the materials and then when you did it?

DAVID: It flew up in the air the last time I tested it. The second time it didn't go very far either, but it did not go face down like the first time. The third one was the best because I figured out what was keeping it down.

MOTHER: What did you learn about building something that flies? What matters?

DAVID: The light weight.

MOTHER: Mhm.

DAVID: And making sure that it's stabilized and not very heavy.

MOTHER: Great!

Families who participated in programmes with an engineering focus talked about engineering more than twice as much when reminiscing about their experiences compared to those who participated in the other programmes. Moreover, analysis of the families' interactions during tinkering revealed that all programmes fostered collaboration, with no differences in joint engagement with objects and materials, or joint talk among families across programmes. On the other hand, families who participated in the engineering-focused programmes engaged in more engineering talk during tinkering than those who participated in the other programmes. In fact, analyses revealed that parent–child engineering talk during tinkering mediated the association between programme and engineering talk during reminiscing. The two engineering-focused programmes were associated with the most engineering talk during tinkering, which in turn, was associated with the greatest amount of engineering talk during reminiscing. This work points to ways to design exhibits to promote engineering talk among families during and after exhibit experiences. It also suggests that engaging families in reminiscing after exhibit experiences can reveal and potentially deepen informal science-related learning opportunities.

Reminiscing style and the development of academic skills

Research on the ways in which autobiographical memory develops in sociocultural contexts has clearly demonstrated the critical role of parental, and especially maternal, reminiscing style in advancing this development. Across multiple cultures studied, mothers and children spontaneously engage in reminiscing conversations about their past experiences, and the ways in which mothers structure these conversations matters for children's developing autobiographical memories. Mothers who are more elaborative, coherent, detailed, and focused on mental states and reactions have children who develop more coherent, detailed autobiographical memories that integrate the inner and outer worlds, creating personal narratives that describe motivations, intentions, and reactions as well as the events that occurred. What we have demonstrated in this chapter is that this kind of elaborated coherent reminiscing also provides a foundation for more decontextualized memory and academic skills. Children of more elaborative mothers develop more complex literacy skills earlier in development than children of less elaborative mothers, as well as better deliberate memory skills. As children mature, reminiscing conversations begin to include more academic content, at least in modern industrialized cultures, and these conversations provide a basis for more abstract semantic knowledge about science and engineering. In all, maternal reminiscing style lays a foundation for the emerging and evolving development of a complex multifaceted memory system that integrates personal memories with abstract skills and knowledge (Baddeley, 2013). This is an exciting set of findings but much remains to be known.

First, much of the research has been conducted within a limited set of cultures, with mostly white industrialized urban families. The autobiographical reminiscing research has been extended into many cultures and we see similar patterns, but the research on ways in which autobiographical reminiscing undergirds more decontextualized skills is still limited in reach. Clearly, certain kinds of academic skills are more or less valued in particular cultures, and in order for this research to be meaningful, deep cultural analysis must determine both what cultural tools and skills are privileged and develop theoretical pathways from early reminiscing contexts to the development of these more abstract skills (Greenfield & Quiroz, 2013; Mistry & Dutta, 2015). Second, and related to the first point, most of the research to date has examined mothers. Increased examination of the role of other family members—fathers, siblings, grandparents, and so on—must be undertaken to fully understand the development of autobiographical memory in sociocultural context especially in cultures where children develop in more extended and communal contexts. Further, in industrialized cultures, where formal education is the norm, the role of teachers as scaffolding personal memories in order to transition children to more abstract skills and knowledge must also be studied. Coffman et al. (2008) have begun to examine how teacher–child conversations begin to build bridges to more abstract knowledge, such as arithmetic, and their findings are provocative. Along similar lines, Andrews and Van Bergen (2020) have demonstrated the role of teachers' elaborative and evaluative questions and comments in children's developing abilities to engage in academic conversations in early day care settings. Intriguingly they find that talk about the future was more predictive of children's developing decontextualized talk than talk about the past. How talk about the past and talk about the future are inter-related and each contribute to children's developing knowledge base is a critical question as the research moves forwards.

Conclusions

We have begun to answer the question that Baddeley posed in 1988. Autobiographical memory serves many functions, including coherence and consistency of identity across time (Conway et al., 2004; McAdams, 1992), the creation of shared history and shared values (Fivush et al., 2011; Wang, 2013), directing future behaviour (Bluck, 2003) and developing empathy and understanding of other (Nelson, 2003). As we have demonstrated in this chapter, autobiographical memory also serves as a foundation for the development of more academic content, by providing a bridge between lived experience and abstract knowledge. Building on coherent elaborated narratives of personal experiences, children develop abstract and academic skills that allow them to become competent members of their culture.

Acknowledgements

Authors are in alphabetical order. CAHs contributions to this chapter were supported by the National Science Foundation under Grant No DRL-1516541.

References

Andrews, R., & Van Bergen, P. (2020). Characteristics of educators' talk about decontextualised events. *Australasian Journal of Early Childhood*, *45*(4), 362–376.

Baddeley, A. (1988). But what the hell is it for? In M. M. Gruneberg, P. E. Morris, & R. N. Sykes (Eds.), *Practical aspects of memory: Current research and issues, Vol. 1. Memory in everyday life* (pp. 3–18). John Wiley & Sons Ltd.

Baddeley, A. (2013). *Essentials of human memory* (classic edition). Psychology Press.

Baker-Ward, L., Ornstein, P., & Holden, D. J. (1984). The expression of memorization in early childhood. *Journal of Experimental Child Psychology*, *37*(3), 555–575.

Benjamin, N., Haden, C. A., & Wilkerson, E. (2010). Enhancing building, conversation, and learning through caregiver-child interactions in a children's museum. *Developmental Psychology*, *46*(2), 502–515.

Bjorklund, D. F. Ornstein, P. A., & Haig, J. R. (1977). Developmental differences in organization and recall: Training in the use of organizational techniques. *Developmental Psychology*, *13*(3), 175–183.

Bjorklund, D. F., Dukes, C., & Brown, R. D. (2009). The development of memory strategies. In M. Courage & N. Cowan (Eds.), *The development of memory in infancy and childhood* (pp. 145–175). Psychology Press.

Bluck, S. (2003). Autobiographical memory: Exploring its functions in everyday life. *Memory, 11*(2), 113–124.

Boland, A. M., Haden, C. A., & Ornstein, P. A. (2003). Boosting children's memory by training mothers in the use of an elaborative conversational style as an event unfolds. *Journal of Cognition and Development, 4*(1), 39–65.

Britto, P. R. (2012). *School readiness: A conceptual framework.* United Nations Children's Fund.

Bruner, J. (1991). The narrative construction of reality. *Critical Inquiry, 18*(1), 1–21.

Coffman, J. L., Ornstein, P. A., McCall, L. E., & Curran, P. J. (2008). Linking teachers' memory-relevant language and the development of children's memory skills. *Developmental Psychology, 44*(6), 1640–1654.

Conway, M. A., & Pleydell-Pearce, C. W. (2000). The construction of autobiographical memories in the self-memory system. *Psychological Review, 107*(2), 261–288.

Conway, M. A., Singer, J. A., & Tagini, A. (2004). The self and autobiographical memory: Correspondence and coherence. *Social Cognition, 22*(5), 491–529.

DeLoache, J. S., Cassidy, D. J., & Brown, A. L. (1985). Precursors of mnemonic strategies in very young children's memory. *Child Development, 56*(1), 125–137.

Donald, M. (2001). *A mind so rare: The evolution of human consciousness.* W. W. Norton & Company.

Dowdall, N., Melendez-Torres, G. J., Murray, L., Gardner, F., Hartford, L., & Cooper, P. J. (2020). Shared picture book reading interventions for child language development: A systematic review and meta-analysis. *Child Development, 91*(2), e383–e399.

Farrant, K., & Reese, E. (2000). Maternal style and children's participation in reminiscing: Stepping stones in children's autobiographical memory development. *Journal of Cognition and Development, 1*(2), 193–225.

Fivush, R. (2011). The development of autobiographical memory. *Annual Reviews of Psychology, 62*(1), 559–582.

Fivush, R. (2019). *Family narratives and the development of an autobiographical self: Social and cultural perspectives on autobiographical memory.* Routledge.

Fivush, R., Habermas, T., Waters, T. E., & Zaman, W. (2011). The making of autobiographical memory: Intersections of culture, narratives and identity. *International Journal of Psychology, 46*(5), 321–345.

Fivush, R., Haden, C. A., & Reese, E. (2006). Elaborating on elaborations: Role of maternal reminiscing style in cognitive and socioemotional development. *Child Development, 77*(6), 1568–1588.

Fivush, R., & Nelson, K. (2006). Parent–child reminiscing locates the self in the past. *British Journal of Developmental Psychology, 24*(1), 235–251.

Fivush, R., & Waters, T. E. (2019). Development and organization of autobiographical memory form and function. In: J. Mace (Ed.), *The organization and structure of autobiographical memory* (pp. 52–71). Oxford University Press.

Fivush, R., & Zaman, W. (2013). Gender, subjective perspective, and autobiographical consciousness. In P. J. Bauer & R. Fivush (Eds.), *The Wiley handbook on the development of children's memory* (Vol. 2, pp. 586–604). John Wiley & Sons Ltd.

Gauvain, M. (2001). *The social context of cognitive development.* Guilford.

Goldstone, R. L., & Sakamoto, Y. (2003). The transfer of abstract principles governing complex adaptive systems. *Cognitive Psychology, 46*(4), 414–466.

Greenfield, P. M., & Quiroz, B. (2013). Context and culture in the socialization and development of personal achievement values: Comparing Latino immigrant families, European American families, and elementary school teachers. *Journal of Applied Developmental Psychology, 34*(2), 108–118.

Gunderson, E. A., & Levine, S. C. (2011). Some types of parent number talk count more than others: Relations between parents' input and children's cardinal-number knowledge. *Developmental Science, 14*(5), 1021–1032.

Gutiérrez, K. D., & Rogoff, B. (2003). Cultural ways of learning: Individual traits or repertoires of practice. *Educational Researcher, 32*(5), 19–25.

Haden, C. A. (2010). Talking about science in museums. *Child Development Perspectives*, 4(1), 62–67.

Haden, C. A., Acosta, D., & Pagano, L. (2021). Making memories in museums. In L. E. Baker-Ward, D. F. Bjorklund, & J. L. Coffman (Eds.), *The development of children's memory: The scientific contributions of Peter A. Ornstein* (pp. 186–202). Cambridge University Press.

Haden, C. A., Cohen, T., Uttal, D., & Marcus, M. (2016). Building learning: Narrating and transferring experiences in a children's museum. In D. Sobel & J. Jipson (Eds.), *Cognitive development in museum settings: Relating research and practice* (pp. 84–103). Psychology Press.

Haden, C. A., Jant, E. A., Hoffman, P. C., Marcus, M., Geddes, J. R., & Gaskins, S. (2014). Supporting family conversations and children's STEM learning in a children's museum. *Early Childhood Research Quarterly*, 29(3), 333–344.

Haden, C. A., Ornstein, P. A., Eckerman, C. O., & Didow, S. M. (2001). Mother–child conversational interactions as events unfold: Linkages to subsequent remembering. *Child Development*, 72(4), 1016–1031.

Harris, P. L., Koenig, M. A., Corriveau, K. H., & Jaswal, V. K. (2018). Cognitive foundations of learning from testimony. *Annual Review of Psychology*, 69(1), 251–273.

Jant, E. A., Haden, C. A., Uttal, D. H., & Babcock, E. (2014). Conversation and object manipulation influence children's learning in a museum. *Child Development*, 85(5), 2029–2045.

Kendeou, P., Van den Broek, P., White, M. J., & Lynch, J. S. (2009). Predicting reading comprehension in early elementary school: The independent contributions of oral language and decoding skills. *Journal of Educational Psychology*, 101(4), 765–778.

Kulkofsky, S., Wang, Q., & Kim Koh, J. B. (2009). Functions of memory sharing and mother–child reminiscing behaviors: Individual and cultural variations. *Journal of Cognition and Development*, 10(1–2), 92–114.

Langley, H. A., Coffman, J. L., & Ornstein, P. A. (2017). The socialization of children's memory: Linking maternal conversational style to the development of children's autobiographical and deliberate memory skills. *Journal of Cognition and Development*, 18(1), 63–86.

Leyva, D., Reese, E., & Wiser, M. (2012). Early understanding of the functions of print: Parent–child interaction and preschoolers' notating skills. *First Language*, 32(3), 301–323.

Leyva, D., Sparks, A., & Reese, E. (2012). The link between preschoolers' phonological awareness and mothers' book-reading and reminiscing practices in low-income families. *Journal of Literacy Research*, 44(4), 426–447.

McAdams, D. P. (1992). Unity and purpose in human lives: The emergence of identity as a life story. In R. A. Zucker, A. I. Rabin, J. Aronoff, & S. J. Frank (Eds.), *Personality structure in the life course: Essays on personology in the Murray tradition* (pp. 323–375). Springer Publishing Company.

Miller, P. H. (1990). The development of strategies of selective attention. In D. F. Bjorklund (Ed.), *Children's strategies: Contemporary views of cognitive development* (pp. 157–184). Lawrence Erlbaum Associates.

Mistry, J., & Dutta, R. (2015). Human development and culture. In R. M. Lerner (Ed.), *Handbook of child psychology and developmental science* (7th ed., Vol. 1, pp. 369–406). John Wiley & Sons Ltd.

National Academies of Sciences, Engineering, and Medicine. (2018). *How people learn II: Learners, contexts, and cultures*. The National Academies Press.

National Early Literacy Panel. (2008). *Developing early literacy: Report of the National Early Literacy Panel*. National Institute for Literacy. https://www.nichd.nih.gov/sites/default/files/publications/pubs/documents/NELPReport09.pdf

National Research Council. (2009). *Learning science in informal environments: People, places, and pursuits* (P. Bell, B. Lewenstein, A. W. Shouse, & M. A. Feder, Eds.). National Academies Press.

National Research Council. (2012). *A framework for K-12 science education: Practices, crosscutting concepts, and core ideas*. National Academies Press.

National Research Council. (2013). *Next Generation Science Standards: For states, by states*. National Academies Press.

Neha, T., Reese, E., Schaughency, E., & Taumoepeau, M. (2020). The role of whānau (New Zealand Māori families) for Māori children's early learning. *Developmental Psychology*, 56(8), 1518–1531.

Nelson, K. (2003). Narrative and self, myth and memory: Emergence of the cultural self. In R. Fivush & C. Haden (Eds.), *Autobiographical memory and the construction of a narrative self: Developmental and cultural perspectives* (pp. 3–28). Lawrence Erlbaum Associates Publishers.

Nelson, K., & Fivush, R. (2004). The emergence of autobiographical memory: A social cultural developmental theory. *Psychological Review, 111*(2), 486–511.

Nelson, K., & Fivush, R. (2020). The development of autobiographical memory, autobiographical narratives, and autobiographical consciousness. *Psychological Reports, 123*(1), 71–96.

Ornstein, P. A., Haden, C. A., & Coffman, J. (2011). Learning to remember: Mothers and teachers talking with children. In N. Stein & S. Raudenbush (Eds.), *Developmental and learning sciences go to school* (pp. 69–83). Routledge/Taylor and Francis.

Ornstein, P. A., Haden, C. A., & Elischberger, H. B. (2006). Children's memory development: Remembering the past and preparing for the future. In E. Bialystok & F. I. M. Craik (Eds.), *Lifespan cognition: Mechanisms of change* (pp. 143–161). Oxford University Press.

Ornstein, P. A., Naus, M. J., & Liberty, C. (1975). Rehearsal and organizational processes in children's memory. *Child Development, 46*(4), 818–830.

Pagano, L. C., Haden, C. A., & Uttal, D. H. (2020). Museum program design supports parent-child engineering talk during tinkering and reminiscing. *Journal of Experimental Child Psychology, 200*, 104944.

Pagano, L. C., Haden, C. A., Uttal, D. H., & Cohen, T. (2019). Conversational reflections about tinkering experiences in a children's museum. *Science Education, 103*(6), 1493–1512.

Peterson, C., Jesso, B., & McCabe, A. (1999). Encouraging narratives in preschoolers: An intervention study. *Journal of Child Language, 26*(1), 49–67.

Piaget, J. (1970). *Science of education and the psychology of the child* (D. Coltman, Trans.). Orion Press.

Pruden, S. M., Levine, S. C., & Huttenlocher, J. (2011). Children's spatial thinking: Does talk about the spatial world matter? *Developmental Science, 14*(6), 1417–1430.

Reese, E. (1995). Predicting children's literacy from mother-child conversations. *Cognitive Development, 10*(3), 381–405.

Reese, E. (2018). Encouraging collaborative reminiscing between young children and their caregivers. In M. L. Meade, C. B. Harris, P. Van Bergen, J. Sutton, & A. J. Barnier (Eds.), *Collaborative remembering: Theories, research, and applications* (pp. 317–333). Oxford University Press.

Reese, E., Hayne, H., & MacDonald, S. (2008). Looking back to the future: Māori and Pakeha mother–child birth stories. *Child Development, 79*(1), 114–125.

Reese, E., Jack, F., & White, N. (2010). Origins of adolescents' autobiographical memories. *Cognitive Development, 25*(4), 352–367.

Reese, E., Leyva, D., Sparks, A., & Grolnick, W. (2010). Maternal elaborative reminiscing increases low-income children's narrative skills relative to dialogic reading. *Early Education and Development, 21*(3), 318–342.

Reese, E., Macfarlane, L., McAnally, H., Robertson, S. J., & Taumoepeau, M. (2020). Coaching in maternal reminiscing with preschoolers leads to elaborative and coherent personal narratives in early adolescence. *Journal of Experimental Child Psychology, 189*, 104707.

Reese, E., & Neha, T. (2015). Let's kōrero (talk): The practice and functions of reminiscing among mothers and children in Māori families. *Memory, 23*(1), 99–110.

Reese, E., & Newcombe, R. (2007). Training mothers in elaborative reminiscing enhances children's autobiographical memory and narrative. *Child Development, 78*(4), 1153–1170.

Reese, E., & Robertson, S. J. (2019). Origins of adolescents' earliest memories. *Memory, 27*(1), 79–91.

Ricoeur, P. (1991). Life in quest of narrative. In D. Wood (Ed.), *On Paul Ricoeur: Narrative and interpretation* (pp. 20–33). Routledge.

Roberts, W. A. (2002). Are animals stuck in time? *Psychological Bulletin, 128*(3), 473–489.

Rogoff, B. (1990). *Apprenticeship in thinking: Cognitive development in social context*. Oxford University Press.

Rogoff, B., Dahl, A., & Callanan, M. (2018). The importance of understanding children's lived experience. *Developmental Review, 50*, 5–15.

Rowe, M. L. (2012). A longitudinal investigation of the role of quantity and quality of child-directed speech in vocabulary development. *Child Development, 83*(5), 1762–1774.

Rubin, D. C. (2006). The basic-systems model of episodic memory. *Perspectives on Psychological Science, 1*(4), 277–311.

Rudek, D. J. (2004). *Reminiscing about past events: Influences on children's deliberate memory and meta-cognitive skills* [Unpublished doctoral dissertation]. Loyola University Chicago.

Rudek, D. J., & Haden, C. A. (2005). Mothers' and preschoolers' mental state language during reminiscing over time. *Merrill-Palmer Quarterly, 51*(4), 523–549.

Salmon, K., & Reese, E. (2016). The benefits of reminiscing with young children. *Current Directions in Psychological Science, 25*(4), 233–238.

Shanahan, T., & Lonigan, C. J. (2010). The National Early Literacy Panel: A summary of the process and the report. *Educational Researcher, 39*(4), 279–285.

Sigel, I. E. (1993). The centrality of a distancing model for the development of representational competence. In R. R. Cocking & K. A. Renninger (Eds.), *The development and meaning of psychological distance* (pp. 141–158). Lawrence Erlbaum Associates.

Sparks, A., & Reese, E. (2013). From reminiscing to reading: Home contributions to children's developing language and literacy in low-income families. *First Language, 33*(1), 89–109.

Suggate, S., Schaughency, E., McAnally, H., & Reese, E. (2018). From infancy to adolescence: The longitudinal links between vocabulary, early literacy skills, oral narrative, and reading comprehension. *Cognitive Development, 47*, 82–95.

Sutton, J., Harris, C. B., Keil, P. G., & Barnier, A. J. (2010). The psychology of memory, extended cognition, and socially distributed remembering. *Phenomenology and the Cognitive Sciences, 9*(4), 521–560.

Tompkins, V. (2015, March). *Mothers' rare words in three contexts: Relations to preschoolers' language skill* [Poster presentation]. Biennial meeting of the Society for Research in Child Development, Philadelphia, PA.

Tompkins, V., Duffy, K., Allen, E., & Smith, R. (2019). Who was Abraham Lincoln anyway? Mother-child reminiscing across levels of abstraction. *Developmental Psychology, 55*(7), 1493–1508.

Tõugu, P., Tulviste, T., Schröder, L., Keller, H., & De Geer, B. (2012). Content of maternal open-ended questions and statements in reminiscing with their 4-year-olds: Links with independence and interdependence orientation in European contexts. *Memory, 20*(5), 499–510.

Tulviste, T., Tõugu, P., Keller, H., Schröder, L., & De Geer, B. (2016). Children's and mothers' contribution to joint reminiscing in different sociocultural contexts: Who speaks and what is said. *Infant and Child Development, 25*(1), 43–63.

Tweed, R. G., & Lehman, D. R. (2002). Learning considered within a cultural context: Confucian and Socratic approaches. *American Psychologist, 57*(2), 89–99.

Wang, Q. (2013). The cultured self and remembering. In P. J. Bauer & R. Fivush (Eds.), *The Wiley handbook on the development of children's memory* (Vol. 2, pp. 605–625). John Wiley & Sons Ltd.

Wang, Q. (2016). Remembering the self in cultural contexts: A cultural dynamic theory of autobiographical memory. *Memory Studies, 9*(3), 295–304.

Wellman, H. M. (2018). Theory of mind: The state of the art. *European Journal of Developmental Psychology, 15*(6), 1–28.

12

Working memory in language learning and bilingual development

Michael F. Bunting and Zhisheng (Edward) Wen

Introduction

The themes of this volume have reminded us of Neisser's (1978) pertinent observation, 'If *X* is an interesting or socially significant aspect of memory, then psychologists have hardly ever studied *X*' (p. 4). Does Neisser's conditional statement apply to the relationship between language and memory? Most certainly. In 1966, the report from the US government's Automatic Language Processing Advisory Committee (ALPAC) compared the difficulty and importance of the scientific study of language to that of particle physics, concluding, 'Language is second to no phenomenon in importance' (ALPAC, 1966, p. 44). The social and scientific importance of language is, it would follow, without match. For a long time, language and memory were studied in parallel, but not always together. They were comrades in the cognitive revolution with a one–two punch to behaviourism from Miller's (1956) paper on short-term memory capacity and Chomsky's (1957) groundbreaking idea that natural languages are innate and syntactically rule-governed. For several ensuing decades, however, language and memory were studied apart as separate cognitive abilities, each with distinct research traditions and methods (Duff & Piai, 2020). It is only in recent decades that language and memory research have become entwined. The *Cambridge handbook of Working Memory and Language*, edited by Schwieter and Wen (2022), provides the most up-to-date and comprehensive reviews of working memory and language.

Indeed, memory, in all its forms and types (be it sensory, short-term/working, and long-term; Waugh & Norman, 1965), is critical to essential aspects of language representation, evolution, acquisition, production, and subskills processing (Wen, 2016; cf. Corballis, 2020). The last 50 years have witnessed diverse theoretical approaches to language: from the early behaviourist view that equates language learning to verbal behaviour (Skinner, 1957); to Chomsky's cognitive view in the 1950s and one that emphasizes the prewired language faculty (aka universal grammar) and the central role of syntactic structure (1957); to the contemporary blossoming of many alternative accounts, particularly the construction-oriented, usage-based functional paradigms (Bybee, 2010; Goldberg, 2003; Tomasello, 2003). For example, if language

Michael F. Bunting and Zhisheng (Edward) Wen, *Working memory in language learning and bilingual development* In: *Memory in Science for Society.* Edited by: Robert H. Logie, Zhisheng (Edward) Wen, Susan E. Gathercole, Nelson Cowan, and Randall W. Engle, Oxford University Press. © Oxford University Press 2023. DOI: 10.1093/oso/9780192849069.003.0012 is viewed as the innate and static representation of linguistic competence, then long-term memory is a prerequisite for storing and sustaining the lexical items (in the mental lexicon) and the grammar rules (e.g. including metalinguistic knowledge). On the other hand, when language is considered as the acquisition of linguistic sequences at different levels (e.g. the connectionist view; Ellis, 1996, 2012), then immediate or working memory must serve as a moving window of consciousness if one is to integrate overtime to make sense of these sequences, such as sentences and discourse (Chafe, 1994; Corballis, 2020).

On other occasions, such as during sentence processing or language comprehension, all listeners must perceive a sentence incrementally and store the first part of it while they take in the rest, with a view to extracting its gist (Clark & Clark, 1977). Anaphors, for example, must be held in storage until they can be bound to appropriate discourse referents, the efficiency of which is likely to be modulated by working memory capacity (e.g. Joseph et al., 2015). Parenthetical asides can disrupt the flow of a sentence and separate the subject and verb of a sentence by several seconds, thus necessitating memory to maintain one's train of thought. In short, given that memory and language are integrated and intertwined to such an extent, it becomes imperative to understand their interactions (Duff & Piai, 2020). In the next sections, we explore the theoretical links between the more specific system of working memory and its relation to language learning and bilingual development. Following these, we trace the conceptions of working memory from Baddeley's multicomponent model to Cowan's embedded-processes model, in tandem with their implications for language learning and bilingualism. We then end the chapter by calling for a paradigm shift from working memory components to executive functions, particularly the emerging construct of attentional control.

Working memory capacity and language structure

Prior to the inception of the concept of working memory in 1960 (Miller et al., 1960), many modern theories of human cognition describe a *primary* memory system (James, 1890; Waugh & Norman,1965) that is (1) dedicated to the temporary processing, maintenance, and holding of information that is relevant to current tasks; and (2) functionally or structurally distinct from secondary (or long-term) memory. This *primary* system is short-term memory, the precursor to what is now known as working memory. As such, the conceptual distinction between long-term memory and short-term or working memory is intuitive to most people. Long-term memory, the repository of a lifetime of memories and the accumulative repertoire of knowledge acquired, is vast and limitless in comparison to the restricted number of thoughts people can keep simultaneously active in immediate consciousness (Cowan, 2008). Cognitive psychologists refer to the active form of short-term memory as working memory, or the small amount of information that can be kept in an *accessible* state in order to be used in ongoing mental tasks (Cowan, 1995). Apt examples of working memory at work in our daily life are dialling numbers to call a friend (which requires us to keep the digits in our head while dialling), mental calculations of arithmetic (say, multiplying 38 by 25), and playing a game of table tennis (keep the previous scores in the head while concentrating on playing).

In the last 60 years, particularly after the advent of Baddeley's seminal model (Baddeley & Hitch, 1974), the concept of working memory has become a buzzword in cognitive science (Conway et al., 2007; Oberauer et al., 2018). Cognitive scientists have approached the *nature, structure*, and *operation* of working memory from multiple disciplines and perspectives spanning psychology, linguistics, neuroscience, as well as archaeology/anthropology, and artificial intelligence or computer modelling (see the variety of opinions offered in Logie et al., 2021; Miyake & Shah, 1999; Schwieter & Wen, 2022). The current proliferation of a dozen or so theoretical models of working memory is a two-edged sword. On the one hand, they have given rise to enormous enthusiasm and a deeper understanding of the construct of working memory, but on the other hand, they have also incurred lingering controversies and debates over its nature, structure, and relationship with long-term memory (Baddeley, 2012; Cowan, 2017). Notwithstanding these seemingly disparate differences in research focus and emphasis (Logie et al., 2021), most working memory theorists would agree on its essential features and functions. These unifying understandings include, first of all, that working memory serves to order, store, and manage immediate sensory details until they can be properly incorporated into the cognitive process that must integrate those data. More importantly, these theorists concur that the amount of data that can be stored for immediate, accurate recall (*availability*) is limited in size, normally ranging between four chunks (Cowan, 2001) to seven units of information (Miller, 1956); and in duration, which usually lasts between a few seconds to 20 seconds (Waugh & Norman, 1965); and the speed with which it can be recalled (*accessibility*) varies.

To begin with, the limited capacity of working memory actually constitutes the signature feature of this central construct of human cognition (Carruthers, 2003; Klingberg, 2009). Perceived this way, the implications of working memory are also far-reaching for language design features, acquisition, and processing, as well as long-term development. For example, as an advocate of the emergentist account of language processing (MacWhinney & O'Grady, 2015), William O'Grady (2017) has cited evidence from English and Korean to argue that working memory limitations constrain and shape essential aspects of language, ranging from phonology to grammar. Similarly, Gomez-Rodriguez et al. (2022), by analysing the big data of cross-linguistic corpora, argue that (working) memory limitations are hidden in grammar. Lu and Wen (2022) also set out to make the distinction between the magical number four (Cowan, 2001) and the magical number seven (Miller, 1956) in that short-term memory limitations of seven refer to the range of the momentary chunk number, while the working memory limitations of four refer to the average mean of momentary chunk number (cf. Liu's mean dependency distance or MDD, 2008) within the 'focus of attention' analogous to Cowan's (1999) embedded-processes model (see also Adams et al., 2022; Cowan et al., 2020).

The multicomponent model and language learning

In terms of working memory and language, the most widely cited framework is the seminal model by Baddeley and Hitch (1974). In this standard model, Baddeley and colleagues first described working memory as having two different subsystems or component buffers: visuospatial working memory for manipulating and briefly maintaining information from the spatial domain; and, verbal working memory for handling verbally mediated representations and processing (see also Baddeley, 1986, 2003, 2007; Baddeley et al., 2021; Baddeley & Logie, 1999). Research on this structural view of working memory has addressed the further subdivision of these two primary components into more subcomponents. For example, the phonological loop can be further demarcated into a phonological short-term store and the articulatory rehearsal mechanism. Given the instrumental roles these two subcomponents play in storing and sustaining the phonological form of language, the phonological loop (also known as phonological working memory; e.g. Pierce et al., 2017; Wen, 2016) has been positioned as a 'language learning device' (Baddeley et al., 1998). To tap into the construct of the phonological loop, some 'simple' storage-focused versions of memory span tasks, such as the digit span, the letter span, and so on, have been used for many years. More recently though, it has been argued that the *non-word repetition span task* emerged as the best proxy (Gathercole et al., 1994, 2006).

Among the multiple components, phonological working memory has received the most attention in language-related research (Baddeley 2003, 2015). Evidence accumulating from developmental and cognitive psychology, neuropsychology, psycholinguistics, and specific language impairments converges on the close associations between phonological working memory and the sound-based and chunking-based aspects of language acquisition and processing (Baddeley, 2003; Gathercole & Baddeley, 1993). Recently, Llompart and Dabrowska (2020) comprehensively demonstrated its close links with lexical knowledge, grammatical knowledge, and collocational knowledge in native language acquisition and processing. These same links are also apparent in bilingualism studies. Empirical studies in both psychology and second language acquisition also point to its associations with the acquisition and development of L2 vocabulary (Atkins & Baddeley, 1998; Cheung, 1996; Service, 1992), formulaic sequences, and/or collocational chunks (Ellis, 2012; Foster et al., 2014), and grammar development, particularly among *ab initio* or young L2 learners (French & O'Brien, 2008) and low-proficiency adult L2 learners (Serafini & Sanz, 2016). Other laboratory-based experimental and longitudinal studies also corroborate the significant role phonological working memory plays in the early grammatical development of oral skills (O'Brien et al., 2006, 2007).

In contrast to phonological working memory, the central executive of the multicomponent model has received much less attention and is thus less understood (Baddeley, 2015). The situation improved with the original tripartite model by Baddeley being further refined and expanded to include a fourth component, namely, the episodic buffer (Baddeley, 2000). As Baddeley (2022) acknowledged recently, language-related research has served as the key catalyst to boosting the refinement of the original tripartite model. In more recent years, Baddeley and colleagues have explored the implications of the episodic buffer and the central executive for language, particularly the function of chunking and binding (Atkinson et al., 2021; Baddeley, 2022; Baddeley et al., 2011).

Executive models of working memory and language processing

In contrast to the multicomponent model, Cowan (1995, 2001; see Cowan, Chapter 9, this volume; see also Cowan et al., 2020) suggests that memory is a singular system consisting of elements at various levels of activation. To understand Cowan's view of the memory system, it is important to distinguish among items (i.e. memories) that are inactive, items that are active above some threshold, and items that are hyperactive. At any moment most elements in the system are in a relatively inactive state and, thus, are conceptually in long-term memory. A subset of those elements, however, may be in a higher state of activation (i.e. above some threshold of activation) but outside the focus of attention and, thus, outside of conscious awareness, at least for the most part. Elements in this state are conceptually in short-term memory. Even information in an activated but unconscious state can influence ongoing processing, such as in subliminal perception or semantic priming. The temporary persistence of recently activated sensory and semantic information in short-term memory is automatic, and there is functionally no capacity limitation on short-term memory as thus described. However, activation is fleeting, and the contents of short-term memory quickly return to an inactive state due to decay, unless they are reactivated and returned to conscious awareness by means of focused attention.

Cowan (1995, 2005; see also Cowan, Chapter 9, this volume; Cowan et al., 2020) refers to elements in a hyperactive state as primary memory (cf. James, 1890), or the focus of attention. Maintaining elements in a hyperactive state requires controlled and effortful attention. Because the capacity for attention is considerably limited, only a number of elements can be maintained within the focus of attention. Following an extensive review of decades of research, Cowan (2001) concluded that the fundamental capacity of the focus of attention is, on average, four chunks (i.e. groups) of items. Because individuals vary in the capacity of the focus of attention, the actual range is from three to five (i.e. the magical number of four plus or minus one; also see Lu & Wen, 2022). The capacity-limited focus of attention is analogous to the central executive in the Baddeley and Hitch (1974; Baddeley, 1986) model of working memory, to the supervisory attentional system in Norman and Shallice's (1986) model, and controlled attention in Posner and Snyder (1975) and Schneider and Shiffrin's (1977) models.

According to Cowan's (1995) model of memory, working memory consists of the contents of short-term memory plus the limited-capacity controlled-attention processes associated with the focus of attention. Working memory functionally serves two purposes: processing control and memory maintenance. The active and attention-demanding control processes include the cross-modal integration of acoustic and visual information, organizing information into meaningful chunks, and binding recently encoded information with existing forms of knowledge representations in long-term memory (Baddeley, 1986; Cowan, 1995). In addition to these supervisory-type functions, working memory serves as a storage function. Attention, in the service of memory, can maintain a limited number of items in the focus of attention. Additionally, without the help of attention, a greater number of sensory and semantic elements are briefly but automatically maintained in the activated portion of memory (i.e. short-term memory) until they are overridden by new and interfering elements or lost due to decay. These two forms of memory maintenance—the effortful versus the automatic—are often called active and passive storage, respectively.

Working memory processes include the automatic, temporary persistence of sensory and semantic information recently activated in the brain and the inclusion of a subset of the activated information in the focus of attention. For example, to comprehend ongoing spoken language, one must hold in mind what has been said so far. Usually, one remembers the meaning of what has been said rather than all of the actual words, especially for the more distant speech in a long conversation (cf. Clark & Clark, 1977). One also has available a fleeting sensory trace of the most recently heard speech. The most recent words or even salient or important words that occurred distantly in the speech may still be in the focus of one's attention. These aspects of comprehending spoken language—maintaining and updating a semantic representation of the speech and processing new sensory information—are the job of working memory.

Dual processes of working memory in language comprehension

Daneman and Carpenter (1980)

Storage-plus-processing or complex span tasks (e.g. counting span, operation span, and reading span) are the gold standard measures of individual differences in working memory capacity (Cowan et al., 2005). Complex span tasks are essentially dual tasks that make concurrent demands on cognitive processing and short-term memory storage. Complex span tasks are the most widely used measures of the supervisory functions of working memory.

The complexity of working memory span tasks (i.e. their dual-task nature) contributes to the difficulty of determining what constructs they assess. Consider, for example, how one complex span task, Daneman and Carpenter's (1980) reading–word span task, works (for an updated version, see Conway et al., 2005). Each reading span trial consists of a short sentence (9–16 words) preceding a to-be-remembered word (the single word memorandum). The sentences are usually unrelated to one another and descriptive, but not overly complex. One task demand is a test of reading comprehension. In one common version of the task, half of the sentences make sense (e.g. 'Carol was so well behaved today that her parents rewarded her with a piece of cake'), while the other half do not (e.g. 'Jacob and Johan drove a yellow light bulb from New York to Los Angeles'). Participants read a sentence, respond 'yes' or 'no' to indicate if it makes sense or not, and remember a single word memorandum. The sentence–word strings are presented in groups, or sets, ranging in size from two up to seven trials. Participants are asked to recall only the words, but their accuracy in reading comprehension is also monitored. Reading span makes concurrent demands on temporary storage and domain-specific knowledge of reading comprehension.

Decades of research support the claims that (1) the attentional-control processes of working memory are domain free and regulate cognition across language, imagery, reasoning, and many other cognitive domains; and (2) individual differences in attentional control processes account for the well-documented high correlations between working memory and intellectual aptitude (Conway et al., 2002, 2005; Kane et al., 2004; Miyake et al., 2000). Although many attentional-control processes have been hypothesized, the clearest distinction is between task-switching processes and updating/inhibition processes (Miyake et al., 2000).

Caplan and Waters (1999)

Of relevance to the language community, Caplan and Waters (1999) suggested that verbal working memory should be differentiated for verbal (but not syntactic) processes for cognitive tasks generally versus syntactic/grammatical processes that support linguistically mediated tasks such as sentence processing and comprehension. Contrary to the approach that describes working memory in multiple task-specific processes, Kane et al. (2004) demonstrated that linguistic and non-linguistic (but still verbally mediated) tasks rely on a single pool of working memory resources and that working memory processes are, by and large, domain general.

Working memory is an important component in many learning processes, including taking notes, following directions, or ignoring distractions (Piolat et al.,

2005; Engle, 2002; Engle et al., 1991). Evidence suggests that it is also an important part of language comprehension. Speakers with larger working memories are also better able to learn vocabulary (in both first and second languages; Atkins & Baddeley, 1998), write more proficiently (Ransdell & Levy, 1996), and have better reading and listening comprehension (Daneman & Merikle, 1996; de Bruïne et al., 2021; Peng et al., 2018).

Understanding how people read and comprehend text in their native language is challenging enough for cognitive science. Second language interpretation and reading of texts add to the level of complexity, both for the reader trying to deal with the text and the scientists trying to understand the processes. There are at least two main reasons for this challenge. For one, defining the locus of the problem is difficult, for example, is it with the reader or with the text? For another, there are many cognitive processes involved. Working with texts involves both language-specific and language-independent comprehension skills.

Reading comprehension is more than the physical act of reading, such as the act of moving one's eyes over a page of printed text and pronouncing words. Skilled readers can 'read' without comprehension. A skilled reader of English could, for example, read aloud this passage from the prologue to *Sir Gawain and the Green Knight* with relative ease while having little or no idea of the meaning of these words written in Middle English:

Middle English	Modern English
And neuenes hit his aune nome, as hit now hat;	And names it with his own name, which it now has;
Tirius to Tuskan and teldes bigynnes,	Tirius turns to Tuscany and founds dwellings,
Langaberde in Lumbardie lyftes vp homes.	Longobard raises home in Lombardy.

Likewise, when lacking sufficient background knowledge or specialized vocabulary, one may also have a difficult time comprehending a passage composed of otherwise familiar words, such as this passage from Einstein's (1905) particle theory of light:

According to the assumption to be contemplated here, when a light ray is spreading from a point, the energy is not distributed continuously over ever-increasing spaces, but consists of a finite number of energy quanta that are localized in points in space, move without dividing, and can be absorbed or generated only as a whole.

It could be said that comprehension is making sense of what is read, but even 'making sense' has many different definitions. Making sense of a text may simply mean being able to extract the main idea of the passage, but it may also mean being able to answer simple questions about it or even drawing inferences from what is read (Schwartz, 1984). Some researchers have suggested that text comprehension serves two primary functions: information transfer and social interaction (Butcher & Kintsch, 2003). Researchers interested in information transfer may focus on story understanding, memory for factual material presented in texts, learning from texts, and drawing inferences from the text. Researchers interested in social interaction focus on the use of language to establish social roles, regulate social interactions, and convey emotion.

Reading for the gist of a text or reading for pleasure is generally less cognitively demanding than reading for detailed understanding and the ability to remember information later. Engle and Conway (1998) offered these predictions for the interplay of memory and language on reading comprehension, surmising that comprehension is least cognitively demanding when written language:

- Is constructed of simple sentences with relatively few words (e.g. 'he ran' or 'language is complex').
- Does not require memory of previous sentences to understand the current text (e.g. the sentence 'the robber jumped from the bank' could refer to a criminal leaving a financial institution, or the edge of a river).
- Avoids ambiguity that could lead to later misinterpretation.
- Uses information that is presented linearly.
- Is presented in a distracter-free environment.

Working memory in second language acquisition and bilingual development

Strong support for working memory playing an important role in second language comprehension comes from the *developmental interdependence hypothesis*, which states that a learner's competence in the second language is at least partially dependent on competence in the first language (Cummins, 1979, 2021). This would be expected if fundamental cognitive abilities play a role in the ability for both first and second languages (Lasagabaster, 2001). Surprisingly, research interest in the second language acquisition field was not strong until the early 1990s, when Harrington and Sawyer (1992) found a positive relationship between L2 working memory reading span and L2 reading comprehension scores. Since then, research into the working memory–second language acquisition nexus (Wen, 2016) was boosted by cognitive-oriented second language acquisition scholars such as Ellis (1996), Skehan (1998), and Williams (2012). For most early studies, the research design has followed the traditions from cognitive psychology, in particular, Baddeley's seminal tripartite model. Similar to research in psychology and psycholinguistics, these emerging second language acquisition studies have demonstrated that working memory, as measured by the non-word repetition span task (Gathercole et al., 1994) or the reading span task (Daneman & Carpenter, 1980), has been shown to predict second language vocabulary acquisition time (Cheung, 1996), English as a second language vocabulary ability (Masoura & Gathercole, 2005), second language implicit grammar teaching (Robinson, 2005), and second language reading (Walter, 2004) and writing abilities (Abu-Rabia, 2003).

In addition to these L2 parallels of L1 language domains, second language-oriented studies have also explored the role of working memory in the acquisition and processing of L2 formulaic sequences or collocations (Ellis, 2012; Foster et al., 2014), in the noticing of the corrective feedback during L2 interactions (Li, 2017; Mackey et al., 2002), and in the linguistic performance dimensions (i.e. fluency, accuracy, and complexity) of task-based speech production (Ahmadian, 2012). Although there have been a number of studies in both psychology and bilingual processing investigating the effects of working memory on translation/interpreting, Wen (2015, 2016) has categorized these specialized strands of empirical studies into the broad realms of second language acquisition.

The growing body of empirical studies in both psychology and language sciences (including second language acquisition) has paved the ground for large-scale meta-analytical studies. On the one hand, Linck et al. (2014) conducted a meta-analysis of data from 79 samples involving 3707 participants providing 748 effect sizes and found a positive, but modest link between working memory and the L2 learning process and products (with $r = 0.25$). Other studies have examined reverse effects, that is, whether speaking or learning a second/foreign language (being bilingual) will have an impact on one's working memory. For this, Grundy and Timmer (2017) in their meta-analysis of 88 effect sizes in 27 independent studies have also found a positive but modest association ($r = 0.2$) between bilingualism effects on working memory. That is, working memory appears to share only a small amount of variance with bilingualism, and it is unclear whether being bilingual enhances working memory, having a good working memory makes it easier to become bilingual, or whether both share variance with some third, common factor.

Executive functions and (second) language use

Further elucidation of the control processes of memory and attention is one of the most significant scientific advances in cognitive psychology in recent decades (Bunting, 2006; Conway et al., 2005; Daneman & Carpenter, 1980; Miyake et al. 2000). General-purpose, attention-controlled executive functions moderate the operation of various cognitive subprocesses and regulate the dynamics of human cognition (Miyake et al., 2000). The best defined and most distinguishable executive functions are task switching and the combined operations of memory updating and inhibition.

The role of task switching in second/foreign language use

Task switching—*the ability to manage attention switching effectively during dual- and multitask situations*—is one of many general-purpose, attention-controlled executive functions that regulate complex cognitive processes, including foreign language use. The ability to switch between a primary and secondary language may be especially important to second/foreign language acquisition. Language learning requires mapping linguistic forms to functions and meaning, and so the ability to switch attention from one to another during language comprehension or use could be highly beneficial.

Demonstrating rapid and efficient task switching is a milestone for advanced-level second language users, requiring their production and comprehension of the second language. Robinson (2004) offered this real-world example: 'taking a telephone call, answering brief questions correctly, and writing down a message in the second language while simultaneously "keeping an eye" on, and monitoring a rapidly changing TV screen to record numerical and written information in the second language (e.g. about changing stock prices, changes to flight times, or other job specific information)' (p. 1).

There are significant individual differences in task-switching ability (Monsell, 2003). Some individuals are markedly better at doing two or three things at once than others, controlling for the extent of prior practice and background knowledge drawn on in the task situation. Task-switching ability also deteriorates with ageing; young adults are typically better at dual-task performance than older and elderly adults (Salthouse et al., 1998). Salthouse et al. (1998) found that switching from one task to another produced meaningful and reliable switch costs, both in the time and error measures. Furthermore, the measures of task switching were positively and significantly related to measures of reasoning ability, speeded pattern matching, and problem-solving (i.e. faster switchers had higher levels of cognitive performance).

In a study with greater implications for language learning, Miyake et al. (2000) found that individual differences in task-switching ability were positively related to performance on the Wisconsin Card Sorting Task. This is a neurological test of abstract reasoning and perseveration errors. The task requires participants to sort cards according to one of three rules that change at regular intervals during the task. Perseveration errors occur when participants fail to realize the rule change and continue to sort the cards according to an old rule. Perseveration errors are common during language learning and may be similarly related to task-switching ability.

The ability to manage complex multitask demands is a domain-free executive function, but language-specific deficits may also exacerbate task-switching costs (Segalowitz & Frenkiel-Fishman, 2005). This would especially be the case when one's knowledge of a foreign language is sparse. Segalowitz and Frenkiel-Fishman (2005) showed that second language proficiency affected performance on a learning task when participants switched from performing the task in their first language, English, to a language they were learning as a foreign language, French.

The role of updating and inhibition in foreign language use

To comprehend ongoing spoken language, one must hold in mind what has been said so far. Usually, one remembers the meaning of what has been said rather than all of the actual words, especially for the more distant speech in a long conversation. One also has available a fleeting sensory trace of the most recently heard speech. The most recent words or even salient or important words that occurred distantly in the speech may still be in the focus of one's attention. All of these aspects of comprehending spoken language—maintaining and updating a semantic representation of the speech and processing new sensory information—occur in working memory.

The process of updating the contents of working memory requires monitoring and coding incoming information for relevance to the task at hand and then appropriately revising the items held in working memory by replacing old, no longer relevant information with newer, more relevant information (Morris & Jones, 1990). Updating is more than a simple maintenance function. That is, the essence of updating lies in the requirement to manipulate relevant information actively in working memory, rather than passively store information.

Inhibition is a highly related, maybe even integral process, of updating; it is one's deliberate ability to inhibit dominant, automatic, or prepotent responses when necessary. The colour naming Stroop task is a prototypical inhibition task. In computerized versions of this task, subjects are asked to name quickly and accurately the colour in which objects, words, and non-words appear. In critical trials, the word is the name of a colour (e.g. the word 'red') appearing in another colour (e.g. written in blue). One needs to inhibit or override the tendency to produce a more dominant or automatic response (i.e. name the word) in order to do the task correctly (i.e. say the colour).

In everyday cognition, people engage in multiple tasks with competing goals, and even when they limit their attention to a single, important task, they often need to ignore outside distractions and their own intruding thoughts. Updating and inhibiting executive functions make this possible. For example, the ability to update the contents of working memory with new information is important for reading and language comprehension, either in a first or foreign language. Working memory is important for language use and comprehension when there is a need to retain a specific word or phrase until disambiguating information comes later in the sentence or paragraph; (e.g. the word 'bank' has multiple meanings, thus context is needed to ensure that the correct interpretation, as in the sentence: 'I'm low on cash, so remind me to stop by the bank later today'; Engle & Conway, 1998). The demands on working memory are even greater when one encounters similar words with different meanings in one's first and second languages. Active working memory control may be important for suppressing first language knowledge when it interferes with second language comprehension (Michael & Gollan, 2005). Research on bilingualism suggests that both languages are always active and thus require the bilingual to use cognitive resources to control the relative levels of activation of the two languages (see Michael & Gollan, 2005, for a review).

Some researchers have argued that updating and inhibition are separate, though related executive functions (Miyake et al., 2000). Although this may be the case, the most common measure of individual differences in working memory capacity, the complex span task, reflects both functions simultaneously (Bunting, 2006; Conway et al., 2005; Miyake et al., 2000).

Bilingualism advantage and working memory

The bilingual advantage is the tantalizing notion that bilinguals, especially language-balanced bilinguals, enjoy certain cognitive advantages relative to monolinguals. The hypothesis underlying the advantage is that the cognitive skills developed to cope with the demands of controlling two languages generalize to more efficient processing in working memory (Blom et al., 2014), attention (Friesen et al., 2015), and executive function (Blumenfeld & Marian, 2014). The mental gymnastics of controlling two languages include suppressing interference from a non-target language(s) while speaking or recognizing a target language (Struys et al., 2018) and producing or recognizing language switches when changing from one language to the other (Abutalebi & Green, 2008).

Some early studies hinted at the promise of bilingual advantage. On tasks requiring cognitive control—*the ability to regulate behaviour and resolve interference among competing representations*—bilinguals outperformed monolinguals selectively on trials inducing conflict (Bialystok et al., 2004). Other evidence reflected broader patterns: bilinguals are better at conflict monitoring during goal-directed tasks, performing faster generally under high, but not low, conflict-monitoring conditions (Costa et al., 2009). However, subsequent mounting evidence pointed away from a global cognitive processing advantage and failed to find significant differences in general working memory between bilinguals and monolinguals (Bialystok, 2010; Bonifacci et al., 2011; Engel de Abreu, 2011; Namazi & Thordardottir, 2010).

Two recent reviews suggest the evidence for the bilingual advantage is, at best, mixed. Van den Noort et al. (2019) conducted an analysis of 2692 bilingual participants, of whom 601 were children and 2091 were adults, across 46 original studies. Although a majority of the studies (54.3%) found a bilingual advantage on cognitive control tasks, many reported mixed (28.3%) or no evidence (17.4%) in support of it. Van den Noort et al. observed a preponderance of methodological issues with many of the studies, including issues with how participants were selected and determined to be bilingual and the use of non-standardized tests of cognitive control and other constructs. They criticized some of the studies for ignoring individual differences or failing to use true longitudinal designs.

Calvo et al. (2016) similarly raised methodological and procedural concerns, but their assessment of the literature is that while bilingualism may not enhance general working memory processing, bilingualism may improve certain aspects of it. Calvo et al. identified several studies that failed to find an overall bilingual benefit, but even among many of these studies, they did find such effects in specific tasks or conditions (e.g. Bialystok et al., 2004). Hence, they called for further research on bilingualism and discrete aspects of working memory and executive function.

Working memory training and language classroom practice

For decades, researchers and educators alike have turned to educational technology in hopes of discovering ways in which to attenuate or even remediate cognitive difficulties associated with poor educational progress. Evidence-based programmes have been developed to promote language-specific skills such as reading (e.g. Fast ForWord; see Tallal et al., 1998) and phonological awareness (O'Connor et al., 1993; Tyler et al., 2011), as well as more general skills such as working memory, self-regulation, and cooperative learning (e.g. Tools of the Mind; see Diamond et al., 2007, 2019). In terms of second language learning and classroom practice, no serious research has been done on the possible effects of working memory interventions on facilitating L2 learning, despite its great potential (Tsai et al., 2016).

The efficacy of some of these regimes is well established. For instance, phonological awareness training has been endorsed by the US Department of Education (US Department of Education, Institute of Education Sciences, & What Works Clearinghouse, 2012). The evidence supporting the efficacy of working memory training, however, is considerably less certain. Working memory training falls in the category of intensive training that focuses on specific core cognitive competencies as opposed to domain-specific skills. Sometimes known as 'brain-training', this sector of the ed-tech market exploded in the first decade of this century fuelled in part by tantalizing findings and positive effects on ageing populations (Buschkuehl et al., 2008), individuals with anxiety disorders (Schmidt et al., 2009), and children and adults with disorders of memory and attention (Klingberg, 2010). However, under increased scrutiny, further research and meta-analyses cast doubt on the efficacy of working memory and some other kinds of general cognitive abilities training. A comprehensive review of the evidence, pro and con, behind working memory training is beyond the scope of this chapter, but interested readers are referred to

Novick et al. (2019) for a comprehensive and balanced review of the literature on working memory and other cognitive training.

Holmes and Gathercole (2014) examined the impact of working memory training for children in classrooms with training administered by teachers. In Trial 1, a whole class of children aged 8–9 years received training and their performance on a range of working memory tasks was assessed before and after training. In Trial 2, the impact of teacher-led training on end-of-year school assessments was evaluated with children with poor academic performance. The students completed a standard protocol for CogMed working memory training (Pearson Education, n.d.). The children in Trial 1 made training gains across a range of untrained standardized measures of working memory. Improvements, which were significant for all assessed aspects of working memory, were greatest for tasks that required children to recall sequences of visuospatial information or simultaneously hold in mind and manipulate sequences of verbal or visuospatial information. Trial 2 with low-achieving children offered evidence that training enhanced academic performance. In comparison to untrained peers, children in the trained group made significantly greater progress across the academic year in English (speaking, listening, reading, and writing skills), and a greater proportion of the trained group reached target levels of attainment in the National Curriculum tests.

Rode et al. (2014) similarly tested the effectiveness of working memory training in a classroom context. They, too, found some evidence for the efficacy of working memory training in the form of correlations between training task performance and pretest and post-test transfer measures of working memory, tests of school achievement, including reading, and teacher ratings. However, the effect sizes were very small and inconsistent across transfer tasks, leading the authors to conclude that they had little evidence in support of the hypothesis that working memory training has generalized effects on intellectual functioning or academic performance. Some strengths of their study were that they used a standard working memory training task over 17 sessions in the classroom, they documented the construct validity of their training and transfer tasks, and they tested a large number of children ($n = 156$ third-grade children participated in the training, while $n = 126$ participated in other school activities). A limitation, however, was that they could not invoke a double-blind training paradigm given the way the training was administered in the classroom.

The role of working memory in following instructions

The ability to follow instructions is important to everyday life. In an academic context, difficulty in following instructions can have negative repercussions on grades, learning, and overall achievement (Dunham et al., 2020). A relationship between working memory and the process of following instructions has been long established. In one early investigation, Brener (1940) tested the ability of participants to follow simple commands, such as, 'put a comma below B'. Brener reported that the ability to follow instructions correlated significantly with the digit span task, a long-used measure of short-term memory.

In a series of studies, Yang (2011; see also Yang et al., 2014) has studied the role of working memory in the act of following instructions. Yang (2011) conducted seven experiments in which a dual-task methodology was used to isolate contributions of the subcomponents of working memory, including the visuospatial sketchpad, phonological loop, and central executive, as specified in the multicomponent model (Baddeley & Hitch, 1974). Articulatory suppression (i.e. participants repeating numbers while listening to instructions) was used to isolate the role of the phonological loop. To disrupt the visuospatial sketchpad, participants performed a simple finger-tapping task while listening to instructions (Experiment 4), while they closed their eyes in order to prevent the encoding of visual and spatial information listening to instructions (Experiment 5). To attenuate the involvement of the central executive during listening, participants performed a backward counting task. According to Yang, the central executive had the most significant role in following instructions, as it supported the encoding and maintenance of sequences of actions. The phonological loop appeared to support the perception and maintenance of verbal instructions, while the visuospatial sketchpad appeared to be to encode and bind visual and spatial cues in an action (e.g. 'push the black pencil, and spin the green eraser, and touch the red pencil, and push the blue ruler, and touch the white eraser'; Yang, 2011, p. 78).

Sometimes called the teach-back or self-enactment method, a strategy for improving children's ability to follow instructions entails having children immediately act out the information received in the instructions. Jaroslawska et al. (2016) found that instruction recall was enhanced by carrying out the instructions as they were presented or immediately afterwards at recall. Conversely, the alternative strategies of the child silently reading along with the instructions or repeating the spoken instructions aloud during presentation did not improve instruction-following over spoken presentation alone. Jaroslawska et al. proposed that the benefits of physically acting out the instructions arise from the existence of a short-term motor store that maintains the temporal, spatial, and motoric features of either planned or already executed actions. Yang et al. (2017) found that children with attention deficit hyperactivity disorder benefit from the self-enactment method, too. They also showed that both children with and without attention deficit hyperactivity disorder benefited when teachers demonstrated the instructions, at least when the instructions were given verbally.

Parallels can be made between children's developing working memory capacities (which are still limited) and older adults' (usually considered) declining performance. The ability to successfully follow instructions is imperative across the lifespan and certainly no less so for older adults. Coats et al. (2021) found that, like children, older adults benefited when instructions were acted or demonstrated by the person giving the instructions. However, unlike the children in Jaroslawska et al.'s (2016) study, older adults did not benefit when asked to self-enact the instructions. For a more comprehensive discussion of research on remembering and acting on instructions see Allen et al. (Chapter 10, this volume).

Improving working memory capacity

Working memory and problem-solving

Individual differences in working memory have been shown to be related to differences in several problem-solving and intelligence tests (Conway et al., 2002), as well as reading comprehension (Just & Carpenter, 1992). Researchers have theorized that these relationships are due to people with lower working memory having difficulty retaining contextual cues in the text (Daneman & Green, 1986), and with higher working memory people retain multiple interpretations of the text during reading, thus improving their pragmatic interpretation (Just & Carpenter, 1992).

It is one thing to recognize that basic abilities such as working memory capacity are important for language ability, but it is quite another thing to use that information to improve performance. The following are two simple strategies that may be used for people to work within their inherent working memory capacity limitations to improve language comprehension performance.

Chunking

Although individuals differ in their memory capacities, all people do possess the same general limitation that they can hold no more than a few (generally considered four plus or minus one) items in working memory at any given time (Cowan, 2001). Despite these limitations, there are means by which working memory capacity can be expanded in a practical sense, and the first is through chunking.

A chunk is defined as a meaningful unit of information, although it may take many forms. A chunk can be a single letter or number, or something larger like a word or complex idea. Take the following instance: when asked to remember a 12-letter sequence, most people fail because the task places too great a demand on working memory.

S N I C I D A I B A F A

Someone might have difficulty with this task until told to divide the items by threes. But what if someone were asked to retain the following items in working memory?

BTWPOSWTHIMO

People familiar with the lingo of email, texting, and instant messaging might recognize these letters as acronyms for, By The Way, Parent Over Shoulder, What The Heck, and In My Opinion. As such, the letters are divided into four chunks, as opposed to 12 (unrelated letters). The letters BTW are a single chunk and require only a single unit of effort to retain, as opposed to three. This example illustrates how chunks are defined by how the mind groups items, rather than some intrinsic size or value.

By chunking, people can increase the amount of information held in working memory for short periods. This can be accomplished by becoming familiar enough with the materials being studied that tricks and strategies become available for grouping items that might otherwise not be apparent. Developing more complex relationships between people and places also allows these relationships to act as binders, consolidating them into chunks. For example, chess masters (but not amateurs) have been shown to chunk relationships between pieces in order to better recall the pieces' positions on a chessboard (Gobet & Clarkson, 2004: Gobet et al., 2001). Just like with chess players, as analysts expand their understanding of the context in which an interaction takes place (e.g. relationships between people, hierarchy, location), working memory for the situation becomes freed up for other things.

Withholding judgement

A second and more complex means by which working memory can be expanded is by withholding judgement about a text during its interpretation. Research has shown that people with low working memory pay more attention to specific details about a text while reading it, while those with higher working memory benefit from taking in more information about the context of the text during the initial reading, saving observational judgement until after the interpretation is complete (Whitney et al., 1991). While we are not able to say whether increasing working memory would therefore improve one's ability to interpret context, the benefits of a 'withhold judgement' strategy are still valid. Working memory is best targeted to taking in information about context regarding the text, at least on first interpretation. Those people with higher working memory, and therefore those who tend to possess the greatest reading comprehension, use this strategy, and all people, regardless of working memory capacity, could benefit from it as well.

Elaboration and refreshing

According to Cowan's (2005) embedded-processes model of working memory, stimuli in the focus of attention are impervious to forgetting due to decay or interference. However, stimuli outside of the focus of attention but still recently active are susceptible to forgetting due to interference or decay. For memory representations outside of the focus of attention and on the verge of forgetting, refreshing is a general attention-based mechanism by which briefly thinking of a stimulus returns it to a state of heightened activation (Bartsch et al., 2018; Cowan, 1999; Johnson, 1992). The benefits of refreshing on delayed item recognition, or long-term memory, have been well established in studies that contrasted the refreshing of a single word to repeated reading of a word (Johnson et al., 2002).

Elaboration or elaborative rehearsal refers to still other general attention-based processes that more deeply encode and store information for later retrieval (Craik & Lockhart 1972; Johnson, 1992; Johnson et al., 2002). By making associations between the new stimuli one is trying to learn and prior knowledge, one can enrich the memory representation of the new stimuli. Elaborative rehearsal can involve thinking of examples, visualizing an image of the stimuli, and using mnemonic devices (for a comprehensive review of mnemonic devices, see Bellezza, 1981). The beneficial effects of elaboration for long-term memory are also well established (Bartsch et al., 2018; Craik & Tulving, 1975).

However, Bartsch et al. (2018) compared the effects of elaboration and refreshing on working memory performance, and their results cast doubt that either of these processes much improves working memory performance. Across four conditions, they either instructed participants to use refreshing or elaboration and contrasted the results to a baseline method of repeating (or rereading) the memory stimuli during the maintenance phase of a working memory task. For the memory task, participants were instructed to view and remember six nouns in serial order. After the list presentation, either the first three words or the last three words were to receive additional processing. In the repeat-without-elaboration condition, the three words reappeared sequentially on the screen, and the participants had to re-read them silently. In the refresh-without-elaboration condition, visual prompts appeared on the screen in place of the words, and participants were instructed to 'think of' the corresponding words as soon as the prompts appeared. In the repeat-with-elaboration condition, the three to-be-processed words appeared on the screen in sequential order, and participants were instructed to generate a vivid mental image of the three objects interacting. In the combined refresh-with-elaboration condition, participants were both prompted to 'think of' the words and form a vivid mental image of the items together. Immediately following one of the four additional processing stages described above, serial-position recognition memory for the original six was tested, thus serving as a test effect of additional processing on working memory. After completing a block of four trials, participants were given a distraction task (solving mental arithmetic problems for 2 minutes) before their recognition memory for the words was tested again. This served as a test of additional processing on long-term memory for the words.

As would be expected from prior research, Bartsch et al. (2018) found that elaboration had a beneficial effect on long-term memory. However, contrary to previous findings, they found no such effect for refreshing, which they hypothesized to mean that refreshing and elaboration are separate processes. Furthermore, and perhaps more surprising, neither refreshing nor elaboration did much to improve working memory performance, compared with a baseline condition that consisted of no processing of the memory representations after encoding.

Conclusion

In this chapter, we have summarized the developments of the conceptions of working memory construct and its assessment procedures in cognitive psychology that have lent theoretical frameworks and methodological ramifications to the language sciences including bilingual development. After some 50 years of intense research from both cognitive psychology and psycholinguistics, theoretical links are emerging between working memory constructs and language learning domains and skills, paving the ground for an integrated framework and overarching model (Wen, 2016; Wen & Schwieter, 2022). More importantly, Baddeley's seminal multicomponent model of working memory has contributed substantial inspiration to language and bilingualism research, while executive models such as Cowan's and Engle's models are gaining increasing credence. Thus, we are witnessing a paradigm shift in key working memory models from the originally dominated components' view to the more recent emphasis on executive and functional models, with the future trends directing towards the new conception and measurement of attention control (Engle, 2018; see Mashburn et al., Chapter 7, this volume). It is conceivable that these trends represent promising future directions for exploring the working memory–language nexus.

Acknowledgements

We would like to thank Alan Baddeley, Nelson Cowan, Robert Logie, Peter Skehan, and Richard Sparks for their constructive comments and suggestions on different aspects of the chapter. Any errors and misinterpretations are our sole responsibility.

References

Abu-Rabia, S. (2003). The influence of working memory on reading and creative writing processes in a second language. *Educational Psychology*, 23(2), 209–222.
Abutalebi, J., & Green, D. W. (2008). Control mechanisms in bilingual language production: Neural evidence from language switching studies. *Language and Cognitive Processes* 23(4), 557–582.
Adams, E., Forsberg, A., & Cowan, N. (2022). The Embedded-Processes Model and Language Use. In J. Schwieter & Z. Wen (Eds.), *The Cambridge Handbook of Working Memory and Language* (Cambridge Handbooks in Language and Linguistics, pp. 73–97). Cambridge University Press.

Ahmadian, M. J. (2012). The effects of guided careful online planning on complexity, accuracy and fluency in intermediate EFL learners' oral production: The case of English articles. *Language Teaching Research, 16*(1), 129–149.

Atkins, P. W. B., & Baddeley, A. D. (1998). Working memory and distributed vocabulary learning. *Applied Psycholinguistics, 19*(4), 537–552.

Atkinson, A. L., Allen, R. J., Baddeley, A. D., Hitch, G. J., & Waterman, A. H. (2021). Can valuable information be prioritized in verbal working memory? *Journal of Experimental Psychology: Learning, Memory, and Cognition, 47*(5), 747–764.

Automatic Language Processing Advisory Committee (ALPAC). (1966). *Languages and machines: Computers in translation and linguistics. A report by the Automatic Language Processing Advisory Committee, Division of Behavioral Sciences, National Academy of Sciences, National Research Council.* National Academy of Sciences, National Research Council. https://www.nap.edu/resource/alpac_lm/ARC000005.pdf

Baddeley, A. D. (1986). *Working memory.* Oxford University Press.

Baddeley, A. D. (2000). The episodic buffer: A new component of working memory? *Trends in Cognitive Sciences, 4*(11), 417–423.

Baddeley, A. D. (2003). Working memory: Looking back and looking forward. *Nature Reviews Neuroscience, 4,* 829–839.

Baddeley, A. D. (2007). *Working memory, thought, and action.* Oxford University Press.

Baddeley, A. D. (2012). Working memory: Theories, models, and controversies. *Annual Review of Psychology, 63,* 1–29.

Baddeley, A. D. (2015). Working memory in second language learning. In Z. E. Wen, M. B. Mota, & A. McNeill (Eds.), *Working memory in second language acquisition and processing* (pp. 17–28). De Gruyter.

Baddeley, A. D. (2022). Working memory and challenges of language. In J. Schwieter & Z. E. Wen (Eds.), *The Cambridge handbook of working memory and language* (pp. 19–28). Cambridge University Press.

Baddeley, A. D., Allen, R. J., & Hitch, G. J. (2011). Binding in visual working memory: The role of the episodic buffer. *Neuropsychologia, 49*(6), 1393–1400.

Baddeley, A. D., & Hitch, G. J. (1974). Working memory. In G. A. Bower (Ed.), *Recent advances in learning and motivation* (Vol. 8, pp. 47–89). Academic Press.

Baddeley, A. D., Hitch, G. J., & Allen, R. (2021). A multicomponent model of working memory. In R. H. Logie, V. Camos, & N. Cowan (Eds.), *Working memory: State of the science* (pp. 10–43). Oxford University Press.

Baddeley, A. D., Gathercole, S. E., & Papagno, C. (1998). The phonological loop as a language learning device. *Psychological Review, 105*(1), 158–173.

Baddeley, A. D., & Logie, R. H. (1999). Working memory: The multiple-component model. In A. Miyake & P. Shah (Eds.), Models of working memory: Mechanisms of active maintenance and executive control (pp. 28–61). Cambridge University Press.

Bartsch, L. M., Singmann, H., & Oberauer, K. (2018). The effects of refreshing and elaboration on working memory performance, and their contributions to long-term memory formation. *Memory & Cognition, 46*(5), 796–808.

Bellezza, F. S. (1981). Mnemonic devices: Classification, characteristics, and criteria. *Review of Educational Research, 51*(2), 247–275.

Bialystok, E. (2010). Global-local and trail-making tasks by monolingual and bilingual children: Beyond inhibition. *Developmental Psychology, 46*(1), 93–105.

Bialystok, E., Craik, F. I. M., Klein, R., & Viswanathan, M. (2004). Bilingualism, aging, and cognitive control: Evidence from the Simon task. *Psychology and Aging, 19*(2), 290–303.

Blom, E., Küntay, A. C., Messer, M., Verhagen, J., & Leseman, P. (2014). The benefits of being bilingual: Working memory in bilingual Turkish-Dutch children. *Journal of Experimental Child Psychology, 128,* 105–119.

Blumenfeld, H. K., & Marian, V. (2014). Cognitive control in bilinguals: Advantages in stimulus-stimulus inhibition. *Bilingualism, 17*(3), 610–629.

Bonifacci, P., Giombini, L., Bellocchi, S., & Contento, S. (2011). Speed of processing, anticipation, inhibition and working memory in bilinguals. *Developmental Science, 14*(2), 256–269.

Brener, R. (1940). An experimental investigation of memory span. *Journal of Experimental Psychology*, *26*(5), 467–482.

Bunting, M. F. (2006). Proactive interference and item similarity in working memory. *Journal of Experimental Psychology: Learning, Memory, and Cognition*, *32*(2), 183–196.

Buschkuehl, M., Jaeggi, S. M., Hutchison, S., Perrig-Chiello, P., Däpp, C., Müller, M., Breil, F., Hoppeler, H., & Perrig, W. J. (2008). Impact of working memory training on memory performance in old-old adults. *Psychology and Aging*, *23*(4), 743–753.

Butcher, K., & Kintsch, W. (2003). Text comprehension and discourse processing. In A. F. Healy, R. W. Proctor, & I. B. Weiner (Eds.), *Handbook of psychology* (pp. 575–595). Wiley.

Bybee, J. (2010). *Language, usage, and cognition*. Cambridge University Press.

Calvo, N., Ibáñez, A., & García, A. M. (2016). The impact of bilingualism on working memory: A null effect on the whole may not be so on the parts. *Frontiers in Psychology*, *7*, 265.

Caplan, D., & Waters, G. S. (1999). Verbal working memory and sentence comprehension. *Behavioral and Brain Sciences*, *22*(1), 77–94.

Carruthers, P. (2003). On Fodor's problem. *Memory & Language*, *18*(5), 502–523.

Chafe, W. (1994). *Discourse, consciousness, and time*. University of Chicago Press.

Cheung, H. (1996). Non-word span as a unique predictor of second-language vocabulary learning. *Developmental Psychology*, *32*(5), 867–873.

Chomsky, N. (1957). *Syntactic structures*. Mouton & Co.

Clark, H. H., & Clark, E. V. (1977). *Psychology and language: An introduction to psycholinguistics*. Harcourt Brace Jovanovich.

Coats, R. O., Waterman, A. H., Ryder, F., Atkinson, A. L., & Allen, R. J. (2021). Following instructions in working memory: Do older adults show the enactment advantage? *Journals of Gerontology: Series B*, *76*(4), 703–710.

Conway, A. R. A., Cowan, N., Bunting, M. F., Therriault, D. J., & Minkoff, S. R. B. (2002). A latent variable analysis of working memory capacity, short-term memory capacity, processing speed, and general fluid intelligence. *Intelligence*, *30*(2), 163–183.

Conway, A. R. A., Jarrold, C., Kane, M. J., Miyake, A., & Towse, J. N. (Eds.). (2007). *Variation in working memory*. Oxford University Press.

Conway, A. R. A., Kane, M. J., Bunting, M. F., Hambrick, D. Z., Wilhelm, O., & Engle, R. W. (2005). Working memory span tasks: A methodological review and user's guide. *Psychonomic Bulletin & Review*, *12*(5), 769–786.

Corballis, M. (2020). One person, two minds. *Brain*, *143*(10), 3164–3167.

Costa, A., Hernández, M., Costa-Faidella, J., & Sebastián-Gallés, N. (2009). On the bilingual advantage in conflict processing: Now you see it, now you don't. *Cognition*, *113*(2), 135–149.

Cowan, N. (1995). *Attention and memory: An integrated framework* [Oxford Psychology Series, No. 26]. Oxford University Press.

Cowan, N. (1999). An embedded-processes model of working memory. In A. Miyake & P. Shah (Eds.), *Models of working memory: Mechanisms of active maintenance and executive control* (pp. 62–101). Cambridge University Press.

Cowan, N. (2001). The magical number 4 in short-term memory: A reconsideration of mental storage capacity. *Behavioral and Brain Sciences*, *24*(1), 87–185.

Cowan, N. (2005). *Working memory capacity*. Psychology Press.

Cowan, N. (2008). What are the differences between long-term, short-term, and working memory? *Progress in Brain Research*, *169*, 323–338.

Cowan, N. (2017). The many faces of working memory and short-term storage. *Psychonomic Bulletin & Review*, *24*(4), 1158–1170.

Cowan, N., Elliott, E. M., Saults, J. S., Morey, C. C., Mattox, S., Hismjatullina, A., & Conway, A. R. A. (2005). On the capacity of attention: Its estimation and its role in working memory and cognitive aptitudes. *Cognitive Psychology*, *51*, 42–100.

Cowan, N., Morey, C. C., & Naveh-Benjamin, M. (2020). An embedded-processes approach to working memory: How is it distinct from other approaches, and to what ends? In R. H. Logie, V. Camos, & N. Cowan (Eds.), *Working memory: State of the science* (pp. 44–84). Oxford University Press.

Craik, F. I. M., & Lockhart, R. S. (1972). Levels of processing: A framework for memory research. *Journal of Verbal Learning and Verbal Behavior, 11*(6), 671–684.

Craik, F. I. M., & Tulving, E. (1975). Depth of processing and the retention of words in episodic memory. *Journal of Experimental Psychology: General, 104*(3), 268–294.

Cummins, J. (1979). Linguistic interdependence and the educational development of bilingual children. *Review of Educational Research, 49*(2), 222–251.

Cummins, J. (2021). *Rethinking the education of multilingual learners: A critical analysis of theoretical concepts.* Channel View Publications.

Daneman, M., & Carpenter, P. A. (1980). Individual differences in working memory and reading. *Journal of Verbal Learning and Verbal Behavior, 19*(4), 450–466.

Daneman, M., & Green, I. (1986). Individual differences in comprehending and producing words in context. *Journal of Memory and Language, 25*(1), 1–18.

Daneman, M., & Merikle, P. M. (1996). Working memory and language comprehension: A meta-analysis. *Psychonomic Bulletin & Review, 3*, 422–433.

de Bruïne, A., Jolles, D., & van den Broek, P. (2021). Minding the load or loading the mind: The effect of manipulating working memory on coherence monitoring. *Journal of Memory & Language, 118*, 104212.

Diamond, A., Barnett, W. S., Thomas, J., & Munro, S. (2007). Preschool program improves cognitive control. *Science, 318*(5855), 1387–1388.

Diamond, A., Lee, C., Senften, P., Lam, A., & Abbott, D. (2019). Randomized control trial of Tools of the Mind: Marked benefits to kindergarten children and their teachers. *PLoS One, 14*(9), e0222447.

Duff, M., & Piai, V. (Eds.). (2020). *Language and memory: Understanding their interactions, interdependencies, and shared mechanisms.* Frontiers Media SA.

Dunham, S., Lee, E., & Persky, A. M. (2020). The psychology of following instructions and its implications. *American Journal of Pharmaceutical Education, 84*(8), 7779.

Einstein, A. (1905). Über einen die erzeugung und verwandlung des lichtes betreffenden heuristischen gesichtspunkt [On a heuristic point of view about the creation and conversion of light]. *Annalen der Physik, 17*(6), 132–148.

Ellis, N. C. (1996). Working memory in the acquisition of vocabulary and syntax: Putting language in good order. *Quarterly Journal of Experimental Psychology Section A: Human Experimental Psychology, 49*(1), 234–250.

Ellis, N. C. (2012). Formulaic language and second language acquisition: Zipf and the phrasal teddy bear. *Annual Review of Applied Linguistics, 32*, 17–44.

Engel de Abreu, P. M. J. (2011). Working memory in multilingual children: Is there a bilingual effect? *Memory, 19*(5), 529–537.

Engle, R. W. (2002). Working memory capacity as executive attention. *Current Directions in Psychological Science, 11*(1), 19–23.

Engle, R. W. (2018). Working memory and executive attention: A revisit. *Perspectives on Psychological Science, 13*(2), 190–193.

Engle, R. W., & Conway, A. R. A. (1998). Working memory and comprehension. In R. Logie & K. Gilhooly (Eds.), *Working memory and thinking* (pp. 67–92). Psychology Press.

Engle, R. W., Carullo, J. J., & Collins, K. W. (1991). Individual differences in working memory for comprehension and following directions. *Journal of Educational Research, 84*(5), 253–262.

Foster, P., Bolibaugh, C., & Kotula, A. (2014). Knowledge of nativelike selections in a L2: The influence of exposure, memory, age of onset, and motivation in foreign language and immersion settings. *Studies in Second Language Acquisition, 36*(1), 101–132.

French, L. M., & O'Brien, I. (2008). Phonological memory and children's second language grammar learning. *Applied Psycholinguistics, 29*(3), 463–487.

Friesen, D. C., Latman, V., Calvo, A., & Bialystok, E. (2015). Attention during visual search: The benefit of bilingualism. *International Journal of Bilingualism, 19*(6), 693–702.

Gathercole, S. E., & Baddeley, A. D. (1993). *Working memory and language.* Lawrence Erlbaum Associates, Inc.

Gathercole, S. E., Alloway, T. P., Willis, C., & Adams, A. M. (2006). Working memory in children with reading disabilities. *Journal of Experimental Child Psychology, 93*(3), 265–281.

Gathercole, S. E., Willis, C. S., Baddeley, A. D., & Emslie, H. (1994). The Children's Test of Nonword Repetition: A test of phonological working memory. *Memory, 2*(2), 103–127.

Gobet, F., & Clarkson, G. (2004). Chunks in expert memory: Evidence for the magical number four . . . or is it two? *Memory, 12*(6), 732–747.

Gobet, F., Lane, P. C. R., Croker, S., Cheng, P. C. H., Jones, G., & Pine, J. M. (2001). Chunking mechanisms in human learning. *Trends in Cognitive Sciences, 5*(6), 236–243.

Goldberg, A. E. (2003). Constructions: A new theoretical approach to language. *Trends in Cognitive Sciences, 7*(5), 219–224.

Gomez-Rodriguez, C., Christiansen, M. H., & Ferrer-i-Cancho, R. (2022). Memory limitations are hidden in grammar. *Glottometrics, 52*, 39–64.

Grundy, J. G., & Timmer, K. (2017). Bilingualism and working memory capacity: A comprehensive meta-analysis. *Second Language Research, 33*(3), 325–340.

Harrington, M., & Sawyer, M. (1992). L2 working memory capacity and L2 reading skill. *Studies in Second Language Acquisition, 14*(1), 25–38.

Holmes, J., & Gathercole, S. E. (2014). Taking working memory training from the laboratory into schools. *Educational Psychology, 34*(4), 440–450.

James, W. (1890). *The principles of psychology*. Holt.

Jaroslawska, A. J., Gathercole, S. E., Logie, M. R., & Holmes, J. (2016). Following instructions in a virtual school: Does working memory play a role? *Memory & Cognition, 44*(4), 580–589.

Johnson, M. K. (1992). MEM: Mechanisms of recollection. *Journal of Cognitive Neuroscience, 4*(3), 268–280.

Johnson, M. K., Reeder, J. A., Raye, C. L., & Mitchell, K. J. (2002). Second thoughts versus second looks: An age-related deficit in reflectively refreshing just-activated information. *Psychological Science, 13*(1), 64–67.

Joseph, H. S., Bremner, G., Liversedge, S. P., & Nation, K. (2015). Working memory, reading ability and the effects of distance and typicality on anaphor resolution in children. *Journal of Cognitive Psychology, 27*(5), 622–639.

Just, M., & Carpenter, P. A. (1992). A capacity theory of comprehension: Individual differences in working memory. *Psychological Review, 99*(1), 122–149.

Kane, M. J., Hambrick, D. Z., Tuholski, S. W., Wilhelm, O., Payne, T. W., & Engle, R. W. (2004). The generality of working memory capacity: A latent-variable approach to verbal and visuospatial memory span and reasoning. *Journal of Experimental Psychology: General, 133*(2), 189–217.

Klingberg, T. (2009). *The overflowing brain: Information overload and the limits of working memory*. Oxford University Press.

Klingberg, T. (2010). Training and plasticity of working memory. *Trends in Cognitive Sciences, 14*(7), 317–324.

Lasagabaster, D. (2001). The effect of knowledge about the L1 on foreign language skills and grammar. *International Journal of Bilingual Education and Bilingualism, 4*(5), 310–331.

Li, S. (2017). Student and teacher beliefs and attitudes about oral corrective feedback. In H. Nassau & E. Kartchava (Eds.), *Corrective Feedback in Second Language Teaching and Learning: Research, Theory, Applications, Implications* (pp. 143–157). Milton: Taylor and Francis.

Linck, J. A., Osthus, P., Koeth, J. T., & Bunting, M. F. (2014). Working memory and second language comprehension and production: A meta-analysis. *Psychonomic Bulletin & Review, 21*(4), 861–883.

Liu, H. (2008). Dependency distance as a metric of language comprehension difficulty. *Journal of Cognitive Science, 9*(2), 159–191.

Llompart, M., & Dabrowska, E. (2020). Explicit but not implicit memory predicts ultimate attainment in the native language. *Frontiers in Psychology, 11*, 569–586.

Logie, R., Camos, V., & Cowan, N. (Eds.). (2021). *Working memory: State of the science*. Oxford University Press.

Lu, B., & Wen, Z. E. (2022). Short-term and working memory capacity and the language device: Chunking and parsing complexity. In J. Schwieter & Z. E. Wen. (Eds.), *The Cambridge handbook of working memory and language* (pp. 393–418). Cambridge University Press.

Mackey, A., Philp, J., Egi, T., Fujii, A., & Tatsumi, T. (2002). Individual differences in working memory, noticing of interactional feedback and L2 development. In P. Robinson (Ed.), *Individual differences and second language instruction* (pp. 181–209). John Benjamins Publishing Company.

MacWhinney, B., & O'Grady, W. (Eds.). (2015). *The handbook of language emergence*. Wiley-Blackwell.

Masoura, E. V., & Gathercole, S. E. (2005). Contrasting contributions of phonological short-term memory and long-term knowledge to vocabulary learning in a foreign language. *Memory, 13*(3/4), 422–429.

Michael, E. B., & Gollan, T. H. (2005). Being and becoming bilingual: Individual differences and consequences for language production. In J. F. Kroll & A. M. B. deGroot (Eds.), *Handbook of bilingualism: Psycholinguistic approaches* (pp. 389–407). Oxford University Press.

Miller, G. A. (1956). The magical number seven, plus or minus two: Some limits on our capacity for processing information. *Psychological Review, 63*(2), 81–97.

Miller, G. A., Galanter, E., & Pribram, K. H. (1960). *Plans and the structure of behavior*. Holt, Rinehart and Winston, Inc.

Miyake, A., Friedman, N. P., Emerson, M. J., Witzki, A. H., Howerter, A., & Wager, T. D. (2000). The unity and diversity of executive functions and their contributions to complex 'front lobe' tasks: A latent variable analysis. *Cognitive Psychology, 41*, 48–100.

Miyake, A., & Shah, P. (Eds.). (1999). *Models of working memory: Mechanisms of active maintenance and executive control*. Cambridge University Press.

Monsell, S. (2003). Task switching. *Trends in Cognitive Sciences, 7*(3), 134–140.

Morris, N., & Jones, D. M. (1990). Memory updating in working memory: The role of the central executive. *British Journal of Psychology, 81*(2), 111–121.

Namazi, M., & Thordardottir, E. (2010). A working memory, not bilingual advantage, in controlled attention. *International Journal of Bilingual Education and Bilingualism, 13*(5), 597–616.

Neisser, U. (1978). Memory: What are the important questions? In M. M. Gruneberg, P. E. Morris, & R. N. Sykes (Eds.). *Practical aspects of memory* (pp. 3–24). Academic Press.

Norman, D. A., & Shallice, T. (1986). Attention to action. In R. J. Davidson, G. E. Schwartz, & D. Shapiro (Eds.), *Consciousness and self-regulation* (pp. 1–18). Springer.

Novick, J. M., Bunting, M. F., Dougherty, M. R., & Engle, R. W. (Eds.). (2019). *Cognitive and working memory training: Perspectives from psychology, neuroscience, and human development*. Oxford University Press.

O'Brien, I., Segalowitz, N., Collentine, J., & Freed, B. (2006). Phonological memory and lexical, narrative, and grammatical skills in second language oral production by adult learners. *Applied Psycholinguistics, 27*(3), 377–402.

O'Brien, I., Segalowitz, N., Freed, B., & Collentine, J. (2007). Phonological memory predicts L2 oral fluency gains in adults. *Studies in Second Language Acquisition, 29*(4), 557–582.

O'Connor, R. E., Jenkins, J. R., Leicester, N., & Slocum, T. A. (1993). Teaching phonological awareness to young children with learning disabilities. *Exceptional Children, 59*(6), 532–546.

O'Grady, W. (2017). Working memory and language: From phonology to grammar. *Applied Psycholinguistics, 38*(6), 1340–1343.

Oberauer, K., Lewandowsky, S., Awh, E., Brown, G., Conway, A., Cowan, N., Donkin, C., Farrell, S., Hitch, G. J., Hurlstone, M. J., Ma, W. J., Morey, C. C., Nee, D. E., Schweppe, J., Vergauwe, E., & Ward, G. (2018). Benchmarks for models of short-term and working memory. *Psychonomic Bulletin, 144*(9), 885–958.

Pearson Education (n. d.). *Cogmed working memory training*. https://www.pearsonassessments.com/store/usassessments/en/Store/Professional-Assessments/Cognition-%26-Neuro/Interventions/Cogmed-Working-Memory-Training/p/100000069.html?tab=overview

Peng, P., Barnes, M., Wang, C., Wang, W., Li, S., Swanson, H. L., Dardick, W., & Tao, S. (2018). A meta-analysis on the relation between reading and working memory. *Psychological Bulletin, 144*(1), 48–76.

Pierce, L. J., Genesee, F., Delcenserie, A., & Morgan, G. (2017). Variations in phonological working memory: Linking early language experiences and language learning outcomes. *Applied Psycholinguistics, 38*(6), 1265–1300.

Piolat, A., Olive, T., & Kellogg, R. T. (2005). Cognitive effort during note taking. *Applied Cognitive Psychology, 19*(3), 291–312.

Posner, M., & Snyder, C. (1975). Attention and cognitive control. In R. L. Solso (Ed.), *Information processing and cognition: The Loyola symposium* (pp. 55–85). Lawrence Erlbaum.

Ransdell, S., & Levy, C. M. (1996). Working memory constraints on writing quality and fluency. In C. M. Levy & S. Ransdell (Eds.), *The science of writing: Theories, methods, individual differences, and applications* (pp. 93–105). Lawrence Erlbaum Associates, Inc.

Robinson, P. (2004). *Candidates for aptitude constructs and measures: Dual and multi-task performance, executive control and ids in task switching.* University of Maryland Center for Advanced Study of Language.

Robinson, P. (2005). Cognitive abilities, chunk-strength, and frequency effects in implicit artificial grammar and incidental L2 learning. *Studies in Second Language Acquisition, 27*, 235–268.

Rode, C., Robson, R., Purviance, A., Geary, D. C., & Mayr, U. (2014). Is working memory training effective? A study in a school setting. *PLoS One, 9*(8), e104796.

Salthouse, T., Friscoe, N., McGuthery, E., & Hambrick, D. Z. (1998). Relation of task switching to speed, age and fluid intelligence. *Psychology and Aging, 13*(3), 445–461.

Schmidt, N. B., Richey, J. A., Buckner, J. D., & Timpano, K. R. (2009). Attention training for generalized social anxiety disorder. *Journal of Abnormal Psychology, 118*(1), 5–14.

Schwartz, S. (1984). *Measuring reading competence: A theoretical-prescriptive approach.* Plenum Press.

Schwieter, J. W., & Wen, Z. E. (Eds.). (2022). *The Cambridge handbook of working memory and language.* Cambridge University Press.

Segalowitz, N., & Frenkiel-Fishman, S. (2005). Attention control and ability level in a complex cognitive skill: Attention shifting and second-language proficiency. *Memory & Cognition, 33*(4), 644–653.

Serafini, E. J., & Sanz, C. (2016). Evidence for the decreasing impact of cognitive ability on second language development as proficiency increases. *Studies in Second Language Acquisition, 38*(4), 607–646.

Service, E. (1992). Phonology, working memory, and foreign-language learning. *Quarterly Journal of Experimental Psychology Section A, 45*(1), 21–50.

Shiffrin, R. M., & Schneider, W. (1977). Controlled and automatic human information processing: II. Perceptual learning, automatic attending and a general theory. *Psychological Review, 84*(2), 127–190.

Skehan, P. (1998). *A cognitive approach to language learning.* Oxford University Press.

Skinner, B. F. (1957). *Verbal behavior.* Appleton-Century-Crofts.

Struys, E., Woumans, E., Nour, S., Kepinska, O., & Van den Noort, M. (2018). A domain general monitoring account of language switching in recognition tasks: Evidence for adaptive control. *Bilingualism: Language & Cognition, 22*(3), 606–623.

Tallal, P., Merzenich, M., Miller, S., & Jenkins, W. (1998). Language learning impairments: Integrating basic science, technology, and remediation. *Experimental Brain Research, 123*(1–2), 210–219.

Tomasello, M. (2003). *Constructing a language: A usage-based theory of language acquisition.* Harvard University Press.

Tsai, N., Au, J., & Jaeggi, S. M. (2016). Working memory, language processing, and implications of malleability for second language acquisition. In G. Granena, D. O. Jackson, & Y. Yilmaz (Eds.), *Cognitive individual differences in second language processing and acquisition* (pp. 69–88). John Benjamins Publishing Company.

Tyler, A. A., Gillon, G., Macrae, T., & Johnson, R. L. (2011). Direct and indirect effects of stimulating phoneme awareness vs. other linguistic skills in preschoolers with co-occurring speech and language impairments. *Topics in Language Disorders, 31*(2), 128–144.

US Department of Education, Institute of Education Sciences, & What Works Clearinghouse. (2012, June). *Early childhood education interventions for children with disabilities intervention report: Phonological awareness training.* https://ies.ed.gov/ncee/wwc/Docs/InterventionReports/wwc_pat_060512.pdf

van den Noort, M., Struys, E., Bosch, P., Jaswetz, L., Perriard, B., Yeo, S., Barisch, P., Vermeire, K., Lee, S. H., & Lim, S. (2019). Does the bilingual advantage in cognitive control exist and if so, what are its modulating factors? A systematic review. *Behavioral Sciences, 9*(3), 27.

Walter, C. (2004). Transfer of reading comprehension skills to L2 is linked to mental representations of text and to L2 working memory. *Applied Linguistics, 25*(3), 315–339.

Waugh, N. C., & Norman, D. A. (1965). Primary memory. *Psychological Review, 72*(2), 89–104.

Wen, Z. (2015). Working memory in second language acquisition and processing: The phonological/executive model. In Z. Wen, M. Borges Mota, & A. McNeill (Eds.), *Working Memory in Second Language Acquisition and Processing.* Bristol, Blue Ridge Summit: Multilingual Matters.

Wen, Z. E. (2016). *Working memory and second language learning: Towards an integrated approach.* Multilingual Matters.

Wen, Z. E., & Schwieter, J. (2022). Towards an integrated account of working memory and language. In Schwieter, J., & Wen, Z. E. (Eds.), *The Cambridge handbook of working memory and language* (pp. 909–927). Cambridge University Press.

Whitney, P., Ritchie, B., & Clark, M. (1991). Working-memory capacity and the use of elaborative inferences in text comprehension. *Discourse Processes*, *14*(2), 133–146.

Williams, J. N. (2012). Working memory and SLA. In S. Gass & A. Mackey (Eds.), *Handbook of second language acquisition* (pp. 427–441). Routledge/Taylor & Francis.

Yang, T. (2011). *The role of working memory in following instructions* [PhD thesis, University of York].

Yang, T., Allen, R. J., & Gathercole, S. E. (2014). Examining the role of working memory resources in following spoken instructions. *Journal of Cognitive Psychology*, *28*(2), 186–198.

Yang, T., Allen, R. J., Holmes, J., & Chan, R. C. K. (2017). Impaired memory for instructions in children with attention-deficit hyperactivity disorder is improved by action at presentation and recall. *Frontiers in Psychology*, *8*, 39.

PART 3

IMPAIRMENTS OF MEMORY

13

Age-related changes in everyday prospective memory

Fergus I. M. Craik and Julie D. Henry

Introduction

Remembering to perform a planned action at a future time is a frequently occurring necessity of everyday life. Researchers have studied this ability under the heading of 'prospective memory' (PM), with a major emphasis on the conditions under which the planned intention is *not* fulfilled. Our daily lives are filled with many trivial tasks that require PM, such as remembering to switch off a light, to post a letter, or to check how much milk is left in the fridge. However, many other planned intentions can have more serious consequences if they are not fulfilled in a timely manner; examples include remembering to take medication, to check food cooking, to turn off appliances, and to pay bills. Such PM failures are obviously important because they compromise a person's ability to live independently, and reduced PM function has been directly linked to poorer everyday functioning in many clinical populations (see, e.g. Pirogovsky et al., 2012; Twamley et al., 2008), and of particular interest in this chapter, late adulthood (Hering et al., 2018; Rose et al., 2015; Sheppard et al., 2019).

Although PM problems can occur at any age, their frequency typically increases in older adulthood (Henry et al., 2004). The evidence for this conclusion is admittedly mixed, and we discuss possible reasons for these discrepancies in a later section of the chapter, but one point to make immediately is that at least some of this variance likely reflects the complexity of PM itself. Although the term PM may imply a single construct or a defined cognitive skill, it is in fact a complex multi-componential process that entails at least four distinct phases (intention formation, retention, initiation, and completion), each of which draws on basic cognitive resources that are differentially affected by ageing (Ellis, 1996; Kliegel et al., 2000). As we will discuss, when PM is considered in this more nuanced way, apparent 'discrepancies' in age effects across individual studies can often be reconciled surprisingly easily.

Other preliminary points to make about PM include the distinction between its 'pure' prospective aspects and the aspects of retrospective memory that each PM task involves. Successful PM task completion is contingent not only on remembering that *something* should be done at a particular time or in response to a particular

Fergus I. M. Craik and Julie D. Henry, *Age-related changes in everyday prospective memory* In: *Memory in Science for Society.*
Edited by: Robert H. Logie, Zhisheng (Edward) Wen, Susan E. Gathercole, Nelson Cowan, and Randall W. Engle, Oxford University Press.
© Oxford University Press 2023. DOI: 10.1093/oso/9780192849069.003.0013

event occurring (the prospective component), but also on accurate recollection of the specific action to be performed or the information to be conveyed (the retrospective component), and there is some evidence to suggest that age-related differences are greater in the former of these two components (Cohen et al., 2001). A final important distinction concerns the different *cues* that should initiate the planned intention; in general these cues are either 'event-based' (e.g. 'Next time you go to the pharmacy remember to pick up some aspirin') or 'time-based' (e.g. 'Remember to take your medication at 6 pm' or 'Remember to phone your mother in half-an-hour'; McDaniel & Einstein, 2000), but many other important distinctions between cue types are now also recognized. We first review some general characteristics and models of PM functioning, and then return to consider the age-related cognitive and motivational differences that may influence PM performance in both laboratory and real-life situations.

Characteristics of prospective memory

How does a person fulfil a delayed intention? If the action and its consequences are sufficiently important and the delay between intention and action sufficiently short, the person is likely to retain the intention in working memory, thereby increasing the chances of fast and effective completion of the task. In most cases, however, the retention period is long enough that the intention drops out of mind, and PM success then depends on monitoring the environment for the appropriate time or place to perform the action, in conjunction with the hope that well-established habits may provide a reminder that an action is required.

Strategic monitoring and spontaneous retrieval

The PM literature from the 1980s to the present has attempted to clarify and illustrate the roles of strategic monitoring on the one hand, and spontaneous retrieval of the intention on the other. Monitoring is a deliberate, controlled process that depends on executive functions mediated by the frontal lobes. It also reflects metacognitive abilities, in the sense that the individual may decide that the target action is important enough to initiate a monitoring plan that will be costly in terms of processing resources (Rummel & Meiser, 2013). This last point implies that monitoring may detract from the performance of other ongoing activities, and this issue has been the focus of considerable research in the PM literature. Initial studies by Einstein and McDaniel (1996) suggested that remembering to execute a planned action could be triggered spontaneously by the target event, and that retaining the intention might therefore be cost-free in terms of processing resources.

Indeed, in their *multiprocess framework*, McDaniel and Einstein proposed a role for *both* strategic monitoring and spontaneous retrieval (Einstein & McDaniel, 2005; McDaniel & Einstein, 2000). They suggested that spontaneous retrieval may occur as a result of the person setting up an associative link between the future cue and the intended action, the 'reflexive-associative theory' (Einstein & McDaniel, 2005). The credibility of spontaneous retrieval is apparently boosted by the everyday occurrence of thoughts occasionally 'popping into mind' (e.g. Kvavilashvili & Mandler, 2004), and by the related phenomenon of 'involuntary memory' (e.g. Berntsen, 2010). However, it seems to us likely that such apparently uncued and 'spontaneous' thoughts and memories are in fact typically triggered by some current thought or environmental event, although the link itself may never enter conscious awareness (see also Hintzman, 2011 for a discussion of the pervasiveness of 'reminding' in our mental lives). Moreover, and as will be discussed later, Rebekah Smith (2003, 2017; Smith et al., 2007) has consistently found that monitoring a PM intention impairs performance on a simultaneously performed ongoing task, and on the basis of these findings, has argued that there are no cost-free automatic processes involved in PM.

Encoding intentions

There is considerable overlap between the factors associated with effective encoding of episodic memories and those that improve encoding of PM intentions. For instance, deeper semantic processing (such as when the PM cue is presented as a picture rather than as a word) has been shown to enhance PM function (Einstein & McDaniel, 1996). Interestingly too, and similar to the distinction that has been made between automatic and strategic retrieval PM processes, recent evidence suggests that while initial encoding of the intention can be supported by deliberative and strategic processes, it may also be supported by more perfunctory processing, occurring in a transient manner, or 'in passing' (Scullin et al., 2018). However, several explicit strategies have been shown to substantially improve the likelihood of prospective remembering. One of the most effective of these is *implementation intentions*, in which the person formulates and repeats a prospective action to be carried out in a specified context, such as '*If* it is sunny tomorrow, *then* I will walk to the office' (Gollwitzer & Brandstätter, 1997). This strategy is also discussed later in the chapter. Martiny-Huenger et al. (2017) proposed a novel simulation account of action planning, which explains how it is possible for conscious thought to elicit behavioural automaticity via the creation of direct stimulus–response links. By this view, the implementation intentions statement elicits sensorimotor simulations, thereby establishing a direct perception–action link. The creation of this direct link then means that encountering the anticipated situation triggers the intended action via relatively automatic, reflexive processes. Such self-regulatory plans appear to be effective in real life as well as in the laboratory, although McDaniel and Scullin (2010) made the point that they do not benefit responses that are already completely automatized.

Retrieving intentions

Although PM consists of four distinct phases, the retrieval phase has conventionally been of greatest interest to theorists, especially when the cues to remember the encoded intention are subtle or absent. Here, a major contribution has been the distinction between time-based and event-based cues, proposed initially by Einstein and McDaniel (1990; Einstein et al., 1995) in the context of attempts to reconcile their 1990 finding of no age-related differences in PM with Craik's (1983, 1986) suggestion that older adults should be especially vulnerable to PM failures. Craik had argued that remembering in the absence of cues or an appropriate context is highly dependent on 'self-initiated mental operations'. That is, contextual and other features of the sought-for memory must be reconstructed mentally based on the available aspects of the triggering question and plausible options from relevant past experiences. Craik also assumed that such self-initiated operations are mediated by frontal lobe processes (e.g. Stuss & Alexander, 2007), and since the frontal lobes are among the first brain regions to be negatively affected by ageing (e.g. Raz, 2000), self-initiated control processes should also be particularly vulnerable to disruption. In their 1995 article, Einstein, McDaniel, and colleagues found age-related differences in time-based but not event-based PM tasks. They concluded that time-based tasks (with no explicit cues) are particularly dependent on self-initiated processing and so should show age-related impairments in line with Craik's earlier suggestions (Einstein et al., 1995). On the other hand, event-based tasks often involve a greater degree of 'environmental support' which reduces the need for self-initiation (Craik, 1983, 1986). These points are taken up again in the section on age-related differences in PM.

Another consideration relevant to PM success concerns the nature of the ongoing background activities, and the relation between these activities and PM cues. The degree to which the background activity demands participants' attention is one obvious factor, as it follows from current models of PM (e.g. McDaniel and Einstein's multiprocess framework and the *preparatory attention and memory processes* model (PAM) proposed by R. E. Smith, 2003) which regard PM and the ongoing task as competing for shared cognitive resources either all of the time (in the PAM model), or where strategic demands of the PM task are high (in the multiprocess framework). Consistent with both models, there is evidence that where PM task demands are greater this incurs costs to ongoing task processing (Rummel et al., 2017). The natural corollary of course, is that where an ongoing task is resource demanding, this should negatively impact PM performance. However, while some studies have shown that high ongoing task demands disrupt PM performance, this relationship appears to be complex (Meier & Zimmermann, 2015), and it has been argued that, for at least some ongoing tasks, there may be *no* capacity sharing between PM and ongoing task performance at all, with mathematical modelling instead pointing to potential alterations in response thresholds (Strickland et al., 2018, 2019). Thus, although there is good evidence that challenging ongoing tasks can and often do disrupt PM performance, there is ongoing debate about the precise mechanism(s) by which this disruption occurs.

Another important and related determinant of PM retrieval success is the individual's level of interest or *absorption* in the background task; a person whose attention is absorbed in an engrossing novel or film is less likely to remember the PM commitment, although the importance of this variable appears to be greater in laboratory than naturalistic settings (Schnitzspahn et al., 2011). The other side of the coin to absorption is *disengagement*—the ability to break the current attentional set, in order to monitor the environment for relevant cues. For time-based PM tasks this monitoring may involve periodic checking of a clock (Einstein & McDaniel, 1996), and specifically, a series of test–wait–test monitoring cycles, which become increasingly frequent as the target time draws closer, resulting in more frequent disengagements from the ongoing task (Harris & Wilkins, 1982). The probability of effective disengagement however depends on several factors, including the importance of the PM task, the person's degree of absorption in the ongoing task, as well as individual difference characteristics such as mental flexibility, a core component of executive control. Indeed, *perseveration*—continuing to engage in a particular activity when it is no longer relevant or appropriate to do so—is a hallmark of frontal dysfunction (Luria, 1980; Sandson & Albert, 1984) and likely plays some part in age-related PM inefficiencies (Scullin et al., 2012).

A further important characteristic of event-based PM tasks is the extent to which the cue forms part of the ongoing task. For example, if the cue is a word and the ongoing task involves verbal processing, the participant's attention will be directed at potential PM cues, albeit with a different processing goal in mind. McDaniel and Einstein (2000; Einstein & McDaniel, 2008) refer to this situation as involving *focal* as opposed to *non-focal* cues, although again it seems that this is a *dimension* of relatedness rather than a categorical distinction. As will be discussed later, this focal/non-focal difference is also relevant to studies of ageing—age differences are typically greater for non-focal cues in normal adult ageing (Kliegel et al., 2008). The nature and location of the event-based cue is another important determinant of PM success; the more distinctive and unusual the cue, the more successful it is likely to be. As Einstein and McDaniel (1990) comment, 'distinctiveness may alert the subject to view the target event as something other than part of the ongoing task—that is, alerting subjects that it is also a cue for remembering to perform an action' (p. 722). Emotionally salient cues, for instance, improve PM function in some clinical populations (Lui et al., 2021), and can eliminate age differences in PM (Altgassen et al., 2010). Two other aspects of PM success that are particularly relevant to understanding PM age effects are first, possible differences in motivation and second, group differences in the likelihood of setting up explicit reminders such as notes, alarms, and objects in atypical places, and we discuss these factors at greater length in the following sections.

Finally, as with other types of memory, the concordance of context between the encoding and retrieval situations is also a major factor in successful PM. An everyday example is finding yourself in an upstairs room to collect something, and then realizing that the specific 'something' has vanished from mind! It can be readily reinstated, however, by going back down to the living room where the intention was formulated. The importance of a close match between the types of processing carried out at encoding and retrieval has long been acknowledged in the broader episodic memory literature, and variously referred to as *encoding specificity* (Tulving & Thomson, 1973), *repetition of operations* (Kolers, 1973), and *transfer-appropriate processing* (Morris et al., 1977). Contextual information makes up a substantial portion of the encoding of any event, and this is clearly the case for PM tasks as well (Einstein & McDaniel, 1996; Ellis, 1996).

Current frameworks for prospective memory effects

Before describing the experimental and practical work on age-related effects in PM, we will bring together some of the previously discussed findings and ideas in two of the main frameworks that guide research at the present time. A central tenet of the *multiprocess framework* (McDaniel & Einstein, 2000) and the *dynamic multiprocess framework* (Scullin et al., 2013) is that either automatic or attention-driven processes can be used to support prospective remembering at the critical retrieval stage. However, the relative importance of these processes varies as a function of task-specific and contextual demands. In contexts where the person's intended activity is simple and the cue is salient (such as responding to an alarm), they may not need to engage in self-initiated switching of attention, as prospective remembering may be supported by the cue eliciting a relatively automatic orienting response. In contrast, in situations where the intended action is complex or the cue is not salient, PM is assumed to involve strategic, controlled processing. In such circumstances, strategic monitoring of the environment for the appropriate cue is required for a PM task to be executed successfully.

Both frameworks predict that people will engage in monitoring when PM cues are expected and disengage when they are not; in the latter case, PM success will therefore depend on the probability of spontaneous retrieval of the intention. When cues are expected, the attentional demands of monitoring will reduce performance on any ongoing task, but such costs are absent (according to the multiprocess theory) when no monitoring takes place. It is worth commenting that in laboratory situations where PM cues are part of participants' expectations, monitoring may be more prevalent than in real-life situations, especially if the real-life intended action is relatively trivial, or the cue is not expected for some time. In a similar way, while laboratory tasks nearly always require the participant to encode a PM intention with conscious effort, in real-life settings this encoding process may often occur quite casually and perfunctorily. A further point about spontaneous retrieval of intentions in real life is that many related objects or events may remind the person of the PM task (Kvavilashvili & Fisher, 2007).

In contrast to both the original and dynamic multiprocess frameworks, a central tenet of the PAM model (R. E. Smith, 2003) is that PM tasks *never* involve cost-free automatic processes. According to R. E. Smith and colleagues (2007), 'resource-demanding processes are required for successful performance of a delayed intention' (p. 743), and attentional processes are switched in whenever cues are expected. They also argue that events in the environment may act as cues to switch on monitoring processes. However, the PAM model acknowledges that specific characteristics of the PM task determine the *degree* to which demands are imposed on top-down self-initiated processes that consume cognitive resources (and therefore that all PM tasks are, in this key respect, not equal).

Memory losses in healthy ageing

In this section we discuss general memory losses and their implications for understanding age effects in PM. Some degree of memory loss does occur in the course of healthy ageing, but these losses are more apparent under some conditions than others. For example, age-related difficulties are substantial in free recall tasks involving no cues or reminders, in working memory tasks in which the material held must be transformed, manipulated, or integrated with further incoming information, and in retrieving highly specific items such as names from semantic memory. In contrast, age-related losses are minimal in 'primary memory' tasks such as digit span, in recall of recency items in free recall lists, in the knowledge and use of vocabulary information, and in the use of habitual procedural knowledge such as driving or playing a musical instrument (reviews in Craik, 2021; Luo & Craik, 2008; McDaniel et al., 2008). While ageing also disrupts recognition memory, and specifically, is associated with reduced discrimination accuracy and a more liberal response criterion (Fraundorf et al., 2019), recognition memory typically shows smaller age-related losses than does free recall performance (Craik & McDowd, 1987; Danckert & Craik, 2013; Rhodes et al., 2019).

What changes might underlie this pattern of strengths and weaknesses? One possibility is that ageing might affect some memory systems (Tulving & Schacter, 1990) more than others, but the findings do not align with this proposal. Both free recall and recognition are measures of episodic memory, yet recall is highly sensitive to ageing whereas recognition is much less so. Also, recall of specific names and vocabulary knowledge reflect access to semantic memory, yet again name retrieval declines in the course of ageing (Burke et al., 2004; Maylor & Valentine, 1992) whereas vocabulary knowledge holds up well with age (Salthouse, 1982).

Craik (2021) has endorsed an alternative framework that appears to provide a better fit to the observed pattern, arguing that processing resources decline with age (Craik & Byrd, 1982) as does the ability to 'self-initiate' necessary mental operations, and that together these changes leave older people more reliant on environmental support (Craik, 1983, 1986). Another proposal, which fits well with these ideas, is that cognitive control processes become less effective with increasing age (Jacoby, 1991; Jennings & Jacoby, 1993). In each of these models, age-related behavioural changes can be most parsimoniously explained by biological changes in the ageing brain. This is because the frontal lobes are among the first brain regions to undergo substantial losses both structurally and functionally in the course of normal ageing (e.g. Raz, 2000), and executive processes impose particular demands on these neural substrates (D'Esposito & Postle, 2015). Because larger prefrontal volume and greater prefrontal cortex thickness are associated with better executive functioning (Yuana & Raz, 2014), it follows that executive functions should become less effective in older adults, and this pattern of decline has been consistently identified (Buckner, 2004; Daniels et al., 2006). Craik's proposal is that self-initiation depends on effective executive functions. One relevant question here is whether the term 'processing resources' is simply another name for cognitive control given that both reflect attentional processes mediated by the frontal lobes. We argue that they are separable constructs on the grounds that machines typically require both a source of power (resources) and a means of managing and directing that power appropriately (control), and that the brain, as a complex biological machine, also requires these two aspects of functioning.

The ability to self-initiate cognitive operations is a functional outcome associated with medial prefrontal regions (Stuss & Alexander, 2007). The relevance to memory is that remembering a prior event is enhanced to the extent that the original contextual details are reinstated at the time of retrieval; this is why cued recall is typically more effective than free recall, and recognition memory is typically best of all. Clearly people can recollect events in the absence of such contextual reinstatement—by reconstructing aspects of the original situation, presumably by alternating processes of reconstruction and verification—but these processes of self-initiation are costly in terms of both resources and control, and are therefore performed less effectively by older adults. Craik (1983, 1986) proposed that remembering incorporates both top-down self-initiation and reconstruction processes, as well as bottom-up processes driven by contextual information from the external environment. Further, these two sets of processes work in a complementary manner; a greater degree of environmental support implies less need for self-initiated activities. One further point in this connection is that a person's accumulated knowledge of the world can also function to reduce the need for self-initiation; this reliance on past experience has been referred to as 'schematic support' (Craik & Bosman, 1992).

Put in a different way, because the efficiency of frontal lobe functioning declines in the course of ageing, older people should be at their greatest disadvantage relative to their younger counterparts when remembering places demands on self-initiation; the corollary, however (and the good news), is that such age-related decrements can be reduced and even eliminated by the provision of more schematic and environmental support. As will be discussed next, this view appears to quite precisely capture what is currently understood about age effects in PM, and the specific circumstances that increase or decrease older adults' vulnerability to PM errors.

Age-related differences in prospective memory

Because intact PM is critical for older adults to continue living independently and to function well in society (Hering et al., 2018; Thöne-Otto & Walther, 2010), it is important to understand how PM processes change in the course of ageing, and how inefficiencies may be remediated or supported. We argued in the previous section that many of the cognitive problems experienced by older adults are attributable to broader age-related declines in processing resources and in cognitive control. In turn, these inefficiencies tend to make the older person more reliant on habitual thoughts and procedures and less able to 'go beyond the information given' in Jerome Bruner's nice phrase. That is, whereas 'proximal' perceptual and motor processes function well, older adults have more difficulty with 'distal' processes such as recollection (Craik, 2021), imagining the future (Addis et al., 2010; Lyons et al., 2014), and mental imagery (Rendell et al., 2012). These inefficiencies have the effect of reducing performance on PM tasks, although many studies have now shown that these age-related difficulties can be reduced in various ways. Here, we first discuss major findings from the experimental literature with some emphasis on how difficulties may be attenuated, and then survey some practical implications of these findings.

Intention formation

In experimental studies of PM, the formation of the intention to carry out the proposed action is necessarily deliberate—it is the prescribed task after all. Yet in many (perhaps most?) everyday situations PM intentions to carry out routine tasks may be formed in a 'perfunctory' manner (Scullin et al., 2018), and this relatively casual form of encoding may be more deleterious to older adults. However, another important consideration at the intention formation stage is whether individuals set up reminders to boost their later chances of fulfilling an intention, and this aspect of strategic thinking may be better developed in older adults given their experience of many previous PM failures. Indeed, some investigators have observed that older adults have been quite reluctant to give up notes, alarms, and other reminders when asked to do so for the purposes of a real-life study (e.g. Moscovitch, 1982). It is certainly the case (as one might expect) that the *importance* of the prospective action is associated with better encoding of the PM intention and thus with higher levels of successful implementation (Kvavilashvili, 1987). In one real-life study, higher levels of importance improved PM completion for older adults resulting in better performance than for a younger group, although only for intentions of low and medium importance; for highly important intentions all participants performed perfectly (Ihle et al., 2012).

Emotionality

Other manipulations of intention formation are known to increase PM performance; one point of present interest is whether such factors increase performance differentially for older adults. One such variable is the emotionality of the target cue; in several laboratory PM studies, age effects have been shown to be eliminated (Altgassen et al., 2010) or reduced (Rendell et al., 2011) by the provision of emotionally valenced cues. In a further study, May et al. (2015) also found that emotional cues increased PM detection in both younger and older adults without disruption of ongoing task performance, and that the use of emotional cues reduced older adults' repetition errors. These effects align with a much broader literature which shows that, as we grow older, there are systematic changes in attention to and prioritization of emotionally valenced information (von Hippel & Henry, 2012).

Implementation intentions

A second manipulation with strong practical implications is the use of implementation intentions, a simple encoding strategy that can be readily applied in many everyday contexts. One study by Chasteen et al. (2001) found that forming such explicit intentions increased older adults' likelihood of completing a PM task, although not on a task requiring a response to a salient cue. The authors concluded that implementation intentions are most effective in situations involving self-initiation. Taking the opposite view, Henry et al. (2020) argued that any benefit of implementation intentions in reducing age-related problems should be greater with event-based PM tasks, since such tasks allow for the formation of a direct perception–action link. The results supported this prediction and also showed that asking participants to form a mental representation of carrying out the task conferred an additional benefit. In general, it appears that the formation of implementation intentions can meaningfully improve the performance of older adults on PM tasks, although the exact nature and extent of the benefit may depend on specific characteristics of the participants and tasks in question.

Retention and the ongoing task

The nature of the background activity (or 'ongoing task' in laboratory situations) affects participants' ability to complete PM tasks, and several factors are relevant to understanding this effect. Clearly there is a necessary trade-off between performance of the ongoing task and monitoring for the PM cue (Schnitzspahn et al., 2011). In a real sense it is a divided attention situation, and there is good evidence that older adults are more penalized than their younger counterparts in such conditions, especially when one of the tasks involves retrieval (Castel & Craik, 2003; Naveh-Benjamin et al., 2005). It also follows that greater degrees of ongoing task difficulty will consume more attentional resources leaving fewer resources for monitoring activities, and in line with this view, Einstein et al. (1997) found that PM errors rose as ongoing task difficulty increased, with this increase especially marked for older adults.

Absorption

In real-life PM situations, the background activity is less likely to be a formal task and more likely to consist of normal daily routines, and an important consideration here is how *absorbed* the individual is in these activities (e.g. Kvavilashvili, 1987; McDaniel & Einstein, 2000; Schnitzspahn et al., 2011). While this can be regarded as the more positive version of ongoing task difficulty, a person's degree of absorption in a book, film, conversation, or many other activities is harder to index objectively and thus its influence on PM performance is not as clearly understood.

Disengagement

Current theories of PM emphasize the flexible switching of attention between monitoring for a PM cue and performing some background activity. At times when the cue is unlikely to occur, it is adaptive to disengage from monitoring and thus have more resources available for other purposes (Henry et al., 2021; Scullin et al., 2013), but PM performance will then necessarily depend more on spontaneous retrieval (Scullin et al., 2013). Given this trade-off, performance levels on the ongoing task can be used to gauge how much attention is devoted to the PM task (R. E. Smith et al., 2017); Einstein and colleagues (1995) found that in time-based tasks requiring explicit monitoring, older adults monitored less often and were less likely to concentrate their monitoring on the period just before the target time (see also Mioni et al., 2020). Similar age effects were also found in a computerized 'cooking breakfast' task in which younger and older adults had to switch between one screen showing cooking progress for various foods, and a second screen (on the same monitor) displaying a table-setting task (Craik & Bialystok, 2006). Older adults checked the cooking screen less often, resulting in poorer cooking performance, and continued to perform the table-setting task at times when the foods needed attention. Interestingly too, older adults show a deficit in global task-switching performance (Wasylyshyn et al., 2011)—that is, their performance is affected simply by the *possibility* of having to switch to a different mode of responding. Consistent with this view, a recent study showed that when monitoring of the time-based cues was entirely prevented (by hiding the timing device), no PM age effects emerged (Varley et al., 2021). However, where the timing device could be accessed, negative age effects on the PM task emerged, and this was true whether the monitoring needed to be completed in an explicit, deliberate manner, or could be completed more perfunctorily. Further, because no negative effects were observed on the ongoing task (which consisted of simple arithmetic problems), the 'costs' of task switching appeared to be quite specific to older adults' PM performance. Taken together, this profile of age-related difficulties fits well with our proposal that much of the PM decline seen in late adulthood stem from inefficiencies in broader executive control processes (Stuss & Craik, 2020; West, 1996).

Initiation of the prospective memory response

Nature of the PM cue

The retrieval (or initiation) phase of PM is obviously critical, since it determines whether the planned action will be successfully carried out. In general, PM cues are either time-based or event-based; as noted, Einstein et al. (1995) found age-related difficulties in the former but not the latter, arguing that only time-based tasks required self-initiated processing. However, in a meta-analysis, Henry and colleagues (2004) found evidence that younger participants outperformed older adults on *both* types of task, and that this could be explained by looking at the characteristics of the event-based tasks, which in the context of the multiprocess framework differed quite meaningfully in the degree to which they place demands on strategic controlled resources. Whereas age effects were smaller for low-demand event-based PM tasks relative to time-based PM tasks, older adults exhibited very comparable levels of impairment on high-demand event-based PM tasks relative to time-based PM tasks.

Haines and colleagues (2020) highlighted a similar, and we would suggest just as important, distinction within time-based cues; those that should be completed at a specific time of day (TOD tasks) and those that should be completed after a specified interval of time (interval tasks). The researchers found that whereas younger and older adults performed similarly on real-life interval tasks, the older adults outperformed their younger counterparts on TOD tasks. They also point out that real-life TOD cues are typically well supported by other environmental events (e.g. a TOD cue of 12.30 pm will be associated with predictable lunch-time occurrences), whereas this is typically not the case for interval cues. This distinction has already added meaningfully to our understanding of when and why older adults can be expected to experience difficulties with time-based PM tasks—and, in particular, why older adults often do surprisingly well on naturalistic tests of PM completed in their everyday lives (as Haines and colleagues note, most naturalistic studies of PM have relied on time-based tasks that involve TOD cues, and which therefore have more environmental support).

The evidence is less clear on how the distinctiveness or 'unusualness' of the PM cue may influence PM age effects. For instance, Mäntylä (1994) found age-related decrements when low-frequency category examples were used as cues, whereas Cherry et al. (2001) found larger age decrements with typical (high-frequency) exemplars.

Moreover, while in one study use of emotional words as PM cues eliminated age differences (Altgassen et al., 2010), in another, emotional words did not eliminate, but only reduced age effects—and this effect was specific to positively valenced cues only (Rendell et al., 2011). These types of mixed findings indicate that the way cue distinctiveness influences age effects is likely to be complex.

By contrast, one task parameter that has been reliably shown to influence age effects is cue focality. As noted earlier, in their multiprocess framework McDaniel and Einstein (2000; Einstein & McDaniel, 2005) made the useful suggestion that PM cues will be more readily detected if they form part of the ongoing task (focal cues), relative to when they are completely separate from the participant's main ongoing task focus (non-focal cues). In real-life situations, this distinction captures the point that many (if not most) PM failures occur when the required action is on a dimension that is very different from the person's ongoing background activity. That is, many daily PM tasks are represented by laboratory situations involving non-focal cues. A meta-analysis of 46 laboratory-based studies conducted by Kliegel et al. (2008) revealed mean age effects of 0.54 and 0.72 (values of Cohen's d statistic) for focal and non-focal tasks respectively, indicating that although older adults generally experience difficulties in relation to both types of PM cue, these difficulties are reliably greater for the latter. Clearly this is an important factor to bear in mind when considering the effects of ageing on PM tasks.

Completing the intention

Effects of delayed implementation

After the PM cue has been successfully detected, are there further age-related problems associated with fulfilling the intention? In real-life PM situations, it is quite often the case that a person remembers a previously formulated intention but needs to delay carrying out the relevant action. Does this influence final completion of the PM intention, and are there age differences in this respect? West and Craik (2001) reported a study in which participants responded to a PM cue either immediately or at the end of the current block of trials. The results showed that correct responding decreased when responses were delayed, but to the same extent in younger and older adults. The effect of delayed responding was also investigated by Einstein et al. (2000). In this experiment the researchers reported an age-related decrement in PM responding after a delay when the cue was easy to detect. With difficult cues, however, there was a general drop in performance with ageing, but neither age group showed a further drop when the delay was increased from 10 to 30 seconds.

Commission errors

Another important consideration at the intention completion stage is the occurrence of commission errors (Figure 13.1). Such errors occur when one fails to 'turn off' a no longer relevant PM intention, and a PM task is erroneously completed a second time.

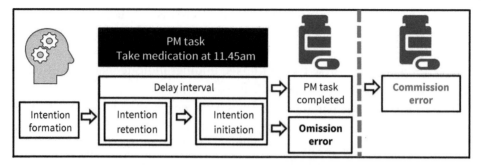

Figure 13.1 Commission errors. Traditionally, PM studies have focused on whether a task has been completed or not, effectively 'stopping' at the dashed red line. Yet beyond this line is another potential type of error known as *commission errors*, which occur *after* the PM task has been completed, or is no longer relevant.

Such errors have been the focus of less empirical study than errors of omission, but are important to consider because they can also lead to problems in everyday life, including embarrassment (such as when one passes a message to a friend twice), financial problems (such as paying the same bill twice), and even a medical emergency (as in the case of accidental medication overdose). It has been argued that PM commission errors are made when residual intention activation co-occurs with a failure in executive control, and thus a failure in retrospective memory for the already-completed intention (Scullin & Bugg, 2013). It is therefore unsurprising that older adults have been shown to be more vulnerable to making these types of errors (Scullin et al., 2012). As noted earlier, the spontaneous retrieval processes that can support PM appear to be preserved in normal adult ageing (Henry et al., 2004). At the same time, relative to young, older adults are more likely to outsource control to the environment. This altered mode of processing creates a shift towards a greater reliance on external relative to internal information (Mayr et al., 2015), which may increase the frequency and/or strength of PM aftereffects elicited by external cues. Importantly (and this is true for PM omission errors at the intention completion stage too), older adults may also be less likely to be *aware* that a PM error has occurred because age-related deterioration in error processing includes reductions in the capacity to consciously detect performance errors (Sim et al., 2020).

Further age-related topics

The age–prospective memory paradox

As discussed earlier, most studies have found an age-related impairment in laboratory-based PM tasks, attributed in part at least to inefficiencies of frontal lobe function. In turn, these normal concomitants of healthy ageing are associated with age-related declines in planning, strategic controlled processing, and self-initiated activities (Henry et al., 2004; Varley et al., 2021). Against this background, it was surprising to find robust age-related *benefits* in PM studies conducted in the naturalistic setting of participants' own lives. This result was reported by Rendell and Thomson (1999), Rendell and Craik (2000), and by several other investigators (e.g. Bailey et al., 2010; Haines et al., 2020; Ihle et al., 2012; Phillips et al., 2008; Schnitzspahn et al., 2011). This 'age–PM paradox' has also been the subject of several reviews (e.g. Haines et al., 2019; Henry et al., 2004; Phillips et al., 2008).

One obvious explanation of the good performance in naturalistic studies is that older adults, aware of the fallibility of their memory systems, take more care to provide notes, alarms, and other reminders in real-life tasks. This is not the whole story, however, as most studies have requested participants to avoid the use of memory aids, and participants typically comply with this request (Rendell & Craik, 2000; Rendell & Thomson, 1999). Further, in a laboratory-based study that presented three counterbalanced reminder conditions (no reminders, self-initiated reminders, and experimenter-initiated reminders), provision of reminder cues benefited younger and older adults' PM performance equivalently (Henry et al., 2012). Other factors associated with older participants' superiority in real-life PM studies include the point that compared to younger adults (especially perhaps undergraduate students!), older adults have more structured and predictable daily routines, and can therefore use this structure to be on the alert for both time-based and event-based PM cues. It also seems likely that older adults will form PM intentions quite deliberately and will regard the fulfilment of such intentions as an important aspect of their current activities (Phillips et al., 2008). This combination of deliberation and importance may well be absent from many fleeting domestic intentions (take a book upstairs, take a cup downstairs, check the milk, take out the rubbish), whose age-related failures give rise to frustration and irritation but not to major feelings of concern.

It seems likely that several different factors are involved in this inversion of performance (and of researchers' expectations); one such factor could be older adults' lack of familiarity with computers used in many laboratory tests—a factor typically absent in studies conducted in everyday life. Henry and colleagues (2004) comment that the age-related PM impairments found in laboratory studies do not apparently translate to deficits in everyday life. This optimistic observation does not accord with older adults' reports of their own current memory difficulties, however (G. Smith et al., 2000). In fact, the contrasting findings of excellent performance by older adults in *studies* involving naturalistic settings, and their experienced difficulties in their day-to-day lives may constitute the *real* age-related PM paradox!

Prospective memory and ageing: practical aspects

It should be clear from the preceding sections that the ability to foresee future events and to direct current action towards appropriate future goals is critical for a great variety of functional behaviours and for independent living in older adults. This includes many of the complex activities of everyday living, such as handling finances, driving, managing medication, shopping, and preparing meals, that are often referred to as instrumental activities of daily living. Highlighting the critical importance of these skills, one of the strongest personal risk factors for nursing home placement, in addition to age and physical health, is a decline in the ability to perform such activities. Indeed, this has been identified as the 'decisive factor' predicting an urgent request for nursing home admission (Van Rensbergen & Pacolet, 2012). In younger adulthood too, failures of PM can have devastating consequences, never more so than when parents or caretakers put children at risk by unintentionally leaving them in cars. This has been identified as the 'most frequently reported occurrence of a catastrophic PM failure' (Diamond, 2019). In the US alone, over 940 children have died in hot cars since 1990 (KidsandCars.org, 2021). Diamond (2019) presents compelling evidence that at the heart of almost all these tragedies is neither negligence nor neglect, but a breakdown of PM that could potentially happen to anyone.

Given the critical importance yet fallibility of PM in so many everyday life contexts and at all stages of the human lifespan, it is surprising that comparatively few research studies have focused on PM in workplace environments. This section provides an overview of what we currently understand about the PM errors most likely to occur in these environments, the circumstances that appear to increase their risk of occurrence, and the importance of these findings for older adults. We will then conclude with an overview of some of the most validated ways to assess PM function, and interventions that can be used to support PM function.

Prospective memory in applied settings

In applied environments, PM has traditionally been examined using analyses of accident reports, ethnographic observations, and diary reports. While such methods lack the statistical power and rigorous control of laboratory studies, they provide insights into the types of PM failures that actually occur in real-life environments, the factors associated with these failures, and how successfully these failures have been detected and responded to. However, simulation-based methods have also been used, and these provide a valuable 'bridge' between controlled laboratory paradigms and naturalistic field studies. A common limitation of many of these simulation studies is a reliance on non-expert participants, which is problematic because the performance of experts often becomes more automatic over time. Yet despite this caveat, because simulation-based methods use paradigms that more closely approximate the demands of many workplace environments, studies using this approach have contributed greatly to our understanding of the mechanisms that underlie real-life PM failures in complex environments.

Here we focus on five contextual variables that have been identified as important in this literature; interruptions, the length of the PM retention interval, the absence of cues that are typically present for habitual tasks, habit intrusions, and multi-tasking. The first of these, interruptions, occur in nearly all occupational settings, and in many are impossible to eliminate completely. Interruptions are problematic because they can lead to a failure to remember to resume the interrupted task at all, or resumption at the incorrect stage. However, in a compelling demonstration of how interruptions vary greatly in their potential impact, Grundgeiger et al. (2010) used mobile eye-tracking data to track how easily nurses could recall an interrupted primary task. The results showed that longer disruptions, or disruptions which led to a physical change in location, were particularly problematic in lengthening the time it took nurses to resume critical care duties. Also, highlighting how not all interruptions are equal in their potential for harm, M. D. Wilson et al. (2020) showed that interruptions during the PM retention interval (the period of time prior to when the PM task actually needed to be completed) did not disrupt later PM task performance in a simulated air traffic control paradigm. It was suggested that when environmental demands provide appropriate PM cuing and offloading, PM may in fact be relatively resistant to interruptions. However, increasing the retention interval of the PM task was detrimental to PM performance—not only by increasing PM failures, but also by incurring greater costs to the ongoing air traffic control tasks. Based on these findings, it was argued that in applied contexts the retention interval of PM tasks should be minimized whenever possible because of their potentially contaminating effect on ongoing tasks.

An absence of cues that are typically available to 'trigger' habitual tasks has been identified as another important factor in understanding real-life PM failures. The sequential steps of highly practised procedures are often executed in a manner that has become habitual in highly skilled professionals. When external circumstances require that one of these actions be performed out of its normal sequence, the usual 'cues' that support task execution are not provided, thereby increasing the risk of PM failure (Dismukes, 2012). In a related manner, it has been argued that in applied settings PM failures may arise because of *habit intrusions*, which occur when there is a need to replace a familiar practised action with an atypical novel one. Habit intrusions have been identified as a specific risk factor for anaesthetic complications (Bosack, 2015), and to account for 16% of the PM errors made by computer programmers (Sanjram & Khan, 2011).

The final key factor to be discussed here that appears to increase vulnerability to PM errors in complex real-life environments is multitasking. In many occupational settings there is a requirement to engage in multiple tasks concurrently, meaning that attention must be appropriately engaged and disengaged in response to changing situational demands. Here, most laboratory paradigms that assess task switching provide a poor approximation of real-life demands, as participants are usually explicitly told that they must engage and disengage from clearly defined task sets. In real life, the ability to multitask needs to be managed spontaneously; also, the precise point at which task switching ought to occur may be ambiguous, requiring judgement and expertise to detect. A recent simulation study using a challenging yet realistic dynamic maritime surveillance task provided evidence of capacity sharing between PM monitoring and ongoing task performance (Strickland et al., 2019). As we discussed earlier, these results are important because an earlier study identified no evidence of capacity sharing in a simpler paradigm (Strickland et al., 2018). These findings reinforce the need for studies whose results may be generalized to complex real-life environments, using more challenging research designs. They also have direct practical implications for work design, suggesting that in order to successfully complete both concurrent PM and ongoing tasks, it is critical to not exceed the operator's capacity to perform within safety thresholds (which Strickland's group refer to as the *red zone*).

Taken together, the current literature shows that many variables influence the likelihood of PM errors occurring in workplace environments. It is therefore striking that even though the ageing population in nearly all developed countries has led to recommendations that the age of retirement be increased, almost nothing is known about PM age effects in workplace environments. Even the simulation studies completed to date have focused predominantly or exclusively on younger adults. This is an important gap that must be addressed, particularly since our earlier discussion of the factors shown to influence age-related PM effects in experimental settings allow for competing possibilities about how older adults' PM performance might fare in applied environments. On the one hand, older adults who have worked in the same profession for many years might be only minimally affected by factors known to increase their younger counterparts' vulnerability to PM errors because of their greater familiarity with specific task demands, and larger repertoire of strategies to draw on in response to changing situational pressures. However, it is also possible that with advancing age older adults place too much reliance on habitual processing, and that this makes them more vulnerable to specific types of PM errors, such as those arising from habit intrusions. An important future research question is therefore to establish whether vulnerability to specific types of PM errors in workplace environments differs between young and older adults. It seems entirely possible that quite different types of workplace support are needed at different stages of the adult lifespan.

Assessment of prospective memory

Finally, as with many other cognitive abilities, the capacity for successful PM functioning varies considerably between individuals, and this variability is typically much greater in older than in younger adults. Where a judgement must be made about the need for PM support, and the type of support if needed, a standardized behavioural assessment is critical to objectively quantify the nature and severity of any PM difficulties, and to identify the residual abilities that might be used to compensate for losses. Although PM can, in theory, be indexed via self-report (and such measures are appealing to many because of their brevity and ease of administration), these assessments are generally weakly related to objective test performance (Fuermaier et al., 2015; Henry, 2021b; Raskin et al., 2018; Thompson et al., 2015). For this reason, objective performance-based assessments are always to be preferred, particularly in clinical or applied environments where the results have implications for the assessed individual.

We have detailed elsewhere an overview of four validated behavioural tests, including their key characteristics and psychometric properties (Henry, 2021a), and how each has unique characteristics that are likely to be more or less desirable, depending on the specific goal of the assessment, and the features of the setting. However, in brief, these are: (1) the *Royal Prince Alfred Prospective Memory Test* (RPA-ProMem; Radford et al., 2011), a four-item behavioural measure, which, because of its brevity, may be of value in particularly time-pressured environments; (2) the *Cambridge Prospective Memory Test* (CAMPROMT; B. A. Wilson et al., 2005), a longer measure that includes a greater number of tasks, and includes the unique feature of standardized prompts to probe for specific types of errors; (3) the *Memory for Intentions Screening Test* (MIST; Raskin et al., 2010), that allows for an even more detailed understanding of the source of any PM impairment, differentiating between multiple PM task parameters and six distinct types of error; and (4) *Virtual Week* (VW; Rendell & Craik, 2000; Rose et al., 2015), which provides a closer approximation to real-life PM tasks that require multitasking, requiring several different intentions to be managed concurrently, with the participant asked to plan and execute multiple PM tasks while involved in an engaging board game setting (Figure 13.2).

Interventions to reduce older adults' prospective memory failures

Once problems with PM have been identified using a validated, behavioural assessment, the natural question that follows is what can be done to reduce these difficulties. The good news is that an extensive literature now exists focused on the development and validation of interventions targeting the many special populations known to struggle with PM, with many of these interventions directed specifically to older adults (for a recent review, see Jones et al., 2021). In the broader cognitive intervention literature, a distinction has been made between *compensatory approaches* which teach new ways to accomplish a cognitive task by working around cognitive weaknesses, and *restorative approaches*, which are based on principles of neural plasticity, and provide cognitive training to try and restore function directly. A range of psycho-educational methods have also been developed, that focus on increasing understanding of cognitive weaknesses, and metacognitive approaches, which target self-monitoring by enhancing both the ability to detect and self-correct errors as well as self-knowledge of strengths and weaknesses. Depending on the nature of the PM impairment, each of these approaches has potential value (Henry, 2021a).

Figure 13.2 In the Virtual Week task, participants move round the board whose squares represent hours of one day from 8 am until 10 pm. Participants must remember to carry out a variety of prospective memory assignments associated with specified times and events. Reproduced from Rose, N. S., Rendell, P. G., Hering, A., Kliegel, M., Bidelman, G. M., & Craik, F. I. M. (2015). Cognitive and neural plasticity in older adults' prospective memory following training with the Virtual Week computer game. *Frontiers in Human Neuroscience, 9*, 592.

In contrast to much of the literature on cognitive interventions, most studies seeking improvements in older adults' PM abilities have used such compensatory approaches as elaborative encoding. This manipulation involves the provision of memory aids that link to-be-remembered information to previously existing memories and knowledge, making it easier to recall the new information in the future. As discussed earlier in the chapter, one of the most successful elaborative encoding strategies is implementation intentions, a simple goal-setting strategy that involves generating and saying aloud a simple statement that has a precise format and structure whereby a specific cue is stated first, followed by an action. Although exceptions have been noted (Fine et al., 2021), most studies show that use of this encoding strategy can meaningfully enhance older adults' PM (Brom & Kliegel, 2014; Bugg et al., 2013; Chasteen et al., 2001; Lee, et al., 2016), particularly when combined with visual imagery (Henry et al., 2021). In addition, older adults have been shown to benefit from external aids, including memory notebooks, content-specific alerts, or one of the best validated internal memory rehabilitation strategies, spaced retrieval, which involves continuously retrieving information over increasingly longer delay intervals (see, e.g. Ozgis et al., 2009). There has also been increased focus on the potential for technology to enhance rehabilitation strategies. Raghunath et al. (2020) developed a digital memory notebook application, to support everyday memory skills in older adults with mild cognitive impairment. This tool integrated features of several validated external compensatory aids such as a calendar function and reminders, but in a user-friendly interface that included large targets, simple icons, and confirmation feedback. A pilot study involving 20 older adults of varying cognitive ability showed that most people found this tool easy to use.

Relative to compensatory approaches, restorative approaches have been the focus of far less study in the PM literature, but these seek to strengthen specific cognitive domains in order to improve functional performance more generally. There is encouraging evidence to support the value of this approach. For instance, a recent meta-analysis of more than 200 randomized controlled trials showed that cognitive training is modestly effective in improving older adults' cognitive function, with an overall small-sized net gain in cognition for healthy older adults ($g = 0.28$; Basak et al., 2020). However, Basak et al. (2020) also showed that, although training a single cognitive ability (single component training) reliably generated benefits on the trained cognitive ability (i.e. near transfer effects), only a few yielded far transfer effects (i.e. benefits in cognitive domains other than the trained one). Additionally, in the largest study to date to test which types of PM interventions might lead to meaningful benefits, a restorative intervention that targeted only PM also failed to elicit far transfer effects (Henry et al., 2021). These data align with a growing literature suggesting that interventions targeting a single cognitive ability do not reliably generate far transfer effects; thus at present the strongest evidence base supports the value of specific compensatory strategies to enhance PM function in late adulthood (see B. A. Wilson, Chapter 16, this volume, for a more general discussion).

Future directions

The general theme of this book is the interaction between theory and practice in various domains of memory research; in particular, the idea that whereas theoretical concepts and frameworks provide a way to understand real-world phenomena, the observations and findings from practical settings are critical to sharpen and appropriately modify the relevant theoretical views. How well is this happening in the area of ageing and PM?

Based on the literature reviewed in this chapter, we believe that a good start has been made. The age-related decline in the efficiency of executive functions is a case in point. As discussed earlier, this change reduces older adults' ability to self-initiate thoughts and actions, making them more dependent on environmental cues and re-minders. This difficulty is already well known to older adults, but there is scope to increase the presence of such support in workplace settings, thereby decreasing the number of PM failures in adults of all ages. The experimental distinction between time-based and event-based forms of PM, and the more recent discovery that older adults are more vulnerable to time-interval errors than to TOD errors, point the way to the provision of relevant interventions at useful times and locations in practical settings.

Other related ways in which declining executive function efficiency can affect PM behaviour include the ease with which information (e.g. task-based intentions) can be dropped from working memory or, equivalently, from conscious awareness. Such momentary lapses are well illustrated by the finding that on over a quarter of occasions in which participants were late to respond to a time-based task, they had checked the clock during the last 10 seconds before the target time (Harris & Wilkins, 1982). Similar lapses have been studied under the heading of mind wan-dering and shown to be linked to working memory capacity (e.g. McVay & Kane, 2009). Further investigations examining the effects of ageing on the interactions among PM, mind wandering, and working memory capacity would appear to be a fruitful direction for both laboratory-based and real-life studies (Kliegel & Jäger, 2006). As we noted earlier, PM errors of commission are particularly concerning in some practical settings—accidentally taking medications twice is a good example. This topic has been somewhat neglected experimentally, and again seems to us a good candidate for future work.

There appears to be a growing consensus that PM is not a distinct form of memory nor yet a different 'memory system'. Instead, recent investigators have suggested that PM 'may be better described as a collection of different memory and decision-making elements' (McBride & Workman, 2017, p. 233). Nonetheless, the relations between PM and other memory processes are of great theoretical interest (Einstein & McDaniel, 1990; Glisky, 1996; Wilkins & Baddeley, 1978). It seems clear that many of the variables associated with good encoding for episodic memory are also effective as ways to boost the effectiveness of PM intentions; we have previously mentioned semantic processing, picture stimuli, and emotionally salient cues in this respect. With regard to retrieval processes, Cohen et al. (2001) found that semantic relatedness predicted good performance on the retrospective component of a PM task, but a change in the study-test format had the greatest negative effect on the prospective component. Such findings highlight how differ-ences between tasks, cues, targets, and subjective intentions all have the potential to affect PM function, and these complexities need to be disentangled in future studies.

Lastly, we are intrigued by the striking differences in the effects of ageing on laboratory-based and real-life PM tasks. The relatively poor performance levels on the former appear to be directly influenced by age-related declines in executive func-tion, whereas we have tentatively attributed the superior performance of older adults on many real-life tasks to their higher levels of motivation, ascribed importance of the studies, and the greater regularity of their lifestyles. However, there seems to be a third set of circumstances, namely older adults who are *not* in experimental studies, but simply living in their own day-to-day environments. In such settings, older adults report having substantial problems with everyday PM (G. Smith et al., 2000), an apparent anomaly in light of older adults' objectively good performance in experimental studies conducted in real-life settings. One way to reconcile this disparity is that older adults experience many annoying but essentially trivial failures of PM in domestic settings, but typically these are instances of low importance. Arguably, when a PM intention refers to an important outcome—either in a study or in their ongoing lives—older adults can be highly successful, partly by using notes and reminders, and partly by periodically refreshing the intention and its target outcome. A final question concerns the relative levels of motivation in *younger* adults; if their levels of motivation are equated to those of older adults, is their level of PM success also equivalent, or are there still age-related differences—positive or negative? By addressing each of these questions in future research, it will be possible to gain a clearer understanding of the circumstances that support PM function in older adults' everyday lives—with direct and important implications for their ability to continue living safely and independently (and with minimal frustration!) in their own homes.

Conclusion

In conclusion, we are delighted to contribute this chapter to a book honouring Alan Baddeley's many ground-breaking contributions to theories of human memory and the application of these theories to real-world problems. It has been a hallmark of Alan's approach that his ideas have always been rooted in everyday experience, and he has similarly stressed the necessary interactions between theory and practice in his many books and journal articles. The first author of the present chapter and Alan have been friends for longer than either of us *can* remember at this stage of our lives; not always in perfect theoretical harmony, but always eager to share and discuss ideas over a pint or two. We also collaborated in writing an obituary for Donald Broadbent—a mutual scientific hero. The present chapter deals with PM—remembering to remember at some future time. This has not been one of Alan's main scientific topics, although he has studied aspects of PM in both healthy adults (Wilkins & Baddeley, 1978) and in individuals with memory impairments (Baddeley, 2004). We hope that this chapter will revive his interest in the findings and ideas.

References

Addis, D. R., Musicaro, R., Pan, L., & Schacter, D. L. (2010). Episodic simulation of past and future events in older adults: Evidence from an experimental recombination task. *Psychology and Aging*, 25(2), 369–376.

Altgassen, M., Phillips, L. H., Henry, J. D., Rendell, P. G., & Kliegel, M. (2010). Emotional target cues eliminate age differences in prospective memory. *Quarterly Journal of Experimental Psychology*, *63*(6), 1057–1064.

Bailey, P. E., Henry, J. D., Rendell, P. G., Phillips, L. H., & Kliegel, M. (2010). Dismantling the 'age-prospective memory paradox': The classic laboratory paradigm simulated in a naturalistic setting. *Quarterly Journal of Experimental Psychology*, *63*(4), 646–652.

Basak, C., Qin, S., & O'Connell, M. A. (2020). Differential effects of cognitive training modules in healthy aging and mild cognitive impairment: A comprehensive meta-analysis of randomized controlled trials. *Psychology and Aging*, *35*(2), 220–249.

Berntsen, D. (2010). The unbidden past: Involuntary autobiographical memories as a basic mode of remembering. *Current Directions in Psychological Science*, *19*(3), 138–142.

Bosack, R. C. (2015). Anesthetic complications-how bad things happen. In R. C. Bosack & S. Lieblich (Eds.), *Anesthesia complications in the dental office* (pp. 3–5). John Wiley & Sons Inc.

Brom, S. S., & Kliegel, M. (2014). Improving everyday prospective memory performance in older adults: Comparing cognitive process and strategy training. *Psychology and Aging*, *29*(3), 744–755.

Buckner, R. L. (2004). Memory and executive function in aging and AD: Multiple factors that cause decline and reserve factors that compensate. *Neuron*, *44*(1), 195–208.

Bugg, J. M., Scullin, M. K., & McDaniel, M. A. (2013). Strengthening encoding via implementation intention formation increases prospective memory commission errors. *Psychonomic Bulletin & Review*, *20*(3), 522–527.

Burke, D. M., Locantore, J. K., Austin, A. A., & Chae, B. (2004). Cherry pit primes Brad Pitt: Homophone priming effects on young and older adults' production of proper names. *Psychological Science*, *15*(3), 164–170.

Castel, A. D., & Craik, F. I. M. (2003). The effects of aging and divided attention on memory for item and associative information. *Psychology and Aging*, *18*(4), 873–885.

Chasteen, A. L., Park, D. C., & Schwarz, N. (2001). Implementation intentions and facilitation of prospective memory. *Psychological Science*, *12*(6), 457–461.

Cherry, K. E., Martin, R. C., Simmons-Gerolamo, S. S., Pinkston, J. B., Griffing, A., & Gouvier, W. D. (2001). Prospective remembering in younger and older adults: Role of the prospective cue. *Memory*, *9*(3), 177–193.

Cohen, A. L., West, R., & Craik, F. I. M. (2001). Modulation of the prospective and retrospective components of prospective remembering in younger and older adults. *Aging, Neuropsychology, and Cognition*, *8*(1), 1–13.

Craik, F. I. M. (1983). On the transfer of information from temporary to permanent memory. *Philosophical Transactions of the Royal Society of London. Series B, Biological Sciences*, *302*(1110), 341–359.

Craik, F. I. M. (1986). A functional account of age differences in memory. In F. Klix et al. (Eds.), *Human memory and cognitive capabilities* (pp. 409–422). Elsevier Science Publishers.

Craik, F. I. M. (2021). *Remembering: An activity of mind and brain*. Oxford University Press.

Craik, F. I. M., & Bialystok, E. (2006). Planning and task management in older adults: Cooking breakfast. *Memory & Cognition*, *34*(6), 1236–1249.

Craik, F. I. M., & Bosman, E. A. (1992). Age-related changes in memory and learning. In H. Bouma & J. A. M. Graafmans (Eds.), *Gerontechnology* (pp. 79–92). IOS Press.

Craik, F. I. M., & Byrd, M. (1982). Aging and cognitive deficits: The role of attentional resources. In F. I. M. Craik & S. E. Trehub (Eds.), *Aging and cognitive processes* (pp. 191–211). Plenum Press.

Craik, F. I. M., & McDowd, J. M. (1987). Age differences in recall and recognition. *Journal of Experimental Psychology: Learning, Memory, and Cognition*, *13*(3), 474–479.

Danckert, S. L., & Craik, F. I. M. (2013). Does aging affect recall more than recognition memory? *Psychology and Aging*, *28*(4), 902–909.

Daniels, K., Toth, J., & Jacoby, L. (2006). The aging of executive functions. In E. Bialystok & F. I. M. Craik (Eds.), *Lifespan cognition: Mechanisms of change* (pp. 96–111). Oxford University Press.

D'Esposito, M., & Postle, B. R. (2015). The cognitive neuroscience of working memory. *Annual Review of Psychology*, *66*, 115–142.

Diamond, D. M. (2019). When a child dies of heatstroke after a parent or caretaker unknowingly leaves the child in a car: How does it happen and is it a crime? *Medicine Science and the Law*, *59*(2), 115–126.

Dismukes, R. K. (2012). Prospective memory in workplace and everyday situations. *Current Directions in Psychological Science*, *21*(4), 215–220.

Einstein, G. O., & McDaniel, M. A. (1990). Normal aging and prospective memory. *Journal of Experimental Psychology: Learning, Memory, and Cognition*, *16*(4), 717–726.

Einstein, G. O., & McDaniel, M. A. (1996). Retrieval processes in prospective memory: Theoretical approaches and some new empirical findings. In M. A. Brandimonte, G. O. Einstein, & M. A. McDaniel (Eds.), *Prospective memory: Theory and applications* (pp. 115–142). Lawrence Erlbaum.

Einstein, G. O., & McDaniel, M. A. (2005). Prospective memory: Multiple retrieval processes. *Current Directions in Psychological Science*, *14*(6), 286–290.

Einstein, G. O., & McDaniel, M. A. (2008). Prospective memory and metamemory: The skilled use of basic attentional and memory processes. In A. S. Benjamin & B. H. Ross (Eds.), *Psychology of learning and motivation* (pp. 145–173). Academic Press.

Einstein, G. O., McDaniel, M. A., Manzi, M., Cochran, B., & Baker, M. (2000). Prospective memory and aging: Forgetting intentions over short delays. *Psychology and Aging*, *15*(4), 671–683.

Einstein, G. O., McDaniel, M. A., Richardson, S. L., Guynn, M. J., & Cunfer, A. R. (1995). Aging and prospective memory: Examining the influences of self-initiated retrieval processes. *Journal of Experimental Psychology: Learning, Memory, and Cognition*, *21*(4), 996–1007.

Einstein, G. O., Smith, R. E., McDaniel, M. A., & Shaw, P. (1997). Aging and prospective memory: Task demands at encoding and retrieval. *Psychology and Aging*, *12*(3), 479–488.

Ellis, J. (1996). Prospective memory or the realization of delayed intentions: A conceptual framework for research. In M. A. Brandimonte, G. O. Einstein, & M. A. McDaniel (Eds.), *Prospective memory: Theory and applications* (pp. 1–22). Lawrence Erlbaum.

Fine, L., Loft, S., Bucks, R. S., Parker, D., Laws, M., Olaithe, M., Pushpanathan, M., Rainey Smith, S. R., Sohrabi, H. R., Martins, R. N., & Weinborn, M. (2021). Improving prospective memory performance in community-dwelling older adults: Goal management training and implementation intentions. *Experimental Aging Research*, *47*(5), 414–435.

Fraundorf, S. H., Hourihan, K. L., Peters, R. A., & Benjamin, A. S. (2019). Aging and recognition memory: A meta-analysis. *Psychological Bulletin*, *145*(4), 339–371.

Fuermaier, A. B. M., Tucha, L., Koerts, J., Aschenbrenner, S., Kaunzinger, I., Hauser, J., Weisbrod, M., Lange, K. W., & Tucha, O. (2015). Cognitive impairment in adult ADHD: Perspective matters! *Neuropsychology*, *29*(1), 45–58.

Glisky, E. L. (1996). Prospective memory and the frontal lobes. In M. Brandimonte, G. O. Einstein, & M. A. McDaniel (Eds.), *Prospective memory: Theory and applications* (pp. 249–266). Erlbaum.

Gollwitzer, P. M., & Brandstätter, V. (1997). Implementation intentions and effective goal pursuit. *Journal of Personality and Social Psychology*, *73*(1), 186–199.

Grundgeiger, T., Sanderson, P., MacDougall, H. G., & Venkatesh, B. (2010). Interruption management in the intensive care unit: Predicting resumption times and assessing distributed support. *Journal of Experimental Psychology: Applied*, *16*(4), 317–334.

Haines, S. J., Randall, S. E., Terrett, G., Busija, L., Tatangelo, G., McLennan, S. N., Rose, N. S., Kliegel, M., Henry, J. D., & Rendell, P. G. (2020). Differences in time-based task characteristics help to explain the age-prospective memory paradox. *Cognition*, *202*, e104305.

Haines, S. J., Shelton, J. T., Henry, J. D., Terrett, G., Vorwerk, T., & Rendell, P. G. (2019). Prospective memory and cognitive aging. In O. Braddick (Ed. in Chief) *Oxford research encyclopedia of psychology* (pp. 1–25). Oxford University Press.

Harris, J. E., & Wilkins, A. J. (1982). Remembering to do things: A theoretical framework and an illustrative experiment. *Human Learning*, *1*(2), 123–136.

KidsandCars.org. (2021). *Heatstroke*. htpps://www.kidsandcars.org/how-kids-get-hurt/heat-stroke/

Henry, J. D. (2021a). Prospective memory impairment in neurological disorders: implications and management. *Nature Reviews Neurology*, *17*(5), 297–307.

Henry, J. D. (2021b). Reply to: Assessing prospective memory beyond experimental tasks. *Nature Reviews Neurology*, *17*(7), 459–460.

Henry, J. D., Hering, A., Haines, S., Grainger, S. A., Koleits, N., McLennan, S., Pelly, R., Doyle, C., Nathan, R. S., Kliegel, M., & Rendell, P. G. (2021). Acting with the future in mind: Testing competing prospective memory interventions. *Psychology and Aging*, 36(4), 491–503.

Henry, J. D., MacLeod, M. S., Phillips, L. H., & Crawford, J. R. (2004). A meta-analytic review of prospective memory and aging. *Psychology and Aging*, 19(1), 27–39.

Henry, J. D., Rendell, P. G., Phillips, L. H., Dunlop, L., & Kliegel, M. (2012). Prospective memory reminders: a laboratory investigation of initiation source and age effects. *Quarterly Journal of Experimental Psychology*, 65(7), 1274–1287.

Henry, J. D., Terrett, G., Grainger, S. A., Rose, N. S., Kliegel, M., Bugge, M., Ryrie, C., & Rendell, P. G. (2020). Implementation intentions and prospective memory function in late adulthood. *Psychology and Aging*, 35(8), 1105–1114.

Hering, A., Kliegel, M., Rendell, P. G., Craik, F. I. M., & Rose, N. S. (2018). Prospective memory is a key predictor of functional independence in older adults. *Journal of the International Neuropsychological Society*, 24(6), 640–645.

Hintzman, D. L. (2011). Research strategy in the study of memory: Fads, fallacies, and the search for the 'coordinates of truth'. *Perspectives on Psychological Science*, 6(3), 253–271.

Ihle, A., Schnitzspahn, K., Rendell, P. G., Luong, C., & Kliegel, M. (2012). Age benefits in everyday prospective memory: The influence of personal task importance, use of reminders and everyday stress. *Aging, Neuropsychology, and Cognition*, 19(1–2), 84–101.

Jacoby, L. L. (1991). A process dissociation framework: Separating automatic from intentional uses of memory. *Journal of Memory and Language*, 30(5), 513–541.

Jennings, J. M., & Jacoby, L. L. (1993). Automatic versus intentional uses of memory: Aging, attention, and control. *Psychology and Aging*, 8(2), 283–293.

Jones, W. E., Benge, J. F., & Scullin, M. K. (2021). Preserving prospective memory in daily life: A systematic review and meta-analysis of mnemonic strategy, cognitive training, external memory aid, and combination interventions. *Neuropsychology*, 35(1), 123–140.

Kliegel, M., & Jäger, T. (2006). Delayed-execute prospective memory performance: The effects of age and working memory. *Developmental Neuropsychology*, 30(3), 819–943.

Kliegel, M., Jäger, T., & Phillips, L. H. (2008). Adult age differences in event-based prospective memory: A meta-analysis on the role of focal versus nonfocal cues. *Psychology and Aging*, 23(1), 203–208.

Kliegel, M., McDaniel, M. A., & Einstein, G. O. (2000). Plan formation, retention, and execution in prospective memory: A new paradigm and age-related effects. *Memory & Cognition*, 28(6), 1041–1049.

Kolers, P. A. (1973). Remembering operations. *Memory & Cognition*, 1(3), 347–355.

Kvavilashvili, L. (1987). Remembering intention as a distinct form of memory. *British Journal of Psychology*, 78(4), 507–518.

Kvavilashvili, L., & Fisher, L. (2007). Is time-based prospective remembering mediated by self-initiated rehearsals? Role of incidental cues, ongoing activity, age, and motivation. *Journal of Experimental Psychology: General*, 136(1), 112–132.

Kvavilashvili, L., & Mandler, G. (2004). Out of one's mind: A study of involuntary semantic memories. *Cognitive Psychology*, 48(1), 47–94.

Lee, J. H., Shelton, J. T., Scullin, M. K., & McDaniel, M. A. (2016). An implementation intention strategy can improve prospective memory in older adults with very mild Alzheimer's disease. *British Journal of Clinical Psychology*, 55(2), 154–166.

Lui, S. Y. S., Leung, S. S. W., Yang, T. X., Ho, K. K. Y., Man, C. M. Y., Leung, K. H. L., Wong, J. O. Y., Wang, Y., Cheung, E. F. C., & Chan, R. K. C. (2021). The benefits of emotionally salient cues on event-based prospective memory in bipolar patients and schizophrenia patients. *European Archives of Psychiatry and Clinical Neuroscience*, 271(8), 1503–1511.

Luo, L., & Craik, F. I. M (2008). Aging and memory: A cognitive approach. *Canadian Journal of Psychiatry*, 53(6), 346–353.

Luria, A. R. (1980). *Higher cortical functions in man*. Springer.

Lyons, A. D., Henry, J. D., Rendell, P. F., Corballis, M. C., & Suddendorf, T. (2014). Episodic foresight and aging. *Psychology and Aging*, 29(4), 873–884.

Mäntylä, T. (1994). Remembering to remember: Adult age differences in prospective memory. *Journal of Gerontology*, *49*(6), P276–P282.

Martiny-Huenger, T., Martiny, S. E., Parks-Stamm, E. J., Pfeiffer, E., & Gollwitzer, P. M. (2017). From conscious thought to automatic action: A simulation account of action planning. *Journal of Experimental Psychology: General*, *146*(10), 1513–1525.

May, C. P., Manning, M., Einstein, G. O., Becker, L., & Owens, M. (2015). The best of both worlds: Emotional cues improve prospective memory execution and reduce repetition errors. *Aging, Neuropsychology, and Cognition*, *22*(3), 357–375.

Maylor, E. A., & Valentine, T. (1992). Linear and nonlinear effects of aging on categorizing and naming faces. *Psychology and Aging*, *7*(2), 317–323.

Mayr, U., Spieler, D. H., & Hutcheon, T. G. (2015). When and why do old adults outsource control to the environment? *Psychology and Aging*, *30*(3), 624–633.

McBride, D. M., & Workman, R. A. (2017). Is prospective memory unique? A comparison of prospective and retrospective memory. *Psychology of Learning and Motivation*, *67*, 213–238.

McDaniel, M. A., & Einstein, G. O. (2000). Strategic and automatic processes in prospective memory retrieval: A multiprocess framework. *Applied Cognitive Psychology*, *14*(7), S127–S144.

McDaniel, M. A., Einstein, G. O., & Jacoby, L. L. (2008). New considerations in aging and memory: The glass may be half full. In F. I. M. Craik & T. A. Salthouse (Eds.), *The handbook of aging and cognition* (3rd ed., pp. 251–310). Academic Press.

McDaniel, M. A., & Scullin, M. K. (2010). Implementation intention does not automatize prospective memory responding. *Memory & Cognition*, *38*(2), 221–232.

McVay, J. C., & Kane, M. J. (2009). Conducting the train of thought: Working memory capacity, goal neglect, and mind wandering in an executive-control task. *Journal of Experimental Psychology: Learning, Memory, and Cognition*, *35*(1), 196–204.

Meier, B., & Zimmermann, T. D. (2015). Loads and loads and loads: The influence of prospective load, retrospective load, and ongoing task load in prospective memory. *Frontiers in Human Neuroscience*, *9*, e322.

Mioni, G., Grondin, S., McLennan, S. N., & Stablum, F. (2020). The role of time-monitoring behaviour in time-based prospective memory performance in younger and older adults. *Memory*, *28*(1), 34–48.

Morris, C. D., Bransford, J. D., & Franks, J. J. (1977). Levels of processing versus transfer appropriate processing. *Journal of Verbal Learning and Verbal Behavior*, *16*(5), 519–533.

Moscovitch, M. (1982). A neuropsychological approach to perception and memory in normal and pathological aging. In F. I. M. Craik & S. Trehub (Eds.), *Aging and cognitive processes* (pp. 55–78). Plenum Press.

Naveh-Benjamin, M., Craik, F. I. M., Guez, J., & Kreuger, S. (2005). Divided attention in younger and older adults: Effects of strategy and relatedness on memory performance and secondary task costs. *Journal of Experimental Psychology: Learning, Memory, and Cognition*, *31*(3), 520–537.

Ozgis, S., Rendell, P. G., & Henry, J. D. (2009). Spaced retrieval significantly improves prospective memory performance of cognitively impaired older adults. *Gerontology*, *55*(2), 229–232.

Phillips, L. H., Henry, J. D., & Martin, M. (2008). Adult aging and prospective memory: The importance of ecological validity. In M. Kliegel, M. A. McDaniel, & G. O. Einstein (Eds.), *Prospective memory: Cognitive neuroscience, developmental and applied perspectives* (pp. 161–186). Erlbaum.

Pirogovsky, E., Woods, S. P., Vincent Filoteo, J., & Gilbert, P. E. (2012). Prospective memory deficits are associated with poorer everyday functioning in Parkinson's disease. *Journal of the International Neuropsychological Society*, *18*(6), 986–995.

Radford, K. A., Lah, S., Say, M. J., & Miller, L. A. (2011). Validation of a new measure of prospective memory: The Royal Prince Alfred Prospective Memory Test. *Clinical Neuropsychologist*, *25*(1), 127–140.

Raghunath, N., Dahmen, J., Brown, K., Cook, D., & Schmitter-Edgecombe, M. (2020). Creating a digital memory notebook application for individuals with mild cognitive impairment to support everyday functioning. *Disability and Rehabilitation: Assistive Technology*, *15*(4), 421–431.

Raskin, S., Buckheit, C., & Sherrod, C. (2010). *Memory for intentions test: Manual*. Psychological Assessment Resources, Inc.

Raskin, S. A., Shum, D. H. K., Ellis, J., Pereira, A., & Mills, G. (2018). A comparison of laboratory, clinical, and self-report measures of prospective memory in healthy adults and individuals with brain injury. *Journal of Clinical and Experimental Neuropsychology*, 40(5), 423–436.

Raz, N. (2000). Aging of the brain and its impact on cognitive performance: Integration of structural and functional findings. In F. I. M. Craik & T. A. Salthouse (Eds.), *The handbook of aging and cognition* (2nd ed., pp. 1–90). Lawrence Erlbaum Associates Publishers.

Rendell, P. G., Bailey, P. E., Henry, J. D., Phillips, L. H., Gaskin, S., & Kliegel, M. (2012). Older adults have greater difficulty imaging future rather than atemporal experiences. *Psychology and Aging*, 27(4), 1089–1098.

Rendell, P. G., & Craik, F. I. M. (2000). Virtual week and actual week: Age-related differences in prospective memory. *Applied Cognitive Psychology*, 14(7), S43–S62.

Rendell, P. G., Phillips, L. H., Henry, J. D., Brumby-Rendell, T., de la Piedad Garcia, X., Altgassen, M., & Kliegel, M. (2011). Prospective memory, emotional valence and ageing. *Cognition and Emotion*, 25(5), 915–925.

Rendell, P. G., & Thomson, D. M. (1999). Aging and prospective memory: Differences between naturalistic and laboratory tasks. *Journals of Gerontology Series B: Psychological Sciences and Social Sciences*, 54(4), 256–269.

Rhodes, S., Greene, N. R., & Naveh-Benjamin, M. (2019). Age-related differences in recall and recognition: A meta-analysis. *Psychonomic Bulletin & Review*, 26(5), 1529–1547.

Rose, N. S., Rendell, P. G., Hering, A., Kliegel, M., Bidelman, G. M., & Craik, F. I. M. (2015). Cognitive and neural plasticity in older adults' prospective memory following training with the Virtual Week computer game. *Frontiers in Human Neuroscience*, 9, 592.

Rummel, J., & Meiser, T. (2013). The role of metacognition in prospective memory: Anticipated task demands influence attention allocation strategies. *Consciousness and Cognition*, 22(3), 931–943.

Rummel, J., Smeekens, B. A., & Kane, M. J. (2017). Dealing with prospective memory demands while performing an ongoing task: Shared processing, increased on-task focus, or both? *Journal of Experimental Psychology: Learning, Memory, and Cognition*, 43(7), 1047–1062.

Salthouse, T. A. (1982). *Adult cognition: An experimental psychology of human aging.* Springer-Verlag.

Sandson, J., & Albert, M. L. (1984). Varieties of perseveration. *Neuropsychologia*, 22(6), 715–732.

Sanjram, P. K., & Khan, A. (2011). Attention, polychronicity, and expertise in prospective memory performance: Programmers' vulnerability to habit intrusion error in multitasking. *International Journal of Human Computer Studies*, 69(6), 428–439.

Schnitzspahn, K. M., Ihle, A., Henry, J. D., Rendell, P. G., & Kliegel, M. (2011). The age-prospective memory-paradox: An exploration of possible mechanisms. *International Psychogeriatrics*, 23(4), 583–592.

Scullin, M. K., & Bugg, J. M. (2013). Failing to forget: Prospective memory commission errors can result from spontaneous retrieval and impaired executive control. *Journal of Experimental Psychology: Learning, Memory, and Cognition*, 39(3), 965–971.

Scullin, M. K., Bugg, J. M., & McDaniel, M. A. (2012). Whoops, I did it again: Commission errors in prospective memory. *Psychology and Aging*, 27(1), 46–53.

Scullin, M. K., McDaniel, M. A., Dasse, M. N., Lee, J. H., Kurinec, C. A., Tami, C., & Krueger, M. L. (2018). Thought probes during prospective memory encoding: Evidence for perfunctory processes. *PLoS One*, 13(6), e0198646.

Scullin, M. K., McDaniel, M. A., & Shelton, J. T. (2013). The dynamic multiprocess framework: Evidence from prospective memory with contextual variability. *Cognitive Psychology*, 67(1–2), 55–71.

Sheppard, D. P., Matchanova, A., Sullivan, K. L., Kazimi, S. I., & Woods, S. P. (2019). Prospective memory partially mediates the association between aging and everyday functioning. *Clinical Neuropsychologist*, 34(4), 755–774.

Sim, J., Brown, F. L., O'Connell, R. G., & Hester, R. (2020). Impaired error awareness in healthy older adults: An age group comparison study. *Neurobiology of Aging*, 96, 58–67.

Smith, G., Della Sala, S., Logie, R. H., & Maylor, E. A. (2000). Prospective and retrospective memory in normal ageing and dementia: A questionnaire study. *Memory*, 8(5), 311–321.

Smith, R. E. (2003). The cost of remembering to remember in event-based prospective memory: Investigating the capacity demands of delayed intention performance. *Journal of Experimental Psychology: Learning, Memory, and Cognition*, 29(3), 347–361.

Smith, R. E. (2017). Prospective memory in context. *Psychology of Learning and Motivation*, 66, 211–249.

Smith, R. E., Hunt, R. R., McVay, J. C., & McConnell, M. D. (2007). The cost of event-based prospective memory: Salient target events. *Journal of Experimental Psychology: Learning, Memory, and Cognition*, *33*(4), 734–746.

Smith, R. E., Hunt, R. R., & Murray, A. E. (2017). Prospective memory in context: Moving through a familiar space. *Journal of Experimental Psychology: Learning, Memory, and Cognition*, *43*(2), 189–204.

Strickland, L., Elliott, D., Wilson, M. D., Loft, S., Neal, A., & Heathcote, A. (2019). Prospective memory in the red zone: Cognitive control and capacity sharing in a complex, multi-stimulus task. *Journal of Experimental Psychology: Applied*, *25*(4), 695–715.

Strickland, L., Loft, S., Remington, R. W., & Heathcote, A. (2018). Racing to remember: A theory of decision control in event-based prospective memory. *Psychological Review*, *125*(6), 851–887.

Stuss, D. T., & Alexander, M. P. (2007). Is there a dysexecutive syndrome? *Philosophical Transactions of the Royal Society of London. Series B, Biological Sciences*, *362*(1481), 901–915.

Stuss, D. T., & Craik, F. I. M. (2020). Alterations in executive functions with aging. In K. H. Heilman & S. E. Nadeau (Eds.)., *Cognitive changes and the aging brain* (pp. 168–187). Cambridge University Press.

Thompson, C. L., Henry, J. D., Rendell, P. G., Withall, A., & Brodaty, H. (2015). How valid are subjective ratings of prospective memory in mild cognitive impairment and early dementia? *Gerontology*, *61*(3), 251–257.

Thöne-Otto, A. I. T., & Walther, K. (2010). Assessment and treatment of prospective memory disorders in clinical practice. In M. Kliegel, M. A. McDaniel, & G. O. Einstein (Eds.), *Prospective memory: Cognitive, neuroscience, developmental and applied perspectives* (pp. 321–345). Lawrence Erlbaum Associates.

Tulving, E., & Schacter, D. L. (1990). Priming and human memory systems. *Science*, *247*(4940), 301–306.

Tulving, E., & Thomson, D. M. (1973). Encoding specificity and retrieval processes in episodic memory. *Psychological Review*, *80*(5), 352–373.

Twamley, E. W., Woods, S. P., Zurhellen, C. H., Vertinski, M., Narvaez, J. M., Mausbach, B. T., Patterson, T. L., & Jeste, D. V. (2008). Neuropsychological substrates and everyday functioning implications of prospective memory impairment in schizophrenia. *Schizophrenia Research*, *106*(1), 42–49.

Van Rensbergen, G., & Pacolet, J. (2012). Instrumental activities of daily living (I-ADL) trigger an urgent request for nursing home admission. *Archives of Public Health*, *70*(1), 1–8.

Varley, D., Henry, J. D., Gibson, E., Suddendorf, T., Rendell, P. G., & Redshaw, J. (2021). An old problem revisited: How sensitive is time-based prospective memory to age effects? *Psychology and Aging*, *36*(5), 616–625.

von Hippel, W., & Henry, J. D. (2012). Social cognitive aging. In S. T. Fiske, & C. N. Macrae (Eds.), *The SAGE handbook of social cognition* (pp. 390–411). Sage Publications.

Wasylyshyn, C., Verhaeghen, P., & Sliwinski, M. J. (2011). Aging and task switching: A meta-analysis. *Psychology and Aging*, *26*(1), 15–20.

West, R. L. (1996). An application of prefrontal cortex function theory to cognitive aging. *Psychological Bulletin*, *120*(2), 272–292.

West, R., & Craik, F. I. M. (2001). Influences on the efficiency of prospective memory in younger and older adults. *Psychology and Aging*, *16*(4), 682–696.

Wilkins, A. J., & Baddeley, A. D. (1978). Remembering to recall in everyday life: An approach to absent-mindedness. In M. M. Gruneberg, P. E. Morris & R. N. Sykes (Eds.), *Practical aspects of memory* (pp. 27–34). Academic Press.

Wilson, B. A., Emslie, H., Foley, J., Shiel, A., Watson, P., Hawkins, K, Groot, Y., & Evans, J. J. (2005). *The Cambridge prospective memory test*. Pearson.

Wilson, M. D., Strickland, L., Farrell, S., Visser, T. A. W., & Loft, S. (2020). Prospective memory performance in simulated air traffic control: Robust to interruptions but impaired by retention interval. *Human Factors*, *62*(8), 1249–1264.

Yuana, P., & Raz, N. (2014). Prefrontal cortex and executive functions in healthy adults: A meta-analysis of structural neuroimaging studies. *Neuroscience & Biobehavioral Reviews*, *42*, 180–192.

14

Mental imagery

Using working memory theory to design behaviour change interventions

Jackie Andrade

This morning I walked on Dartmoor, a wild and hilly area in the south-west of England. Now that I'm at my desk, I can vividly recall the skylark singing overhead, the warmth of the May sun after a frosty start, and the view across farmland spotted with bright yellow fields of oilseed rape. I can also imagine a scene where I arrive at the top of the hill to find the aftermath of a party, with litter strewn around and loud music playing over still-sleeping bodies. This violation has never happened here, but my semantic memory contains all the ingredients to create this image. I feel joy imagining the first scene, anger imagining the second. The images are not merely static. I can animate them, creating a version where the revellers quietly wake up, pack up their things, and leave everything looking as pristine as it did before.

These examples illustrate that mental imagery is supported by episodic and semantic memory, that it is connected to emotions, and that it is a dynamic, creative process requiring temporary storage and manipulation of sensory information. In this chapter, I describe theoretical research on the cognitive processes underpinning mental imagery, and explore how this research has guided research on a range of problems including mental health and weight loss. This applied research in turn raises theoretical questions that promise to shape how we conceptualize the relationship between working memory, conscious experience, and long-term memory.

Working memory and mental imagery

In his 1986 monograph *Working Memory*, Alan Baddeley described a tripartite model of working memory containing three modular sets of cognitive processes. In this model, the articulatory loop (later called the phonological loop; Baddeley, 1997) temporarily stores auditory and speech-based information, keeping representations active via a subvocal rehearsal process. The visuospatial sketchpad temporarily stores pictorial information, keeping it active via an 'inner scribe' sequential rehearsal or rewriting process (Logie, 1995). The central executive controls

Jackie Andrade, *Mental imagery* In: *Memory in Science for Society*. Edited by: Robert H. Logie, Zhisheng (Edward) Wen, Susan E. Gathercole, Nelson Cowan, and Randall W. Engle, Oxford University Press. © Oxford University Press 2023. DOI: 10.1093/oso/9780192849069.003.0014 allocation of attentional resources to these 'slave systems' and manipulates representations held in them.

The working memory model has undergone further development since then, notably with the addition of a fourth component called the episodic buffer. The episodic buffer serves as a store for multimodal or 'bound' representations that may include information from perception, the visuospatial sketchpad and phonological loop, and long-term memory. The episodic buffer thus serves as an interface between the central executive and other memory systems (Baddeley et al., 2011, 2021). Evidence supporting this model has been reviewed extensively elsewhere (e.g. Baddeley, 2002). Figure 14.1 illustrates its structure and components.

Throughout the development of this working memory model, there has been a focus on working memory serving complex cognition. This focus meant that the model lent itself to guiding research into abilities that are important for life outside the laboratory. In the decade or so that followed the publication of *Working Memory* (Baddeley, 1986), evidence accumulated for the hypothesis that being able to temporarily store an unfamiliar sequence of speech sounds is an essential component of native and second language acquisition. Reviewing this evidence, Baddeley et al. (1998) concluded that the key role of the phonological loop is to support language acquisition.

Baddeley (1986) also suggested a function for the visuospatial sketchpad, arguing that there 'appears to be good evidence for the occurrence of a temporary visuospatial store . . . that is capable of retaining and manipulating images' (p. 143). In support of this hypothesis, Logie et al. (1990) showed that visuospatial imagery and memory tasks interfere more with each other than they do with verbal tasks of similar complexity. They asked participants to remember a grid of black and white squares as the visual short-term memory task, or a sequence of letters as the verbal short-term memory task, at the same time as performing another task. These

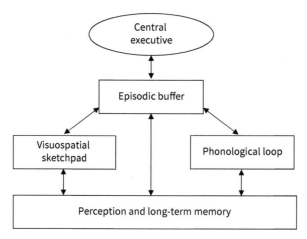

Figure 14.1 The four-component model of working memory (based on Baddeley et al., 2011, 2021).

competing tasks involved predominantly visuospatial or verbal processing. For example, participants were asked to imagine a path through a matrix constructed by following a series of instructions such as, 'in the starting square put a 1, in the next square UP put a 2' (Brooks, 1967) or they performed a verbal equivalent of this task by remembering a set of similar but meaningless sentences, 'in the starting square put a 1, in the next square to the GOOD put a 2'. Logie et al. found that the imagery version of Brooks' task reduced performance more on the concurrent visual grid-memory task than the verbal letter-memory, and vice versa for the verbal version of the task. This finding supports the idea that visuospatial working memory supports imagery as well as temporary storage functions, that it involves representations that contain visual perceptual information rather than purely abstract or propositional representations, and that it has limited capacity.

This chapter takes the idea of a visuospatial sketchpad supporting visual imagery as a starting point, and extends it to argue that working memory supports imagery in different sensory modalities and with different emotional qualities. By doing this, working memory underpins recollection of memories, craving and desire, and anticipation of the future, making it a key determinant of how we behave and what we experience.

Mental imagery as mental experience

Performing a Brooks' (1967) matrix task and imagining countryside scenes differ in an important way. The imagery I described at the start of this chapter is much richer with sensory details and emotion—delight at the lark's song, dismay at the mess from the party. It is an episodic *experience* rather than a cold representation in working memory. Important early studies of mental imagery focused on imagery as mental representation, for example, revealing how different parts of the brain supported imagery (Farah, 1989) and whether its content was analogous to its real-world counterpart in the sense of preserving distance, space, and time (Kosslyn, 1996), and whether brain injury affected attention to perceived and imagined stimuli in similar ways (Bisiach & Luzzatti, 1978). Interest in the relationship between imagery and perception has continued, with functional brain imaging studies revealing similarities in brain activation when visual stimuli are perceived and imagined (Ganis et al., 2004). Studies of imagery in other modalities show an association between imagery and activation of sensory brain areas supporting perception in the same modality (e.g. Kobayashi et al., 2004; Zvyagintsev et al., 2013).

There was less focus on imagery as experience, on what it felt like to imagine something. Although the cognitive revolution in psychology (Neisser, 1967) moved researchers' focus from behaviourist input–output relationships to intervening mental processes, there was still a wariness of researching mental experiences in their own right. In contrast, in clinical psychology, there had been long-standing interest in symptoms of mental illness that could be classified as imagery experiences, for example, auditory hallucinations in schizophrenia and flashbacks in post-traumatic stress disorder, and in the use of imagery as a therapeutic tool. For example, exposure treatments for anxiety and phobia can effectively use imagined contact with the feared situation as well as *in vivo* experiences. In cognitive behavioural therapy for addiction, imagery has been used to induce drug cravings prior to practising strategies for coping with such cravings. Experiences and treatments of mental illness have provided a rich source of ideas and a testing ground for hypotheses about the cognitive psychology of mental imagery.

In general, these ground-breaking studies of imagery measured performance on tasks requiring imagery, or reduction in distressing symptoms, rather than imagery experience. Exceptions to this trend were work on imagery vividness that assessed whether being good at generating vivid images related to other mental abilities (Marks, 1973) and what sensory features characterized vivid imagery (Cornoldi et al., 1992). There were some puzzling findings from this work, namely observations of weak or negative associations between participants' ratings of imagery vividness and their performance on tasks assumed to involve imagery, such as remembering colours (Reisberg et al., 1986). These studies told us about individual differences in imagery ability and about the content of vivid images, but they said little about the cognitive processes that need harnessing to create a vivid image, in the sense of a mental experience that feels veridical, analogous to the equivalent perceptual experience.

What makes an image vivid?

Alan Baddeley and I wanted to know how people create vivid images (Baddeley & Andrade, 2000). We used the working memory model to design a set of experiments to test the extent to which temporary storage of sensory information and attentional or executive processing were important factors. Rather than measuring image content or performance on imagery tasks, we asked participants to report their imagery experience, that is, the vividness of their image. We largely used repeated measures designs to avoid the problem, suggested by research on individual differences in imagery vividness, that individuals might vary hugely in how they rated their images yet these differences might tell us little about underlying differences in cognitive functions.

We asked participants to hold in mind a recently presented and meaningless tune or pattern, or to create an image from a written cue, for example to imagine the sound of people laughing or the appearance of children playing. Participants generated and maintained each image for a few seconds either unhampered by a secondary task or while performing a task designed to load the visuospatial sketchpad, such as tapping a pattern on a keypad, or the phonological loop, such as counting aloud. They then rated the vividness of their image on a scale of 0, indicating they were trying to think

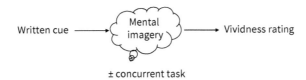

Figure 14.2 Experimental protocol in Baddeley and Andrade's (2000) study of working memory and imagery.

about the scene but were experiencing no image at all, to 10, indicating an image that was as clear and vivid as normal hearing or vision (Figure 14.2). As with studies of short-term memory such as the one by Logie et al. (1990), we looked for a crossover interaction as a signature of modality-specific working memory involvement. That is what we found. Visual images were less vivid with the concurrent visuospatial task than with the concurrent verbal task, while the converse was true for auditory images (Figure 14.3). For images in both modalities, there was also a general effect of cognitive load: vividness was lower in both concurrent task conditions than it was in the control condition.

There were also effects that implicated long-term memory in imagery vividness. Participants anticipated that they would be able to imagine a bizarre situation, such as trees marching, more vividly than a familiar one, such as a ship coming into a harbour, perhaps drawing on metacognitive knowledge that bizarre images tend to be memorable. However, their ratings showed the opposite pattern when they imagined these scenes in our experiments with a fixed time period. Bizarre images were

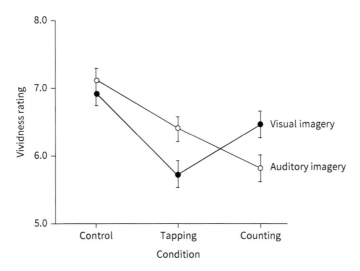

Figure 14.3 Combined data from five experiments showing effects of visuospatial and verbal concurrent tasks (pattern tapping and counting respectively) on mean rated vividness of visual and auditory imagery, with standard error bars. Reprinted from Andrade, J. (2001). The contribution of working memory to conscious experience. In J. Andrade (Ed.), *Working Memory in Perspective* (pp. 60–78). Psychology Press.

rated as less vivid than more familiar images. Participants also rated static or simple images (the appearance of a rose garden or the sound of a hairdryer) as more vivid than dynamic, changing images (the appearance of a cat climbing a tree or the sound of someone singing in the shower). These findings suggested that images were more vivid when a ready-made 'snapshot' was available in long-term memory. Consistent with this interpretation, we found that participants with greater knowledge of a subject imagined exemplars with greater vividness than those with less knowledge. It appeared that images were more vivid when long-term memory provided plenty of experiential detail and ready-made examples. Images were less vivid when information needed to be combined or continually updated to generate novel or dynamic mental experiences.

We interpreted these findings as evidence that vivid imagery requires the temporary storage of sensory information in the phonological loop and visuospatial sketchpad of working memory. It also requires central executive resources, to retrieve information from long-term memory, manipulate information to create novel images, and update information to create animated images. It seemed likely that the processing part of this executive role would proceed most efficiently if the information retrieved from long-term memory were stored temporarily in working memory as a bound representation, for example, as a single representation of an elephant eating peas rather than separate representations drawn from disparate semantic memories of peas and elephants. This assumption fits the phenomenology of imagery, outside of the constraints of laboratory experiments, as a multimodal experience in which, for example, imagining the scent of a rose automatically and seemingly simultaneously involves imagining the visual appearance of the rose. Studies of desire imagery discussed later in the chapter corroborate this introspection. For example, participants beginning treatment for alcohol addiction reported that their craving for alcohol frequently involved imagery of the smell and taste of alcohol and the sensation of swallowing it, with a mean of 2.3 sensory modalities reported overall (Kavanagh et al., 2009). Baddeley (2000) introduced the episodic buffer, described earlier, to the working memory model as a separate storage module for multimodal representations like these images.

Working memory loads reduce emotionality as well as vividness of imagery

Life at the Applied Psychology Unit in Cambridge, UK, during the 1990s involved institutionalized coffee and tea breaks, twice a day, every day. Attendance was expected. It was at one of these breaks that David Kavanagh, visiting on sabbatical from the University of Sydney, asked me what I thought of a puzzling new approach to treating post-traumatic stress disorder, called eye-movement desensitization and reprocessing or EMDR (Shapiro, 1989). EMDR had no clear theoretical basis but incorporated elements of exposure treatments and cognitive behavioural therapy. The attention-grabbing aspect was that clients were asked to make rapid side-to-side eye movements, tracking the therapist's moving finger, while recalling memories of trauma. I had not heard of EMDR but, buoyed by the early results of our studies of image vividness, I suggested that requiring someone to make side-to-side eye movements while recalling a traumatic experience would interfere with visuospatial working memory and make their recollection less vivid and less distressing. Perhaps this was a useful thing, therapeutically.

We tested this idea using a similar procedure to the theoretical studies described above. For mundane reasons, these experiments were published before the theoretical work that inspired them (Andrade et al., 1997). We asked non-clinical participants to view newspaper photos of distressing events or to recall emotional autobiographical memories, and then to hold an image or memory in mind for a few seconds while performing no other task, a computerized side-to-side eye movement task modelled on the one used in EMDR, or another comparison task such as tapping a pattern or counting aloud. Although the working memory model did not encompass emotion, an application of the model to trauma memories clearly needed to assess whether loading working memory could reduce the strength of emotion experienced by someone recalling a happy or upsetting event. We therefore asked participants to rate their emotional response to the image on a scale similar to the one we used to measure vividness, ranging from −10 (extremely negative), through 0 (neutral), to +10 (extremely positive).

As predicted, visuospatial tasks reduced the vividness of these visual images more than a verbal task. Eye movements had a somewhat stronger effect than pattern tapping or visual interference (Kavanagh et al., 2001), which we attributed to eye movements imposing visual interference, from shifting one's gaze across a scene, as well as spatiomotor control whereas the other tasks involved only one of these elements. Subsequent research confirmed that side-to-side eye movements are a particularly effective means of loading the visuospatial sketchpad (Lawrence et al., 2004; Pearson & Sahraie, 2003; Postle et al., 2006).

The effects of eye movements on participants' ratings of emotion were very similar, in time course and extent, to their effects on the vividness of their images and recollections. This finding was important for several reasons. One reason was that it suggested that the therapeutic role of eye movements in EMDR might be something to do with the load they imposed on working memory. Prompted by a reviewer pointing out that our results were counterintuitive as an explanation of EMDR, because a less vivid recollection should impede rather than facilitate imaginal desensitization, we hypothesized that working memory loads might act as a treatment aid rather than having a direct impact on symptoms. By helping clients recollect a traumatic memory in a less vivid, less distressing form, eye movements might increase the person's tolerance for holding the memory in mind while habituating to the emotion it engenders and re-evaluating its meaning (Kavanagh et al., 2001;

Lilley et al., 2009). This mechanism of therapeutic effect remains to be established but the general approach, of treating the eye movement component of EMDR as a working memory load, has generated considerable research and debate (see review by Landin-Romero et al., 2018).

A second reason that the results were important is that they showed how the working memory model could be used to design studies of emotion as well as image vividness. Kemps and Tiggemann (2007a, experiment 2) demonstrated this point nicely by asking participants to focus exclusively on either the auditory or the visual aspects of emotional autobiographical memories. They found the predicted crossover interaction that is the signature of modality-specific working memory involvement. Eye movements reduced the vividness and emotionality of visual auto-biographical images relative to counting aloud, whereas counting aloud reduced the vividness and emotionality of auditory autobiographical images relative to eye movements.

The third reason is that finding that the impact of working memory loads on emotion mirrored those on vividness influenced our view of imagery as an affectively charged mental experience, with emotion being embodied in the image rather than triggered by it. In addition to finding that vivid images were more emotive, we also found the converse, that more emotive stimuli were imagined more vividly than neutral stimuli (Bywaters et al., 2004). This view of imagery as embodied experience is consistent with Barsalou's (2008) grounded cognition approach, in which action-related, sensory and emotional aspects of concepts are an integral element of concepts rather than mere associates of them. It is corroborated by Holmes' work on the closer association between imagery and emotion than between verbal or propositional thinking and emotion. For example, Holmes and Mathews (2005) demonstrated that imagining anxiety-provoking scenarios increased state anxiety more than verbally thinking about the same scenarios. These ideas on the association between imagery and emotion were important when we moved to studying desire.

Mental imagery and cravings

Drug and food cravings are strange. They can maintain addictions or unhealthy eating habits even when the individual wants to quit or lose weight (Tiffany & Wray, 2012). The really strange part is that these desires require cognitive resources that are also needed for working towards the goal to quit, thus cravings impair cognitive performance (Sayette et al., 1994; Zwaan & Truitt, 1998). Subsequent research, discussed in more detail below, shows that working memory loads weaken cravings (Skorka-Brown et al., 2014, 2015) and increase healthy rather than indulgent food choices (van Dillen & Andrade, 2016). Given a limited supply of cognitive resources, why should we use those resources to feed desires that contravene our valued goals?

David Kavanagh, Jon May, and I wanted to solve this conundrum. Introspecting our desires for pre-work coffee and post-work beer, we postulated that mental imagery might be the driving force of cravings. That beer seemed more tempting, more refreshing, more *necessary*, when we imagined it vividly. Embellishing the image with details of the smell, the coldness, the feel of the little bubbles, and the taste of the first sip made things worse, strengthening the desire to stop work and go to the pub. Imagery seemed at first too obvious an answer but a review of the literature found no tests of the role of imagery in desire and no theories of desire that included imagery, despite evidence that guided imagery could trigger drug cravings (Tiffany & Hakenewerth, 1991).

We developed a theory of desire, called elaborated intrusion theory (EI theory; Kavanagh et al., 2005), with imagery as the engine (Figure 14.4). At the heart of EI theory was the idea that desire, or craving in a drug-use context, resulted from the operation of working memory processes that maintained information about the desired activity or substance in consciousness. In contrast to most other theories of craving, environmental, physiological, cognitive, and mood cues served to increase the likelihood of experiencing an apparently spontaneous, intrusive thought about the activity ('I need some coffee' or an image of a coffee cup, for example) but did not directly cause the craving. We subsequently reported evidence for this assumption: most respondents with alcohol addiction reported that they sometimes experienced thoughts about alcohol that went of their own accord, without leading to a craving (Kavanagh et al., 2009). In EI theory, these thoughts become desires when they are elaborated, a process that loads heavily on working memory. Working memory resources are used to retrieve desire-related information from long-term memory, and to use that information to construct richly sensory mental images of achieving one's desire. This sensory imagery is affectively charged in the same way as the actual experience. This assumption is central to answering the question we set ourselves, which was why do we devote cognitive resources to craving even when doing so is counterproductive to our goals to abstain? The EI theory answer to this question is that we do so because desire images embody the same pleasure or relief as actual consumption. The positive feeling is short-lived, because our current state of deprivation compares poorly with our imagined satisfaction. The more vividly we imagine satisfying our desire, the starker this contrast becomes, fuelling a cycle of more vivid, positively charged imagery leading to increased awareness of deficit and motivating us to acquire the substance despite a conflicting goal to avoid it.

Studies of the phenomenology of drug and food cravings support the hypothesis that they involve multisensory imagery. An early study by May et al. (2004) asked people to complete a questionnaire while they were experiencing a craving. They typically reported that the craving began when they 'suddenly thought about it' or imagined having the substance, and that the craving episode involved imagining the appearance and taste of the substance (but not the sound of it). Subsequent questionnaire studies confirmed the association between imagery, intrusive thoughts,

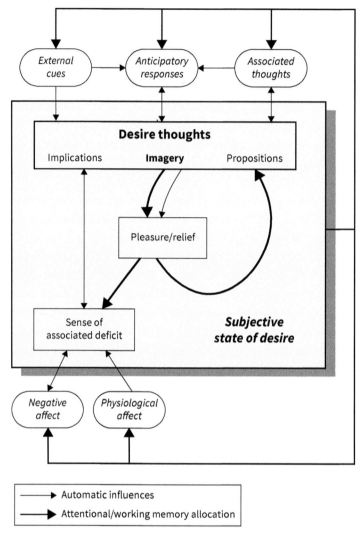

Figure 14.4 The elaborated intrusion theory of desire, showing the contribution of triggers (rounded external boxes), intrusive thoughts ('desire thoughts'), and sensory imagery to desire (central square box). Reprinted with permission Kavanagh, D. J., Andrade, J., & May, J. (2005). Imaginary relish and exquisite torture: The elaborated intrusion theory of desire. Psychological Review, 112(2), 446–467.

and desire strength (e.g. Kavanagh et al., 2009) and led to the development of new tools for measuring craving as an episodic mental experience that varied in strength and frequency. Psychometric studies of these measures confirmed the involvement of three correlated but separable factors of desire strength, imagery, and intrusiveness (May et al., 2014; Statham et al., 2011). Recent research has extended these scales to behavioural addictions, showing similar results in gambling (Cornil et al., 2018, 2019).

Experimental tests of EI theory built on observations that craving for drugs such as alcohol and nicotine, and for everyday substances like food and drink, involved visual imagery as well as imagery of taste, smell, and bodily sensations, but did not involve auditory imagery (Kavanagh et al., 2009; May et al., 2004, 2008). In working memory terms, we hypothesized that craving imagery loads the visuospatial sketchpad more than the phonological loop. Tests of this hypothesis compared the effects of visuospatial loads, such as clay modelling, or competing visual imagery, against auditory imagery or verbal task loads. Relative to auditory loads, visuospatial tasks reduce craving for cigarettes (May et al., 2010; Versland & Rosenberg, 2007) and food (Andrade, Pears, et al., 2012; Kemps & Tiggemann, 2007b). Kemps and Tiggemann (2009) showed that cravings for coffee involved visual, olfactory, and taste imagery. A competing olfactory task reduced coffee craving strength, as did a visuospatial task, relative to an auditory task.

According to the working memory model, this interference should be mutual. If craving involves mental imagery, which we have established uses visuospatial and central executive working memory resources (and, by definition, episodic buffer resources), then craving should impair performance on other tasks that require those resources. There is some evidence that this is the case (e.g. Sayette et al., 1994; Zwaan & Truitt, 1998), but it is not always clear-cut. For example, Kemps et al. (2008) showed that participants performed more poorly on an operation span task after abstaining from chocolate for 24 hours and completing the task in the presence of chocolate, than in a no-craving control condition. However, this was only the case for those who reported habitually craving chocolate. Meule et al. (2012) found slower and more errorful performance on an N-back task that involved pictures of high-calorie foods than one with neutral stimuli. However, this was equally the case for participants with higher trait and state food cravings as for those with lower cravings. These findings support the hypothesis that craving involves working memory resources but suggest that the contribution of working memory to individual differences in cravings requires further investigation.

Mental imagery and desire for functional goals

Overall, the research presented above shows that cravings for substances fit the predictions of EI theory. Cornil's work on gambling provides evidence that they extend to compulsive behaviours. EI theory was intended as a general theory of desire, not merely of dysfunctional craving, therefore its predictions should apply to motivations for functional behaviours. Applications of EI theory to functional desires have focused on developing a psychometrically robust measure of motivation based on the theory. Robinson et al. (2016) adapted the theory-based craving experience questionnaire that had been developed for measuring substance cravings (May et al., 2014; Statham et al., 2011), to assess desire to reduce alcohol consumption. Initial analysis suggested that desire to reduce alcohol consumption contained the same three factors as desire for alcohol, namely intensity, imagery, and intrusiveness or availability. Subsequent analysis of the same dataset showed a better fit with a four-factor model that split the imagery factor into incentives imagery (e.g. 'imagine how good it would be to do it') and self-efficacy imagery (e.g. 'imagine succeeding at it') (Kavanagh et al., 2018). Studies of motivation for diabetes self-management (Parham et al., 2017) and motivation for healthy eating and physical activity (Kavanagh et al., 2020) replicated this finding and support the hypothesis that imagery is a core component of desire for functional goals as well as of unwanted cravings.

Functional imagery training: using imagery to change behaviour

Evidence that EI theory applies to functional as well as dysfunctional desires raises the question of how we might use the theory to strengthen desires for healthy goals and behaviours. According to the theory, encouraging vivid, positively affective imagery of goal achievement will make the goal more accessible and serve as a standard against which to mentally contrast the current reality. Keeping this imagined goal in mind will motivate behaviour towards it. Research shows that imagining future episodes increases future-oriented choices and decreases impulsive decision-making oriented to short-term rewards (Daniel et al., 2013). We predict that imagining a specific, goal-related episode will be maximally effective, not only shifting attention to the future episode but increasing desire for it and confidence that it can be achieved. Likewise, encouraging imagery of the benefits of working towards the goal is predicted to increase desire for the required behaviours, and confidence that they can be performed. Consistent with these predictions, there is evidence from sport and exercise research that mental practice increases performance (Driskell et al., 1994) and that imagery supports motivation (Hall et al., 1998), and from findings that imagery also helps convert good intentions to actual behaviour, increasing the efficacy of 'if . . . then' plans or implementation intentions (Knäuper et al., 2009, 2011).

Our research on working memory and vividness of imagery showed that images were more vivid when they were better supported by long-term memory. Motivationally effective images should therefore be as personal as possible, drawing on autobiographical memories as well as general knowledge. According to EI theory, episodes of desire begin with a spontaneous thought or image about the goal that intrudes into awareness. For example, one's focus on a piece of work might be interrupted by a thought about one's children. This thought may be elaborated as an image that includes catching an early train to get home to be with them, motivating one to stop working and pack up. As Hofmann et al. (2012) showed in their ecological momentary assessment study of desire, people frequently experience desires that conflict with their goals. Sometimes these conflicts can be between an immediate reward and a longer-term goal, for example, to stop and get some coffee or to complete a piece of work, but they can also be between conflicting longer-term goals as in the example above where the person has a goal to complete the work but also to spend time with family. We hypothesize that one way to shift the balance between different goals is to mentally rehearse the one that is the priority. If the person in the example above wants to spend more time with their family, then practising imagery of enjoying that time should help it intrude more readily. It is important that this imagery is personal, using sensory details from memories of specific episodes when they have enjoyed time with their family. Imagery of spending time with children today is predicted to be a more effective motivator than imagery of a more abstract concept of being a good parent.

The working memory model leads to another prediction, which is that conflicting goal images will compete for limited capacity resources. According to EI theory, the concrete, practised image of drinking some coffee will more effectively grab working memory resources than a more abstract image of success at work. The image is ready-made in episodic memory and requires less executive resource for its retrieval and maintenance. Developing and practising imagery of functional goals should help them compete more effectively for resources. May et al. (2010) showed that smokers reported stronger cigarette craving after generating neutral auditory images than after neutral visual imagery designed to compete with their visual craving images. We propose that goal imagery should be even more effective at blocking craving imagery, because it is affectively charged and therefore more likely to be given attentional priority.

We developed an intervention called functional imagery training (FIT; or 'functional decision-making' in its initial phases; Andrade, May, & Kavanagh, 2012) that brings together these ideas from EI theory and working memory. FIT encourages individuals to imagine the potential immediate benefits of changing their behaviour as well as the longer-term benefits of attaining their goal. To ensure that the imagery is personal, important, and relevant, guided imagery exercises take place in the context of a directed, person-centred conversation. As in motivational interviewing (Miller & Rollnick 2012), this conversation follows a structure that begins with weighing up the general pros and cons of behaviour change versus no change and moves towards creating a specific action plan. Each step of the conversation is accompanied by personalized, multisensory guided imagery to amplify the emotional impact of the conversation and to teach the use of imagery for self-motivation. A FIT session therefore begins by eliciting and imagining the benefits of behaviour change for the person as they might experience them at a specific episode in the future, and contrasting that image with imagery of the same future episode but where no change has happened (Figure 14.5). This phase of FIT is designed to strengthen desire for the end goal. Here and throughout FIT, the practitioner uses active listening skills, and particularly reflection, to draw attention to what is most meaningful, important, and motivating for the person and to highlight any points where imagery has affected how

Figure 14.5 Structure of a typical FIT session. The imagery work shown here happens in a context of active listening where the practitioner helps the individual identify aspects of their goal that are motivating because they link to their core values. Booster sessions help develop skills in updating motivational imagery based on recent experiences or new goals.

motivated or confident they feel. Next, the practitioner asks the person about their ideas for working towards their goal, and helps them mentally develop and practise those ideas in imagery. They discuss past successes and anticipated obstacles, and use imagery to apply strategies that were successful in the past to overcoming future obstacles, focusing on recollecting feelings of success. By the end of a typical initial FIT session, the person will have developed a vivid image of taking the first steps towards their goal, playing forward to successfully staying on track for a few days or weeks and experiencing the benefits of working towards their goal, and ultimately achieving their goal. They are encouraged to practise that imagery frequently by pairing it with an everyday behaviour—for example, practising imagery while waiting for the kettle to boil—so that it becomes really vivid and accessible. The idea is to create goal-relevant imagery that comes to mind readily at moments of decision and attracts working memory resources because it is positively emotionally charged. In effect, FIT is creating personal, episodic memories of the future that motivate behaviour towards that imagined future.

As an example, take someone who would like to reduce their alcohol consumption because their drinking habit has been causing arguments with their spouse and they would like to save their marriage. It is Friday evening and their friends are suggesting going for a few beers after work, as they always do. 'Saving a marriage' is a hard goal to imagine, being abstract, uncertain, and distant in the future. In a conflict between that goal and the more concrete, immediate, and familiar goal of going for a beer, the Friday drinks habit is likely to win and abstinence be put off until tomorrow. A FIT practitioner would help the person identify concrete, proximal, *imageable* benefits of staying sober tonight. The person might decide to go to the pub but drink less, motivated by an image of their spouse's pleasure at them arriving home sober or of enjoying taking their children to football on Saturday morning without a hangover. Practising that image will help it come to mind more readily and vividly at the moment of decision when they are offered another beer.

As discussed earlier, the most vivid and emotive images are likely to be those that are retrieved in a coherent form from long-term memory rather than created *de novo*, just as familiar images were more vivid than bizarre images in Baddeley and Andrade (2000). FIT therefore typically progresses as initial, longer sessions to develop imagery and train imagery practice, and shorter booster sessions to support the person to incorporate memories of recent successes and new goals into their imagery as they make progress or encounter obstacles. These boosters are important, because images based on recent episodic memories are assumed to be richer in sensory detail and therefore more vivid and emotive than images based on scenarios generated *de novo* from semantic memory. By guiding participants to use episodic memories to strengthen their imagery, FIT teaches a habit of generating powerful motivational images of behaviour change. In our laboratory work on working memory, participants spontaneously retrieved information from long-term memory in order to satisfy the task demands. In FIT, we explicitly teach people to do this because initially it is harder to generate vivid imagery of novel behaviours and familiar habits have a cognitive advantage. The aim is that, with time, functional imagery becomes a habitual way of thinking about the future, especially when faced with a decision between an immediate reward and a conflicting, more distant goal.

Testing functional imagery training for health behaviour change

In a first test of FIT in this form, Andrade et al. (2016) used a stepped wedge design to test the effect of FIT versus waiting list on consumption of high-calorie snacks over two 2-week periods. The FIT intervention comprised a single FIT session lasting no more than 45 minutes, and a brief booster call a week later. FIT reduced snacking and led to a small amount of weight loss. Both changes were associated with an increase in motivation following FIT. This initial study showed that FIT could support motivation in a way that produced a short-term change in a habitual behaviour.

In an interesting extension of this study, Robinson et al. (2020) used a social robot to deliver FIT in a 'conversation' where the participant was encouraged to answer the robot's pre-programmed questions aloud. The robot guided a set of personalized imagery exercises as in human-delivered FIT. Robinson used a similar stepped wedge design but with 4 weeks rather than only 2 weeks between sessions. Participants reduced their snacking by more than half after receiving robot-delivered FIT, with large effect sizes. This study supports the assumption that the imagery and imagery training components of FIT are effective, and not just the supportive relationship with a human therapist.

Considerably longer-term changes are required for new behaviours to become habitual (Lally et al., 2010) and result in health improvements. In a randomized controlled trial, Solbrig et al. (2019) tested a more intensive version of FIT delivered by human over 6 months to participants with overweight or obesity who wished to lose weight. Participants received an initial face-to-face session of FIT lasting 1 hour and a second by telephone lasting up to 45 minutes and delivered 1 week later. Thereafter, there were 10-minute telephone booster calls every 2 weeks for 3 months and then every month for the next 3 months. In total, participants received no more than

4 hours of contact time and this was matched across the FIT condition and the control. Participants in the control condition received motivational interviewing, that is, a person-centred conversation directed towards their behaviour change goal but without guided imagery or imagery training. In contrast to other weight loss interventions, we provided no lifestyle advice or education.

Because we had chosen an active, supportive control condition that is known to benefit weight loss attempts (Armstrong et al., 2011), we were surprised by the extent to which FIT provided additional benefit. Participants in the control condition lost an average of 0.74 kg over the first 6 months of the trial and maintained most of that weight loss over the next 6 months with no additional support, giving a mean weight loss at 12 months of 0.67 kg and mean reduction in waist circumference of 2.5 cm. Participants randomized to FIT lost considerably more weight—4.11 kg—in the first 6 months. Typically in weight loss studies, by 1 year after the start of the study participants will have regained around half of the weight they lost during the intervention phase (Jolly et al., 2011). The most exciting finding from our trial was that this did not happen: participants in the FIT condition finished the study having lost an average of 6.44 kg at 12 months and 9.1 cm of waist circumference. These findings support the use of personalized imagery during a motivational interview and of training participants to use imagery to stay motivated after the intervention ends.

Analysing qualitative data from trial participants, Solbrig (2018, Chapter 7) found that they often described how FIT changed their mindset, from one where healthy living was an effort, to one where it was enjoyable and they felt happy about the changes they were seeing well before they reached their ultimate goal: 'I am in a different frame of mind about stuff now, I can continue on and enjoy what I am doing and not worry'. Participants also reported benefits for their mental health problems, for example, 'mood has really improved with all the positive thoughts and images in my head', and some adapted FIT to address other problems they were experiencing, for example, to deal with work-related stress or family arguments.

Imagery and mental health

These responses suggest that FIT can benefit mental health. Consistent with this suggestion, Di Simplicio and colleagues (2020) found that FIT produced larger reductions in self-harming than usual care or waiting list in a pilot trial with young adults who self-harmed. Rhodes developed a version of FIT for delivering to groups, including sports teams. His research focuses on reducing performance anxiety by enhancing mental resilience or 'grit'. Rhodes et al. (2018) showed that this group FIT increased professional footballers' grit. A subsequent study showed lasting improvements in penalty kicks (Rhodes et al., 2020). Following on from these studies, we are now developing FIT as an intervention for specific and generalized anxiety disorders.

On the interplay between theory and application

The title of this chapter implies that the working memory model has guided the development of behaviour change interventions. I have focused on FIT as an intervention with potential application across many different behaviours to improve mental as well as physical health. FIT emerged from a line of research that began with studies of modality-specific and modality-general processes in mental imagery, guided by the tripartite working memory model, and continued with applications to problems of trauma memory and craving, and thereafter to the question of whether one could use imagery to create 'cravings' for functional behaviours. Having arrived at this point, one could argue that FIT could have been developed without ever considering working memory. There was evidence that the client-centred, directed active listening approach of motivational interviewing was an effective brief intervention for a wide range of problems including addiction and weight management (e.g. Rubak et al., 2005). There was also evidence that imagining emotional situations led to greater change in emotional experience than verbally thinking about the same situations (Holmes & Mathews, 2005) and that imagery could convert plans into actions (Knäuper et al., 2009). This evidence would be sufficient to justify adding an imagery component to motivational interviewing, as FIT has done.

However, this argument overlooks the real value of a theory, which is to help frame questions and guide the design of research to answer them. The working memory model guided the design of experiments to test the importance of sensory short-term storage versus general attentional processes in emotional autobiographical recall and in cravings for food and drugs. From previous studies of short-term memory, the signature finding to suggest a role of modality-specific processes rather than central executive processes was a crossover interaction, with auditory interference impacting auditory short-term memory performance more than visual short-term memory, and vice versa for visuospatial interference. Alan Baddeley and I wanted to see if the same interaction emerged when the variable was image vividness rather than short-term memory performance. This meant comparing the effects of auditory and visual interference on auditory imagery as well as visual. Using the working memory model therefore prompted us to consider imagery in another sensory modality, whereas almost all studies of mental imagery up to that that point had only considered visual imagery. The focus on specific senses led us and others working on mechanisms of EMDR (e.g. Lilley et al., 2009) to ask participants to rate how the extent to which their emotional memories involved visual information rather than other senses, or to instruct them only to focus on and rate the visual components. The distinction between modality-general and modality-specific working memory resources led researchers to establish the working memory load imposed by their experimental tasks, by testing the extent to which they interfered with performance on a working memory task unrelated to autobiographical recall (Engelhard et al., 2010). The working memory model thus helped researchers specify their hypotheses about the role of eye movements in EMDR more precisely and to design better experiments to test them.

In extending this research to drug and food cravings, and then to motivation for functional goals, we broadened our focus to the full range of senses, including olfaction, taste, bodily sensations, and emotions. The working memory model says nothing about these other senses, although there has been some work guided by the model on olfactory memory (Andrade & Donaldson, 2007; Dade et al., 2001). However, the fact that experiments testing the model needed to contrast visual with auditory tasks meant that the mould was broken—visual imagery was no longer the only interesting aspect of mental sensory experience. Furthermore, researchers could adopt the standard working memory-guided experimental designs to explore the role of other senses in cravings, as Versland and Roseberg (2007) and Kemps and Tiggemann (2009) did when they included a concurrent olfactory task in their studies of cigarette and coffee cravings respectively. In these ways, the working memory model has shaped research on multisensory imagery that extends beyond the auditory and visual modalities covered by the model.

A core assumption is that the different modular systems in working memory have limited capacity for storing and processing information. This assumption underpinned the design of experiments that compare the effects of competing loads in different modalities. It allowed us and others to test the hypothesis from EI theory that craving depended on sensory imagery. It is embedded in FIT as the assumption that vivid accessible goal imagery will not only strengthen desire for the goal but also, through competition for limited working memory capacity, weaken cravings for an immediate temptation that would violate the goal. FIT includes a 'cravings buster' exercise to demonstrate this. Participants generate a vivid, multisensory image of something they often crave, say chocolate, and which conflicts with their chosen goal, say weight loss. They are then instructed to switch their attention to their goal image, to imagine working towards and achieving their goal. Having generated that image as vividly as possible, they are asked, 'What happened to the chocolate?' People typically reply that their craving image disappeared or faded. This exercise demonstrates to individuals how they have limited capacity to imagine two different scenarios at once and how they can consciously control how to allocate their limited cognitive resources.

Good theories don't stand still. Although much of the application of the working memory theory has used the tripartite model from 1986, itself a revision of the original Baddeley and Hitch (1974) model, theoretical work on working memory has added a fourth component to the model, the episodic buffer, as mentioned earlier (Baddeley, 2000). This more recent version of the working memory model has already been applied to understanding developmental disorders, Parkinson's disease, and language processing, demonstrating its continued potential to guide research (see Baddeley, 2019; Baddeley et al., 2021 for recent reviews). The quadripartite working memory model has not yet been applied to problems of craving and desire but it has the potential to shift focus. Perhaps the key requirement of motivational imagery is not the maintenance of sensory information in modality-specific stores but rather that of a bound, multisensory image in an episodic buffer.

This example shows the potential of the working memory model to guide a new phase of research on desire, but applied research is not a one-way street where theory guides the search for solutions to real-world problems and gets nothing in return. The process of researching problems that are messier, more complex, and less constrained than those tackled in the laboratory raises new challenges for theory, feeding its evolution. An important question that arises from researching interventions for trauma memory or drug and food cravings is how can we make the benefits stick, to bring about lasting improvements? Work on memory consolidation and reconsolidation has shaped one way of conceptualizing this problem. Studies of the neural basis of memory encoding have shown that it takes up to 6 hours for a memory to be fully encoded or consolidated (Walker et al., 2003). Until this consolidation is complete, the memory is labile and vulnerable to interference or suggestion. Reconsolidation refers to the finding that memories become labile again when they are reactivated, that is, recalled under specific conditions (see Meir et al., 2018), providing an opportunity to rewrite troubling memories. There is evidence that fear memories, for example, can be permanently extinguished if the extinction takes place during the reconsolidation window, after the fear memory has been reactivated (Schiller et al., 2010). Applying this finding to a mental health problem, Soeter and Kindt (2015) briefly introduced spider-fearful participants to a tarantula and then administered a beta-blocker, propranolol, or placebo. Propranolol caused a marked reduction in fear responses towards spiders. Merely using propranolol without using the tarantula to reactivate spider memories had no effect, supporting the interpretation that memory reconsolidation is the mechanism underpinning this behaviour change.

Holmes and her team applied these ideas to trauma memory. They used the working memory model to design experiments to test whether visuospatial loads would influence how distressing material was encoded in long-term memory. As predicted, visuospatial interference during a 'trauma film' reduced the extent to which memories of the film intruded in the days after watching (Stuart et al., 2006). They argued that selectively blocking the encoding of visual information produced memories that were less rich in the sensory detail that rendered them distressing and liable to intrude (note, however, that Pearson & Sawyer, 2011, reported evidence that effects were due to general rather than modality-specific working memory load). Combining this idea with research on memory consolidation, they went on to show that visuospatial interference shortly after memory encoding had a similar effect (Holmes et al., 2010), showing that the consolidation period provides a window of opportunity for reducing the impact of trauma. Recent research develops these ideas to test whether working memory loads during a reconsolidation period can alter existing trauma memories. Kessler et al. (2020) exposed healthy participants to a trauma film and then, 3 days later, induced a reconsolidation phase for the memory of the film with a visual reminder followed by a 10-minute delay. Participants then played a 15-minute verbal 'Pub Quiz' game or visuospatial Tetris game, or did neither. They reported intrusions of memories of the film for the next 3 days. Playing Tetris during the reconsolidation period reduced intrusions more than the Pub Quiz, which reduced them more than the no-task control. This study raises the exciting promise that rendering trauma memories labile using a simple reminder-and-wait process, and then weakening them with modality-specific working memory interference, could produce permanent reductions in trauma symptoms, specifically intrusive re-experiencing.

These studies, bringing together the working memory model with work on memory consolidation, hold the promise of using memory research to design interventions that are maximally long-lasting. They suggest the possibility that disrupting cravings for drugs or food under reconsolidation conditions could permanently weaken the power of those cravings. If we can help people create and elaborate affectively charged images of succeeding at their goals, as FIT aims to do, perhaps we can not only give them a strategy for staying motivated but permanently change how they think about those goals, resulting in the 'mindset' shift that some of our weight loss trial participants reported. In other words, there is a possibility that changing what information people temporarily store in working memory and how they elaborate it can affect how that information is consolidated or reconsolidated in long-term memory. Very recent theoretical research on the relationship between working memory and long-term memory supports this possibility. Maintaining stimuli in working memory for longer produces better incidental retention in long-term memory (Cotton & Ricker, 2021) and this benefit of time in working memory may be due to enhanced elaboration of stimuli (Loaiza & Lavilla, 2021).

In conclusion, research guided by the working memory model has revealed that conscious experiences involve active attention to information, thus mental images are more vivid and more highly emotionally charged when they can be temporarily stored and manipulated in working memory without competition for processing or storage capacity. There is evidence that changing what we experience, by interfering with distressing images or elaborating motivational ones in working memory, can bring about long-lasting changes in well-being. The findings and ideas from applications of the working memory model in turn raise questions about exactly how working memory and the conscious experiences it supports influence our long-term episodic memories.

References

Andrade, J. (2001). The contribution of working memory to conscious experience. In J. Andrade (Ed.), *Working memory in perspective* (pp. 60–78). Psychology Press.

Andrade, J., & Donaldson, L. (2007). Evidence for an olfactory store in working memory? *Psychologia*, *50*(2), 76–89.

Andrade, J., Kavanagh, D., & Baddeley, A. (1997). Eye movements and visual imagery: A working memory approach to the treatment of post-traumatic stress disorder. *British Journal of Clinical Psychology, 36*(2), 209–223.

Andrade, J., Khalil, M., Dickson, J., May, J., & Kavanagh, D. J. (2016). Functional imagery training to reduce snacking: Testing a novel motivational intervention based on elaborated intrusion theory. *Appetite, 100*, 256–262.

Andrade, J., May, J., & Kavanagh, D. J. (2012). Sensory imagery in craving: From cognitive psychology to new treatments for addiction. *Journal of Experimental Psychopathology, 3*(2), 127–145.

Andrade, J., Pears, S., May, J., & Kavanagh, D. J. (2012). Use of a clay modeling task to reduce chocolate craving. *Appetite, 58*, 955–963.

Armstrong, M. J., Mottershead, T. A., Ronksley, P. E., Sigal, R. J., Campbell, T. S., & Hemmelgarn, B. R. (2011). Motivational interviewing to improve weight loss in overweight and/or obese patients: A systematic review and meta-analysis of randomized controlled trials. *Obesity Reviews, 12*(9), 709–723.

Baddeley, A. D. (1986). *Working memory*. Oxford University Press.

Baddeley, A. D. (1997). *Human memory: Theory and practice* (revised edition). Psychology Press.

Baddeley, A. (2000). The episodic buffer: A new component of working memory? *Trends in Cognitive Sciences, 4*(11), 417–423.

Baddeley, A. D. (2002). Is working memory still working? *European Psychologist, 7*(2), 85.

Baddeley, A. (2019). *Working memories: Postmen, divers and the cognitive revolution*. Routledge.

Baddeley, A. D., Allen, R. J., & Hitch, G. J. (2011). Binding in visual working memory: The role of the episodic buffer. *Neuropsychologia, 49*(6), 1393–1400.

Baddeley, A. D., & Andrade, J. (2000). Working memory and the vividness of imagery. *Journal of Experimental Psychology: General, 129*(1), 126–145.

Baddeley, A., Gathercole, S., & Papagno, C. (1998). The phonological loop as a language learning device. *Psychological Review, 105*, 158–173.

Baddeley, A. D., & Hitch, G. J. (1974). Working memory. In G. A. Bower (Ed.), *Recent advances in learning and motivation* (Vol. 8, pp. 47–89). Academic Press.

Baddeley, A. D., Hitch, G. J., & Allen, R. (2021). A multicomponent model of working memory. In R. H. Logie, V. Camos, & N. Cowan (Eds.), *Working memory: State of the science* (pp. 10–43). Oxford University Press.

Barsalou, L. W. (2008). Grounded cognition. *Annual Review of Psychology, 59*, 617–645.

Bisiach, E., & Luzzatti, C. (1978). Unilateral neglect of representational space. *Cortex, 14*(1), 129–133.

Brooks, L. R. (1967). The suppression of visualisation by reading. *Quarterly Journal of Experimental Psychology, 19*(4), 289–299.

Bywaters, M., Andrade, J., & Turpin, G. (2004). Determinants of the vividness of visual imagery: The effects of delayed recall, stimulus affect and individual differences, *Memory, 12*(4), 479–488.

Cornil, A., Long, J., Rothen, S., Perales, J. C., de Timary, P., & Billieux, J. (2019). The gambling craving experience questionnaire: Psychometric properties of a new scale based on the elaborated intrusion theory of desire. *Addictive Behaviors, 95*, 110–117.

Cornil, A., Lopez-Fernandez, O., Devos, G., de Timary, P., Goudriaan, A. E., & Billieux, J. (2018). Exploring gambling craving through the elaborated intrusion theory of desire: A mixed methods approach. *International Gambling Studies, 18*(1), 1–21.

Cornoldi, C., de Beni, R., Cavedon, A., Mazzoni, G., et al. (1992). How can a vivid image be described? Characteristics influencing vividness judgments and the relationship between vividness and memory. *Journal of Mental Imagery, 16*(3–4), 89–107.

Cotton, K., & Ricker, T. J. (2021). Working memory consolidation improves long-term memory recognition. *Journal of Experimental Psychology: Learning, Memory, and Cognition, 47*(2), 208–219.

Dade, L. A., Zatorre, R. J., Evans, A. C., & Jones-Gotman, M. (2001). Working memory in another dimension: Functional imaging of human olfactory working memory. *Neuroimage, 14*(3), 650–660.

Daniel, T. O., Stanton, C. M., & Epstein, L. H. (2013). The future is now: Reducing impulsivity and energy intake using episodic future thinking. *Psychological Science, 24*(11), 2339–2342.

Di Simplicio, M., Appiah-Kusi, E., Wilkinson, P., Watson, P., Meiser-Stedman, C., Kavanagh, D. J., & Holmes, E. A. (2020). Imaginator: A proof-of-concept feasibility trial of a brief imagery-based psychological intervention for young people who self-harm. *Suicide and Life-Threatening Behavior*, *50*(3), 724–740.

Driskell, J. E., Copper, C., & Moran, A. (1994). Does mental practice enhance performance? *Journal of Applied Psychology*, *79*(4), 481–492.

Engelhard, I. M., van Uijen, S. L., & van den Hout, M. A. (2010). The impact of taxing working memory on negative and positive memories. *European Journal of Psychotraumatology, 1*, 5623–5630.

Farah, M. J. (1989). The neural basis of mental imagery. *Trends in Neurosciences, 12*(10), 395–399.

Ganis, G., Thompson, W. L., & Kosslyn, S. M. (2004). Brain areas underlying visual mental imagery and visual perception: An fMRI study. *Brain Research. Cognitive Brain Research, 20*(2), 226–241.

Hall, C. R., Mack, D. E., Paivio, A., & Hausenblas, H. A. (1998). Imagery use by athletes: Development of the Sport Imagery Questionnaire. *International Journal of Sport Psychology, 29*(1), 73–89.

Hofmann, W., Baumeister, R. F., Förster, G., & Vohs, K. D. (2012). Everyday temptations: An experience sampling study of desire, conflict, and self-control. *Journal of Personality and Social Psychology, 102*(6), 1318–1335.

Holmes, E. A., James, E. L., Kilford, E. J., & Deeprose, C. (2010). Key steps in developing a cognitive vaccine against traumatic flashbacks: Visuospatial Tetris versus verbal Pub Quiz. *PLoS One, 5*(11), e13706.

Holmes, E. A., & Mathews, A. (2005). Mental imagery and emotion: A special relationship? *Emotion, 5*(4), 489–497.

Jolly, K., Lewis, A., Beach, J., Denley, J., Adab, P., Deeks, J. J., Daley, A., & Aveyard, P. (2011). Comparison of range of commercial or primary care led weight reduction programmes with minimal intervention control for weight loss in obesity: Lighten Up randomised controlled trial. *BMJ, 343*, d6500.

Kavanagh, D. J., Andrade, J., & May, J. (2005). Imaginary relish and exquisite torture: The elaborated intrusion theory of desire. *Psychological Review, 112*(2), 446–467.

Kavanagh, D. J., Freese, S., Andrade, J., & May, J. (2001). Effects of visuospatial tasks on desensitization to emotive memories. *British Journal of Clinical Psychology, 40*(3), 267–280.

Kavanagh, D. J., May, J., & Andrade, J. (2009). Tests of the elaborated intrusion theory of craving and desire: Features of alcohol craving during treatment for an alcohol disorder. *British Journal of Clinical Psychology, 48*(3), 241–254.

Kavanagh, D., Robinson, N., Connolly, J., Connor, J., Andrade, J., & May, J. (2018). The revised four-factor motivational thought frequency and state motivation scales for alcohol control. *Addictive Behaviours, 87*, 69–73.

Kavanagh, D. J., Texeira, H., Connolly, J., Andrade, J., May, J., Godfrey, S., Carroll, A., Taylor, K., & Connor, J. P. (2020). The motivational thought frequency scales for increased physical activity and reduced high-energy snacking. *British Journal of Health Psychology, 25*(3), 558–575.

Kemps, E., & Tiggemann, M. (2007a). Reducing the vividness and emotional impact of distressing auto-biographical memories: The importance of modality-specific interference. *Memory, 15*(4), 412–422.

Kemps, E., & Tiggemann, M. (2007b). Modality-specific imagery reduces cravings for food: An application of the elaborated intrusion theory of desire to food craving. *Journal of Experimental Psychology: Applied, 13*(2), 95–104.

Kemps, E., & Tiggemann, M. (2009). Competing visual and olfactory imagery tasks suppress craving for coffee. *Experimental and Clinical Psychopharmacology, 17*(1), 43–50.

Kemps, E., Tiggemann, M., & Grigg, M. (2008). Food cravings consume limited cognitive resources. *Journal of Experimental Psychology: Applied, 14*(3), 247–254.

Kessler, H., Schmidt, A. C., James, E. L., Blackwell, S. E., von Rauchhaupt, M., Harren, K., Kehyayan, A., Clark, I. A., Sauvage, M., Herpertz, S., Axmacher, N., & Holmes, E. A. (2020). Visuospatial computer game play after memory reminder delivered three days after a traumatic film reduces the number of intrusive memories of the experimental trauma. *Journal of Behavior Therapy and Experimental Psychiatry, 67*, 101454.

Knäuper, B., McCollam, A., Rosen-Brown, A., Lacaille, J., Kelso, E., & Roseman, M. (2011). Fruitful plans: Adding targeted mental imagery to implementation intentions increases fruit consumption. *Psychology and Health, 26*(5), 601–617.

Knäuper, B., Roseman, M., Johnson, P. J., & Krantz, L. H. (2009). Using mental imagery to enhance the effectiveness of implementation intentions. *Current Psychology, 28*(3), 181–186.

Kobayashi, M., Takeda, M., Hattori, N., Fukunaga, M., Sasabe, T., Inoue, N., Nagai, Y., Sawada, T., Sadato, N., & Watanabe, Y. (2004). Functional imaging of gustatory perception and imagery: 'top-down' processing of gustatory signals. *Neuroimage*, *23*(4), 1271–1282.

Kosslyn, S. M. (1996). *Image and brain: The resolution of the imagery debate*. MIT Press.

Lally, P., Van Jaarsveld, C. H., Potts, H. W., & Wardle, J. (2010). How are habits formed: Modelling habit formation in the real world. *European Journal of Social Psychology*, *40*(6), 998–1009.

Landin-Romero, R., Moreno-Alcazar, A., Pagani, M., & Amann, B. L. (2018). How does eye movement desensitization and reprocessing therapy work? A systematic review on suggested mechanisms of action. *Frontiers in Psychology*, *9*, 1395.

Lawrence, B. M., Myerson, J., & Abrams, R. A. (2004). Interference with spatial working memory: An eye movement is more than a shift of attention. *Psychonomic Bulletin & Review*, *11*(3), 488–494.

Lilley, S., Andrade, J., Turpin, G., Sabin-Farrell, R., & Holmes, E. A. (2009). Visuo-spatial working memory interference with recollections of trauma. *British Journal of Clinical Psychology*, *48*(Pt. 3), 309–321.

Loaiza, V. M., & Lavilla, E. T. (2021). Elaborative strategies contribute to the long-term benefits of time in working memory. *Journal of Memory and Language*, *117*, 104205.

Logie, R. H. (1995). *Visuo-spatial working memory*. Lawrence Erlbaum Associates.

Logie, R. H., Zucco, G. M., & Baddeley, A. D. (1990). Interference with visual short-term memory. *Acta Psychologica*, *75*, 55–74.

Marks, D. F. (1973). Visual imagery differences in the recall of pictures. *British Journal of Psychology*, *64*(1), 17–24.

May, J., Andrade, J, Panabokke, N., & Kavanagh, D. (2004). Images of desire: Cognitive models of craving, *Memory*, *12*(4), 447–461.

May, J., Andrade, J., Kavanagh, D. J., Feeney, G. F. X., Gullo, M., Statham, D. J., Skorka-Brown, J., Connolly, J. M., Cassimatis, M., Young, R. McD., & Connor, J. P. (2014). The Craving Experience Questionnaire: A brief, theory-based measure of consummatory desire and craving. *Addiction*, *109*(5), 728–735.

May, J., Andrade, J., Kavanagh, D., & Penfound, L. (2008). Imagery and strength of craving for eating, drinking and playing sport. *Cognition and Emotion*, *22*(4), 633–650.

May, J., Andrade, J., Panabokke, N., & Kavanagh, D. (2010). Visuospatial tasks suppress craving for cigarettes. *Behaviour Research and Therapy*, *48*, 476–485.

Meir Drexler, S., & Wolf, O. T. (2018). Behavioral disruption of memory reconsolidation: From bench to bedside and back again. *Behavioral Neuroscience*, *132*(1), 13–22.

Meule, A., Skirde, A. K., Freund, R., Vögele, C., & Kübler, A. (2012). High-calorie food-cues impair working memory performance in high and low food cravers. *Appetite*, *59*(2), 264–269.

Miller, W. R., & Rollnick, S. (2012). *Motivational interviewing: Helping people change*. Guilford Press.

Neisser, U. (1967). *Cognitive psychology*. Appleton-Century-Crofts.

Parham, S. C., Kavanagh, D. J., Gericke, C. A., King, N., May, J., & Andrade, J. (2017). Assessment of motivational cognitions in diabetes self-care: The motivation thought frequency scales for glucose testing, physical activity and healthy eating. *International Journal of Behavioral Medicine*, *24*(3), 447–456.

Pearson, D. G., & Sawyer, T. (2011). Effects of dual task interference on memory intrusions for affective images. *International Journal of Cognitive Therapy*, *4*(2), 122–133.

Pearson, D., & Sahraie, A. (2003). Oculomotor control and the maintenance of spatially and temporally distributed events in visuo-spatial working memory. *Quarterly Journal of Experimental Psychology Section A*, *56*(7), 1089–1111.

Postle, B. R., Idzikowski, C., Sala, S. D., Logie, R. H., & Baddeley, A. D. (2006). The selective disruption of spatial working memory by eye movements. *Quarterly Journal of Experimental Psychology*, *59*(1), 100–120.

Reisberg, D., Culver, L. C., Heuer, F., & Fischman, D. (1986). Visual memory: When imagery vividness makes a difference. *Journal of Mental Imagery*, *10*(4), 51–74.

Rhodes, J., May, J., Andrade, J., Kavanagh, D.J. (2018). Enhancing grit through functional imagery training in professional soccer. *Sports Psychologist*, *32*, 220–225.

Rhodes, J., May, J., & Booth, A. (2020). Penalty success in professional soccer: A randomised comparison between imagery methodologies. *Journal of Imagery Research in Sport and Physical Activity*, *15*(1), 20200014.

Robinson, N. L., Connolly, J., Hides, L., & Kavanagh, D. J. (2020). Social robots as treatment agents: Pilot randomized controlled trial to deliver a behavior change intervention. *Internet Interventions, 21*, 100320.

Robinson, N., Kavanagh, D.J., Connor, J. May, J., & Andrade, J. (2016). Assessment of motivation to control alcohol use: The motivational thought frequency and state motivation scales for alcohol. *Addictive Behaviors, 59*, 1–6.

Rubak, S., Sandbæk, A., Lauritzen, T., & Christensen, B. (2005). Motivational interviewing: A systematic review and meta-analysis. *British Journal of General Practice, 55*(513), 305–312.

Sayette, M. A., Monti, P. M., Rohsenow, D. J., Gulliver, S. B., Colby, S. M., Sirota, A. D., Niaura, R., & Abrams, D. B. (1994). The effects of cue exposure on reaction time in male alcoholics. *Journal of Studies on Alcohol, 55*(5), 629–633.

Schiller, D., Monfils, M. H., Raio, C. M., Johnson, D. C., LeDoux, J. E., & Phelps, E. A. (2010). Preventing the return of fear in humans using reconsolidation update mechanisms. *Nature, 463*(7277), 49–53.

Shapiro, F. (1989). Eye movement desensitisation: A new treatment for post-traumatic stress disorder. *Journal of Behaviour Therapy and Experimental Psychiatry, 20*(3), 211–217.

Skorka-Brown, J., Andrade, J., & May, J. (2014). Playing 'Tetris' reduces the strength, frequency and vividness of naturally occurring cravings. *Appetite, 76*, 161–165.

Skorka-Brown, J., Andrade, J., Whalley, B., & May, J. (2015). Playing Tetris decreases drug and other cravings in real world settings. *Addictive Behaviors, 51*, 165–170.

Soeter, M., & Kindt, M. (2015). An abrupt transformation of phobic behavior after a post-retrieval amnesic agent. *Biological Psychiatry, 78*(12), 880–886.

Solbrig, L. (2018). *Functional Imagery Training, a novel, theory-based motivational intervention for weight-loss* [Unpublished doctoral dissertation]. University of Plymouth.

Solbrig, L., Whalley, B., Kavanagh, D. J., May, J., Parkin, T., Jones, R., & Andrade, J. (2019). Functional imagery training versus motivational interviewing for weight loss: A randomised controlled trial of brief individual interventions for overweight and obesity. *International Journal of Obesity, 43*(4), 883–894.

Statham, D. J., Connor, J. P., Kavanagh, D. J., Feeney, G. F. X., May, J., Andrade, J., & Young, R. McD. (2011). Measuring alcohol craving: Development of the Alcohol Craving Experience questionnaire. *Addiction, 106*, 1230–1238.

Stuart, A. D., Holmes, E. A., & Brewin, C. R. (2006). The influence of a visuospatial grounding task on intrusive images of a traumatic film. *Behaviour Research and Therapy, 44*(4), 611–619.

Tiffany, S. T., & Hakenewerth, D. M. (1991). The production of smoking urges through an imagery manipulation: Psychophysiological and verbal manifestations. *Addictive Behaviors, 16*(6), 389–400.

Tiffany, S. T., & Wray, J. M. (2012). The clinical significance of drug craving. *Annals of the New York Academy of Sciences, 1248*, 1–17.

Van Dillen, L. F., & Andrade, J. (2016). Derailing the streetcar named desire. Cognitive distractions reduce individual differences in cravings and unhealthy snacking in response to palatable food. *Appetite, 96*, 102–110.

Versland, A., & Rosenberg, H. (2007). Effect of brief imagery interventions on craving in college student smokers. *Addiction Research & Theory, 15*(2), 177–187.

Walker, M. P., Brakefield, T., Hobson, J. A., & Stickgold, R. (2003). Dissociable stages of human memory consolidation and reconsolidation. *Nature, 425*(6958), 8–12.

Zvyagintsev, M., Clemens, B., Chechko, N., Mathiak, K. A., Sack, A. T., & Mathiak, K. (2013). Brain networks underlying mental imagery of auditory and visual information. *European Journal of Neuroscience, 37*(9), 1421–1434.

Zwaan, R. A., & Truitt, T. P. (1998). Smoking urges affect language processing. *Experimental and Clinical Psychopharmacology, 6*(3), 325–330.

15

Neuropsychology of working memory

From theory to clinic and from clinic to theory

Roberto Cubelli, Robert H. Logie, and Sergio Della Sala

Definition of neuropsychology

Neuropsychology is 'the discipline which investigates, with the methods of experimental psychology, the disorders of higher mental functions following acquired brain lesions' (De Renzi, 1967, p. 422). The investigation of the relationship between mind and brain began within the phrenological framework (Bouillaud, 1825), developed in France and Italy (see Zago et al., 2015) and established with Broca (1865; see Cubelli & Montagna, 1994). However, it is with Wernicke (1874) that neuropsychology was initiated as a scientific discipline. We would argue that to understand contemporary findings it is crucial to set them in the context of the development of neuropsychology over many decades. Therefore, we provide a review both of the historical progression and of recent developments in theory and findings regarding impairments of the moment-to-moment use of human memory, known as working memory (WM).

Models in neuropsychology

Based on the observation of three different clinical profiles, Wernicke (1874) proposed the first neuropsychological model depicting the functional and anatomical architecture of the mechanisms underlying the task of oral repetition of single words. The model, later revised by Lichtheim (1885), criticized by Freud (1891), and reformulated by Wernicke (1906), represented a landmark for the use of conceptual models in neuropsychology. Ever since, each neuropsychological model typically includes three features: (1) the description of the different cognitive operations involved in performing tasks; (2) the prediction of distinctive patterns of selective impairments due to focal brain damage; and, in many cases, (3) the localization of the processing stages in cerebral areas or networks. Crucial evidence for building the cognitive architecture of neuropsychological models is the *double dissociation* between tasks. This refers to the observations that when performing each of two tasks,

Roberto Cubelli, Robert H. Logie, and Sergio Della Sala, *Neuropsychology of working memory* In: *Memory in Science for Society*.
Edited by: Robert H. Logie, Zhisheng (Edward) Wen, Susan E. Gathercole, Nelson Cowan, and Randall W. Engle, Oxford University Press.
© Oxford University Press 2023. DOI: 10.1093/oso/9780192849069.003.0015

two patients (or groups of patients) show the opposite patterns of impairment and sparing in cognitive performance. That is, one patient shows impairments in performance of task A and unimpaired performance on task B, whereas another patient shows impairments in performance of task B and unimpaired performance of task A. This contrasting pattern across different patients is argued to indicate that the neural and cognitive mechanisms underlying tasks A and B are at least partly different and each includes at least one aspect of cognition that is involved in one task but not in the other.

Wernicke–Lichtheim's model, reproposed by Geschwind (1965), has proved extremely useful in clinical settings (Goodglass & Kaplan, 1972). However, the model was too coarse to allow a detailed appraisal of single cases or to allow the precise identification of symptoms or syndromes. In particular, it was not well enough specified to derive information aimed at understanding the cognitive and linguistic mechanisms and the representations underlying verbal processing.

Investigating the cognitive impairments due to acquired cerebral lesions requires in-depth knowledge of human cognitive function. Clinical intuitions are not sufficient; an interdisciplinary approach based on the advancement of cognitive sciences is vital to assemble theoretical models describing the functional architecture of cognitive functions in order to understand the nature of the impairments that may arise in those functions (see Coltheart et al., 1980).

Cognitive neuropsychology is a branch of cognitive psychology (Coltheart, 2001). It investigates the impairments of cognitive processes following brain damage and accounts for the observed patterns of spared and impaired abilities 'in terms of damage to one or more components of a theory or model of normal cognition' (Ellis and Young 1996, p. 4). In turn, this refined diagnosis has a reciprocal benefit in allowing researchers to develop, test, or modify the models of intact cognitive functioning (Coltheart, 2003; see also Cubelli et al., 2016).

The reciprocal interaction between cognitive psychology and clinical neuropsychology in using and informing theoretical models is well represented by the development and use of the multicomponent model of WM, first proposed in outline by Baddeley and Hitch (1974), then developed in greater detail (e.g. Baddeley, 1986, 2000; Baddeley et al., 2021; Baddeley & Logie, 1999; see Baddeley, Chapter 2, this volume). WM refers to the temporary retention of a small amount of information in support of the performance of current tasks. As described by Baddeley (1986), WM was thought to comprise an articulatory loop (later renamed phonological loop; Vallar & Baddeley, 1984a) that supports the temporary retention of a small amount of verbal material, a visuospatial scratchpad that supports retention of a small amount of visual and spatial information, and a central executive that coordinates the functioning of the other two systems and is the host for higher-level cognition such as reasoning, problem-solving, and the control of attention. Subsequently, a range of conceptual, computational, and neuroscience models of WM in healthy individuals have been developed (e.g. Barrouillet & Camos, 2015; Cowan, 1995, 1999,

2017; Cowan, Chapter 9, this volume; Logie, 1995, 2011, 2016; Logie, Belletier, & Doherty, 2021; O'Reilly et al., 2016; for recent reviews see Logie, Camos, & Cowan, 2021). However, variations of the multiple component model, and in particular the phonological loop concept, have been used most extensively in cognitive neuro-psychology (for recent reviews, see Baddeley & Hitch, 2019; Logie, 2019; Papagno & Shallice, 2019), and it is this model on which we will focus in this chapter.

The multiple component model of working memory: a synergy between cognitive psychology and cognitive neuropsychology

Neuropsychology has played a crucial role in the development of the multiple component WM model, and in turn the development of the theoretical model has enhanced the understanding of cognitive impairments in brain-damaged individuals. More detailed early reviews of the neuropsychology of WM are presented in Della Sala and Logie (1993), Logie and Della Sala (2005), and Vallar and Shallice (1990). The concept of a multiple component WM is extraordinary in its simplicity. It was founded and developed on the basis that the success of a model derives from how fruitful it is in generating hypotheses to be tested empirically and how useful it is in accounting for the patterns of performance observed both in brain-damaged individuals and in healthy individuals.

Developing a conceptual model of cognitive function relies on the collection of experimental evidence for the assumed characteristics of that model. Just as those characteristics should predict patterns of performance in healthy adults, they should also predict the pattern of cognitive impairments that result from selective brain damage. If the predictions from the model match the patterns of data observed, this increases confidence in the model. If the data do not match the predictions, this can then lead to refinements in the model. This approach allows for advances in theoretical understanding from studying cognitive impairments in patients, and those advances can enhance understanding of healthy cognition. In this regard, the subsequent data-driven fractionation of the multiple components of WM has been singularly successful.

The ideas that led to current understanding of WM have their origins in the writings of Locke (1690), James (1890), Hebb (1949), and Broadbent (1958), each of whom argued for the distinction between information that is currently in mind and required for current tasks, often referred to as short-term memory (STM), and stored knowledge that has accumulated over the lifetime of the individual, often referred to as long-term memory (LTM) (for an historical overview, see Logie, 1996). This distinction has been supported by a large amount of experimental evidence from cognitive psychology (e.g. Klatzky, 1975). Baddeley and Hitch (1974) noted the additional evidence for a separation between STM and LTM arose from observations of double dissociations in neuropsychological patients; specifically, that there were patients with STM impairments but largely intact LTM, and other patients with LTM impairments but intact STM (e.g. Baddeley & Warrington, 1970; Drachman & Arbit, 1966; Milner, 1966; Warrington & Shallice, 1969; for recent reviews see MacPherson & Della Sala, 2019; Shallice & Papagno, 2019). That is, STM and LTM can be damaged independently, suggesting that they are separate systems that nevertheless interact.[1]

Baddeley and Hitch (1974) conducted a series of experiments with healthy adults to investigate whether STM might comprise multiple components rather than a unitary system, using the dual-task paradigm. This paradigm involves comparing performance on tasks performed on their own with performance when participants are asked to perform two tasks at the same time. The logic is that if task A and task B both rely on one set of cognitive functions then performance should drop substantially when performing the tasks together compared with performing any one task on its own. In contrast, if task A and task B rely on different cognitive functions, then performing them together should result in little, if any drop in performance. Multiple experiments (for early reviews see Baddeley, 1986; Baddeley & Logie, 1999; Logie, 1995) demonstrated this kind of double dissociation, and led to the proposal of a multicomponent structure of STM, described earlier, including an articulatory loop or phonological loop, a visuospatial scratchpad, and a central executive.

The multicomponent model of WM proved consistent with neuropsychological clinical data (Della Sala & Logie, 1993). At a gross level, the pathological reduction of STM capacity (span) following brain damage can be modality specific and linked with specific brain areas (De Renzi & Nichelli, 1975): it is associated with left parietal lesions for verbal stimuli (e.g. Tree & Playfoot, 2019; Vallar, 2019; see early reviews in Vallar & Shallice, 1990) and with right parietal lesions for visuospatial information (e.g. Farah et al., 1988; Hanley et al., 1991; for an early review see Logie & Della Sala, 2005). Similar dissociations have been reported in people affected by early-stage dementia (Baddeley, Della Sala, & Spinnler, 1991; De Tollis et al., 2021). This double dissociation between verbal STM and visuospatial STM can be accounted for by assuming selective impairments of the different storage components of the model. With the increased sophistication of neuroimaging techniques and the advancement of their analyses, and the accrued knowledge from lesion studies (see Lugtmeijer et al., 2021), it became apparent that this gross, dichotomous dissociation should be further fractionated, considering their relative task demand of the span tasks (see e.g. Geva et al., 2021; Ghaleh et al., 2020) as well as the issue of functional connectivity across hemispheres (Siegel et al., 2016).

---

[1] Short-term memory and its abbreviation STM typically, but not consistently, refers to temporary memory for verbal material comprising lists of unrelated letters, words, or digits (1–9). We shall note where the term STM is used more broadly to include verbal and non-verbal material.

The case of the phonological loop

The verbal component of the multiple component model of WM was initially labelled the articulatory loop (Baddeley, 1986) to designate the presumed central role of articulation in storing and maintaining information. This hypothesis was supported by experimental evidence: (1) sequences of words of short spoken duration are better recalled than words of long spoken duration, even when the number of syllables and phonemes is held constant (Baddeley et al., 1975; (2) STM verbal span is positively related to reading fluency (Hulme et al., 1984); and (3) verbal STM span is reduced when rehearsal is prevented by asking participants to repeat an irrelevant word, a technique known as articulatory suppression (Baddeley et al., 1984; Murray, 1965). Based on these findings, it was suggested that the same mechanisms necessary for fluent speech are involved in verbal STM tasks (Baddeley, 1986).

Articulation was assumed to be critical for the process of subvocal rehearsal of the material to be memorized. It followed that if STM patients have an impaired articulatory loop, the process of fluent speech production should also be defective. However, patients showing a clear selective deficit of verbal STM, like JB (Shallice & Butterworth, 1977) and PV (Basso et al., 1982), did not present with any articulatory deficits in spontaneous speech. Similarly, in healthy participants, articulatory suppression was expected to disrupt the effects that are thought to be characteristic of a normally functioning articulatory loop (Baddeley et al., 1984; Murray, 1965, 1968): the phonological similarity effect, specifically lists of similar sounding letters or words are less well remembered than lists of letters or words that sound different (Conrad & Hull, 1964), and the word length effect, namely that lists of words that take longer to pronounce are less well remembered than lists of words that take less time to pronounce (Baddeley et al., 1975). If these effects reflect the same articulatory-based mechanism, they should be equally sensitive to articulatory suppression. Yet, when rehearsal is prevented by articulatory suppression, the word length effect disappears with both visual (Baddeley et al., 1975) and auditory presentation (Baddeley et al., 1984), whereas the phonological similarity effect is abolished only with visual presentation (Murray, 1968; Levy, 1971).

Taken together, these findings called for a revision of the articulatory loop hypothesis. The original WM model was modified by separating the unitary verbal subsystem into two components: a phonological short-term store and an articulatory rehearsal mechanism (Salamè & Baddeley, 1982; Vallar & Baddeley, 1984a). The phonological store was conceived as a passive buffer capable of storing phonologically coded material for a limited time: auditory stimuli enter it directly, whereas visual presentation of verbal material requires subvocal rehearsal to gain access to the store. In contrast, the rehearsal mechanism was thought to be an active process with two functions: to refresh information within the phonological store, thus preventing decay of the information it contains, and to allow visual stimuli to enter the phonological store. Articulatory suppression was thought to prevent the latter, and to inhibit rehearsal, but not to prevent phonological encoding of aurally pre-sented verbal material. The phonological similarity effect was thought to reflect the content of the store; the word length effect revealed the activity of the articulatory re-hearsal. Following this fractionation, the verbal subsystem was renamed the phono-logical loop (Vallar & Baddeley, 1984a).

The revised model (illustrated in Figure 15.1) allowed a better description of the impairments shown by patients presenting with selective STM disorders. A patient, known by the initials PV, who had a short-term verbal memory impairment, pre-sented with reduced verbal span but no deficits in spontaneous speech. Therefore, she was assumed to suffer from a reduced capacity of the phonological store, which impaired her ability to comprehend longer sentences (Vallar & Baddeley, 1984b) and to learn novel words (Baddeley et al., 1998). Since her performance mirrored that of normal participants under articulatory suppression, that is, absence of the phono-logical similarity effect with visual presentation, and absence of the word length ef-fect with both auditory and visual input (Vallar & Baddeley, 1984a), she might have suffered from a secondary functional disengagement, or disconnection, of the ar-ticulatory mechanism due to the defective phonological store. In this way, the ar-ticulatory rehearsal activity can be viewed as a motor-based strategic control process

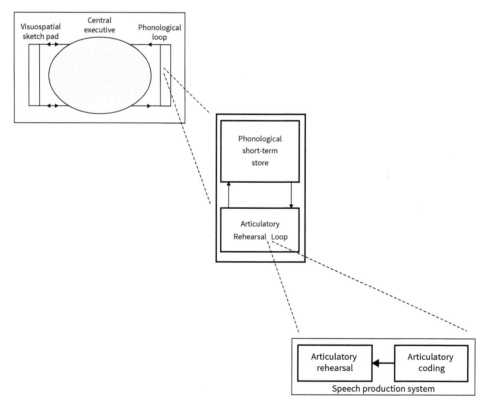

Figure 15.1 Fractionations of the articulatory loop (Baddeley, 1986; Salamè & Baddeley, 1982; Cubelli & Nichelli, 1992).

that can enhance the STM capacity only when the phonological store is unimpaired (Shallice & Vallar, 1990).

Patients with anarthria, that is, completely unable to articulate overt speech but with normal language comprehension and production as indicated by their written or typed output, are likely suffering from a primary deficit at the level of the rehearsal mechanism (see Logie et al., 1989). Indeed, in a very simple counting task used with healthy participants (Logie & Baddeley, 1987), anarthric patients behaved like normal participants under articulatory suppression with an increment in error rate as a function of the increase in the number of elements to be counted (Cubelli et al., 1993). In verbal STM tasks, however, their performance replicated neither that of normal participants who are free to rehearse subvocally nor that of normal participants suppressing articulation. Rather, they showed an abnormal pattern of experimental effects, which varied according to the cerebral lesion site (Cubelli & Nichelli, 1992; Vallar & Cappa, 1987). Patients with lesions in a brain area known as the pons (pontine lesions) showed the phonological similarity effect with both auditory and visual stimuli, but not the word length effect with both modalities of presentation, thus revealing a deficit in rehearsing information within the phonological store. In contrast, patients with bilateral opercular lesions in an area of the frontal lobes showed both phonological similarity and word length effects but only with auditory stimuli, revealing an inability to transfer visually presented stimuli into the phonological store.

Cubelli and Nichelli (1992) suggested a fractionation of the articulatory rehearsal mechanism in two different processes, one involved in articulatory recoding, and the second devoted to the articulatory rehearsal of information within the phonological store. Both are relevant for overt speech (the impairment of one of them is sufficient to cause anarthria) but they each have a different role in verbal STM. Articulatory recoding feeds visual information into the phonological store, while the articulatory rehearsal process recycles stored stimuli. Once again, two functions previously attributed to a unitary mechanism were assigned to two different theoretical components of cognition. In healthy adult participants, articulatory suppression is assumed to engage both the articulatory recoding and the articulatory rehearsal processes; by contrast, a cerebral lesion can impair one component selectively, determining different patterns of the experimental effects in STM tasks. Pontine lesions cause an impairment of the rehearsal mechanism, whereas frontal opercular lesions result in an impairment of the articulatory recoding process. Consistent with this hypothesis, auditory digit span appears to be more impaired in pontine patients than in frontal opercular patients (Cubelli & Nichelli, 1992; see also Papagno et al., 2008).

The model outlined by Baddeley and Hitch (1974) and more fully developed in Baddeley (1986) has been effective in orienting basic research and clinical investigations. It has become more complex by means of progressive adjustments and successive fractionation (Figure 15.1) (See also Baddeley, Chapter 2, this volume). Its development serves as a clear example of how the interaction between experimental psychology and cognitive neuropsychology can advance both theoretical understanding of an important cognitive function and the understanding of selective cognitive deficits associated with diverse focal brain lesions.

A theory can offer a reasonable interpretation of data (Coltheart, 2001). However, a model aimed at accounting for empirical data must be consistent with the accrued knowledge. Adding new modules or processing stages (or boxes in Figure 15.1) to any theoretical model, and particularly a multicomponent model, must be motivated by robust and replicable data patterns. This is particularly important when using data patterns from neuropsychological single cases to inform cognitive models to avoid creating a new component of the model for each single case studied. Any refinement of a model should at least consider whether there are multiple single cases showing consistent, similar patterns of cognitive impairment and sparing, a principle that governs the theoretical developments discussed in this chapter.

The visuospatial sketchpad and further fractionations

The first detailed description of the multicomponent model (Baddeley, 1986) included a unitary visuospatial component. However, subsequently there were reports of a number of neuropsychological double dissociations between visual WM and spatial WM tasks, suggesting the existence of separate visual and spatial memory components of visuospatial WM. For example, Farah et al. (1988) reported patient LH who, as a result of a closed head injury in an automobile accident, suffered damage in both temporal/occipital areas, right temporal lobe, and right inferior frontal lobe. He performed well on spatial tasks concerned with memory for locations and for pathways, such as letter rotation, three-dimensional form rotation, mental scanning, and recalling a recently described pathway, but was severely impaired in visual ability such as remembering colours and the relative size of objects and shapes of states in the map of the US. A similar case was that of LE (Wilson et al., 1999), a professional sculptress who, following systemic lupus erythematosus, resulting in diffuse damage to both the cortex and the white matter, was unable to generate visual images of possible sculptures, and had a severe visual STM deficit, including very poor performance on the Visual Pattern Test, a test that requires retention of black and white checkerboard patterns (Wilson et al., 1987), and on the Four Doors Test (Baddeley et al., 2019), a recognition memory task among pictures of doors that are similar in appearance. However, she could draw complex figures that did not rely on memory, and performed within the normal range for spatial tasks such as the Corsi blocks test that requires recalling sequences of spatial locations of blocks in a random array (Corsi, 1972; Milner, 1971).

Contrasting cases, that is, patients with the opposite pattern of performance, were also reported. Patient EP was affected by a slowly progressive deterioration of the brain and showed a focal atrophy of the anterior part of the right temporal lobe, including the hippocampus (Luzzatti et al., 1998). Her performance was flawless on visual imagery tasks, such as making judgements about relative animal size, or the relative shapes or colours of objects. On the other hand, she was impaired on a range of spatial, topographical tasks such as describing from memory the relative locations of landmarks in her home town. Mr Smith, the patient reported by Hanley and Davies (1995), suffered from a right internal carotid artery stenosis. He was 'terrible with maps' (p. 197) to the point that he was unable to find his way around his own house. He also had difficulties in getting dressed with a mismatch between orientation of the clothing (e.g. sleeves) and the position of his body parts. The patient reports were confirmed with formal testing in that his spatial knowledge and ability to manipulate objects mentally were impaired. However, his ability to perceive and represent visual features of objects and scenes was intact. For example, he had no difficulty in comparing the colours or forms of objects, and had no difficulty in making mental size comparisons between objects and animals when presented with their names. He could readily identify the shapes of countries from silhouettes, but was unable to move these silhouettes into their correct relative geographical position. Finally, he performed very poorly on the Corsi blocks test, and on a series of mental rotation tasks. Della Sala et al. (1999) described two brain injured patients, one of whom presented with specific deficits on the Corsi blocks task, while performing normally on the Visual Pattern Test. In contrast, the other patient presented with the opposite profile, being severely impaired on memory for the Visual Pattern Test without showing a corresponding impairment on the Corsi blocks task. Darling et al. (2006) presented participants with a single probe, the capital letter P, which appeared in one of 30 squares. The task was to identify whether the probe letter P was in the same or in a different location or had the same or a different appearance (typeface). Among the brain-damaged patients assessed, one showed a clear deficit of spatial location memory while performing normally on the visual appearance memory task. Another patient showed the opposite profile.

Similar dissociations supporting the fractionation between a visual and a spatial component of WM have been observed in group studies of healthy participants using interference methods (Della Sala et al., 1999), with older participants (Beigneux et al., 2007), or children born preterm (Retzler et al., 2022). Logie and Pearson (1997) tested groups of children aged 5, 8, and 11 years on their memory span for a version of Corsi block task, and on their memory span for visually presented visual matrix patterns. Performance on both tasks improved across age groups. However, performance between these tasks correlated very poorly within each age group, and memory span for the more static visual matrix patterns increased with age much more rapidly than did memory span for the sequence of movements to random blocks. Similar results from a developmental study with different spatial and visual tasks were reported by Pickering et al. (2001). This technique, known as 'developmental fractionation' (Hitch, 1990) indicates that the cognitive systems responsible for the two tasks seem to develop at different rates and to have little overlap within a given age group.

Earlier studies with healthy adults by Logie and Marchetti (1991) and by Tresch et al. (1993) both showed that retention of visual appearance such as colour shade or geometric form was disrupted by concurrent presentation of irrelevant visual input or colour discrimination, while retention of the location of objects presented at different positions on a screen was disrupted by a concurrent arm movement or a movement discrimination task (for further reports of this distinction in healthy adults see Baddeley & Lieberman, 1980; Della Sala et al., 1999; Hecker & Mapperson, 1997). The pattern of dissociations provided evidence consistent with fractionation of visuospatial WM (Logie, 1995) into a limited capacity visual store, referred to as the visual cache, and a rehearsal component, referred to as the inner scribe, involved in retaining movement sequences as well as refreshing the contents of the visual cache.

Dysexecutive syndrome: the relevance of non-anatomical terminology

The term 'dysexecutive' is now common parlance in neuropsychology. It denotes a series of cognitive and behavioural symptoms, encompassing dysfunctions of planning, inhibition, attention, emotion regulation, abstract thinking, response to novelty, and goal keeping (MacPherson & Della Sala, 2015). We owe the term to Alan Baddeley (1986; Baddeley & Wilson, 1988), who proposed it to identify deficits in the functioning of the central executive component of WM (Baddeley & Della Sala, 1996). The term also was a substitute for the label 'frontal syndrome' which defines a behavioural syndrome in terms of neuroanatomy (Baddeley, 2021), unlike other classic neuropsychological syndromes, such as amnesia, agnosia, aphasia, dyslexia, dysgraphia, that refer to the type of cognitive impairment. Indeed, not all brain-damaged patients with a lesion in the frontal lobes are dysexecutive (e.g. Anderson et al., 1991; Baddeley et al., 1997; Vataja et al., 2003). Moreover, not all dysexecutive patients have frontal lobe damage (e.g. Burgess et al., 1998; Chan et al., 2015; Godefroy et al., 2010; Goldenberg et al., 2007; Mandonnet et al., 2017). Historical examples (see Baddeley & Della Sala, 1996; Della Sala & Logie, 1998) include the behaviour shown by Nietzsche, which was distinctly 'frontal', but that was shown to be due to a syphilis, or to a generalized infection, not as a result of damage specifically to the frontal lobes of the brain (Kaufmann, 1974), or the 'madness' of the UK King George III (who reigned as monarch from 1760 to 1820), affected by porphyria, a metabolic disease (Bennett 1995).

Central executive and dual-tasking

As reported by Baddeley, Bressi, et al. (1991) and Spinnler et al. (1988), patients suffering from Alzheimer's disease (AD), over and above their impairments in episodic LTM, also have clear dysexecutive deficits (see, e.g. Bergeron et al., 2020;

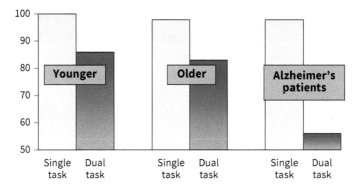

Figure 15.2 Typical outcome from dual-task studies based on the multiple component WM framework (Baddeley et al., 1986). Once individual performance is equated across groups for single task performance, the impact of dual-tasking on younger and older healthy adults does not differ, but there is a large impact on individuals with AD.

Ossenkoppele et al., 2015). In the Baddeley version of the multiple component WM model, the central executive is conceived as coordinating the operation of subsidiary slave systems. Therefore, the theory predicts that patients with dysexecutive impairments due to a defective central executive would have difficulty in coordinating the simultaneous operation of the phonological loop and the visuospatial sketch pad. Accordingly, concurrent performance of two tasks, each of which is thought to engage a different component of the WM system, should be particularly susceptible to the effect of AD. Therefore, a deficit in the ability to perform two such tasks concurrently compared with the ability found in healthy adults could be a cognitive marker of the disease.[2] Crucially, in the paradigms used, each task is selected to rely primarily on a different component of WM. Under these conditions, and when the demands of the two individual tasks are titrated to individual level of ability, there are no or very small, yet statistically significant, effects of healthy ageing on dual-task performance (Baddeley et al., 1986; Baddeley, Bressi, et al., 1991; Logie et al., 2004; Rhodes et al., 2019; for reviews, see Jaroslawska & Rhodes, 2019; Logie, Horne, & Pettit, 2015). However, several studies of our own (e.g. Baddeley et al., 1986, 1991; Kaschel et al., 2009; Della Sala, Cocchini, et al., 2010; Della Sala, Foley, et al., 2010; MacPherson et al., 2012) and from other laboratories (e.g. Holtzer et al., 2004; Ramsden et al., 2008; Sebastian et al., 2006) have reported a dual-tasking deficit in AD, compared with healthy age-matched, and education-matched controls (Figure 15.2). The observed impairment is not simply due to the impact of overall cognitive demand on a damaged brain, because AD patients show a dual-task impairment even when combining two tasks that they find very easy when performed as single tasks, and show no greater drop in performance than healthy controls when the demands of a single

---

[2] Some degree of performance decrement is observed in healthy older and younger participants when they are asked to simultaneously carry out two independent tasks (Baddeley et al., 1986; Doherty et al., 2019; Logie et al., 1990; Rhodes et al., 2019). A cost of coordination, albeit minimal, is inevitable.

task are increased (Logie et al., 2004). The AD dual-task deficit also predicts behavioural problems in everyday life (Alderman, 1996). As a spin-off from these studies, a version of dual-task has been developed that is not dependent on use of language. This version, which is based on temporary feature binding in memory, requiring the integration of two different simultaneous sources of information (such as shape and colour) on a temporary basis (Figure 15.3), proved highly sensitive and specific to AD (Della Sala et al., 2012, 2018; Logie, Parra, & Della Sala, 2015; Parra et al., 2009, 2010). It has also been shown to predict the development of AD in individuals who are asymptomatic, but are known to carry a gene mutation that will result in the onset of AD 10 years later (Parra et al., 2010).

Revising the model

A number of findings during the 1990s subsequently proved fundamental in driving a further revision of the multiple component WM model (Baddeley, 2000). This was particularly relevant in stressing the relationship between WM and LTM. For example, patient PV, described earlier as having a verbal STM deficit assumed to reflect an impairment of the phonological loop, nevertheless showed normal recall of prose passages and her everyday conversation was unimpaired (Vallar & Baddeley, 1984b). This suggested that memory for prose does not rely on the phonological loop. Patients with dense episodic amnesia often perform very poorly on prose recall, both if tested immediately after presentation and, even more so, after a delay (e.g. Della Sala et al., 1997; Isaac & Mayes, 1999; Wilson et al., 1995). However, this is not always the case. For example, Baddeley and Wilson (2002; see also Wilson & Baddeley, 1988) reported the cases of two severely amnesic patients who scored well within the normal range on immediate recall of a passage of prose. To account for these apparently anomalous findings, Baddeley (2000) postulated the existence of another component to the WM model, namely a limited capacity storage system, which could hold information for a limited time, allowing the integration of information coming from various sources, including language and other information activated from LTM, as well as multiple sensory inputs. He named this system the episodic buffer, and viewed it as separate from the central executive and from the phonological loop or visuospatial sketch pad.

Subsequent studies have provided supportive evidence for the episodic buffer concept. For example, immediate memory for feature binding in healthy adults is no more affected by a demanding secondary task than is immediate memory for individual features (Allen et al., 2006, 2012), nor is binding across modalities (e.g. a sound and a visual feature) any more affected by a demanding concurrent task than binding within modality (e.g. colour and shape; Baddeley et al., 2011). Moreover, although the beneficial effects of chunking in prose recall had been traditionally attributed to the functions of the central executive, Baddeley et al. (2009) reported

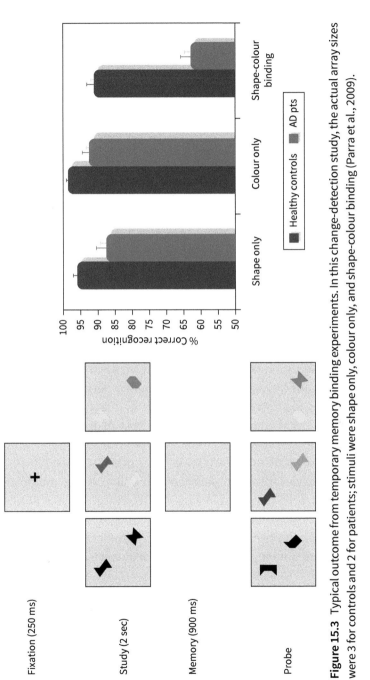

Figure 15.3 Typical outcome from temporary memory binding experiments. In this change-detection study, the actual array sizes were 3 for controls and 2 for patients; stimuli were shape only, colour only, and shape-colour binding (Parra et al., 2009).

that an unrelated concurrent task reduced the retrieval of proper sentences and unrelated words by the same amount, and postulated that the benefit of chunking in prose came from LTM activation of language rather than the intervention of the central executive. This view is consistent also with the tenet that a deficit of the episodic buffer may explain the feature binding deficit in people with AD, described earlier.

This revised version of the WM model proposed by Baddeley (2000), also addressed the issue of the relationship between STM and LTM. Whereas in the original model (Baddeley, 1986), the central executive allowed access to LTM information, in the revised model each STM buffer is directly linked to the LTM component storing the same type of information: for instance, the phonological loop interacts directly with the language component of the LTM. This bidirectional relationship between the phonological loop and the lexicon offers an account for cases AB (Martin et al., 1994) and ML (Martin & He, 2004), who presented with a challenging combination of spared and impaired memory functions. They both had a reduced memory span, but unlike verbal STM patient PV (see above), this impairment was coupled with the standard phonological similarity and word length effects with both auditory and visual stimuli. A major feature of their STM performance was that they did not have the typical lexical effect, that is, they had no advantage for remembering real word lists compared with non-word sequences. They were also impaired in sentence comprehension tasks and their errors increased with the increasing number of words presented (Martin & Romani, 1994). The neuropsychological profile of these patients challenged the original WM model, but it can be tentatively explained by the revised version that postulates direct links between the STM subsystems and the corresponding LTM components. The impaired performance of these patients would then be due to a disconnection between an intact phonological loop and an unimpaired verbal LTM component. Verbal STM span does not benefit from lexical and semantic information; hence, words are treated as unfamiliar words with only phonological and articulatory information being processed.

In a parallel development, an alternative revision of the multiple component model of WM has been proposed by Logie (2011, 2016; Logie et al., 2021). The argument here is that the concept of the central executive acted as a placeholder for a range of high-level cognitive functions that were not explained by the characteristics of the phonological loop and the visuospatial sketch pad. Baddeley (1996) himself described the central executive as an 'area of residual ignorance'. Logie (2016) noted that the concept of a central executive implies some form of homunculus of the mind that acts as a control mechanism, but offers no explanation of how it functions, and raises the questions as to whether there is a higher-level homunculus, what controls that homunculus, and so on to an infinite number of homunculi each controlling the one below it in the hierarchy. This also raises the question of how a central executive might be implemented in the brain. It has been variously linked with the frontal lobes, although as noted earlier, damage to the frontal lobes does not invariably lead to impairments in executive functioning. Cowan (see Chapter 9, this volume; see also Cowan et al., 2011, 2021) has provided evidence that the executive function of control of attention is linked with activation in the anterior cingulate. However, it seems unlikely that a cognitive function as complex as executive control would be supported by a single brain structure or brain area.

The arguments above led Logie (2016) to suggest that executive functions arise from the interaction of multiple cognitive systems, and that executive control is an emergent property of integrated and effective communication across networks of brain structures. So, control arises from multiple local interactions rather than a centralized controller. This concept of 'self-organizing systems' without a central controller has previously been used to account for control of biological systems (e.g. Willshaw, 2006). An analogy can be seen in the behaviour of crowds of people, which arises from the interactions between the many individuals in the crowd (e.g. Morgan, 1986). Self-organizing principles have been shown to account for the coordinated action of colonies of insects through to groups of baboons (e.g. Eisenreich et al., 2017), and flocking behaviour in birds (e.g. Ling et al., 2019; Reynolds, 1987). The principles have also been applied to control within the 'internet of things' (e.g. Nascimento & Lucena, 2017), a development for domestic technology that does not lend itself to centralized control, and is being applied to the design of autonomous control in self-driving cars (Hayashi & Namerikawa, 2016). A similar concept of a self-organizing system could offer a framework for the concept of an episodic buffer, not as a single system or brain structure, but rather the emergent property of concurrently activated functioning of multiple systems as well as activated knowledge in LTM relevant for the current task. Similar ideas have been incorporated in computational models of WM (Barnard, 1999; Vandierendonck, 2016, 2021) and in at least one neurobiological model (Hazy et al., 2007; O'Reilly et al., 2016).

This approach of assuming executive control as an emergent property of the interaction among multiple systems also gains support from studies of AD patients, and provides insight into the nature of the cognitive impairment in these patients. Unlike patients with more focal damage from a stroke or neurosurgery, the damage in AD is more widespread, and spreads further as the disease progresses, so progressively damages the communication between different structures in the brain, and does so more rapidly than it damages the structures themselves (e.g. Charlton & Morris, 2015). In most cases, the initial damage is in the temporal cortex, associated with episodic memory impairments, although prefrontal areas and the occipital cortex can also be affected in some patients.

The findings, described earlier, of specific deficits in dual-task performance and in feature binding for AD patients, but much less impaired performance on single tasks or with immediate memory for single features, can be explained by the degrading of communication between systems that support performance on each task or that allow binding between features to occur. If the primary damage is to the communication network and less to individual structures, then this would allow for much less impaired single task performance or memory for single features.

The studies of patients with more focal lesions likewise show that there are impairments of specific cognitive functions linked with localized damage to specific structures, and only when damage is very extensive is there complete breakdown of cognitive control, such as in the very advanced stages of AD. We can account most readily for these observations by assuming that control is distributed across the brain and arises from effective communication within and between networks, so is neither 'central', nor a single 'executive'. There is always a risk that if a psychological concept is given a name such as 'episodic buffer' or 'central executive' this might lead to an assumption that this reflects a specific region of interest in the brain that might be revealed by specific damage to that region, or by activation of that region in healthy participants performing tasks that are thought to require the episodic buffer or executive control. On the basis of our current arguments, it might be better to think of 'episodic buffering' as a function that reflects currently activated information from multiple sources, and executive control reflects the communication and interactions between multiple different components of the neural architecture.

Similar arguments could be applied to the psychological concepts of the phonological loop and visuospatial scratchpad, suggesting that a more detailed analysis is required of what kinds of cognitive functions might support performance of tasks that are used to explore the characteristics of these concepts (for more detailed discussions, see Logie, 2016, 2018; Logie et al., 2021).

Cognitive models of working memory and methodological concerns in cognitive neuropsychology

WM models should not derive only from studies on healthy volunteers. Often, these models ignore evidence gleaned from sibling disciplines, making them less useful in accounting for the full gamut of data. We argue instead that each theoretical model of WM should generate predictions that are explicit as to the pathological profile that should result from impaired functioning of a specific component or subcomponent of the model. These predictions should be testable, leading to the observation of cases hitherto unrecognized.

The pattern of performance presented by a given neuropsychological patient is not less interpretable than that from a group of healthy participants. In both cases, we need an explicit model of the cognitive function under investigation, and a complete profile of spared and impaired functions of the assessed individual, quite independently of the precise anatomical lesion site, which bears little explanatory power in terms of cognition. The availability of a model offering interpretations of the observed cognitive profile makes any discussion about the need for a 'pure' case rather moot.

If a model allows defined predictions, the presence of associated deficits does not prevent interpretation (as an example, consider the case of deep dyslexia, observed in patients with large lesions and complex neuropsychological patterns, yet accountable by models of reading; Coltheart et al., 1980; Cubelli et al., 2016).

According to Hécaen (1972), neuropsychology is 'the discipline which deals with mental functions and their relationship with cerebral structures [. . .]. Springing from clinical observation, Neuropsychology does not deny it: each individual case could become the object of a study, insofar as each peculiar impairment noted by the clinicians would allow them to postulate a valuable interpretation' (translation by the current authors from the original French). It may be true that every patient could be the source of theoretical analyses and possible theoretical interpretations. However, we would disagree that all patients are good candidates for testing models and specific predictions. Before running experiments and ascertaining the presence of the expected effects, a precise functional diagnosis of the patient is required, as is a detailed analysis of how a cognitive task may be performed by patients and by healthy controls.

In studies with healthy participants, the performance of any one individual may differ from that of other participants for idiosyncratic, uncontrollable, or unpredictable factors. However, it may be misleading to treat this variability as random noise that can be handled by statistical averaging. For example, participants may develop an alternative strategy such as remembering colours and shapes by the verbal labels rather than their visual characteristics, so what is assumed to be a test of visual STM is actually a test of verbal STM. Remembering a sequence of words might rely on the word meanings, the shape of the letters in the word, and the first letter of each word, in addition to word phonology. Consequently, the observed pattern of performance from the group aggregate data might not reflect the cognitive functions of all individual participants, or even how individual participants perform the same tasks under different experimental conditions or on different occasions (e.g. Belletier et al., 2022; Logie et al., 1996). In cognitive neuropsychology, the single case approach is very common, with a strong emphasis on accumulating evidence from multiple single cases rather than from aggregating group data (Caramazza & McCloskey, 1988). Moreover, a clinical label is not necessarily helpful, because the same clinical label may designate patients with different cognitive impairments and the same observed patterns of impairment may be due to different impaired functions. Consider, for example the clinical label AD. It is important to specify the degree of severity, whether it is AD rather than vascular dementia, frontotemporal dementia, or even healthy ageing. Moreover, when testing groups of healthy older people, it is important to assess whether any of them might be in the early stages of hitherto undiagnosed dementia or have undetected brain pathology such as a tumour. A further example is case GB (Baddeley & Wilson, 1985) who showed a post-traumatic mutism and normal capability of active subvocal rehearsal. The inability to speak following a severe closed-head injury usually is due to aphonia, which is a pathological loss of vocalization whereby the person can articulate but only in whisper, rather than to anarthria (see Cubelli & Nichelli, 1992). This being the case, these kinds of patients would not be suitable to investigate the role of articulation in STM tasks.

In neuropsychological single cases, multiple symptoms and deficits may have different causes and interfere with each other in performing single tasks. As a group, the patients with apraxia of speech reported by Rochon et al. (1991) and Waters et al. (1992) showed the same pattern of performance seen in normal participants under articulatory suppression. At the individual level, however, none of them showed the expected pattern (Caplan et al., 2012). In apraxia of speech, the ability to produce articulate speech is impaired but not abolished. These patients vary in severity and clinical picture but they do try to use articulation even if their production is cumbersome and effortful, reduced, and distorted. Hence, their performance in verbal STM tasks is difficult to interpret. The experimental effects typically observed when people are free to rehearse may be present because of the preserved articulation, even if defective, or may be absent because articulation is impaired, even partially. In both cases, results are uninterpretable within a theoretical framework unless there is a clear understanding of the precise nature of the impairment, and how the patient is attempting the tasks they are set.

Describing and interpreting a clinical picture are two different processes. Observing that patient PV had an impaired verbal memory span, coupled with the absence of the recency effect[3] in a supra-span free recall task (Vallar & Papagno, 1986) does not necessarily imply that the loss of the recency effect was due to the defective phonological store. Since span and recency are affected by different factors in healthy participants and doubly dissociate in neuropsychological patients (Della Sala et al., 1998), it is more conservative to conclude that PV had two independent deficits, one leading to a reduced span and the other responsible for the lack of recency (for a discussion, see Caplan et al., 2012). A 'pure' deficit does not mean that a patient shows only one type of impairment, but that the observed pattern of performance corresponds to the clinical picture predicted from cognitive theory, assuming the selective damage of a cognitive function identified in the theory. This does not mean that some other cognitive function, associated with some other aspect of the theory, is not also impaired. From this point of view, a patient with a 'pure' deficit can have more than one independent deficit. This possibility, together with the possible use of an alternative strategy by the patient to help compensate for their impairment, means that the site of a lesion does not necessarily reflect the operation of the impaired function in a healthy brain. Likewise, the possibility that healthy participants may use alternative strategies to perform a cognitive task means that a theoretical understanding of how each participant is performing a given task

---

[3] When asked to recall, in any order (immediate free recall), a list of items that is longer than typical STM capacity or span (supra-span), the recency effect refers to the tendency to remember the last few items in the list better than items earlier in the list (e.g. Glanzer & Cunitz, 1966).

in a brain imaging study is essential if the brain activation patterns can be interpreted as reflecting a given cognitive function. A more fruitful approach to theory development as well as to cognitive neuropsychology may be to explore the range of cognitive functions, or 'cognitive toolbox' that a healthy participant or a neuropsychological patient has available, and then to explore how each individual selects from the cognitive tools available to them when they attempt to perform the tasks they are set (see Logie, 2018, for a detailed discussion).

Conclusion

The development of knowledge in cognitive sciences requires an interaction: from theory to clinical observations and back. We have tried to illustrate in this chapter why it is important for cognitive psychology to recognize the heuristic value for theory development that may be obtained from neuropsychological observations, as well as from studies of healthy individuals. We have argued that such theoretical development should not ignore these observations, or avoid the challenges to theory that might arise from empirical, neuropsychological data, when those data do not fit with the preferred theoretical framework of a given researcher. Rather, researchers should attempt to account for data from patients by testing available frameworks, and if necessary, modify those frameworks according to the data observed from patients as well as from healthy adults. Genuine advances in scientific as well as clinical understanding requires a reciprocal relationship.

References

Alderman, N. (1996). Central executive deficit and response to operant conditioning methods. *Neuropsychological Rehabilitation*, 6(3), 161–186.

Allen, R. J., Baddeley, A. D., & Hitch, G. J. (2006). Is the binding of visual features in working memory resource-demanding? *Journal of Experimental Psychology: General*, 135(2), 298–313.

Allen, R. J., Hitch, G. J., Mate, J., & Baddeley, A. D. (2012). Feature binding and attention in working memory: A resolution of previous contradictory findings. *Quarterly Journal of Experimental Psychology*, 65(12), 2369–2383.

Anderson, S. W., Damasio, H., Jones, R. D., & Tranel, D. (1991). Wisconsin card sorting test performance as a measure of frontal lobe damage. *Journal of Clinical Experimental Neuropsychology*, 13(6), 909–922.

Baddeley, A. D. (1986). *Working memory*. Oxford University Press.

Baddeley, A. D. (1996). Exploring the central executive. *Quarterly Journal of Experimental Psychology*, 49(1), 5–28.

Baddeley, A. D. (2000). The episodic buffer: A new component of working memory? *Trends in Cognitive Sciences*, 4(11), 417–423.

Baddeley, A. D. (2003). Working memory: Looking back and looking forward. *Nature Reviews Neuroscience*, 4(10), 829–839.

Baddeley, A. D. (2021). Developing the concept of working memory: The role of neuropsychology. *Archives of Clinical Neuropsychology*, 36(6), 861–873.

Baddeley, A. D., Allen, R. J., Atkinson, A. L., & Kemp, S. (2019). The problem of detecting long-term forgetting: Evidence from the Crimes Test and the Four Doors Test. *Cortex, 110*, 69–79.

Baddeley, A. D., Allen, R. J., & Hitch, G. J. (2011). Binding in visual working memory: The role of the episodic buffer. *Neuropsychologia, 49*(6), 1393–1400.

Baddeley, A. D., Bressi, S., Della Sala, S., Logie, R., & Spinnler, H. (1991). The decline of working memory in Alzheimer's disease. A longitudinal study. *Brain, 114*(Pt. 6), 2521–2542.

Baddeley, A. D., & Della Sala, S. (1996). Working memory and executive control. *Philosophical Transactions of the Royal Society of London. Series B, Biological Sciences, 351*(1346), 1397–1404.

Baddeley, A. D., Della Sala, S., Papagno, C., & Spinnler, H. (1997). Dual-task performance in dysexecutive and nondysexecutive patients with a frontal lesion. *Neuropsychology, 11*(2), 187–194.

Baddeley, A., Della Sala, S., & Spinnler, H. (1991). The two component hypothesis of memory deficit in Alzheimer's disease. *Journal of Clinical and Experimental Neuropsychology, 13*(2), 341–349.

Baddeley, A. D., Gathercole, S., & Papagno, C. (1998). The phonological loop as a language learning device. *Psychological Review, 105*(1), 158–173.

Baddeley, A. D., & Hitch, G. (1974). Working memory. In G. H. Bower (Ed.), *The psychology of learning and motivation* (Vol. 8, pp. 47–89). Academic Press.

Baddeley, A. D., & Hitch, G. J. (2019). The phonological loop as a buffer store: An update. *Cortex, 112*, 91–106.

Baddeley, A. D., Hitch, G. J., & Allen, R. J. (2009). Working memory and binding in sentence recall. *Journal of Memory and Language, 61*(3), 438–456.

Baddeley, A. D., Hitch, G., & Allen, R. (2021). A multicomponent model of working memory. In R. H. Logie, V. Camos, & N. Cowan (Eds.), *Working memory: State of the science* (pp. 10–43). Oxford University Press.

Baddeley, A. D., Lewis, V. J., & Vallar, G. (1984). Exploring the articulatory loop. *Quarterly Journal of Experimental Psychology, 36*(2), 233–252.

Baddeley, A. D., & Lieberman, K (1980). Spatial working memory. In R. S. Nickerson (Ed.), *Attention and performance VIII* (pp. 521–539). Hillsdale, NJ: Erlbaum.

Baddeley, A. D., & Logie, R. H. (1999). Working memory: The multiple component model. In A. Miyake & P. Shah (Eds.), *Models of working memory* (pp. 28–61). Cambridge University Press.

Baddeley, A. D., Logie, R. H., Bressi, S., Della Sala, S., & Spinnler, H. (1986). Dementia and working memory. *Quarterly Journal of Experimental Psychology, 38*(4), 603–618.

Baddeley, A. D., Thomson, N., & Buchanan, M. (1975). Word length and the structure of short-term memory. *Journal of Verbal Learning and Verbal Behavior, 14*(6), 575–589.

Baddeley, A. D., & Warrington, E. K. (1970). Amnesia and the distinction between long- and short-term memory. *Journal of Verbal Learning & Verbal Behavior, 9*(2), 176–189.

Baddeley, A. D., & Wilson, B. A. (1985). Phonological coding and short-term memory in patients without speech. *Journal of Memory and Language, 24*(4), 490–502.

Baddeley, A. D., & Wilson, B. A. (1988). Frontal amnesia and the dysexecutive syndrome. *Brain and Cognition, 7*(2), 212–230.

Baddeley, A., & Wilson, B. A. (2002). Prose recall and amnesia: Implications for the structure of working memory. *Neuropsychologia, 40*(10), 1737–1743.

Barnard, P. J. (1999). Interacting cognitive subsystems: Modelling working memory phenomena within a multiprocessor architecture. In A. Miyake & P. Shah (Eds.), *Models of working memory* (pp. 298–339). Cambridge University Press.

Barrouillet, P., & Camos, V. (2015). *Working memory. Loss and reconstruction.* Hove: Psychology Press.

Basso, A., Spinnler, H., Vallar, G., & Zanobio, M. E. (1982). Left hemisphere damage and selective impairment of auditory verbal short-tern memory. *Neuropsychologia, 20*(3), 263–274.

Beigneux, K., Plaie, T., & Isingrini, M. (2007). Aging effect on visual and spatial components of working memory. *International Journal of Aging & Human Development, 65*, 301–314.

Belletier, C., Doherty, J. M., Jaroslawska, A. J., Rhodes, S., Cowan, N., Naveh-Benjamin, M., Barrouillet, P., Camos, V., & Logie, R. H. (2022). Strategic adaptation to dual-task in verbal working memory: Potential routes for theory integration. *Journal of Experimental Psychology: Learning, Memory, and Cognition.* Advance online publication.

Bennett, A. (1995). *The madness of King George.* Faber and Faber.

Bergeron, D., Sellami, L., Poulin, S., Verret, L., Bouchard, R. W., & Laforce, R. Jr. (2020). The behavioral/dysexecutive variant of Alzheimer's disease: A case series with clinical, neuropsychological, and FDG-PET characterization. *Dementia and Geriatric Cognitive Disorders, 49*(5), 518–525.

Bouillaud, J. B. (1825). Recherches cliniques propres à démontrer que la perte de la parole correspond à la lésion des lobules antérieurs du cerveau, et à confirmer l'opinion de M. Gall, sur le siège de l'organe du langage articulé. *Archives Générales de Médecine, 8,* 25–45.

Broadbent, D. E. (1958). *Perception and communication.* Pergamon Press.

Broca, P. (1865). Sur le siège de la faculté du langage articulé. *Bulletin de la Société d'Anthropologie, 6*(1), 377–393.

Burgess, P. W., Alderman, N., Evans, J., Emslie, H., & Wilson, B. A. (1998). The ecological validity of tests of executive function. *Journal of the International Neuropsychological Society, 4*(6), 547–558

Caplan, D., Waters, G., & Howard, D. (2012). Slave systems in verbal short-term memory. *Aphasiology, 26*(3–4), 279–316.

Caramazza, A., & McCloskey, M. (1988). The case for single-patient studies. *Cognitive Neuropsychology, 5*(5), 517–527.

Chan, E., MacPherson, S. E., Robinson, G., Turner, M., Lecce, F., Shallice, T., & Cipolotti, L. (2015). Limitations of the trail making test part-B in assessing frontal executive dysfunction. *Journal of International Neuropsychological Society, 21*(2), 169–174.

Charlton, R. A., & Morris, R. G. (2015). Associations between working memory and white matter integrity in normal ageing. In R. H. Logie & R. G. Morris (Eds.), *Working memory and aging* (pp. 97–128). Hove, UK: Psychology Press.

Coltheart, M. (2001). Assumptions and methods in cognitive neuropsychology. In B. Rapp (Ed.), *The handbook of cognitive neuropsychology: What deficits reveal about the human mind.* (pp. 3–21). Psychology Press.

Coltheart, M. (2003). Contributions of experimental psychology to neuropsychology. *Japanese Journal of Psychonomic Science, 22*(1), 58–66.

Coltheart, M., Patterson, K., & Marshall, J. C. (1980). *Deep dyslexia.* Routledge & Kegan Paul.

Conrad, R., & Hull, A. J. (1964). Information, acoustic confusion and memory span. *British Journal of Psychology, 55,* 429–432.

Corsi, P. M. (1972). Human memory and the medial temporal region of the brain. *Dissertation Abstracts International, 34*(02), 819B. (University Microfilms No. AAI05-77717).

Cowan, N. (1995). *Attention and Memory: An Integrated Framework.* New York: Oxford University Press.

Cowan, N. (1999). An Embedded-Processes Model of working memory. In A. Miyake & P. Shah (Eds.), *Models of working memory: Mechanisms of active maintenance and executive control* (pp. 62–101). Cambridge: Cambridge University Press.

Cowan, N., Li, D., Moffitt, A., Becker, T. M., Martin, E. A., Saults, J. S., & Christ, S. E. (2011). A neural region of abstract working memory. *Journal of Cognitive Neuroscience, 23*(10), 2852–2863.

Cowan, N., Morey, C. C., & Naveh-Benjamin, M. (2021). An embedded-processes approach to working memory: How is it distinct from other approaches, and to what ends? In R. H. Logie, V. Camos, & N. Cowan (Eds.), *Working memory: State of the science* (pp. 44–84). Oxford University Press.

Cubelli, R., Pedrizzi, S., & Della Sala, S. (2016). The role of cognitive neuropsychology in clinical settings: The example of a single case of deep dyslexia. In J. Macniven (Ed.), *Neuropsychological formulation: A clinical casebook* (pp. 15–27). Springer.

Cubelli, R., & Montagna, C. G. (1994). A reappraisal of the controversy of Dax and Broca. *Journal of the History of the Neurosciences, 3*(4), 215–226.

Cubelli, R., & Nichelli, P. (1992). Inner speech in anarthria: Neuropsychological evidence of differential effects of cerebral lesions on subvocal articulation. *Journal of Clinical and Experimental Neuropsychology, 14*(4), 499–517.

Cubelli, R., Nichelli, P., & Pentore, R. (1993). Anarthria impairs subvocal counting. *Perceptual and Motor Skills, 77*(3), 971–978.

Darling, S., Della Sala, S., Logie, R. H., & Cantagallo, A. (2006). Neuropsychological evidence for separating components of visuo-spatial working memory. *Journal of Neurology, 253*(2), 176–180.

De Renzi, E. (1967). Caratteristiche e problemi della neuropsicologia. *Archivio di Psicologia, Neurologia e Psichiatria, 28*(5), 422–440.

De Renzi, E., & Nichelli, P. (1975). Verbal and nonverbal short term memory impairment following hemispheric damage. *Cortex, 11*(4), 341–353.

Della Sala, S., Cocchini, G., Logie, R. H., & MacPherson, S. E. (2010). Dual task during encoding, maintenance and retrieval in Alzheimer disease and healthy ageing. *Journal of Alzheimer's Disease, 19*(2), 503–515.

Della Sala, S., Foley, J. A., Beschin, N., Allerhand, M., & Logie, R. H. (2010b). Assessing dual task performance using a paper-and-pencil test: Normative data. *Archives of Clinical Neurology, 25*(5), 410–419.

Della Sala, S., Gray, C., Baddeley, A. D., Allamano, N., & Wilson, L. (1999). Pattern span: A tool for unwelding visuo-spatial memory. *Neuropsychologia, 37*(10), 1189–1199.

Della Sala, S., Kozlova, I., Stamate, A., & Parra, M. (2018). Temporary memory binding: A transcultural cognitive marker of Alzheimer's disease. *International Journal of Geriatric Psychiatry, 33*(6), 849–856.

Della Sala, S., & Logie, R. H. (1998). Dualism down the drain, thinking in the brain. In R. H. Logie & K. J. Gilhooly (Eds.), *Working memory and thinking* (pp. 45–66). Psychology Press,

Della Sala, S., & Logie, R. H. (1993). When working memory does not work: The role of working memory in neuropsychology. In F. Boller & H. Spinnler (Eds.), *Handbook of neuropsychology* (Vol. VIII, pp. 1–61). Elsevier Science Publisher B. V.

Della Sala, S., Logie, R. H., Trivelli, C., Cubelli, R., & Marchetti, C. (1998). Dissociation between recency and span: Neuropsychological and experimental evidence. *Neuropsychology, 12*(4), 533–545.

Della Sala, S., Parra, M. A., Fabi, K., Luzzi, S., & Abrahams, S. (2012). Short-term memory binding is impaired in AD but not in non-AD dementias. *Neuropsychologia, 50*(5), 833–840.

Della Sala, S., Spinnler, H., & Venneri, A. (1997). Persistent global amnesia following right thalamic stroke: An 11-year longitudinal study. *Neuropsychology, 11*(1), 90–103.

De Tollis, M., De Simone, M. S., Perri, R., Fadda, L., Caltagirone, C., & Carlesimo, G. A. (2021). Verbal and spatial memory spans in mild cognitive impairment. *Acta Neurologica Scandinavica, 144*(4), 383–393.

Doherty, J. M., Belletier, C., Rhodes, S., Jaroslawska, A., Barrouillet, P., Camos, V., Cowan, N., Naveh-Benjamin, M., & Logie, R. H. (2019). Dual-task costs in working memory: An adversarial collaboration. *Journal of Experimental Psychology: Learning, Memory, and Cognition, 45*(9), 1529–1551.

Drachman, D. A., & Arbit, J. A. (1966). Memory and the hippocampal complex: II. Is memory a multiple process? *Archives of Neurology, 15*(1), 52–61.

Ellis, A. W., & Young, A. W. (1996). *Human cognitive neuropsychology: A textbook with readings.* Lawrence Erlbaum.

Eisenreich, B. R., Akaishi, R., & Hayden, B. Y. (2017). Control without controllers: Toward a distributed neuroscience of executive control. *Journal of Cognitive Neuroscience, 29*, 1684–1698.

Farah, M. J., Hammond, K. M., Levine, D. N., & Calvanio, R. (1988). Visual and spatial mental imagery: Dissociable systems of representation. *Cognitive Psychology, 20*(4), 439–462.

Freud, S. (1891). *Zur Auffassung der Aphasien: Eine kritische Studie.* Franz Deuticke.

Geschwind, N. (1965). Disconnexion syndromes in animals and man. I. *Brain, 88*(2), 237–294.

Geva, S., Truneh, T., Seghier, M. L., Hope, T. M. H., Leff, A. P., Crinion, J. T., Gajardo-Vidal, A., Lorca-Puls, D. L., Green, D. W., PLORAS Team, & Price, C. J. (2021). Lesions that do or do not impair digit span: A study of 816 stroke survivors. *Brain Communications, 3*(2), fcab031.

Ghaleh, M., Lacey, E. H., Fama, M. E., Anbari, Z., DeMarco, A. T., & Turkeltaub, P. E. (2020). Dissociable mechanisms of verbal working memory revealed through multivariate lesion mapping. *Cerebral Cortex, 30*(4), 2542–2554.

Glanzer, M., & Cunitz, A. R. (1966). Two storage mechanisms in free recall. *Journal of Verbal Learning and Verbal Behavior, 5*, 351–360.

Godefroy, O., Azouvi, P., Philippe, R., Roussel, M., LeGall, D., & Meulemans, T., & Groupe de Réflexion sur l'Evaluation des Fonctions Exécutives Study Group. (2010). Dysexecutive syndrome: Diagnostic criteria and validation study. *Annals of Neurology, 68*(6), 855–864.

Goldenberg, G., Hartmann-Schmid, K., Sürer, F., Daumüller, M., & Hermsdörfer, J. (2007). The impact of dysexecutive syndrome on use of tools and technical devices. *Cortex, 43*(3), 424–435

Goodglass, H., & Kaplan, E. (1972). *Assessment of aphasia and related disorders.* Lea Febiger.

Hanley, J. R., & Davies, A. D. (1995). Lost in your own house. In: R. Campbell & M. A. Conway (Eds.), *Broken memories: Case studies in memory impairment* (pp. 195–208). Oxford: Blackwell.

Hanley, J. R., Young, A. W., & Pearson, N. (1991). Impairment of the visuo-spatial sketch pad. *Quarterly Journal of Experimental Psychology, 43A*, 101–125.

Hayashi, Y., & Namerikawa, T. (2016). Flocking algorithm for multiple nonholonomic cars. In *55th Annual Conference of the Society of Instrument and Control Engineers of Japan, SICE* (pp. 1660–1665). Institute of Electrical and Electronics Engineers Inc.

Hazy, T. E., Frank, M. J., & O'Reilly, R. C. (2007). Towards an executive without a homunculus: Computational models of the prefrontal cortex/basal ganglia system. *Philosophical Transactions of the Royal Society of London. Series B, Biological Sciences, 362*(1485), 1601–1613.

Hebb, D. O. (1949). *The organization of behavior*. Wiley.

Hecaen, H. (1972). *Introduction à la neuropsychologie*. Larousse.

Hecker, R., & Mapperson, B. (1997). Dissociation of visual and spatial processing in working memory. *Neuropsychologia, 35*(5), 599–603.

Hitch, G. J. (1990). Developmental fractionation of working memory. In G. Vallar & T. Shallice (Eds.), *Neuropsychological impairments of short-term memory* (pp. 221–246). Cambridge: Cambridge University Press.

Holtzer, R., Burright, R. G., & Donovick, P. J. (2004). The sensitivity of dual-task performance to cognitive status in aging. *Journal of the International Neuropsychological Society, 10*(2), 230–238.

Hulme, C., Thomson, N., Muir, C., & Lawrence, A. (1984). Speech rate and the development of short-term memory span. *Journal of Experimental Child Psychology, 38*(2), 241–153.

Isaac, C. L., & Mayes, A. R. (1999). Rate of forgetting in amnesia: I. Recall and recognition of prose. *Journal of Experimental Psychology: Learning, Memory, and Cognition, 25*(4), 942–962.

James, W. (1890). *The principles of psychology*. Henry Holt and Company

Jaroslawska, A. J., & Rhodes, S. (2019). Adult age differences in the effects of processing on storage in working memory: A meta-analysis. *Psychology and Aging, 34*(4), 512–530.

Kaschel, R., Logie, R., Kaze´n, M., & Della Sala, S. (2009). Alzheimer's disease, but not ageing or depression, affects dual-tasking. *Journal of Neurology, 256*, 1860–1868.

Kaufmann, W. (1974). *Nietzsche: Philosopher, psychologist, antichrist*. Princeton University Press.

Klatzky, R. L. (1975). *Human memory: Structures and processes*. W. H. Freeman.

Levy, B. A. (1971). The role of articulation in auditory and visual short-term memory. *Journal of Verbal Learning and Verbal Behavior, 10*(2), 123–132.

Lichtheim, L. (1885). On aphasia. *Brain, 7*(4), 433–484.

Ling, H. L., Mclvor, G. E., van der Vaart, K., Vaughan, R. T., Thornton, A., & Ouellette, N. T. (2019). Local interactions and their group-level consequences in flocking jackdaws. *Proceedings of the Royal Society of London B: Biological Sciences, 286*(1906), 20190865.

Locke, J. (1690). *An essay concerning humane understanding*. Thomas Basset.

Logie, R. H. (1995). *Visuo-spatial working memory*. Hove: Lawrence Erlbaum.

Logie, R. H. (1996). The seven ages of working memory. In J. T. E. Richardson, R. W. Engle, L. Hasher, R. H. Logie, E. R. Stoltzfus, & R. T. Zacks (Eds.), *Working memory and human cognition* (pp. 31–65). Oxford University Press.

Logie, R. H. (2011). The functional organisation and the capacity limits of working memory. *Current Directions in Psychological Science, 20*, 240–245.

Logie, R. H. (2016). Retiring the central executive. *Quarterly Journal of Experimental Psychology, 69*(10), 2093–2109.

Logie, R. H. (2018). Human cognition: Common principles and individual variation. *Journal of Applied Research in Memory and Cognition, 7*(4), 471–486.

Logie, R. H. (2019). Converging sources of evidence and theory integration in working memory: A commentary on Morey, Rhodes, & Cowan. *Cortex, 112*, 162–171.

Logie, R. H., & Baddeley, A. D. (1987). Cognitive processes in counting. *Journal of Experimental Psychology: Learning, Memory, and Cognition, 13*(2), 310–326.

Logie, R. H., Belletier, C., & Doherty, J. D. (2021). Integrating theories of working memory. In R. H. Logie, V. Camos, & N. Cowan, (Eds.), *Working memory: State of the science* (pp. 389–429). Oxford University Press.

Logie, R. H., Camos, V., & Cowan, N. (Eds.). (2021). *Working memory: State of the science*. Oxford University Press.

Logie, R. H., Cocchini, G., Della Sala, S., & Baddeley, A. D. (2004). Is there a specific executive capacity for dual task co-ordination? Evidence from Alzheimer's Disease. *Neuropsychology, 18*, 504–513.

Logie, R. H., Cubelli, R., Della Sala, S., Alberoni, M., & Nichelli, P. (1989). Anarthria and verbal short-term memory. In J. Crawford, D. Parker (Eds.), *Developments in clinical and experimental neuropsychology* (pp. 203–211). Plenum.

Logie, R. H., & Della Sala, S. (2005). Disorders of visuo-spatial working memory. In P. Shah & A. Miyake (Eds.), *Handbook of visuospatial thinking* (pp. 81–120). Cambridge University Press.

Logie, R. H., Della Sala, S. D., Laiacona, M., Chalmers, P., & Wynn, V. (1996). Group aggregates and individual reliability: The case of verbal short-term memory. *Memory & Cognition, 24*(3), 305–321.

Logie, R. H., Horne, M. J., & Pettit, L. D. (2015). When cognitive performance does not decline across the lifespan. In R. H. Logie & R. G. Morris (Eds.), *Working memory and ageing* (pp. 21–47). Psychology Press.

Logie, R. H., & Marchetti, C. (1991). Visuo-spatial working memory: Visual, spatial or central executive. In: R. H. Logie & M. Denis (Eds.), *Mental images in human cognition* (pp. 105–115). North Holland Press.

Logie, R. H., Parra, M. A., & Della Sala, S. (2015). From cognitive science to dementia assessment. *Policy Insights from the Behavioral and Brain Sciences, 2*(1), 81–91.

Logie, R. H., & Pearson, D. G. (1997). The inner eye and the inner scribe of visuo-spatial working memory: Evidence from developmental fractionation. *European Journal of Cognitive Psychology, 9*(3), 241–257.

Logie, R. H., Zucco, G., & Baddeley, A. D. (1990). Interference with visual short-term memory. *Acta Psychologica 75*(1), 55–74.

Lugtmeijer, S., Lammers, N. A., de Haan, E. H. F., de Leeuw, F. E., & Kessels, R. (2021). Post-stroke working memory dysfunction: A meta-analysis and systematic review. *Neuropsychological Review, 31*(1), 202–219.

Luzzatti, C., Vecchi, T., Agazzi, D., Cesa-Bianchi, M., & Vergani, C. (1998). A neurological dissociation between preserved visual and impaired spatial processing in mental imagery. *Cortex, 34*(3), 461–469.

MacPherson, S. E., & Della Sala, S. (2015). *The handbook of frontal assessment*. Oxford University Press.

MacPherson, S. E., & Della Sala, S. (2019). *Cases of amnesia*. Routledge.

MacPherson, S. E., Parra, M. A., Moreno, S., Lopera, F., & Della Sala, S. (2012). Dual Task Abilities as a Possible Preclinical Marker of Alzheimer's Disease in Carriers of the E280A Presenilin-1 Mutation. *JINS, 18*, 234–241.

Mandonnet, E., Cerliani, L., Siuda-Krzywicka, K., Poisson, I., Zhi, N., Volle, E., & de Schotten, M. T. (2017). A network-level approach of cognitive flexibility impairment after surgery of a right temporo-parietal glioma. *Neurochirurgie, 63*(4), 308–313.

Martin, R. C., & He, T. (2004). Semantic short-term memory and its role in sentence processing: A replication. *Brain and Language, 89*(1), 76–82.

Martin, R. C., & Romani, C. (1994). Verbal working memory and sentence comprehension: A multiple-components view. *Neuropsychology, 8*(4), 506–523.

Martin, R. C., Shelton, J. R., & Yaffee, L. S. (1994). Language processing and working memory: Neuropsychological evidence for separate phonological and semantic capacities. *Journal of Memory and Language, 33*(1), 83–111.

Milner, B. (1966). Amnesia following operation on the temporal lobes. In C. W. M. Whitty & O. Zangwill (Eds.), *Amnesia* (pp. 109–133). Butterworths.

Milner, B. (1971). Interhemispheric differences in the localization of psychological processes in man. *British Medical Bulletin, 27*, 272–277.

Morgan, G. (1986). *Images of organization*. London, UK: Sage.

Murray, D. (1965). Vocalization-at-presentation, with varying presentation rates. *Quarterly Journal of Experimental Psychology, 17*(1), 47–56.

Murray, D. (1968). Articulation and acoustic confusability in short-term memory. *Journal of Experimental Psychology, 78*(4, Pt. 1), 679–684.

Nascimento, N. M., & Lucena, C. J. P. (2017). FIoT: An agent-based framework for self-adaptive and self-organizing applications based on the Internet of Things. *Information Sciences*, *378*, 161–176.

O'Reilly, R. C., Hazy, T. E., & Herd, S. A. (2016). The leabra cognitive architecture: How to play 20 principles with nature (2016). In S. E. F. Chipman (Ed.), *The Oxford handbook of cognitive science* (pp. 91–116). Oxford University Press.

Ossenkoppele, R., Pijnenburg, Y. A., Perry, D. C., Cohn-Sheehy, B. I., Scheltens, N. M, Vogel, J. W., Kramer, J. H., van der Vlies, A. E., La Joie, R., Rosen, H. J., van der Flier, W. M., Grinberg, L. T., Rozemuller, A. J., Huang, E. J., van Berckel, B. N., Miller, B. L., Barkhof, F., Jagust, W. J., Scheltens, P., Seeley, W. W., . . . Rabinovici, G. D. (2015). The behavioural/dysexecutive variant of Alzheimer's disease: Clinical, neuroimaging and pathological features. *Brain*, *138*(9), 2732–2749.

Papagno, C., Lucchelli, F., & Vallar, G. (2008). Phonological recoding, visual short-term store and the effect of unattended speech: Evidence from a case of slowly progressive anarthria. *Cortex*, *44*(3), 312–324.

Papagno, C., & Shallice, T. (2019). Introduction to impairments of short-term memory buffers: Do they exist? *Cortex*, *112*, 1–4.

Parra, M. A., Abrahams, S., Fabi, K., Logie, R. H., Luzzi, S., & Della Sala, S. (2009). Short term memory binding deficits in Alzheimer's disease. *Brain*, *132*(4), 1057–1066.

Parra, M. A., Abrahams, S., Logie, R. H., Méndez, L. G., Lopera, F., & Della Sala, S. (2010). Visual short-term memory binding deficits in familial Alzheimer's disease. *Brain*, *133*(9), 2702–2713.

Parra, M. A., Della Sala, S., Logie, R. H., & Abrahams, S. (2009). Selective impairment in visual short-term memory binding. *Cognitive Neuropsychology*, *26*(7), 583–605.

Pickering, S. J., Gathercole, S. E., Hall, M., & Lloyd, S. A. (2001). Development of memory for pattern and path: Further evidence for the fractionation of visuo-spatial memory. *Quarterly Journal of Experimental Psychology*, *54A*, 397–420.

Ramsden, C. M., Kinsella, G. J., Ong, B., & Storey, E. (2008). Performance of everyday actions in mild Alzheimer's disease. *Neuropsychology*, *22*(1), 17–26.

Retzler, J., Johnson, S., Groom, M. J., & Cragg, L. (2022). A comparison of simultaneous and sequential visuo-spatial memory in children born very preterm. *Child Neuropsychology*, *28*(4), 496–509.

Reynolds, C. W. (1987). Flocks, herds and schools: A distributed behavioral model, SIGGRAPH. *Computers and Graphics*, *21*(4), 25–34.

Rhodes, S., Jaroslawska, A. J., Doherty, J. M., Belletier, C., Naveh-Benjamin, M., Cowan, N., Camos, V., Barrouillet, P., & Logie, R. H. (2019). Storage and processing in working memory: Assessing dual task performance and task prioritization across the adult lifespan. *Journal of Experimental Psychology: General*, *148*(7), 1204–1227.

Rochon, E., Caplan, D., & Waters, G. S. (1991). Short-term memory processes in patients with apraxia of speech: Implications for the nature and structure of the auditory verbal short-term memory system. *Journal of Neurolinguistics*, *5*(2–3), 231–264.

Salamè, P., & Baddeley, A. D. (1982). Disruption of short-term memory by unattended speech: Implications for the structure of working memory. *Journal of Verbal Learning and Verbal Behavior*, *21*(2), 150–164.

Sebastian, M. V., Menor, J., & Elosua, M. R. (2006). Attentional dysfunction of the central executive in AD: Evidence from dual task. *Cortex*, *42*(7), 1015–1020.

Shallice, T., & Papagno, C. (Eds.). (2019). Impairments of short-term memory buffers: Do they exist? [Special issue] *Cortex*, *112*, 1–181.

Shallice, T., & Butterworth, B. (1977). Short term memory impairment and spontaneous speech. *Neuropsychologia*, *15*(6), 729–735.

Shallice, T., & Vallar, G. (1990). The impairment of auditory-verbal short-term storage. In G. Vallar, T. Shallice (Eds.), *Neuropsychological impairments of short-term memory* (pp. 11–53). Cambridge University Press.

Siegel, J. S., Ramsey, L. E., Snyder, A. Z., Metcalf, N. V., Chacko, R. V., Weinberger, K., & Corbetta, M. (2016). Disruptions of network connectivity predict impairment in multiple behavioral domains after stroke. *Proceedings of the National Academy of Sciences of the United States of America*, *113*(30), E4367–E4376.

Spinnler, H., Della Sala, S., Bandera, R., & Baddeley, A. D. (1988). Dementia, ageing and the structure of human memory. *Cognitive Neuropsychology, 5*(2), 193–211.

Tree, J. J., & Playfoot, D. (2019). How to get by with half a loop: An investigation of visual and auditory codes in a case of impaired phonological short-term memory. *Cortex, 112*, 23–36.

Tresch, M. C., Sinnamon, H. M., & Seamon, J. G. (1993). Double dissociation of spatial and object visual memory: Evidence from selective interference in intact human subjects. *Neuropsychologia, 31*(3), 211–219.

Vallar, G. (2019). A 'purest' impairment of verbal short-term memory. The case of PV and the phonological short-term store. In S. MacPherson & S, Della Sala (Eds.), *Cases of amnesia: Contributions to understanding memory and the brain* (pp. 261–291). Routledge.

Vallar, G., & Baddeley, A. D. (1984a). Fractionation of working memory: Neuropsychological evidence for a phonological short-term store. *Journal of Verbal Learning & Verbal Behavior, 23*(2), 151–161.

Vallar, G., & Baddeley, A. D. (1984b). Phonological short-term store, phonological processing and sentence comprehension: A neuropsychological case study. *Cognitive Neuropsychology, 1*(2), 121–141.

Vallar, G., & Cappa, S. F. (1987). Articulation and verbal short-term memory: Evidence from anarthria. *Cognitive Neuropsychology, 4*(1), 55–78.

Vallar, G., & Papagno, C. (1986). Phonological short-term store and the nature of the recency effect: Evidence from neuropsychology. *Brain and Cognition, 5*(4), 428–442.

Vallar, G., & Shallice, T. (1990). *Neuropsychological impairments of short-term memory.* Cambridge University Press.

Vandierendonck, A. (2016). A working memory system with distributed executive control. *Perspectives on Psychological Science, 11*(1), 74–100.

Vandierendonck, A. (2021). Multi-component working memory system with distributed executive control. In R. H. Logie, V. Camos, & N. Cowan (Eds.), *Working memory: State of the science* (pp. 150–174). Oxford University Press.

Vataja, R., Pohjasvaara, T., Mäntylä, R., Ylikoski, R., Leppävuori, A., Leskelä, M., Kalska, H., Hietanen, M., Aronen, H. J., Salonen, O., Kaste, M., & Erkinjuntti, T. (2003). MRI correlates of executive dysfunction in patients with ischaemic stroke. *European Journal of Neurology, 10*(6), 625–631.

Warrington, E. K., & Shallice, T. (1969). The selective impairment of auditory verbal short-term memory. *Brain, 92*(4), 885–896.

Waters, G. S., Rochon, E., & Caplan, D. (1992). The role of high-level speech planning in rehearsal: Evidence from patients with apraxia of speech. *Journal of Memory and Language, 31*(1), 54–73.

Wernicke, C. (1874). *Der Aphasische Symptomenkomplex.* Cohn & Weigart.

Wernicke, C. (1906). Der aphasische Symptomenkomplex. In E. Leyden (Ed.), *Deutsche Klinik am Eingang des zwanzigsten Jahrhunderts in akademischen Vorlesungen* (pp. 487–556). Urban & Schwarzenberg.

Willshaw, D. (2006). Self-organization in the nervous system. In R. G. M. Morris, L. Tarrassenko, & M. Kenward (Eds.), *Cognitive systems: Information processing meets brain science* (pp. 5–33). London, UK: Academic Press.

Wilson, B. A., & Baddeley, A. D. (1988). Semantic, episodic and autobiographical memory in a post-meningitic amnesic patient. *Brain and Cognition, 8*(1), 31–46.

Wilson, B. A., Baddeley, A. D., & Kapur, N. (1995). Dense amnesia in a professional musician following herpes simplex virus encephalitis. *Journal of Clinical and Experimental Neuropsychology, 17*(5), 668–681.

Wilson, B. A., Baddeley, A. D., & Young, A. W. (1999). LE, a person who lost her "mind's eye." *Neurocase, 5*(2), 119–127.

Wilson, J. T. L., Scott, J. H., & Power, K. G. (1987). Developmental differences in the span of visual memory for pattern. *British Journal of Developmental Psychology, 5*(3), 249–255.

Zago, S., Lorusso, L., Porro, A., Franchini, A. F., & Cubelli, R. (2015). Between Bouillaud and Broca: An unknown Italian debate on cerebral localization of language. *Brain and Cognition, 99*(1), 87–96.

16

Memory rehabilitation

To what extent does theory influence clinical practice?

Barbara A. Wilson

> In theory there is no difference between theory and practice. In practice there is.
>
> Yogi Berra (2012, http://www.brainyquote.com)

Introduction

Memory rehabilitation is part of cognitive rehabilitation, which in turn is part of neuropsychological rehabilitation. Neuropsychological rehabilitation is concerned with the amelioration of cognitive, emotional, psychosocial, and behavioural deficits caused by an insult to the brain. It is not synonymous with recovery (if by this we mean getting back to what one was like before the injury or illness) and it is not synonymous with treatment (this is something we do to people or give to people such as surgery or drugs). Rehabilitation *is* a two-way interactive process. The main purposes of neuropsychological rehabilitation are to enable people with disabilities to achieve their optimum level of well-being, to reduce the impact of their problems on everyday life, and to help them return to their own most appropriate environments. It is *not* to teach them to score better on tests or learn lists of words or be faster at detecting stimuli (Wilson et al., 2009). Of the cognitive problems experienced by those who have survived an insult to the brain, memory disorders are common. Almost all people with dementia (approximately 10% of people over the age of 65 years) are affected together with some 36% of survivors of traumatic brain injury and some 70% of survivors of encephalitis, not to mention those with hypoxic brain damage following cardiac or pulmonary arrest, attempted suicide or near drowning, Parkinson's disease, multiple sclerosis, AIDS, Korsakoff's syndrome, epilepsy, cerebral tumours, and so forth (Wilson, 2009). Most people seen for rehabilitation, however, are likely to have sustained a traumatic brain injury, stroke, encephalitis, hypoxic brain damage, or some other non-progressive condition. Although at present there is no effective way to restore lost memory functioning, we can help people to compensate for their problems and to learn more efficiently.

Barbara A. Wilson, *Memory rehabilitation* In: *Memory in Science for Society*. Edited by: Robert H. Logie, Zhisheng (Edward) Wen, Susan E. Gathercole, Nelson Cowan, and Randall W. Engle, Oxford University Press. © Oxford University Press 2023. DOI: 10.1093/oso/9780192849069.003.0016

For those with very severe and widespread cognitive difficulties, it may be that the best we can do is to modify, structure, or rearrange the environment to help them manage without a memory.

Have models, theories, and frameworks enhanced our attempts at rehabilitating those who have sustained an organic memory deficit? A theory is 'a supposition or system of ideas explaining something, especially one based on general principles independent of the particular thing to be explained' (Allen, 1990, p. 1266); a model is 'a representation that can help us to understand and predict related phenomena' (Baddeley, 1992); while a framework is a basic structure underlying a system or concept. To what extent can a theory, a model, or a framework help promote our understanding and explanation of memory disorders, the assessment of memory deficits, the design of interventions, and the evaluation of our memory therapy interventions? This is the concern of this chapter. Although it is true that in rehabilitation, models, theories, and frameworks are useful for facilitating thinking about assessment and treatment, for explaining deficits to therapists, relatives, and patients, and for enabling us to conceptualize outcomes, they are not the only ways to help and some approaches have developed without theories or models through observations and common sense, as we will see.

To what extent have theories and models helped to explain memory disorders?

Although we may talk about having a good, poor, or photographic memory as if it were one thing, one skill, or one system only, there are, of course, several ways of understanding and explaining memory disorders. Bergson, for example, pointed out way back in 1896 (translated by Paul & Palmer, 1896/1994), that memory is not a store but a selection machine. In fact, memory deficits in patients can be classified in a number of ways including the amount of time for which information is stored, the type of information stored, the type of material to be remembered, the modality being employed, the stages involved in the memory process, conscious and unconscious recall (explicit and implicit memory), recall and recognition, retrospective and prospective memory, and anterograde and retrograde amnesia. When attempting to understand the way memory can break down, models and theories have proved to be really helpful. Squires and Knowlton (1995), for example, in their 'Systems model of long-term memory and associated brain structures' have addressed these issues. For clinicians trying to understand the difficulties experienced by patients, we can think of time-based memory as described by Baddeley and Hitch's working memory model (1974; and updated by Baddeley in 2000); or we can consider semantic and episodic memory, in other words memory for general knowledge or for personal experiences, which has been influenced by Tulving's (1972) theory of memory. Modality-specific memory (having problems with verbal or non-verbal memory) has been known for many years (Milner, 1965, 1968). Problems with encoding, storage, or retrieval (the stages involved in memory) is yet another way to understand the breakdown of memory (e.g. Melton, 1963). We can also contemplate whether our patients have problems with conscious or unconscious recall: that is to say, with explicit or implicit memory, as proposed by Warrington and Weiskrantz (1968) and with Glisky and Schacter (1988) being among those who have influenced the field. Yet another way to understand the difficulties experienced by memory-impaired patients is to think about recall and recognition (Kintsch, 1968); or retrospective (remembering past events) versus prospective memory (remembering to do things), discussed by Baddeley (2004); or finally by distinguishing between anterograde (remembering things which have happened since the brain insult), and retrograde amnesia (remembering things that happened before the brain insult), some of the main proponents being Ribot (1881), Kapur (1993, 1999), and Markowitsch (2003). Wilson (2009) discusses these issues in greater detail.

To end this section, it can be acknowledged that the great variety of models and theories of memory discussed here have greatly helped us to pinpoint the nature of impairment manifested by a memory-impaired individual following damage to the brain.

To what extent have theories and models helped in the assessment of memory deficits?

Before planning treatment for someone with memory difficulties, a detailed assessment should take place, which should include a formal neuropsychological assessment of all cognitive abilities including memory in order to build up a picture of a person's cognitive strengths and weaknesses. In addition, assessment of emotional and psychosocial functioning should be carried out. Standardized tests should be complemented with observations, interviews, and self-report measures. The main theoretical approaches to assessment come from both neuropsychological testing and behavioural psychology. Approaches to assessment from neuropsychological procedures include (1) psychometric assessments, (2) localization considerations, (3) theoretical models from cognitive psychology, (4) exclusion models, and (5) ecological models. Behavioural assessments include (1) observations, (2) self-report measures, and (3) interviews.

Let us look at each of these in a little more detail. Psychometric assessments are those which are based on statistical analysis, have established a procedure for administration, collected norms from a representative sample, developed a scoring procedure, and determined the reliability and validity of the test. An example in memory assessment, is the Wechsler Memory Scale Fourth Edition (Wechsler, 2008), which, it could be argued, is a framework rather than a theory or a model. While, in very general terms, this test can discover whether a person is likely to have problems or not and may tell us if a person is better at verbal or perceptual tasks, it does not inform us as to *what* particular problems a person needs help with.

At one time, localization considerations were the main purpose of a neuropsychological assessment. This is less true now with the advent of brain imaging although tests of frontal lobe damage are still widely used (Stuss & Knight, 2013); and those interested in anatomy may still use tests of localization (Jones-Gotman, 1991).

Some theoretical models of cognitive functioning have had a great influence in recent assessment procedures, particularly models of language and reading such as the Psycholinguistic Assessment of Language Processing in Aphasia (PALPA: Kay et al., 1992). In 1985, Coltheart published his dual route model of reading (Coltheart, 1985) which revolutionized our assessment of reading disorders (Wilson et al., 1994). Theories of memory, perception, and attention have also led to new assessments. The Doors and People Test of visual and verbal recall and recognition (Baddeley et al., 1994) was developed from the Baddeley and Hitch (1974) working memory model.

One of the earliest approaches to neuropsychological assessment was the exclusion model whereby one needed to exclude other possible explanations for the disorder (e.g. Liepmann, 1900/1977; Lissauer, 1890). Thus, in order to diagnose 'apraxia', it was necessary to exclude paralysis, weakness, and poor comprehension as explanations for the movement disorder. Similarly, in order to diagnose 'visual object agnosia', one has to exclude poor eyesight, naming difficulties, and so forth. Exclusion can also be used indirectly: for example, one is sometimes asked if memory problems can be explained by anxiety, poor attention, or depression and it is necessary to exclude these or recognize that they cannot be excluded.

Finally, in the taxonomy of neuropsychological tests, there are the ecological measures, which are assessments designed to predict everyday, real-life problems, usually not well delineated by poor scores on a traditional test. Examples of ecological tests include the Rivermead Behavioural Memory Test (Wilson et al., 1985), the Test of Everyday Attention (Robertson et al., 1994), the Behavioural Assessment of the Dysexecutive Syndrome (Wilson et al., 1996), and the Wessex Head Injury Matrix (Shiel et al., 2000).

Standardized neuropsychological tests have relied on theories, models, and frameworks in their development and are useful at answering some of the questions we need to answer in our assessments such as 'How does this person compare with others of the same age?' or 'What kind of memory deficits does this person have?' They are less good, however, at answering many treatment-related questions such as 'What is most distressing for this person?' or 'How do the problems manifest themselves in everyday life?' In order to answer these latter questions we need to turn to the functional or behavioural assessment procedures such as observations, interviews, and self-report measures. It could be argued that these are not theoretically driven and come about through observations and talking to patients and their families. However, behavioural theories have influenced these approaches too. As Wilson (1987) pointed out, the distinction between behavioural assessment and behavioural treatment is artificial as it is difficult to decide where one ends and the other begins. Until the late 1960s, most behavioural assessments were informal as the proponents such as Hersen and Bellack (1978) were more concerned with treating problems. Since the late 1960s, however, behavioural assessments have become more formalized (Baer et al., 1968; Kanfer & Saslow, 1969).

So, what models from behavioural assessment have influenced memory assessment procedures? An early, simplistic model was the stimulus–response model which argued that behaviour (in our case forgetting and remembering behaviour) is influenced both by environmental factors and organismic factors (Mischel, 1968). This is, perhaps, not very helpful when determining the nature of everyday problems in memory-impaired people. More relevant for our purposes is the SORC model of Kanfer and Saslow (1969). This stands for Stimulus, Organism, Response, and Consequences. Wilson (2009) discusses the behavioural aspects in more detail.

In short, when considering to what extent theories and models have influenced our assessment of memory deficits, it is probably true to say that standardized neuro-psychological procedures have indeed been influenced by them but it is less clear with the functional or behavioural approaches which rely, in part, on such models as the SORC model, and also on direct observation which is obviously less model or theory based.

To what extent have theories and models helped rehabilitation programmes aimed at reducing memory problems in everyday life?

Although there is slight evidence that it might be possible to achieve some degree of restoration of working memory in children with attention deficit hyperactivity disorder (Klingberg, 2006; Klingberg et al., 2005), this has been disputed (Melby-Lervåg et al., 2016). Furthermore, no evidence exists for recovery of episodic memory in survivors of brain injury. Thus, restoration of memory functioning is currently an unrealistic goal. The typical person referred for memory rehabilitation is young, most likely to be male, and will have sustained a traumatic brain injury. In addition to memory, they are likely to have problems with attention, planning, and organization. They will probably have emotional problems such as anxiety, depression, and mood swings and may have behaviour problems such as poor self-control and verbal aggression. The patient may want to return to work or may have tried to return and failed. The family is likely to need help and, after a few months, the young person's friends may well start to drift away leaving him or her socially isolated. All these problems may need addressing in rehabilitation.

How can we help and what theories, models, and frameworks have been useful? Diller (1976), in New York, was the first person to set up a programme called 'Cognitive Rehabilitation'. A programme like this would give patients a set of exercises to work through. The rationale being that this would remediate the underlying deficit or teach the patient how to deal with cognitive problems. Not only did early programmes follow this approach, some still believe this is the way to carry out cognitive rehabilitation (Oltra-Cucarella, 2013). This approach shows little evidence of achieving its goal of improving cognitive function (Ponsford et al., 1995; Robertson, 1990), although, of course, it is usual for people to improve on the tasks they practise. There has been some success with attention (Cicerone et al., 2011) and language disorders (Tallal et al., 1998). To date, however, no studies have improved *episodic* memory functioning through exercises or a training regime. Those following the exercise approach, typically do not address the emotional, social, and behavioural consequences of cognitive impairment caused by an insult to the brain nor do they, as a rule, plan for generalization or transfer of learning to the real world.

In the late 1970s, there were very few published papers or chapters on memory rehabilitation and those few that existed tended to teach lists of words, which is not what memory-impaired people need to know. Some tried to teach mnemonics (e.g. Gianutsos & Gianutsos, 1979) in the hope that this would generalize to real-life difficulties but none were specifically concerned with reducing the impact of the problems in everyday life. Wilson (1981) published her first paper on memory rehabilitation addressing the problem of teaching people's names to a man who had a left temporal lobe tumour removed. This man was embarrassed that he could not recall the names of one of his neighbours as well as the staff and patients at the rehabilitation centre he attended. He chose the names he wanted to learn and a visual imagery procedure was used to teach them. A single-case experimental, multiple-baseline-across-behaviours design was employed (discussed in a little more detail later) to determine whether the approach was successful. It was, and the man learned all ten names selected despite learning almost none during the baseline periods.

Although, as far as possible, one should address real-life problems, there are occasions when it is necessary to ask a particular question which has to be answered in a more traditional experimental way (see Baddeley & Wilson, 1994). However, even then, the results should be applicable to real-life problems. An example of this is 'errorless learning' (EL) which is described later in the section on improving learning.

Within the field of memory rehabilitation, there would appear to be three main approaches we can employ to help improve daily functioning. The first is to modify the environment in order to reduce cognitive demands (e.g. labelling drawers and cupboards so people do not have to remember what is contained in them; another example would be to avoid a particular phrase that triggers a repetitive response). The second main approach is to teach or encourage the use of a compensatory memory aid such as a wall calendar or personal organizer (this is one of the most productive approaches to memory rehabilitation and the most likely to lead to greater independence); the third approach is to help people learn more efficiently though vanishing cues, spaced retrieval, or EL (see Wilson, 2009, for full discussion of each of these methods). Currently, it could be argued that external aids and methods for improving learning have been most beneficial in the rehabilitation of memory deficits. Models and theories for environmental modifications do not appear to exist. Instead, common sense and observations seem to offer the most effective solutions.

This approach is used mostly for people with very severe and widespread problems. For these, our only hope of improving quality of life and giving them as much independence as possible is probably to organize or structure the environment so they can function without the need for memory. People with severe memory problems may not be handicapped in environments where there are no demands made on memory. If doors, closets, drawers, and storage jars are clearly labelled, if rooms are cleared of dangerous equipment, and if someone appears to remind or accompany the memory-impaired person when it is time to attend a session or to have a meal, the person may cope reasonably well.

The environmental approach can be used for those without severe problems, too. Hospitals and nursing homes, in addition to other public spaces such as shopping centres, may use colour coding, signs, and other warning systems to reduce the chances of getting lost. 'Smart houses' are already in existence to help 'disable the disabling environment' described by Wilson and Evans (2000). Some of the equipment used in smart houses can be employed to help the severely impaired memory patient survive more easily. For example, 'photo phones' are telephones with large buttons (available from services for visually impaired people), each button can be programmed to dial a particular number and a photo of the person who owns that number can be pasted on to the large button. So, for example, if a memory-impaired person wants to dial her daughter or her district nurse, she simply presses the right photograph and the number is automatically dialled.

People with very severe memory difficulties and widespread cognitive problems may be unable to learn compensatory strategies and have major difficulties learning new episodic information. However, they may be able to learn things implicitly. CW, for example, the musician who has one of the most severe cases of amnesia on record (Wearing, 2005; Wilson, Kopelman, & Kapur, 2008) is unable to lay down new episodic memories but has learned certain things implicitly. Thus, if he is asked 'Where is the kitchen?' he says he does not know but if asked if he would like to go and make himself some coffee, he will go to the kitchen without error. He has implicit but no explicit memory of how to find the kitchen. For people with very severe problems such as CW, our only hope of improving quality of life and giving them as much independence as possible is probably to organize or structure the environment so they can function without the need for memory. As mentioned above, this approach does not, so far, appear to be influenced by models, theories, or frameworks.

A second main strategy, teaching or encouraging the use of external memory aids, is known as the compensatory approach. People with memory problems may forget to use memory aids, and with electronic aids they may have difficulty programming them; they may use aids in an unsystematic or disorganized way and they may be embarrassed by them. The people who need memory aids the most have the greatest difficulty in using them because the use of a memory aid involves memory. Nevertheless, some memory-impaired people use compensatory aids and strategies efficiently; age, severity, widespread cognitive problems, and premorbid use of aids may play a part (Evans et al., 2003; Kime et al., 1996).

One of the best evaluated of the compensatory electronic memory aids is a paging system called NeuroPage developed by an American neuropsychologist and a British engineer living in California (Hersch & Treadgold, 1994). NeuroPage works by programming reminder messages into a central computer which are then sent to the client's pager at fixed times of the day. It was developed for Treadgold's adult son who sustained a severe traumatic brain injury in a road traffic accident and who needed to return to college. It was evaluated in the UK soon after it was first developed.

Following a pilot study (Wilson et al., 1997), a randomized controlled trial (RCT) was carried out (Wilson et al., 2001). People were randomly allocated to the pager first or to the waiting list first. Most of the 143 participants who completed the trial had memory or planning deficits due to acquired brain injury. Each chose what target behaviours they wanted to work on. Baselines were taken and a randomized control crossover design was used to monitor success. A relative or carer also monitored whether the target behaviour was achieved or not. The results showed that less than 50% of the target activities were completed without the pager and this rose to 76% when reminders were sent via the pager. More than 80% of the participants were more successful at carrying out the tasks with the pager than without. Thus, NeuroPage significantly reduced the everyday failures of memory and planning in people with brain injury. This study also demonstrated that it is possible to combine scientific methodology and clinical relevance. Did this very successful memory compensation rely on theory? It would seem not, as the engineer father and neuropsychologist who developed it simply worked out what was needed to help one particular memory-impaired person. Even though Hersch and Treadgold may not have been influenced by theory, one theoretical model has played a part in compensatory behaviour as we see later.

In 2002, Wilson and Evans discussed the cost implications of NeuroPage. They showed that, for at least some of the clients, NeuroPage saved money for health and social services.

Particularly pleasing was that the 2001 study influenced clinical practice as the local health authority set up a clinical service for people throughout the UK and in 2003 a report on the first 40 clients using this service was published (Wilson et al., 2003). A 10-year follow-up study showing how the service changed was also published (Saez et al., 2011). The main changes were (1) from 2007 onwards, people could choose to receive their messages via their mobile telephones (the younger people with traumatic brain injury tended to choose mobile phones whereas the older people with other diagnoses tended to choose the pager); and (2) although messages to do with medication were by far the most frequent in both the 2003 and 2011 studies, there were differences in the types of messages sent. In the later study, more messages were sent regarding cognitive rehabilitation in general, as well as being more targeted for safety and managing mood problems.

External memory aids are now widely available, can be very inexpensive, and have the potential to be highly effective in compensating for prospective memory problems (see Craik & Henry, Chapter 13, this volume). There can be difficulty in both learning and remembering to use electronic reminders, but even in densely amnesic patients, electronic devices can sometimes be used to aid prospective memory. Can theoretical models influence the development of new compensatory aids? On one level, one could argue they are driven more by observation and common sense, but there is a helpful theoretical framework proposed by Bäckman and Dixon (1992) who distinguish four steps in the evolution of compensatory behaviour: (1) origins, (2) mechanisms, (3) forms, and (4) consequences. Although this framework is useful in understanding compensation in neurologically impaired adults, other factors need to be taken into account. Wilson and Watson (1996) considered the framework and showed that variables such as age, severity, and premorbid use of aids, needed to be taken into account when predicting independence and use of compensations several years post rehabilitation.

A review of electronic aids (de Joode et al., 2010) pointed out that the NeuroPage study was the only RCT (at that time) to have been carried out with electronic aids. Furthermore, it is one of the few studies to focus on real-life everyday targets selected by patients rather than using experimental or laboratory-type material such as remembering hypothetical goals. In 2011, another study, albeit with far fewer patients ($N = 12$), was completed with Google Calendar (McDonald et al., 2011). This also used patient-selected targets. There is plenty of evidence for the success of compensations for reducing everyday problems in memory-impaired people including a television assisted prompting device, mobile phones, smart phones, Google Calendar, wearable cameras, and so forth. Jamieson et al. (2014) published a systematic review of external memory aids. A more recent study has been published by Jamieson et al. (2017). Kapur et al. (2019) discuss some recent apps and voice assistant technology. Jamieson et al. (2022) describe the development of a simplified mobile phone for memory-impaired people. With the rapid development of technology, more and more apps are likely to appear. The use of compensatory aids is an area where it is harder to see the influence of theories and models: on the one hand, many of the compensations are driven by observation and common sense but on the other hand, there is Bäckman and Dixon's (1992) theoretical model (or framework), which lists four steps necessary for compensatory behaviour to occur.

Another method to improve learning is spaced retrieval, also known as expanded or expanding rehearsal (Landauer & Bjork, 1978). This strategy involves the presentation of material to be remembered, followed by immediate testing, then a very gradual lengthening of the retention interval. Spaced retrieval may work because it is a form of distributed practice, that is, distributing the learning trials over a period of time rather than massing them together in one block. Distributed practice is known to be more effective than massed practice (Baddeley, 1999). The method has been used to help people with traumatic brain injury, stroke, encephalitis, and dementia. The combination of spaced retrieval and EL which is discussed below, would appear to be a powerful learning strategy for people with progressive conditions in addition to those with non-progressive conditions (Wilson, 2009).

Those of us without severe memory difficulties can benefit from trial-and-error learning. We are able to remember our mistakes and thus can avoid making the same mistake in future attempts. As memory-impaired people have difficulty with this, any erroneous response may be strengthened or reinforced. This is the rationale behind EL, a teaching technique whereby the likelihood of mistakes during learning is minimized as far as possible. Errors can be avoided through the provision of spoken or written instructions, guiding someone through a particular task, or modelling the steps of a procedure little by little. There is now considerable evidence that EL is superior to trial-and-error learning for people with severe memory deficits. In a meta-analysis of EL, Kessels and de Haan (2003) found a large and statistically significant effect size of this kind of learning for those with severe memory deficits. As mentioned above, the combination of EL and spaced retrieval would appear to be a powerful learning strategy for people with progressive conditions in addition to those with non-progressive conditions (Wilson, 2009).

The principle of EL is, today, an important component of memory rehabilitation. It grew from two distinct theoretical backgrounds. First, was errorless discrimination learning from behavioural psychology (Terrace, 1963, 1966) which was soon taken up in developmental learning disability (Cullen, 1976; Sidman & Stoddard, 1967; Walsh & Lamberts, 1979). The second impetus came from implicit learning (or learning without conscious recollection) from cognitive psychology (e.g. Brooks & Baddeley, 1976; Graf & Schacter, 1985). We know that people with amnesia *can* learn normally or nearly normally under some circumstances but that anomalies are sometimes seen. Patients may, for example, 'get stuck' on an incorrect response. After considering these anomalies, Baddeley and Wilson (1994) posed the question 'Do people with amnesia learn better if prevented from making mistakes while learning?' An experiment was carried out in order to answer this question. Participants were given a stem completion task; there were three groups of people (young participants, elderly participants, and people with very severe memory impairment). Both the young and elderly control participants did a little better under the EL condition but every single one of the 16 densely amnesic people did better when prevented from making mistakes. The conclusions to this study were (1) EL was more effective than errorful learning; (2) this advantage was greater for the people with amnesia; (3) the amnesic people showed less forgetting under EL; and (4) we should not ask people with amnesia to guess. The results were so striking that it immediately influenced clinical practice. We should never ask people with amnesia or severe memory problems to guess unless they are being given a test where it is important for them to guess. One could instead say, for example, 'Only tell me if you are sure'. Thus, this method of improving learning is influenced by two theoretical backgrounds. However, although this worked experimentally, the next step was to discover if the principle of EL could be applied to real-life tasks. Wilson et al. (1994) demonstrated that patients with severe memory impairment could learn some practical tasks using EL principles. They also showed that when employing such principles, it was necessary to ensure active participation and incorporate other approaches from learning theory and memory rehabilitation such as spaced retrieval and learning one thing at a time.

Since the mid 1990s, many people have used EL to teach several everyday tasks to people from different diagnostic groups, of different ages, and at different times post insult. EL is superior to trial-error learning for people with severe memory problems (Wilson, 2009). It is less clear whether it is effective for people with language problems (although see Lambon Ralph & Fillingham, 2007). It would appear that although people with language problems seem to benefit from both approaches, they prefer the EL approach. Conroy and Lambon Ralph (2012) discuss this issue in more detail.

Why does EL work? Baddeley and Wilson (1994) believed that it depended on implicit memory and this system is poor at eliminating errors: episodic memory is the system which does this. In order to benefit from our mistakes, we need to be able to remember them. So if people whose episodic memory is almost non-existent (and who only have implicit memory working for them) make an incorrect response, this response may be strengthened. Implicit memory has no way of selecting a correct from an incorrect response. There has been some debate in the literature as to whether EL depends on explicit or implicit memory. Baddeley and Wilson (1994) argued that implicit memory was responsible for the efficacy of such learning: amnesic patients had to rely on implicit memory, a system which is poor at eliminating errors (this is not to say that EL is a *measure* of implicit memory). Nevertheless, there are alternative explanations. For example, the EL advantage could be due to residual explicit memory processes or to a combination of both implicit and explicit systems. Hunkin et al. (1998) argued that the benefit of EL is due entirely to the effects of error prevention on the residual explicit memory capacities, and not to implicit memory at all. Tailby and Haslam (2003) also believe that the benefits of EL are due to residual explicit concurrent memory processes although they do not rule out implicit memory processes altogether. They say the issue is a complex one and that different individuals may rely on different processes. Support for this view can also be found in a paper by Kessels et al. (2005).

Page et al. (2006) claim, however, that preserved implicit memory in the absence of explicit memory is sufficient for EL to occur. They challenge the Hunkin et al. (1998) paper's conclusions as the design of their implicit task was such that it was unlikely to be sensitive to implicit memory for prior errors. Furthermore, there was an element of errorful learning in both the errorless and errorful explicit memory conditions. They also challenge the Tailby and Haslam (2003) paper as it conflates two separate questions. First, whether or not the *advantage* of EL is due to the contribution of implicit memory; and second, whether *learning* under errorless conditions is due to implicit memory? Perhaps some people do use both implicit and explicit systems when *learning* material but this does not negate the argument that the *advantage* of EL is due to implicit memory, particularly implicit memory for prior errors following errorful learning. This point was made clear in the Page et al. (2006) paper. Some people with no or very little explicit recall can learn under certain conditions such as EL. For example, the Baddeley and Wilson (1994) study included 16 very densely amnesic participants with extremely little explicit memory capacity, yet nevertheless, every single one of them showed an errorless over errorful advantage.

Another model which has proved useful in helping to understand why EL is so effective is the Hebbian learning rule (Hebb, 1949). This rule says that 'Neurons that fire together wire together'. Thus, if an input elicits a pattern of neural activity, then Hebbian learning will strengthen the tendency to activate the same pattern on subsequent occasions.

This means that learning will increase the likelihood of making the same response in the future, whether correct or incorrect (McClelland et al., 1999). So if one sees BR and guesses BRING, one is more likely to say BRING the next time BR is seen.

The conclusion to this section on improving learning is that vanishing cues, spaced retrieval, and EL have all been heavily influenced by theories and models.

To what extent have models and theories helped us to plan our rehabilitation?

In 1993, Caramazza and Hillis wrote a paper entitled 'For a theory of remediation of cognitive deficits'. They said they were not concerned with the question of whether cognitive models are helpful in rehabilitation for 'surely they are, it is hard to imagine that efforts at therapeutic intervention would not be facilitated by having the clearest possible idea of what needs to be rehabilitated' (p. 218). This poses the question: do models of cognitive functioning really tell us 'what needs to be rehabilitated'? They are certainly helpful in identifying cognitive strengths and weaknesses and in explaining phenomena and in making predictions about behaviour. They tell us what the cognitive constraints are on any programmes we wish to implement but, it is suggested, they do not tell us 'what needs to be rehabilitated'. Does this mean theories of cognitive functioning are not important? Of course not. We have already seen plenty of examples of how models and theories have helped our clinical practice. The influential working memory model of Baddeley and Hitch (1974) enables us to understand why a patient can have a normal immediate memory with major problems after a delay or distraction; why we might see a difference between visual and verbal memory; and that problems with executive functioning can exist despite a normal phonological loop and visuospatial sketchpad.

Once we have established the 'theoretical problem' faced by an individual, for example, a damaged central executive or an inefficient delayed verbal memory, we are unlikely to set out to rehabilitate the central executive or the delayed verbal memory system. Instead, we would aim to treat accompanying impairments such as 'problems planning a meal' or 'failure to use a notebook'. These are the kinds of things that 'need to be rehabilitated'. 'What needs to be rehabilitated' is *not* an impairment identified by a theoretically informed model but a real-life problem identified by the patient and his or her family. Such problems may well be caused by cognitive deficits but in most cases we do not try to rehabilitate the deficit so much as the everyday problems seen as important by the patient/family.

One attempt to produce a model to help with the understanding of the complexities of rehabilitation and to help us design intervention programmes, is Wilson's (2002) overarching model entitled 'Towards a comprehensive model of cognitive rehabilitation'. This was published in an endeavour to refute those who believed that in order to do good cognitive rehabilitation, one simply needed one good model of the cognitive problem to be treated (Coltheart, 1991). In addition to a good model of cognitive functioning, one needs to understand many other points such as what does the patient find distressing, are there emotional and behavioural difficulties to be addressed, what was the patient's premorbid personality, and how do we know if we have succeeded? Wilson (2002) tried to include all of these issues and this framework was later adapted for use with neuropsychiatric patients (Loschiavo-Alvares et al., 2018). There are other models and theories of neurorehabilitation but the Wilson (2002) model is probably the most comprehensive.

In 2008, Wilson, Rous, and Sopena interviewed 54 psychologists working in brain injury rehabilitation for adults in the UK and asked which models most influenced their work. In total, 57 different models were reported. Because different terms were sometimes used to describe the same model (e.g. 'neuropsychology' and 'cognitive neuropsychology'), the models and theories were grouped into eight categories. Published literature and discussions with other psychologists were used to determine the categories. The top models and theories reported as most influencing clinical practice show considerable overlap with those described as being most important by others (Prigatano, 1987; Wilson, 1987). Cognitive behaviour therapy was endorsed more than any other single model. Indeed, this is one of the most influential theories of emotion. Systemic and family therapy models also come under theories of emotion as do a number of others. The frequency of references to emotion probably reflects the importance of these in the generic clinical psychology training completed by all participants. Cognitive and neuropsychological models come in second place and in this category are more specific cognitive neuropsychological models such as Baddeley and Hitch's (1974) 'working memory model' and 'the supervisory attentional system model' (Norman & Shallice, 1986). Behavioural models and theories including those of learning come third. Again, this is likely to reflect the fact that clinicians use what they are comfortable with and apply this to the different fields in which they work. Other models mentioned included holistic and other broad-based treatments; health, general adjustment, and coping; anatomy and neuroplasticity; development, such as life development and child development; and finally other models, mentioned once only, including organizational models, risk assessment, insight, and ecotherapy (an approach believing that people have a deep connection to their environment and to the earth itself). It should be made clear that not everyone understood what a model or theory was and some reported a therapeutic approach rather than a model or theory. This was accepted for the purposes of the interview.

Interviewees made comments about models and theories such as 'We can't meet patients' needs without drawing on a number of models' and 'We need to develop services with a strong neuropsychological theory'. Another said 'This has been thought provoking and made me think about what I do'.

The study was also criticized, however, because of having too many models. One can ask if this is, in fact, a bad thing? In assessment, psychologists might refer to the working memory model, or models of reading, attention, language, and perception, for example. While in treatment, we tend to draw on a number of models and theories for ideas and understanding, such as cognitive behaviour therapy, learning, plasticity, family therapy, and so forth. We also need to know about models and theories of recovery and evaluation among others, so perhaps one should not criticize the fact that psychologists working in brain injury rehabilitation use many models in their clinical practice. Of course, the results were not necessarily an accurate record of clinical practice but of the participant's *perception* of what they did.

In short, theories and models would appear to have a big influence on helping us to plan our rehabilitation.

To what extent have theories and models helped to evaluate memory rehabilitation?

One of the problems we constantly face in rehabilitation is determining whether or not we are succeeding in our efforts. Because of the great heterogeneity of patients receiving rehabilitation and because of the variety of aims and methods required to achieve success, the measurement of treatment effectiveness is difficult to evaluate (Hart et al., 2008). As I have argued elsewhere (Wilson, 2009), there is no point in doing rehabilitation if it is not effective. In the medical world, RCTs are considered the gold standard for determining the efficacy of any treatment. In rehabilitation, RCTs are possible but they are not easy and need to be carefully thought out. Wilson et al. (2001) reported an RCT. This was not, however, a double-blind, or even a single-blind RCT. It is impossible to do a double-blind RCT in rehabilitation where neither the person giving nor the person receiving the treatment knows whether 'real' treatment or a placebo control has been administered. Psychologists and therapists *have* to know what they are doing. It *is* possible to carry out a single-blind RCT where one party is blind to what is happening. Some studies now use a 'blind' assessor to rate a patient's behaviour and the assessor does not know whether the patient has been treated or not (see Clare et al., 2010). There is, however, an increasing recognition that RCTs are *not* the only way to evaluate rehabilitation. Andrews (1991) says the RCT 'is a tool to be used not a god to be worshipped'. He continues by saying the RCT is excellent where (1) the design is simple, (2) marked changes are expected, (3) the factors involved are relatively specific, and (4) the number of additional variables likely to affect the outcome are few and can be balanced out. This is completely unlike the situation in rehabilitation. We can use a within-subjects design where each patient is seen under two or more conditions or we can use a group study where we are looking at two or more groups under the same conditions. This is the design used in the Baddeley and Wilson (1994) study when we compared young non-brain damaged participants, elderly non-brain damaged participants, and densely amnesic participants. Frameworks for evaluation such as RCTs and other group designs can sometimes be applied to the evaluation of rehabilitation programmes.

Frequently, however, we need to know whether an individual patient is benefiting from our intervention and this is where we can use one of the single case experimental designs. Single case experimental designs were mentioned above. They came originally from behavioural psychology (Nelson & Hayes, 1979) and are very valuable in evaluating an individual's response to treatment. For every patient we see, we should ask ourselves, 'Is this patient changing and, if so, is the change due to what we are doing (or have done) or would it have happened anyway?' Group studies can answer questions about groups (e.g. 'How many patients with amnesia benefit from an EL approach?') but if we want to know whether the man or woman or child we are treating is benefiting from this approach, we have to look at the response of that particular person. The value of single case experimental designs is that they allow us to evaluate an individual's response to treatment, to see if there is change over time, and to find out whether any changes are due to natural recovery or to the intervention itself. In other words, we can tease out the effects of treatment from the effects of spontaneous recovery and other non-specific factors. Given that rehabilitation is planned for individuals, evaluation should take place at the individual as well as the group level.

More recently, the importance of single case experimental designs has been increasingly recognized. This is, in large part, due to the work of Tate, Perdices, and others in Sydney, Australia. In 2008 and 2009, they published a scale for evaluating the methodological quality of single case experimental designs (Perdices et al., 2009; Tate et al., 2008) which lists 11 areas which should be included or addressed in any single case experimental designs. This was later refined and extended to 14 areas (Tate et al., 2011). Then in 2016, Tate et al. published the SCRIBE paper. The acronym stands for the Single-Case Reporting guidelines In BEhavioural interventions 2016. The paper describes 26 items to guide and structure the reporting of single case experimental designs research. A rationale is provided for each item. In addition, examples from the literature are included, again for each item. The authors recommend that these items are used by authors, reviewers, and editors considering manuscripts for publication. Thus, a methodology for judging single case experimental designs is provided. The senior author, Tate, was able to publish the paper in ten journals in the same year, an extraordinary feat in anyone's eyes! Although this cannot be described as a theory or a model, it *is* a framework which provides a structure for evaluating an individual's response to treatment. Once again, we can argue that models, theories, and frameworks have helped us to evaluate our treatment and rehabilitation programmes.

Conclusion

Rehabilitation should focus on improving aspects of everyday life. In addition to emotional and psychosocial problems, neuropsychologists in rehabilitation are likely to treat cognitive difficulties including memory deficits. The main purposes of rehabilitation are to enable people with disabilities to achieve their optimum level of well-being, to reduce the impact of their problems in everyday life, and to help them return to their own most appropriate environments. It is *not* to teach them to score better on tests, to learn lists of words, or to be faster at detecting stimuli. We need to refer to several models and theories to understand, assess, treat, and evaluate our work with impaired people. In particular, models and theories of cognitive functioning, learning, assessment, emotion, and recovery have proved useful. Observations of patients' needs and abilities as well as common sense also play a part in the rehabilitation of such people. This would appear to be particularly true in behavioural assessment and environmental modifications.

Memory disorders can be classified in a number of ways including the amount of time for which information is stored, the type of information stored, the type of material to be remembered, the modality being employed, the stages involved in the memory process, explicit and implicit memory, recall and recognition, retrospective and prospective memory, and anterograde and retrograde amnesia. Here, models and theories have been fundamental.

Before planning treatment for someone with memory difficulties, a detailed assessment should take place. This should include a formal neuropsychological assessment of all cognitive abilities including memory in order to build up a picture of a person's cognitive strengths and weaknesses. In addition, assessment of emotional and psychosocial functioning should be carried out. Standardized tests should be complemented with observations, interviews, and self-report measures. Assessment has been aided both by theories and models and by observations and non-theoretical approaches.

Once the assessment has been carried, out one can design a rehabilitation programme. One of the major ways of helping people with memory problems cope in everyday life is to enable them to compensate through the use of external aids; we can also help them learn more efficiently and, for those who are very severely impaired, we may need to structure or organize the environment to help these people function without a memory. Models and theories have helped most in the development of ways to help people learn more efficiently.

Both the planning and evaluation of the efficacy of memory rehabilitation have also been aided by models, theories, and frameworks. The Wilson (2002) framework to help planning, and evaluation has been helped both by group studies and by single case experimental designs.

References

Allen, R. E. (Ed.). (1990). *The concise Oxford dictionary of current English* (8th ed.). Clarendon Press.

Andrews, K. (1991). The limitations of randomized controlled trials in rehabilitation research. *Clinical Rehabilitation, 5*(1), 5–8.

Bäckman, L., & Dixon, R. A. (1992). Psychological compensation: A theoretical framework. *Psychological Bulletin, 112*(2), 259–283.

Baddeley, A. D. (1992). Memory theory and memory therapy. In B. A. Wilson & N. Moffat (Eds.), *Clinical management of memory problems* (2nd ed., pp. 1–31). Chapman & Hall.

Baddeley, A. D. (1999). *Essentials of human memory*. Psychology Press.

Baddeley, A. D. (2000). The episodic buffer: A new component of working memory? *Trends in Cognitive Sciences, 4*(11), 417–423.

Baddeley, A. D. (2004). The psychology of memory. In A. D. Baddeley, M. D. Kopelman, & B. A. Wilson (Eds.), *The essential handbook of memory disorders for clinicians* (pp. 3–15). John Wiley & Sons.

Baddeley, A. D., Emslie, H., & Nimmo-Smith, I. (1994). *The doors and people test: A test of visual and verbal recall and recognition*. Thames Valley Test Company.

Baddeley, A. D., & Hitch, G. J. (1974). Working memory. In G. H. Bower (Ed.), *The psychology of learning and motivation: Advances in research and theory* (Vol. 8, pp. 47–89). Academic Press.

Baddeley, A. D., & Wilson, B. A. (1994). When implicit learning fails: Amnesia and the problem of error elimination. *Neuropsychologia, 32*(1), 53–68.

Baer, D. M., Wolf, M. M., & Risley, T. R. (1968). Some current dimensions of applied behavior analysis. *Journal of Applied Behavior, 1*(1), 91–97.

Bergson, H. (1994). *Matter and memory* (N. M. Paul & W. S. Palmer, Trans.). Zone Books. (Original work published 1896)

Brooks, D. N., & Baddeley, A. D. (1976). What can amnesic patients learn? *Neuropsychologia, 14*(1), 111–22.

Caramazza, A., & Hillis, A. (1993). For a theory of remediation of cognitive deficits. *Neuropsychological Rehabilitation, 3*(3), 217–234.

Cicerone, K. D., Langenbahn, D. M., Braden, C., Malec, J. F., Kalmar, K., Fraas, M., Felicetti, T., Laatsch, L., Harley, J. P., Bergquist, T., Azulay, J., Cantor, J., & Ashman, T. (2011). Evidence-based cognitive rehabilitation: Updated review of the literature from 2003 through 2008. *Archives of Physical & Medical Rehabiltation, 92*(4), 519–530.

Clare, L., Linden, D. E., Woods, R. T., Whitaker, R., Evans, S. J., Parkinson, C. H., van Paasschen, J., Nelis, S. M., Hoare, Z., Yuen, K. S., & Rugg, M. D. (2010). Goal-oriented cognitive rehabilitation for people with early-stage Alzheimer disease: A single-blind randomized controlled trial of clinical efficacy. *American Journal of Geriatric Psychiatry, 18*(10), 928–939.

Coltheart, M. (1985). Cognitive neuropsychology and the study of reading. In M. I. Posner, & O. S. M. Marin (Eds.), *Attention and performance XI* (pp. 3–37). Lawrence Erlbaum Associates.

Coltheart, M. (1991). Cognitive psychology applied to the treatment of acquired language disorders. In P. Martin (Ed.), *Handbook of behaviour therapy and psychological science: An integrative approach* (pp. 216–226). Pergamon Press.

Conroy, P., & Lambon Ralph, M. A. (2012). Overview and ways forward for future research. *Neuropsychological Rehabilitation*, *22*(2), 319–328.

Cullen, C. N. (1976). Errorless learning with the retarded. *Nursing Times*, 72(12, Suppl.), 45–47.

de Joode, E., van Heugten, C., Verhey, F., & van Boxtel, M. (2010). Efficacy and usability of assistive technology for patients with cognitive deficits: A systematic review. *Clinical Rehabilitation*, *24*(8), 701–714.

Diller, L. (1976). A model for cognitive retraining in rehabilitation. *Clinical Psychologist*, *29*, 13–15.

Evans, J. J., Wilson, B. A., Needham, P., & Brentnall, S. (2003). Who makes good use of memory aids? Results of a survey of people with acquired brain injury. *Journal of the International Neuropsychological Society*, *9*(6), 925–935.

Gianutsos, R., & Gianutsos, J. (1979). Rehabilitating the verbal recall of brain injured patients by mnemonic training: An experimental demonstration using single case methodology. *Journal of Clinical Neuropsychology*, *1*(2), 117–135.

Glisky, E. L., & Schacter, D. L. (1988). Long-term retention of computer learning by patients with memory disorders. *Neuropsychologia*, *26*(1), 173–178.

Graf, P., & Schacter, D. L. (1985). Implicit and explicit memory for new associations in normal and amnesic subjects. *Journal of Experimental Psychology: Learning, Memory, and Cognition*, *11*(3), 501–518.

Hart, T., Fann, J. R., & Novack, T. A. (2008). The dilemma of the control condition in experience-based cognitive and behavioural treatment research. *Neuropsychological Rehabilitation*, *18*(1), 1–21.

Hebb, D. O. (1949). *The organization of behavior: A neuropsychological theory*. Wiley.

Hersch, N., & Treadgold, L. (1994). NeuroPage: The rehabilitation of memory dysfunction by prosthetic memory and cuing. *NeuroRehabilitation*, *4*(3), 187–197.

Hersen, M., & Bellack, A. S. (1978). *Behavioral assessment*. Pergamon Press.

Hunkin, N. M., Squires, E. J., Parkin, A. J., & Tidy, J. A. (1998). Are the benefits of errorless learning dependent on implicit memory? *Neuropsychologia*, *36*(1), 25–36.

Jamieson, M., Cullen, B., Lennon, M., Brewster, S., & Evans, J. (2022). Designing ApplTree: Usable scheduling software for people with cognitive impairments. *Disability and Rehabilitation: Assistive Technology*, *17*(3), 338–348.

Jamieson, M., Cullen, B., McGee-Lennon, M., Brewster, S., & Evans, J. J. (2014). The efficacy of cognitive prosthetic technology for people with memory impairments: A systematic review and meta-analysis. *Neuropsychological Rehabilitation*, *24*(3–4), 419–444.

Jamieson, M., Cullen, B., McGee-Lennon, M., Brewster, S., & Evans, J. (2017). Technological memory aid use by people with acquired brain injury. *Neuropsychological Rehabilitation*, *27*(6), 919–936.

Jones-Gotman, M. (1991). Localization of lesions by neuropsychological testing. *Epilepsia*, 32(Suppl. 5), S41–S52.

Kanfer, F. H., & Saslow, G. (1969). Behavioral diagnosis. In C. Frasnks (Ed.), *Behavior therapy: Appraisal and status* (pp. 417–444). McGraw-Hill.

Kapur, N. (1993). Focal retrograde amnesia in neurological disease: A critical review. *Cortex*, *29*(2), 217–234.

Kapur, N. (1999). Syndromes of retrograde amnesia: A conceptual and empirical synthesis. *Psychological Bulletin*, *125*(6), 800–825.

Kapur, N., Watson, C., Parmar, H., & Watts, A. (2019). Voice assistants in neurorehabilitation. *Neuropsychologist*, *8*, 24–27.

Kay, J., Coltheart, M., & Lesser, R. (1992). *Psycholinguistic assessments of language processing in aphasia*. Psychology Press.

Kessels, R. P. C., & de Haan, E. H. F. (2003). Implicit learning in memory rehabilitation: A meta-analysis on errorless learning and vanishing cues methods. *Journal of Clinical and Experimental Neuropsychology*, *25*(6), 805–814.

Kime, S. K., Lamb, D. G., & Wilson, B. A. (1996). Use of a comprehensive program of external cuing to enhance procedural memory in a patient with dense amnesia. *Brain Injury*, *10*(1), 17–25.

Kintsch, W. (1968). Recognition and free recall of organized lists. *Journal of Experimental Psychology*, *78*(3), 481–487.

Klingberg, T. (2006). Development of a superior frontal-intraparietal network for visuo-spatial working memory. *Neuropsychologia*, *44*(11), 2171–2177.

Klingberg, T., Fernell, E., Olesen, P., Johnson, M., Gustafsson, P., Dahlström, K., Gillberg, C. G., Forssberg, H., & Westerberg, H. (2005). Computerized training of working memory in children with ADHD—a randomized, controlled trial. *Journal of the American Academy of Child and Adolescent Psychiatry*, *44*(2), 177–186.

Lambon Ralph, M. A., & Fillingham, J. K. (2007). The importance of memory and executive function in aphasia: Evidence from the treatment of anomia using errorless and errorful learning. In A. S. Meyer, L. R. Wheeldon, & A. Krott (Eds.), *Automaticity and control in language processing* (Advances in behavioural brain science) (pp. 193–216). Psychology Press.

Landauer, T. K., & Bjork, R. A. (1978). Optimum rehearsal patterns and name learning. In M. M. Gruneberg, P. Morris, & R. N. Sykes (Eds.), *Practical aspects of memory* (pp. 625–632). Academic Press.

Liepmann, H. (1977). Das Krankheitsbild der Apraxie ('motorischen Asymbolie') auf Grund eines Falles von einseitiger Apraxie [The syndrome of apraxia (motor asymbolia) based on a case of unilateral apraxia]. Translated by Bohne WHO, Liepmann K, Rottenberg DA from Monatsschrift für Psychiatrie und Neurologie, 1900, 8, 15–44. In D. A. Rottenberg, & F. H. Hochberg (Eds.), *Neurological classics in modern translation*. Hafner Press. (Original work published 1900)

Lissauer, H. (1890). Ein Fall von Seelenblindheit nebst einem beitrage zue Theorie derselben. *Archiv für Psychiatrie und Nervenkrankheiten*, *21*, 220–270. [English translation in Jackson, M. (1988). Lissauer on agnosia. *Cognitive Neuropsychology*, *5*, 155–192.]

Loschiavo-Alvares, F., Fish, J., & Wilson, B. A. (2018). Applying the comprehensive model of neuropsychological rehabilitation to people with psychiatric conditions. *Clinical Neuropsychiatry*, *15*(2), 83–93.

McClelland, J. L., Thomas, A., McCandliss, B. D., & Fitz, J. A. (1999). Understanding failures of learning: Hebbian learning, competition for representational space, and some preliminary experimental data. In J. Reggia, E. Ruppin, & D. Glanzman (Eds.), *Brain, behavioral, and cognitive disorders: The neurocomputational perspective* (pp. 75–80). Elsevier.

McDonald, A., Haslam, C., Yates, P., Gurr, B., Leeder, G., & Sayers, A. (2011). Google Calendar: A new memory aid to compensate for prospective memory deficits following acquired brain injury. *Neuropsychological Rehabilitation*, *21*(6), 784–807.

Melby-Lervåg, M., Redick, T. S., & Hulme, C. (2016). Working memory training does not improve performance on measures of intelligence or other measures of 'far transfer': Evidence from a meta-analytic review. *Perspectives on Psychological Science*, *11*(4), 512–534.

Markowitsch, H. J. (2003). Functional neuroanatomy of learning and memory. In P. W. Halligan, U. Kischka, & J. C. Marshall (Eds.), *Handbook of neuropsychology* (pp. 724–730). Oxford University Press.

Melton, A. W. (1963). Implications of short-term memory for a general theory of memory. *Journal of Verbal Learning and Verbal Behavior*, *2*(1), 1–21.

Milner, B. (1965). Visually-guided maze learning in man: Effects of bilateral hippocampal, bilateral frontal, and unilateral cerebral lesions. *Neuropsychologia*, *3*(4), 317–338.

Milner, B. (1968). Visual recognition and recall after right temporal lobe excision in man. *Neuropsychologia*, *6*(3), 191–209.

Mischel, W. (1968). *Personality and assessment*. Wiley.

Nelson, R. O., & Hayes, S. C. (1979). The nature of behavioral assessment: A commentary. *Journal of Applied Behavior Analysis*, *12*(4), 491–500.

Norman, D. A., & Shallice, T. (1986). Attention to action: Willed and automatic control of behavior. In R. Davidson, R. Schwartz, & D. Shapiro (Eds.), *Consciousness and self-regulation* (Advances in research and theory IV) (pp. 1–18). Plenum Press.

Oltra-Cucarella, J. (2013). Enhancing memory rehabilitation: New approaches for clinicians. In F. Metzger (Ed.), *Neuropsychology: New research* (pp. 79–100). Nova Science Publishers.

Page, M., Wilson, B. A., Shiel, A., Carter, G., & Norris, D. (2006). What is the locus of the terrorless-learning advantage? *Neuropsychologia*, *44*(1), 90–100.

Perdices, M., & Tate, R. L. (2009). Single-subject designs as a tool for evidence-based clinical practice: Are they unrecognised and undervalued? *Neuropsychological Rehabilitation, 19*(6), 904–927.

Ponsford, J., Sloan, S., & Snow, P. (1995). *Traumatic brain injury: Rehabilitation for adaptive everyday living*. Psychology Press.

Prigatano, G. P. (1987). Neuropsychological rehabilitation after brain injury: Some further reflections. In J. M. Williams & C. J. Long (Eds.), *The rehabilitation of cognitive disabilities* (pp. 29–41). Springer.

Ribot, T. (1881). *Les maladies de la memoire [Diseases of memory]*. Appleton-Century-Crofts.

Robertson, I. (1990). Does computerised cognitive rehabilitation work? A review. *Aphasiology, 4*(4), 381–405.

Robertson, I., Ward, T., Ridgeway, V., & Nimmo-Smith, I. (1994). *The test of everyday attention*. Pearson assessment.

Saez, M. M., Deakins, J., Winson, R., Watson, P., & Wilson, B. A. (2011). A ten-year follow up of a paging service for people with memory and planning problems within a healthcare system: How do recent users differ from the original users? *Neuropsychological Rehabilitation, 21*(6), 769–783.

Shiel, A., Wilson, B. A., McLellan, L., Horn, S., & Watson, M. (2000). *The Wessex Head Injury Matrix (WHIM)*. Thames Valley Test Company.

Sidman, M., & Stoddard, L. T. (1967). The effectiveness of fading in programming a simultaneous form discrimination for retarded children. *Journal of the Experimental Analysis of Behavior, 10*(1), 3–15.

Squire, L. R., & Knowlton, B. J. (1995). Learning about categories in the absence of memory. *Proceedings of the National Academy of Sciences of the United States of America, 92*(26), 12470–12474.

Stuss, D. T., & Knight, R. T. (Eds.). (2013). *Principles of frontal lobe function* (2nd ed.). Oxford University Press.

Tailby, R., & Haslam, C. (2003). An investigation of errorless learning in memory-impaired patients: Improving the technique and clarifying theory. *Neuropsychologia, 41*(9), 1230–1240.

Tallal, P., Merzenich, M., Miller, S., & Jenkins, W. (1998). Language learning impairments: Integrating basic science, technology, and remediation. *Experimental Brain Research, 123*(1–2), 210–219.

Tate, R. L., McDonald, S., Perdices, M., Togher, L., Schultz, R., & Savage, S. (2008). Rating the methodological quality of single-subject designs and n-of-1 trials: Introducing the Single-Case Experimental Design (SCED) scale. *Neuropsychological Rehabilitation, 18*(4), 385–401.

Tate, R. L., Rosenkoetter, U., Vohra, S., Kratochwill, T., Sampson, M., Togher, L., Backman, C., Evans, J. J., Manolov, R., Nickels, L., Ownsworth, T., Schmid, C. H., Perdices, M., Shadish, W., Barlow, D. H., Kazdin, A., McDonald, S., Shamseer, L., Albin, R., Douglas, J., Gast, D., Mitchell, G., . . . Wilson, B. A. (2016). The Single-Case Reporting guideline In BEhavioural Interventions (SCRIBE) 2016 statement. *Archives of Scientific Psychology, 4*(1), 1–9.

Tulving, E. (1972). Episodic and semantic memory. In E. Tulving & W. Donaldson (Eds.), *Organization of memory* (pp. 381–403). Academic Press.

Walsh, B. F., & Lamberts, F. (1979). Errorless discrimination and picture fading as techniques for teaching sight words to TMR students. *American Journal of Mental Deficiency, 83*(5), 473–479.

Warrington, E. K., & Weiskrantz, L. (1982). Amnesia: A disconnection syndrome? *Neuropsychologia, 20*(3), 233–248.

Wearing, D. (2005). *Forever today: A memoir of love and amnesia*. Doubleday.

Wechsler, D. (2008). *Wechsler memory scale–IV*. Psychological Corp.

Wilson, B. A. (1981). Teaching a man to remember names after removal of a left temporal lobe tumour. *Behavioural Psychotherapy, 9*(4), 338–344.

Wilson, B. A. (1987). *Rehabilitation of memory*. Guilford Press.

Wilson, B. A. (2002). Towards a comprehensive model of cognitive rehabilitation. *Neuropsychological Rehabilitation, 12*(2), 97–110.

Wilson, B. A. (2009). *Memory rehabilitation: Integrating theory and practice*. Guilford Press.

Wilson, B. A., Alderman, N., Burgess, P., Emslie, H., & Evans, J. (1996). *Behavioural assessment of the dysexecutive syndrome*. Thames Valley Test Company.

Wilson, B. A., Baddeley, A. D., Evans, J. J., & Shiel, A. (1994). Errorless learning in the rehabilitation of memory impaired people. *Neuropsychological Rehabilitation, 4*(3), 307–326.

Wilson, B. A., Cockburn, J., & Baddeley, A. D. (1985). *The Rivermead behavioural memory test*. Thames Valley Test Company.

Wilson, B. A., Emslie, H. C., Quirk, K., & Evans, J. J. (2001). Reducing everyday memory and planning problems by means of a paging system: A randomised control crossover study. *Journal of Neurology, Neurosurgery, and Psychiatry, 70*(4), 477–482.

Wilson, B. A., & Evans, J. J. (2000). Practical management of memory problems. In G. E. Berrios & J. R. Hodges (Eds.), *Memory disorders in psychiatric practice* (pp. 291–310). Cambridge University Press.

Wilson, B. A., & Evans, J. J. (2002). Does cognitive rehabilitation work? Clinical and economic considerations and outcomes. In G. Prigatano & N. H. Pliskin (Eds.), *Clinical neuropsychology and cost-outcome research: An introduction* (pp. 329–345). Psychology Press.

Wilson, B. A., Evans, J. J., Emslie, H., & Malinek, V. (1997). Evaluation of NeuroPage: A new memory aid. *Journal of Neurology, Neurosurgery, and Psychiatry, 63*(1), 113–115.

Wilson, B. A., Evans, J. J., Gracey, F., & Bateman, A. (2009). *Neuropsychological Rehabilitation: Theory, models, therapy and outcomes.* Cambridge University Press.

Wilson, B. A., Kopelman, M., & Kapur, N. (2008). Prominent and persistent loss of self-awareness in amnesia: Delusion, impaired consciousness or coping strategy? *Neuropsychological Rehabilitation, 18*(5–6), 527–540.

Wilson, B. A., Rous, R., & Sopena, S. (2008). The current practice of neuropsychological rehabilitation in the United Kingdom. *Applied Neuropsychology, 15*(4), 229–240.

Wilson, B. A., Scott, H., Evans, J., & Emslie, H. (2003). Preliminary report of a NeuroPage service within a health care system. *Neurorehabilitation, 18*(1), 3–9.

Wilson, B. A., & Watson, P. C. (1996). A practical framework for understanding compensatory behaviour in people with organic memory impairment. *Memory, 4*(5), 465–486.

Author Index

For the benefit of digital users, indexed terms that span two pages (e.g., 52–53) may, on occasion, appear on only one of those pages.

Subject Index

For the benefit of digital users, indexed terms that span two pages (e.g., 52–53) may, on occasion, appear on only one of those pages.